MECHANICS
for Advanced Level

L. Bostock B.Sc.

S. Chandler B.Sc.

OXFORD
UNIVERSITY PRESS

Great Clarendon Street, Oxford, OX2 6DP, United Kingdom

Oxford University Press is a department of the University of Oxford.
It furthers the University's objective of excellence in research, scholarship,
and education by publishing worldwide. Oxford is a registered trade mark of
Oxford University Press in the UK and in certain other countries

First published by Stanley Thornes (Publishers) Ltd in 1996
Reprinted by Nelson Thornes Ltd in 2002
This edition published by Oxford University Press in 2015

British Library Cataloguing in Publication Data
Data available

978-0-7487-2596-0

10 9 8 7 6 5 4

Printed by Multivista Global Pvt. Ltd

Acknowledgements

Page make-up: Tech-Set Ltd., Tyne & Wear

Although we have made every effort to trace and contact all
copyright holders before publication this has not been possible in all
cases. If notified, the publisher will rectify any errors or omissions at
the earliest opportunity.

Links to third party websites are provided by Oxford in good faith
and for information only. Oxford disclaims any responsibility for
the materials contained in any third party website referenced in
this work.

CONTENTS

NOTES FOR USERS OF MECHANICS AT A-LEVEL

This book, a companion to Core Maths for A-level, covers all the work necessary for the mechanics component of the A-level syllabuses of most Examining Boards where Mechanics is combined with Pure Mathematics, Statistics etc. It also provides fully for an A/S course in Mechanics.

The aim of the book is to provide a sound but simple treatment of mechanics as an experimental science whose 'laws' are based on deductive and experimental evidence. These basic principles are then used as models for real-life situations. The various simplifying assumptions that are needed to form a suitable model are introduced at an early stage and used wherever appropriate throughout the book so that modelling becomes an integral part of the work, along with the appreciation that 'answers' to real problems can only be estimates. As a result, the degree of accuracy to which the answers to real situations are given varies according to the context; the reader should be prepared to decide upon the appropriate accuracy. Later on, the concept of testing and improving models is introduced.

No previous knowledge of the subject is required and we have arranged the topics carefully so that when pure mathematics is needed it is unlikely to be beyond the level then reached in a parallel Pure Mathematics course.

There are many worked examples and, as each topic is introduced, exercises are provided that always begin with straightforward questions which can be solved from an understanding of the basic techniques. Towards the end of some exercises, questions are set (indicated by an asterisk) that are a little more demanding and, as the topic develops, further exercises contain questions that require more thought.

At intervals through the book there are consolidation sections which contain a summary of the work covered in the preceding chapters and an exercise mainly comprising specimen and past examination questions. It is not intended that these questions be used immediately after a topic has been studied; they are of more value if used for revision later on, when confidence and some sophistication of style have been acquired.

We are grateful to the following examination boards for permission to reproduce questions from their past examination papers and specimen papers. Specimen questions are indicated by the suffix s and it should be noted that they have not been subjected to the rigorous checking and moderation procedure by the Boards that their examination questions undergo. (Any answers included have not been provided by the examining boards; they are the responsibility of the authors and may not necessarily constitute the only possible solutions.)

University of London Examinations and Assessment council (ULEAC)
Northern Examinations and Assessment Board (NEAB, SMP)
University of Cambridge Local Examinations Syndicate (UCLES, MEI)
The Associated Examining Board (AEB)
Welsh Joint Education Committee (WJEC)
University of Oxford Delegacy of Local Examinations (OUDLE)

L. Bostock
S. Chandler

1996

USEFUL INFORMATION

ABBREVIATIONS

=	is equal to		$\Rightarrow$	giving, gives or implies
$\equiv$	is equivalent to		+ve	positive
$\approx$	is approximately equal to		−ve	negative
2sf	corrected to 2 significant figures	3dp	corrected to 3 decimal places
A⟲	taking clockwise moments about an axis through A
B⟳	taking anticlockwise moments about an axis through B

NOTATION USED IN DIAGRAMS

Force ——▷——   Acceleration ——▷▷——
Velocity ——▷——   Dimensions ◄————►

Where components and resultant are shown in one diagram the resultant is denoted by a larger arrow-head e.g.   ——▷——

THE VALUE OF *g*

Throughout this book, unless a different instruction is given, the acceleration due to gravity is taken as 9.8 metres per second per second, i.e. $g = 9.8$

ACCURACY OF ANSWERS

Practical problems rarely have exact answers. Where numerical answers are given they are usually corrected to two or three significant figures or decimal places, depending on their context.

Answers found from graphs may not even be reliable beyond the first significant figure.

CHAPTER 1

MOTION

DISTANCE, SPEED AND TIME

Mechanics is the study of how and why objects move in various ways, or do not move at all.

Everyone is familiar with time, distance and speed and we begin this book by recalling the relationships between these quantities.

If an object is travelling with constant speed, the distance it covers is given by

$$\textbf{distance } = \textbf{ speed} \times \textbf{time}$$

hence

$$\textbf{speed } = \frac{\textbf{distance}}{\textbf{time}}$$

Note also that

$$\textbf{average speed } = \frac{\textbf{total distance}}{\textbf{total time}}$$

When these formulae are used the three quantities must be measured in units that are consistent,

- e.g. if distance is measured in kilometres and time is measured in hours then speed must be measured in kilometres per hour (km/h),
- or if distance is measured in metres and time is measured in seconds then speed must be measured in metres per second (m/s).

DISTANCE–TIME GRAPHS

Suppose that an object is moving in a straight line and that its distances from a fixed point on the line are recorded at various times. By plotting corresponding values, a distance–time graph can be drawn to illustrate the motion of the object.

Consider this situation.

A cyclist travelling along a straight road, covers the 18 km between two points A and B in $1\frac{1}{2}$ hours and the next 35 km, from B to C, in $2\frac{1}{2}$ hours. The graph illustrating this journey is given below.

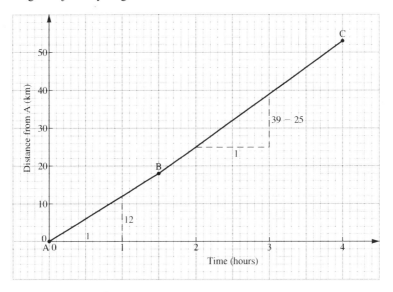

A useful property can be deduced from the graph.

The cyclist's speed from A to B is given by distance ÷ time,

i.e. $18 \div 1\frac{1}{2}$ km/h = 12 km/h

The gradient of the graph for the section from A to B is $12 \div 1 = 12$

Also, the cyclist's speed from B to C is $35 \div 2\frac{1}{2}$ km/h $= 14$ km/h

and the gradient of the graph for the section BC is $(39 - 25) \div 1 = 14$

These two results are examples of the general fact that

the gradient of the distance–time graph gives the speed

Curved Distance–Time Graphs

In practice there are many situations where the speed of a moving object is not constant. In such cases the graph of distance plotted against time is not a straight line but a curve.

As an example, consider the distance, d metres, of a car from a set of traffic lights as the car pulls away from the lights.

This table shows the values of d after t seconds.

t	0	1	2	3	4	5
d	0	2	8	18	32	50

The corresponding distance–time graph is:

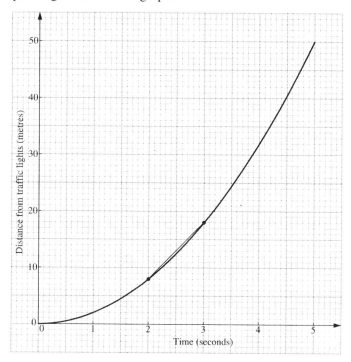

This time the speed of the car cannot be found immediately from the gradient of the graph as the graph is not a straight line. What we can do however is to find the average speed over a chosen interval of time.

Consider, for·example, the motion during the third second (i.e. from $t = 2$ to $t = 3$).

From the values in the table, the average speed during this second is

$$\frac{18 - 8}{3 - 2} \text{ m/s} = 10 \text{ m/s}$$

The gradient of the line joining the points on the graph where $t = 2$ and $t = 3$ is also 10.

So the gradient of a line joining two points on a curved distance–time graph gives the average speed in that time interval. (A line joining two points on a curve is called a *chord*.)

Now suppose that we want to find the speed of the car *at the instant* when $t = 2.5$.

We can choose a shorter interval of time, say from $t = 2.2$ to $t = 2.8$ and find the gradient of this chord which is closer to the curve than the first one was. This value gives a better approximation to the speed in the region of $t = 2.5$ and an even better approximation is given by using a yet shorter time interval.

As the ends of the chord get closer to each other the chord becomes nearer and nearer to the tangent to the curve at the point where $t = 2.5$ so we deduce that the speed at the instant when $t = 2.5$ is given by the gradient of the tangent to the curve at this value of t.

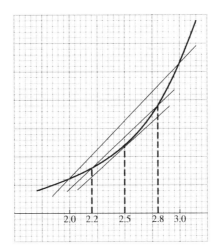

In general

the speed at a particular value of t is given by
the gradient of the tangent to the distance–time graph
at the point where t has that value

Now that this property is established there is no need to find a succession of average speeds by drawing a number of chords; we can go straight to drawing the tangent at the required point. At this stage this is done by judging the position of the tangent visually, so the result can only be a rough approximation to the speed.

EXERCISE 1a

1.

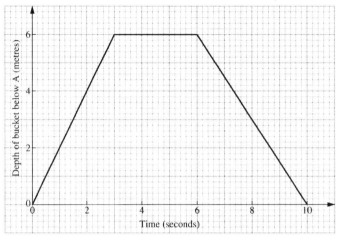

The graph illustrates the motion of a bucket being lowered into a well from the top, A, down to the water level, B, filled with water and drawn up again.

(a) What is the depth of the water level below A?

Find

(b) the speed of the bucket as it descends from A to B

(c) the speed of the bucket as it ascends from B back to A

(d) the time taken to fill the bucket

(e) the average speed for the whole operation including the filling of the bucket.

2. A cyclist rides at 5 m/s along a straight road for 25 minutes. She then dismounts and pushes the bicycle for 5 minutes at 1.7 m/s. Draw a distance–time graph and find the average speed for the whole journey.

3. A goods wagon is shunted 60 m forward in 12 seconds, then 24 m back in 8 seconds and finally 44 m forward in 11 seconds.

(a) Draw the distance–time graph.

(b) Write down the speed in each of the three sections of the motion.

(c) Find the average speed for
   (i) the first 20 seconds
   (ii) the whole journey.

4. In an experiment to measure the viscosity of a lubricant, a ball-bearing was allowed to fall through the lubricant contained in a cylinder. The distance, *d* centimetres, of the ball-bearing from the bottom of the cylinder was measured after *t* seconds, for a series of values of *t* and the results illustrated by this distance–time graph.

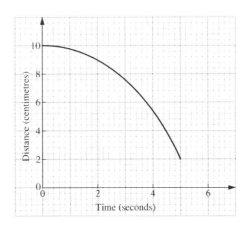

(a) After how long do you think the ball-bearing will reach the bottom?

(b) Find the average speed during (i) the third second   (ii) the fifth second

(c) Estimate the speed when (i) $t = 2.5$ (ii) $t = 4$.

5. A particle P is moving in a straight line from a point A. This table gives the distance, *s* metres, of P from A after *t* seconds.

t (seconds)	0	1	2	3	4	5	6	7	8
s (metres)	0	0.2	0.8	1.8	3.2	5.0	7.2	9.8	12.8

Plot the points and draw a distance–time graph. Hence

(a) find the average speed over the first 4 seconds

(b) estimate the speed when (i) $t = 4.5$ (ii) $t = 6$

(c) find for how long the distance of P from A is less than 6 m.

6. A point of light moving in a straight line on the screen of an oscilloscope is at a distance *s* millimetres from O at a time *t* seconds where $s = \dfrac{20}{t+1}$.

Draw a distance–time graph for values of *t* from 0 to 5 and use it to find

(a) the speed, in mm/s, when $t = 2.5$

(b) the time when the point of light is 8 mm from O

(c) the average speed of the light during the five seconds.

SPEED–TIME GRAPHS

This graph illustrates the motion of a dog chasing a rabbit. The dog starts off enthusiastically, then tires a little and finally gives up hope.

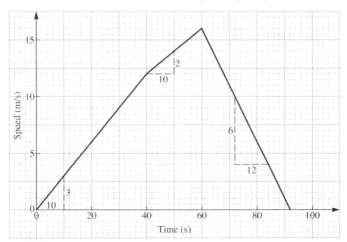

In the early stages of the chase the dog's speed is increasing; we say that the dog is *accelerating*. Acceleration tells us by how much the speed increases in one unit of time and it is measured in a unit of speed per unit of time, e.g. m/s per second, which is written m/s^2. When a speed is going down, the rate at which it decreases is known as *deceleration* or *retardation*.

The following observations can be made from the graph.

(i)   In the first section of the motion the dog's speed increases steadily from zero to 12 m/s in 40 seconds, i.e. the dog accelerates at $\frac{12}{40}$ $m/s^2 = 0.3 \text{ m/s}^2$.
The gradient of the graph for this section is $\frac{3}{10} = 0.3$.

For the next 20 seconds the dog's speed increases steadily from 12 m/s to 16 m/s, i.e. the dog accelerates at $\frac{2}{10}$ $m/s^2 = 0.2 \text{ m/s}^2$.

In this section the gradient of the graph is $\frac{2}{10} = 0.2$

Finally, in slowing down steadily from 16 m/s to rest in 32 seconds, the dog decelerates at $\frac{16}{32}$ $m/s^2 = 0.5 \text{ m/s}^2$
The corresponding gradient is $-\frac{6}{12} = -0.5$

Each section of the motion illustrates the following general facts.

**The gradient of the speed–time graph gives the acceleration;
a negative gradient indicates deceleration.**

(ii) For the first 40 seconds

the average speed is $\frac{1}{2}(0 + 12)$ m/s $= 6$ m/s,
the distance covered is therefore 6×40 m $= 240$ m,

the area under the graph (a triangle) is $\frac{1}{2} \times 40 \times 12 = 240$
(using measurements from the scales on the axes).

For the next 20 seconds

the average speed is $\frac{1}{2}(12 + 16)$ m/s $= 14$ m/s,
the distance covered is therefore 14×20 m $= 280$ m,

the area under the graph (a trapezium) is $\frac{1}{2}(12 + 16) \times 20 = 280$

For the final 32 seconds

the average speed is $\frac{1}{2}(16 + 0)$ m/s $= 8$ m/s,
the distance covered is therefore 8×32 m $= 256$ m,

the area under the graph is $\frac{1}{2} \times 32 \times 16 = 256$

In each section the area under the graph represents the distance covered in that section, illustrating this general fact.

The area under a speed–time graph gives the distance covered.

Curved Speed–Time Graphs

In the example above the speed changes steadily in each section of the motion so the graph for each section is a straight line. The gradient of that line gives the acceleration during that section of motion and it follows that this acceleration is constant within that interval of time.

On the other hand, the graph of the motion of an object whose speed changes in a variable way is a curve, so there is no section where the gradient of a straight-line graph can be used to find the acceleration.

A different approach is therefore needed to find the acceleration in this case.

Finding the Acceleration

This graph shows the speed of a roller-coaster as it goes from the top of the first climb to the top of the next rise.

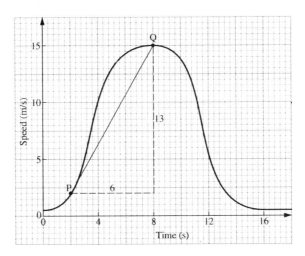

Because the graph of its speed is not made up of straight lines, the acceleration of the roller-coaster at a particular instant cannot be found immediately. However, by using the gradient of a chord joining two points on the graph we can find the *average acceleration* over that part of the motion.

For example the average acceleration, in m/s^2, between the instants when $t = 2$ and $t = 8$ is given by the gradient of PQ, i.e. the average acceleration is $\frac{13}{6}$ $m/s^2 \approx 2.1$ m/s^2.

When the ends of a chord are brought closer together, the average acceleration gives a better approximation to the actual acceleration within the time interval. Ultimately, when the ends of the chord coincide, the chord becomes a tangent to the curve at a point where t has a particular value, i.e.

> **the acceleration at a particular value of t is given by**
> **the gradient of the tangent to the speed–time graph**
> **at the point where t has that value**

Drawing a tangent to a curve is, as we mentioned on page 4, a matter of visual judgement, so an acceleration found in this way is only an approximation.

Note that, to the right of $t = 8$, the gradient of any tangent is negative verifying that the roller-coaster is slowing down, i.e. decelerating.

Finding the Distance Covered

We have seen that if a speed–time graph consists of one or more straight lines, i.e. for each part of the motion the acceleration is constant, the area under the graph gives the distance covered. We showed this, for each separate section of the graph, by using the average speed which, for a straight-line graph, is the speed half-way through the time interval.

When the graph is curved, however, the average speed cannot be found directly in this way, but the method can be adapted to give a good approximation.

If we regard the curve as a series of short straight lines, under each of these lines we have a trapezium whose area can be found.

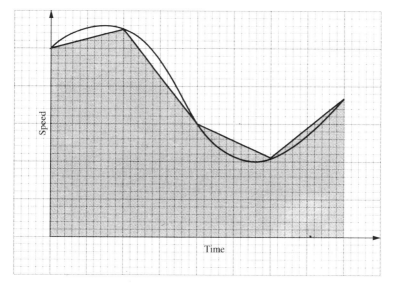

Looking at one of these lines, it is clear that the area underneath it is approximately equal to the area under the corresponding part of the curve. So the sum of the areas beneath all the short lines gives an approximate value for the area under the whole curve.

By making each line successively shorter, so that it gets closer to the curve, the approximation becomes so good that we can now say, whatever the shape of the graph,

the distance covered in an interval of time
is given by the area under the speed–time graph for that interval

When dividing a graph into sections to find an approximation to its area, the arithmetic is simpler if strips of equal width are used.

Example 1b

This graph represents the motion of a cyclist during the first five seconds of a ride.

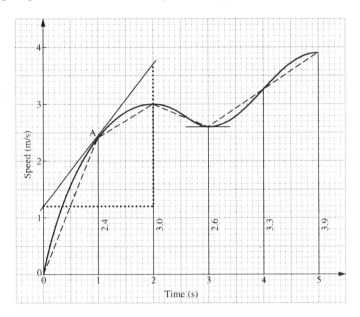

(a) **Find an approximate value for the acceleration after 1 second.**

(b) **For what period of time did the cyclist decelerate?**

(c) **Find an approximate value for the distance travelled by the cyclist in the five seconds. Is this value an under or over estimate?**

(d) **At which time(s) was the acceleration zero?**

(a) To find the acceleration after 1 second, we draw the tangent at A where $t = 1$.

The gradient of this tangent is approximately $(3.7 - 1.2) \div 2 = 1.25$
∴ the acceleration after 1 second is about 1.3 m/s².

(b) When the cyclist decelerates, his speed goes down.

The cyclist decelerates during the third second (from $t = 2$ to $t = 3$).

(c) As the motion spans 5 s, it makes sense to divide the curve into five strips each of width 1 unit. The area of each strip is approximately equal to that of a trapezium.

The area of each strip is approximately $\frac{1}{2}$(sum of parallel sides) × 1
∴ the total area under the curve is given approximately by

$$\tfrac{1}{2}(0 + 2.4) + \tfrac{1}{2}(2.4 + 3) + \tfrac{1}{2}(3 + 2.6) + \tfrac{1}{2}(2.6 + 3.3) + \tfrac{1}{2}(3.3 + 3.9)$$

$$= 13.25$$

Therefore the area under the curve is approximately 13 square units.

2 significant figures is sufficient for an approximation.

The distance covered is approximately 13 metres.

This value is less than the actual value for two reasons:
(i)   we have rounded the calculated value down
(ii) the area of all but one of the trapeziums is a little less than the
      corresponding area under the curve

(d)   The acceleration is zero when the tangent to the curve is horizontal.

The acceleration is zero when $t = 2$ and $t = 3$.

EXERCISE 1b

1.

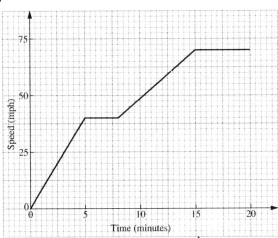

This speed–time graph shows the journey of a train as it moves off from the
platform at a station. Find

(a)   the acceleration during the first 2 minutes, in mph per minute

(b)   the greatest speed of the train

(c)   the time for which the train travels at constant speed between its periods of
      acceleration

(d)   the acceleration during the tenth minute.

2. The speed of a motor cycle increases steadily from 12 m/s to 20 m/s in
    10 seconds. The rider then immediately brakes and brings the vehicle steadily to
    rest in 8 seconds. Draw the graph of speed against time for this journey and find

(a)   the acceleration

(b)   the deceleration

(c)   the distance travelled.

3. This graph shows the speed of a bus as it travels between two stops.

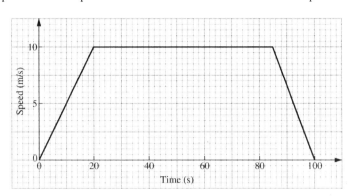

(a) Over what period of time is the bus

(i) accelerating (ii) decelerating (iii) travelling at constant speed?

(b) Find (i) the acceleration (ii) the deceleration.

(c) Find the distance between the two bus stops.

4. A train travelling at 36 m/s starts up an inclined section of track and loses speed steadily at 0.4 m/s². How long will it be before the speed drops to 30 m/s and how far up the incline will the train have travelled by then?

5. This graph shows the speed of a ball that is rolled across a lawn.

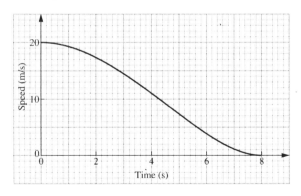

(a) Write down the speed of the ball after

(i) 2 seconds (ii) 7 seconds.

(b) Is the ball accelerating or decelerating when $t = 3$?

(c) Find an approximation for the deceleration during the fifth second.

(d) Find an approximation for the total distance travelled by the ball.

6. A balloon was released into the air on a calm day. Its speed in metres per second was noted at 1 second intervals after release and the results are given in the table.

$t(s)$	0	1	2	3	4	5	6	7	8
$v(m/s)$	0	7	11	13	14	13	11	7	0

Draw the speed–time graph illustrating this data and use it to find
(a) the acceleration of the balloon after (i) 1.5 s (ii) 6 s
(b) the time when the acceleration was zero
(c) the distance travelled by the balloon.
(d) State, with reasons, whether your answer to (c) is an under- or over-estimate.

7. A rocket is fired and its speed t minutes after firing is v km/minute. This graph shows the corresponding values of t and v during the first 4 seconds of the flight. Find
(a) the acceleration, in km/minute^2, 3 minutes after firing
(b) the distance covered in the first 3 minutes
(c) the distance covered in the third minute.

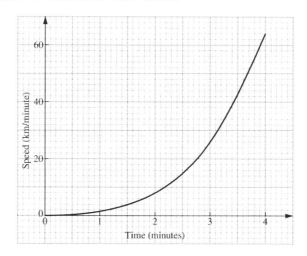

Each question from 8 to 13 is followed by several suggested answers. Some of the questions concern distance–time graphs and others involve speed–time graphs. Read them carefully and choose the correct answer, giving your reasons for rejecting the other answers if you can.

8. This graph shows the motion of an object that starts at O and

A   has a constant speed

B   is at rest when $t = 4$

C   starts with zero speed

D   travels a distance of 6 m.

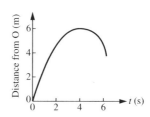

9. A girl throws a ball straight up into the air and catches it on its way down again. This graph represents the motion of the ball.

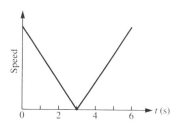

A   The ball goes up with constant speed

B   The acceleration increases as the ball falls

C   The ball is not moving when $t = 3$

D   The distance that the ball rises is negative.

10. A bullet is fired into a block of wood. This graph illustrates the motion of the bullet inside the block.

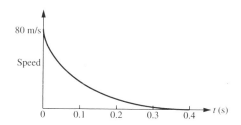

A   The average speed is 40 m/s

B   The bullet has a constant deceleration

C   The bullet stops after 0.4 seconds

D   The bullet penetrates a distance of 16 m into the block.

Use this graph to answer questions 11 and 12.

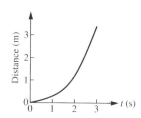

11. The average speed over the 3 seconds of motion is

A $1\frac{1}{2}$ m/s   B   1 m/s   C   3 m/s

12. The speed when $t = 2$ is about

A   0.5 m/s   B   1.5 m/s   C   2 m/s

13.

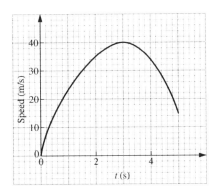

This graph of part of a car journey shows that

A   the car comes to rest when $t = 5$

B   the car changes direction after 3 seconds

C   the acceleration when $t = 2$ is about 1 m/s^2

D   the speed increases for 3 seconds.

MOTION IN A STRAIGHT LINE

If a particle is moving in a straight line it can be moving in either direction along
the line. Rather than try to describe these directions in words in every case, we
can distinguish between them by giving a positive sign to one of the directions;
the other direction is then negative.

DISPLACEMENT

Consider a model engine which starts from a point O and moves in the direction
OA along the straight section of track as shown in the diagram.

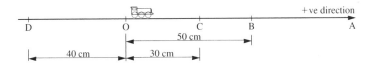

When the engine reaches B, it has travelled a distance of 50 cm and it is also
50 cm from O. However, if the engine then reverses its direction and moves
20 cm back towards O, i.e. to point C,

the total distance that the engine has travelled is 70 cm

but the distance of the engine from O is 30 cm (to the right).

If the engine now continues moving towards O and carries on to point D, beyond O,

the total distance that it has travelled is 140 cm
but the distance of the engine from O is 40 cm (to the left).

If we specify the direction from O to A as the positive direction then
the *distance from O in the direction from O to A* is called the *displacement*.

Hence, from O, the displacement of B is   50 cm

the displacement of C is   30 cm

the displacement of D is −40 cm

Displacement is a quantity which has *both* magnitude (size) *and* direction.
Quantities of this type are *vectors*.

Distance, on the other hand, only has size – the direction doesn't matter.
Distance is a *scalar* quantity.

Example 1c

Starting from floor 4, a lift stops first at floor 11, then at floor 1 and finally at floor 6. The distance between each floor and the next is 4 m.
Taking the upward direction as positive write down, for each of the stops,

(a)   the displacement, *s* metres, of the lift operator from floor 4

(b)   the distance the lift operator has travelled since first leaving floor 4.

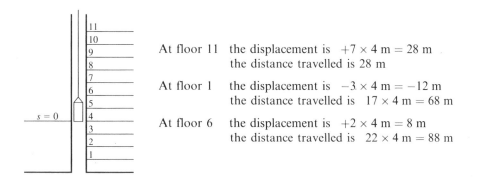

At floor 11   the displacement is $+7 \times 4$ m $= 28$ m
                        the distance travelled is 28 m

At floor 1   the displacement is -3×4 m $= -12$ m
                        the distance travelled is 17×4 m $= 68$ m

At floor 6   the displacement is $+2 \times 4$ m $= 8$ m
                        the distance travelled is 22×4 m $= 88$ m

EXERCISE 1c

1. Which of the following quantities are vector and which are scalar?

(a) 5 km due south   (e) A temperature of 25°C

(b) 6 miles   (f) A force of 8 units vertically downwards

(c) A speed of 4 m/s   (g) A mass of 6 kg

(d) 200 miles north-east   (h) A time of 7 seconds.

2. A spider is at a point A on a smooth wall, 1.2 m above the floor. This graph represents the motion of the spider as it tries (unsuccessfully!) to climb vertically up the wall.

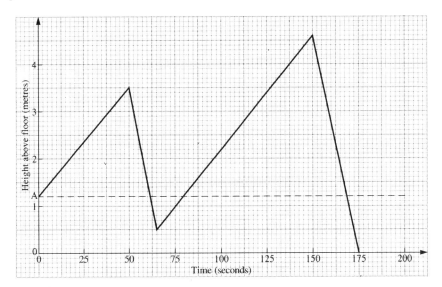

Taking the upward direction as positive,

(a) write down the displacement of the spider from A after

    (i) 50 seconds

    (ii) 80 seconds

    (iii) 175 seconds

(b) find (i) the maximum displacement of the spider from A

        (ii) the total distance the spider travels

(c) find the average speed over the 175 s.

3. A fly is free to move to and fro inside a narrow straight glass tube, closed at both ends. The table below gives the displacement of the fly from the centre of the tube at each instant when it reverses direction. Assuming that the fly moves all the time with a constant speed, illustrate the motion of the fly by drawing a graph of displacement against time.

Time (s)	0	2	5	6	9	10	12
Displacement from O (m)	0	4	−2	0	−6	−4	−8

VELOCITY

If we say that an object is moving with a certain speed *in a particular direction,* we are giving the *velocity* of the object, i.e. velocity is a vector quantity.
Speed is the magnitude of velocity.

Suppose that a body is moving with a *uniform velocity,* i.e. a constant velocity. This means not only that the body has a constant speed, but also that its direction is constant, i.e. that it is moving one way along a straight line.

In order to define which way the body is moving, we specify a velocity in one direction along the line as positive; a velocity in the opposite direction is then negative.

Consider a particle P moving with a constant speed of 5 m/s along the line shown in the diagram.

If P starts from O and moves to the right, which is the chosen positive direction, then its velocity is +5 m/s.
After 1 second the displacement of P from O is +5 m
and after 2 seconds it is +10 m,
and so on.

The displacement is increasing at a rate of +5 m/s.

Now if P starts from A and moves to the left with the same speed, its velocity is −5 m/s.

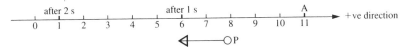

The initial displacement of P from O is +11 m
and after 1 second the displacement is 6 m.
The displacement has decreased by 5 m, i.e. it has increased by −5 m.

Similarly, after 2 seconds the displacement is 1 m and has increased by −10 m.

The displacement is increasing at a rate of −5 m/s

i.e.   **velocity is the rate at which the displacement increases**

ACCELERATION

So far we have associated acceleration only with changing speed. Now that we have defined velocity, however, we must describe acceleration more carefully,

i.e.   **acceleration is the rate at which *velocity* is increasing**

So if the velocity of a particle moving in a straight line increases steadily from 3 m/s to 11 m/s in 4 seconds, the acceleration is +2 m/s².

On the other hand, if the velocity goes down from 14 m/s to 5 m/s in 3 seconds (the velocity has increased by −9 m/s) the acceleration is −3 m/s².

EXERCISE 1d

1. Decide whether each of the following statements is correct or incorrect.
If you think it is incorrect, give your reason.

(a)   A car driving due north at 40 mph has a constant velocity.

(b)   A toy train runs round a circular track with constant velocity 2 m/s.

(c)   A plane flies in a straight path from London to Newcastle so its velocity is constant.

2. A particle is moving along the straight line shown in the diagram. It passes through A, travels to B, then moves from B to C, from C to D and finally from D to E.

This table gives the value of t at each point, where t is the number of seconds that have elapsed since the particle first passed through A.

t	B	C	D	E′
	5	8	15	19

Find the velocity of the particle, constant in each section, in travelling from

(a) A to B   (b) B to C   (c) C to D   (d) D to E.

(Remember to give magnitude and direction.)

3. The velocity of a particle changes steadily from -5 m/s to -21 m/s in 4 seconds. What is the acceleration?

4. A particle moving in a straight line with a constant acceleration has a velocity u m/s at one instant and t seconds later the velocity is v m/s. Find the acceleration of the particle if

(a) $u = 8$, $v = 2$, $t = 3$ (b) $u = 4$, $v = -11$, $t = 5$.

5. A body moving initially at 5 m/s has a constant acceleration of a m/s^2. After 6 seconds its velocity is v m/s. Find v if

(a) $a = 3$ (b) $a = -2$ (c) $a = 0$.

DISPLACEMENT–TIME GRAPHS

For an object P moving in a straight line, a displacement–time graph shows how the distance of P *in a specified direction from a fixed point* varies with time.

Consider an object P which moves in a straight line, travelling through points O, A, B and C on the line and covering each section at a constant speed.
This table gives the displacement, s centimetres, of each of these points from O, and the time, t seconds after leaving O, when P is at each point.

	O	A	B	C
s	0	10	6	-9
t	0	5	7	12

The displacement–time graph can now be drawn.

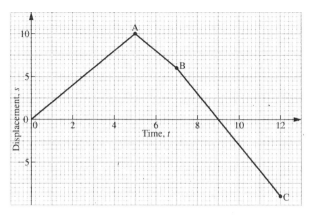

For the section from O to A,
P has travelled a distance of 10 cm in the positive direction in 5 seconds,
i.e. the velocity is 2 cm/s and the gradient of the graph is 2.

For the section from A to B,
P has travelled a distance of 4 cm in the negative direction in 2 seconds,
i.e. the velocity is -2 cm/s and the gradient of the graph is -2.

For the section from B to C,
P has travelled a distance of 15 cm in the negative direction in 5 seconds,
i.e. the velocity is -3 cm/s and the gradient of the graph is -3.

In each case the gradient of the displacement–time graph represents the velocity.

The *average velocity* is the constant velocity that would produce the final increase
in displacement in the total time interval, e.g.

the average velocity from O to B is $\frac{6-0}{7-0}$ cm/s $= \frac{6}{7}$ cm/s

(this is equal to the gradient of the chord OB)

the average velocity from A to C is $\frac{-9-10}{12-5}$ cm/s $= -\frac{19}{7}$ cm/s

(this is equal to the gradient of the chord AC)

the average velocity from O to C is $\frac{-9-0}{12-0}$ cm/s $= -\frac{3}{4}$ cm/s

(this is equal to the gradient of the chord OC).

Note that in moving from O to C, the total *distance* that P has moved is

$(10 + 4 + 15)$ cm,   i.e. 29 cm.

So P's average *speed* is $29 \div 12$ cm/s, i.e. 2.4 cm/s (2 sf), showing that

the average speed is *different* from the average velocity.

For a curved displacement–time graph also, the average velocity over a time
interval is given by the gradient of the chord corresponding to that interval.
So the same reasoning as we used for a distance–time graph shows that the
gradient of the tangent to the curve at a particular value of t, represents the
velocity at that instant.

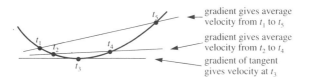

In general, for any type of motion over a given interval of time,

$$\text{average speed} = \frac{\text{distance covered in the time interval}}{\text{the time interval}}$$

$$\text{average velocity} = \frac{\text{increase in displacement over the time interval}}{\text{the time interval}}$$

**the velocity at any instant is represented by
the gradient of the displacement–time graph at the corresponding point**

Examples 1e

1. A particle P moves along a straight line, starting from a fixed point O on that line. The displacement–time graph for its motion over the first six seconds is given below.

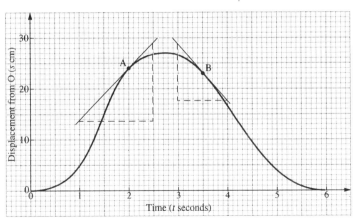

(a) Find the average velocity from

   (i) $t = 0$ to $t = 2$ (ii) $t = 4$ to $t = 6$ (iii) $t = 0$ to $t = 6$.

(b) Estimate the velocity of P at the instant when

   (i) $t = 2$ (ii) $t = 3.5$.

(a) Average velocity is represented by the gradient of a chord.

   (i) average velocity $= \dfrac{24 - 0}{2 - 0}$ cm/s $= 12$ cm/s

   (ii) average velocity $= \dfrac{0 - 16}{6 - 4}$ cm/s $= -8$ cm/s

   (iii) average velocity $= \dfrac{0 - 0}{6 - 0}$ cm/s, i.e. zero

(b)   Velocity is represented by the gradient of a tangent.

(i)   velocity at A $\approx \dfrac{29 - 13.5}{2.5 - 1}$ cm/s,   i.e. 10 cm/s (2 sf)

(ii)   velocity at B $\approx \dfrac{18 - 29}{4 - 3}$ cm/s,   i.e. −11 cm/s (2 sf)

2. **A and O are two fixed points on a straight line. A particle P moves on the line so that, at time t seconds its displacement, s metres, from O is given by $s = (t - 1)(t - 5)$. When $t = 0$, the particle is at point A. Draw a displacement–time graph for values of t from 0 to 6.**

(a)   At what times does the particle pass through O?

(b)   What is the average speed over the 6-second time interval?

(c)   What is the average velocity over the 6-second time interval?

(d)   At what time is the velocity zero?

Using $s = (t - 1)(t - 5)$ for values of t from 0 to 6 gives

t (seconds)	0	1	2	3	4	5	6
s (metres)	5	0	−3	−4	−3	0	5

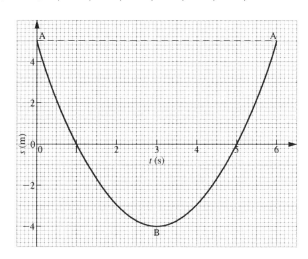

(a)   When the particle is at O its displacement from O is zero, i.e. $s = 0$.
The particle passes through O when $t = 1$ and $t = 5$.

(b)   P starts at A where $s = +5$. P then covers 5 m to O and continues beyond O for a further 4 m
to B where $s = -4$. Then P goes back to O, i.e. 4 m back, and a further 5 m to A.

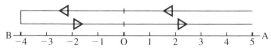

average speed $= \dfrac{\text{total } distance \text{ covered}}{\text{time interval}} = \dfrac{5 + 4 + 4 + 5}{6} = 3$ m/s.

(c) The average velocity is given by the gradient of the chord joining the points where $t = 0$ and $t = 6$, i.e. the average velocity is $\dfrac{5-5}{6-0} = 0$

(d) The velocity is zero when the gradient of the tangent is zero, i.e. when the tangent is parallel to the time axis.

The velocity is zero when $t = 3$.

EXERCISE 1e

1. A boy is practising kicking a football straight towards a wall. For each kick he observes how far the ball rebounds. This graph shows, for one kick, the displacement towards the wall of the ball from the boy.

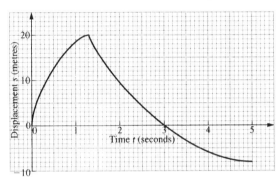

(a) Use the graph to estimate the velocity of the football
   (i) after half a second   (ii) when $t = 3$ (iii) when $t = 5$.

(b) (i) State the velocity when $t = 1.3$.
   (ii) Explain why the graph has a 'point' at this time.

(c) At what time does the football pass the boy when rebounding?

2. A particle P is moving along a straight line. The displacement of P from O, a fixed point on the line, after t seconds is s metres. The table gives some corresponding values of s and t.

t (seconds)	0	1	2	3	4	5	6
s (metres)	0	3	4	3	0	-5	-12

Draw a displacement–time graph and use it to answer the following questions.
(a) Find the average velocity from
   (i) $t = 0$ to $t = 2$ (ii) $t = 2$ to $t = 4$
   (iii) $t = 0$ to $t = 4$ (iv) $t = 2$ to $t = 6$.

(b) Find the average speed for each of the time intervals in part (a).

(c) What do you think the velocity is when $t = 0$?

(d) At what time is the velocity zero?

3. A is a fixed point on a straight line and P is moving on the line. The displacement, s metres, of P from A after t seconds is given by $s = 5t - t^2$.

(a) Copy and complete the following table.

t	0	1	2	3	4	5	6
s	0		6				

(b) Choose suitable scales and draw a displacement–time graph and use it to answer the following questions.

(c) At what time is the velocity zero?

(d) Estimate the velocity when $t = 5$.

(e) Find, for the 6-second journey,
(i) the average velocity   (ii) the average speed.

4. This graph illustrates the motion of a ball bouncing vertically.

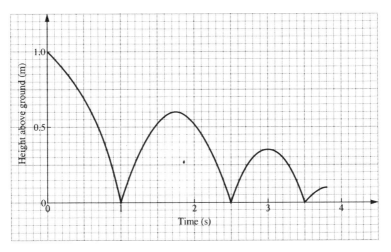

(a) State the times at which the graph shows that the velocity of the ball is zero. (Think carefully before you answer.)

(b) Find the average speed during
(i) the first second
(ii) the first full bounce (i.e. from $t = 1$ to $t = 2.5$)
(iii) the second full bounce.

(c) Find the average velocity during each of the time intervals specified in part (b).

(d) What is the average velocity over the first 3 seconds.

(e) Estimate the velocity of the ball after (i) 0.5 s   (ii) 1.5 s.

VELOCITY–TIME GRAPHS

For an object P travelling in a straight line, a velocity–time graph shows how the speed of P *in a particular direction* varies with time.

This graph shows the variation in the velocity of P during a 20-second period of motion, starting from a point O on the straight line.

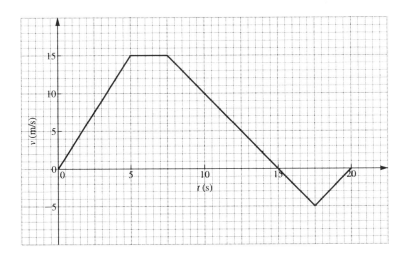

During the first 5 seconds the velocity increases steadily from zero to 15 m/s so the acceleration in this section is 3 m/s² and the gradient of this section of the graph is 3.

For the next 2.5 seconds the velocity is constant, i.e. the acceleration is zero, and the gradient the graph is also zero.

Then the velocity decreases until after another 7.5 seconds the graph crosses the time axis. This shows that the velocity has reduced to zero and that P has momentarily come to rest.

For this section the acceleration is $-\frac{15}{7.5}$ m/s² and the gradient is -2. Note that, although the velocity is decreasing, it is still positive, so P is still moving forward.

When $t = 15$ the velocity is zero and immediately after that the velocity becomes negative, i.e. P *stops* going forward and begins to move in the opposite direction with an acceleration of $-\frac{5}{2.5}$ m/s² (the gradient is -2).

For the last 2.5 seconds the velocity is becoming less negative, i.e. it is increasing. The acceleration is $\frac{0-(-5)}{2.5}$ m/s^2 = 2 m/s^2 and the gradient is 2. Again note that the velocity is still negative so P is still moving backwards until, when $t = 20$, P comes to rest.

Each section demonstrates that, for motion with constant acceleration, the gradient of the velocity–time graph represents the acceleration.

Now consider the distance moved by P in each section, remembering that P moves forward for 15 seconds and then moves in the reverse direction.

Using average velocity × time gives the following results.

Time interval (s)	0–5	5–7.5	7.5–15	15–17.5	17.5–20
Distance moved (m)	37.5	37.5	56.25	6.25	6.25
Direction moved	fwrd	fwrd	fwrd	bkwrd	bkwrd

As P moves 131.25 m forward and then 12.5 m back, the *displacement* of P from O at the end of the 20 seconds is 118.75 m.

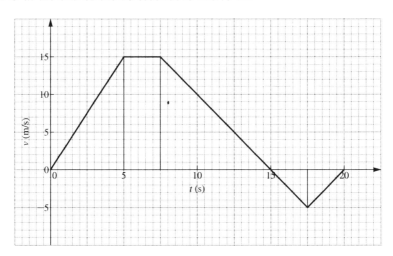

Now we can see that the distance moved in each section is represented by the area between that section of the graph and the time axis. If we take any area that is below this axis as negative, then the displacement of an object moving with constant acceleration is represented by the area between the velocity–time graph and the time axis.

When the acceleration of a moving object is not constant the velocity–time graph is curved. However the arguments we used earlier can be applied again to show that, for a general velocity–time graph:

The average acceleration over a time interval is represented by the gradient of
the corresponding chord. A negative gradient gives a negative acceleration,
indicating that the velocity is decreasing.

The acceleration at a given instant is represented by the gradient of the tangent to
the curve at that particular point on the curve.

The displacement is represented by the area between the curve and the time axis,
regions below that axis being negative.
An approximation to this area can be found
by dividing it into trapezium-shaped strips.

EXERCISE 1f

1.

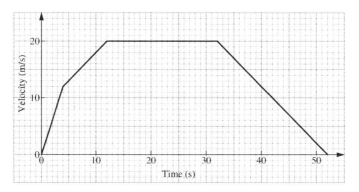

The graph shows the velocity of a car as it moves along a straight road, starting
from a lay-by at a point A.

Find

(a)   the acceleration during the first 4 seconds

(b)   the deceleration during the final 10 seconds

(c)   the distance travelled
      (i)   while accelerating
      (ii)   at constant speed
      (iii) while decelerating.

Explain why the displacement of the car from A at the end of the 52-second
journey is equal to the total *distance* travelled from A.

2. A train is brought to rest from a velocity of 24 m/s with a constant
acceleration of -0.8 m/s^2. Draw a velocity–time graph and find the distance
covered by the train while it is decelerating.

3. A particle is moving in a straight line with a velocity of 10 m/s when it is given an acceleration of -2 m/s^2 for 8 seconds. Draw a velocity–time graph for the eight-second time interval and use it to find

(a) the time when the direction of motion of the particle is reversed

(b) the increase in displacement during the 8 seconds

(c) the total distance travelled in this time.

4. The velocity of a runner was recorded at different times and the resulting velocity–time graph is given below.

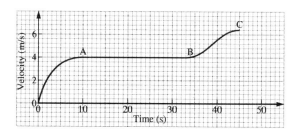

(a) Explain what is happening between
(i) O and A   (ii) A and B   (iii) B and C.

(b) Estimate the length of the race explaining whether your answer is more or less than the actual length.

5. A girl is taking part in an 'It's a Knock Out' game. Starting from her team's base, she has to run forwards to a row of buckets of water, pick one up and run back, trying not to spill any water, to a large cylinder which is at a distance behind the base. Her turn ends when she pours the water into the cylinder. This is the graph of her velocity plotted against time.

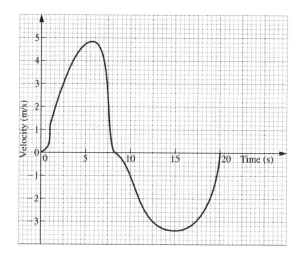

(a) At what time does the girl
   (i) pick up the bucket   (ii) empty the bucket?

(b) Estimate the acceleration of the girl when $t = 3$ and when $t = 12.5$

(c) At what times is the acceleration zero?

(d) Explain what happens after about 6 seconds.

Each question from 6 to 9 is followed by several suggested answers. State which is the correct answer.

6. The diagram shows the displacement–time graph for a particle moving in a straight line.

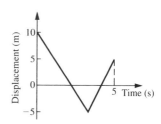

The average velocity for the interval from $t = 0$ to $t = 5$ is

A   0   B   6 m/s   C   −1 m/s   D   2 m/s

7. The diagram shows the displacement–time graph for a particle moving in a straight line.

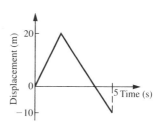

The distance covered by the particle in the interval from $t = 0$ to $t = 5$ is

A   20 m   B   50 m   C   15 m   D   5 m

8. The diagram shows the velocity–time graph for a particle moving in a straight line. The sum of the two shaded areas represents

A   the increase in displacement of the particle

B   the average velocity of the particle

C   the distance moved by the particle

D   the average speed of the particle.

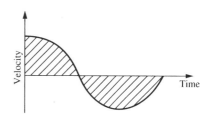

9. A particle moving in a straight line with a constant acceleration of 3 m/s^2 has an initial velocity of −1 m/s. Its velocity 2 seconds later is

A   5 m/s   B   6 m/s   C   4 m/s   D   −7 m/s

CHAPTER 2

CONSTANT ACCELERATION

MOTION WITH CONSTANT ACCELERATION

In Chapter 1 a number of relationships were observed linking the acceleration and velocity of a moving body with the displacement after any time interval.

In the particular case when the acceleration is uniform (i.e. constant) these relationships can be expressed as simple formulae which are known as the equations of motion with constant acceleration.

Consider first the velocity–time graph of an object moving for t seconds with constant acceleration a units.

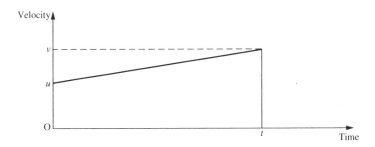

Suppose that at the beginning of the time interval the velocity is u units and at the end it is v units.

The velocity increases by a units each second so after t seconds the increase in velocity is at units.

$$\therefore \qquad v = u + at \qquad\qquad [1]$$

This formula can be used in solving a problem on motion with uniform acceleration, provided that three out of the four quantities u, v, a and t are known, so that the fourth quantity can be calculated. If this is not the case we need another relationship; this can be found if we consider the displacement, s units, of the object from its starting point after t seconds.

The area under the velocity–time graph represents the displacement. This is the area of a trapezium and is $\frac{1}{2}(u+v) \times t$.

$$\therefore \qquad s = \tfrac{1}{2}(u+v)t \qquad\qquad [2]$$

Now we have a formula to use in those problems where three out of the four quantities u, v, s and t are known.

There are however other possibilities. We could, for example, be given information on the values of u, a and t and have to find the displacement. Neither of the formulae found above link these four quantities but we can use them to deduce another relationship if we eliminate v.

From [1] $v = u + at$

Substituting in [2] gives $s = \frac{1}{2}(u+u+at)t$

i.e. $$s = ut + \tfrac{1}{2}at^2 \qquad\qquad [3]$$

In a similar way eliminating u gives

$$s = vt - \tfrac{1}{2}at^2 \qquad\qquad [4]$$

Lastly, a link between u, v, a and s can be found if t is eliminated from [1] and [2].

From [1] $$t = \frac{v-u}{a}$$

Substituting in [2] gives $s = \frac{1}{2}(v+u)\dfrac{(v-u)}{a}$

$$= \frac{1}{2a}(v^2 - u^2)$$

i.e. $$v^2 - u^2 = 2as \qquad\qquad [5]$$

With these formulae established, we are in a position to tackle, *by calculation*, any problem on motion with constant acceleration.

Each formula contains four quantities, but not the fifth, from u, v, a, s and t so it is easy to identify the one to use by noting which quantity is *not* involved.

However always remember that, as we have already seen, many problems can be solved quickly and easily from a velocity–time graph using only the two basic facts that the gradient gives the acceleration and the area under the graph gives the displacement. Solution by graphical methods should not be neglected because calculation is now an alternative. In fact, even when using the formulae, a velocity–time sketch graph often makes the solution clearer.

Choosing the Positive Direction

Displacement, velocity and acceleration are all vectors and therefore have a direction as well as a magnitude. Because we are considering only motion where the acceleration is constant (its direction is constant as well as its magnitude), it follows that all the motion takes place along a straight line. The object can, however, move either way along the line, so it is necessary to decide which is the positive direction; the opposite direction is then negative.

A good way to start is to make a list of the given information, and what is required, using the standard notation for initial and final velocities, displacement and time, giving each value its correct sign. Most motion problems are made clearer if a simple diagram is drawn, using different arrow heads to indicate different quantities. In this book we use:

$\longrightarrow\!\!\triangleright$	$\longrightarrow\!\!\gg$	$\longleftarrow\!\!\longrightarrow$
for velocity,	for acceleration,	for a length

and, later on, $\qquad\longrightarrow\!\!\!>\qquad$ for a force.

Examples 2a

1. **A particle starts from a point A with velocity 3 m/s and moves with a constant acceleration of $\frac{1}{2}$ m/s^2 along a straight line AB. It reaches B with a velocity of 5 m/s.**

   Find (a) the displacement of B from A   (b) the time taken from A to B.

Given: $u = 3,\ v = 5,\ a = \frac{1}{2}$

(a) Required: s
   t is not involved so the formula to use is $v^2 - u^2 = 2as$

   $$5^2 - 3^2 = 2(\tfrac{1}{2})s$$
   $$\therefore \qquad s = 16$$

   The displacement of B from A is 16 m.

(b) Required: t (s is not involved).
   The formula linking $u,\ v,\ a$ and t is $v = u + at$

   $$5 = 3 + \tfrac{1}{2}t$$
   $$\therefore \qquad t = 4$$

   The time taken from A to B is 4 seconds.

Note that, as the value of s was found in part (a), in part (b) we could have used the formula linking u, v, s and t, i.e. $s = \frac{1}{2}(u+v)t$, giving $16 = \frac{1}{2}(3+5)t$ and hence $t = \frac{16}{4} = 4$. Remember though that if you use a *calculated* value there is always the risk that it is not correct, making anything found from it wrong as well. Whenever possible it is best to use *given* values.

An alternative method of solution makes use of the velocity–time graph sketched from the given information.

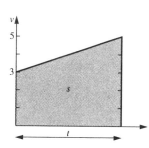

The gradient gives the acceleration.

$$\text{Gradient} = \frac{5-3}{t} = a = \frac{1}{2}$$

$$\therefore \quad t = 4 \quad \Rightarrow \quad \text{the time is 4 seconds.}$$

The area gives the displacement.

$$\text{Area} = \frac{1}{2}(3+5) \times t = \frac{1}{2}(8)(4) = 16$$

$$\therefore \quad s = 16 \quad \Rightarrow \quad \text{the displacement is 16 m.}$$

2. **A cyclist starts riding up a straight steep hill with a velocity of 8 m/s. At the top of the hill, which is 96 m long, the velocity is 4 m/s. Assuming a constant acceleration, find its value.**

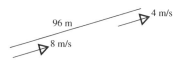

Given: $u = 8$, $v = 4$, $s = 96$

Required: a

The formula without t in it is $v^2 - u^2 = 2as$

$$4^2 - 8^2 = 2a(96)$$

$$\Rightarrow \qquad a = \frac{-48}{192} = \frac{-1}{4}$$

The acceleration is $-\frac{1}{4}$ m/s².

Remember that a negative acceleration is called a deceleration (or a retardation) and it indicates that velocity is decreasing.

A sketch of the velocity–time graph shows that this is the case.

3. The driver of a train begins the approach to a station by applying the brakes to produce a steady deceleration of 0.2 m/s² and brings the train to rest at the platform in 1 minute 30 seconds.

   (a) Find the speed of the train in kilometres/hour at the moment when the brakes were applied,

   (b) the distance then travelled before stopping.

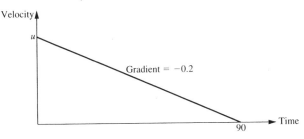

Working in metres and seconds we have:

Given: $a = -0.2$, $v = 0$, $t = 90$

Required: (a) u (b) s

(a) Using $v = u + at$ gives

$$0 = u + (-0.2)(90)$$

$$\therefore \qquad u = 18$$

The speed of the train was 18 m/s,

i.e. $\dfrac{18}{1000} \times 60 \times 60$ km/h $= 64.8$ km/h.

(b) Using $s = vt - \tfrac{1}{2}at^2$ gives

$$s = (0)(90) - \tfrac{1}{2}(-0.2)(90)^2 = 810$$

This is the displacement of the station from the point where the brakes were applied.

$\therefore$ the distance of the train from the station was 810 m.

In a situation where more than one object is involved, or where one object moves in different ways in various sections of the motion, it may be necessary to use 'link' quantities. These are unknown quantities that occur in the motion of more than one object or in more than one section. Whether the problem is tackled by using formulae or by referring to a graph, these quantities enable two incomplete pieces of information to be merged to produce a result.

4. At the same instant two children, who are standing 24 m apart, begin to cycle directly towards each other. James starts from rest at a point A, riding with a constant acceleration of 2 m/s² and William rides with a constant speed of 2 m/s. Find how long it is before they meet.

All quantities will be measured using metres and seconds.

James and William start together so they meet after riding for the same time.

For James

    Given: $u = 0$ Involved: time, t

              $a = 2$ distance, s_1

    Using $s = ut + \frac{1}{2}at^2$ gives $s_1 = \frac{1}{2}(2)t^2$ $\Rightarrow$ $s_1 = t^2$

For William

    Given: constant velocity, 2   Involved: time, t

                                        distance, s_2

    Using   distance = velocity × time gives $s_2 = 2t$

The sum of the distances they run is 24 m, i.e. $s_1 + s_2 = 24$

$\therefore$ $t^2 + 2t = 24$ $\Rightarrow$ $t^2 + 2t - 24 = 0$

                             $\Rightarrow$ $(t + 6)(t - 4) = 0$

                             $\Rightarrow$ $t = -6$ or $t = 4$

Only a positive value for t has any meaning.

The boys meet after 4 seconds.

Note that in this problem the link quantity is t. Neither of the two separate equations gives an actual value of anything, but each gives a different distance in terms of t. These expressions added are equal to the known total distance.

5. **A particle A starts from rest at a point O and moves on a straight line with constant acceleration 2 m/s². At the same instant another particle B, 12 m behind O, is moving with velocity 5 m/s and has a constant acceleration of 3 m/s². How far from O are the particles when B overtakes A?**

A and B travel for different distances. We will choose s to represent the shorter distance, i.e. the distance that A travels; B then travels a distance $(s + 12)$. A and B move for the same time.

For A

    Given: $u = 0$, $a = 2$ Involved: distance, s, and time, t

             $s = ut + \frac{1}{2}at^2$ $\Rightarrow$ $s = \frac{1}{2}(2)t^2$ [1]

For B

    Given: $u = 5$, $a = 3$ Involved: distance, $(s + 12)$, and time, t

             $s = ut + \frac{1}{2}at^2$ $\Rightarrow$ $s + 12 = 5t + \frac{1}{2}(3)t^2$ [2]

Although we want the distance we will eliminate s first and find t, because t appears to powers 1 and 2 and is difficult to eliminate.

$$[2] - [1] \quad \text{gives} \quad 12 = 5t + \tfrac{1}{2}t^2 \quad \Rightarrow \quad t^2 + 10t - 24 = 0$$
$$\Rightarrow \quad (t + 12)(t - 2) = 0$$
$$\Rightarrow \quad t = 2 \text{ or } -12$$

Using the positive value of t in [1] gives $s = 4$.

B overtakes A at a distance 4 m from O.

EXERCISE 2a

In questions 1 to 10 an object is moving with constant acceleration a m/s^2 along a straight line. The velocity at the initial point O is u m/s and t seconds after passing through O the velocity is v m/s. The displacement from O at this time is s m. For each question select the appropriate formula.

1. $u = 0,$ $a = 3,$ $v = 15;$ find t.

2. $t = 10,$ $s = 24,$ $u = 6;$ find v.

3. $a = 5,$ $u = 4,$ $s = 2;$ find v.

4. $u = 16,$ $v = 8,$ $t = 5;$ find s.

5. $t = 7,$ $u = 3,$ $v = 17;$ find a.

6. $t = 7,$ $u = 17,$ $v = 3;$ find a.

7. $u = 5,$ $t = 3,$ $a = -2;$ find s.

8. $a = -4,$ $u = 6,$ $v = 0;$ find t.

9. $v = 3,$ $t = 9,$ $a = 2;$ find s.

10. $v = 7,$ $t = 5,$ $a = 3;$ find u.

The remaining questions can be solved either by calculation from formulae or by reference to a velocity–time graph.

11. The driver of a car travelling on a motorway at 70 mph suddenly sees that the traffic is stationary at an estimated distance of 60 m ahead. He immediately applies the brakes which cause a deceleration of 6 m/s^2. Can a collision be avoided? (70 mph $\approx$ 32 m/s)

12. A bowls player projects the jack along the green with a speed of 4 m/s. It comes to rest 'short' at a distance of 25 m. What is the retardation caused by the surface of the green? With what speed should the jack be projected to reach a length of 30 m?

13. A particle is moving in a straight line with a constant acceleration of 2 m/s^2. If it was initially at rest find the distance covered

(a) in the first four seconds   (b) in the fourth second of its motion.

14. A particle moving in a straight line with a constant acceleration of -3 m/s^2, has an initial velocity at point A of 10.5 m/s.

    (a) Show that the times when the displacement from A is 15 m are given by $t^2 - 7t + 10 = 0$ and find these times.

    (b) Find the times when the displacement from A is -15 m.

15. A racing car is travelling at 130 mph when the driver sees a broken-down car on the track $\frac{1}{10}$ of a mile ahead. Slamming the brakes on he achieves his maximum deceleration of 24.5 mph per second. How far short of the broken-down car does he stop?

16. A body moving in a straight line with constant acceleration takes 3 seconds and 5 seconds to cover two successive distances of 1 m. Find the acceleration. (Hint: use distances of 1 m and 2 m from the start of the motion.)

17. The displacements from a fixed point O, of an object moving in a straight line with constant acceleration, are 10 and 14 metres at times of 2 and 4 seconds respectively after leaving O. Find

    (a) the initial velocity   (b) the acceleration

    (c) the time interval between leaving O and returning to O.

18. A particle starts from a point O with an initial velocity of 2 m/s and travels along a straight line with constant acceleration 2 m/s^2. Two seconds later another particle starts from rest at O and travels along the same line with an acceleration of 6 m/s^2. Find how far from O the second particle overtakes the first.

19. Starting from rest at one set of traffic lights, a car accelerates from rest to a velocity of 12 m/s. It maintains this speed for 42 seconds, until it decelerates to rest at the next set of red lights 60 seconds after leaving the first set. If the acceleration and deceleration are equal, find the distance between the two sets of lights.

20. A stolen car, travelling at a constant speed of 40 m/s, passes a police car parked in a lay-by. The police car sets off three seconds later, accelerating uniformly at 8 m/s^2. How long does the police car take to intercept the stolen vehicle and how far from the lay-by does this happen?

21. A particle P, moving along a straight line with constant acceleration 0.3 m/s^2, passes a point A on the line with a velocity of 20 m/s. At the instant when P passes A, a second particle Q is 20 m behind A and moving with velocity 30 m/s. Prove that, unless the motion of P and/or Q changes, the particles will collide.

22. A bus pulls away from a stop with an acceleration of 1.5 m/s^2 which is maintained until the speed reaches 12 m/s. At the same instant a girl who is 5 m away from the bus stop starts to run after the bus at a constant 7 m/s. Will the girl catch the bus?

FREE FALL MOTION UNDER GRAVITY

In the early days of the study of moving objects it was thought that if two objects with different masses were dropped, the heavier one would fall faster than the lighter. This idea was proved to be false by Galileo in a famous series of experiments; it is alleged that he dropped various objects from the top of the leaning tower of Pisa and timed their descents by the Cathedral clock opposite. Whether this anecdote is fact or fiction, the experiments showed that, regardless of their mass, all objects which gave rise to negligible air resistance had the same acceleration vertically downward when falling freely.

This *acceleration due to gravitational attraction* is represented by the letter g. Its value varies marginally in different parts of the world but it is acceptable to take 9.8 m/s^2 as a good approximation. In some circumstances 10 m/s^2 is good enough.

An object falling completely freely travels in a vertical line, so problems on its motion can be solved by using the equations for uniform acceleration in a straight line.

For bodies that are dropped, the downward direction is usually taken as positive, i.e. we take the positive value of g for the acceleration.

If, on the other hand, a ball is thrown vertically upwards, the upward direction could be chosen as positive; in this case the acceleration is $-g$.

Examples 2b

Take the value of g as 9.8 unless otherwise instructed.

1. **A brick is dropped from a scaffold board and hits the ground 3 seconds later. Find the height of the scaffold board.**

Anything that is 'dropped' from a stationary base is not thrown but released from rest, i.e. its initial velocity is zero.

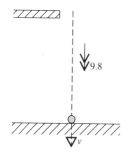

Taking the downward direction as positive:

Known: $u = 0$ Required: s

$a = 9.8$

$t = 3$

Using $s = ut + \frac{1}{2}at^2$ gives

$s = 0 + \frac{1}{2}(9.8)(3^2)$

$\therefore \quad s = 44.1$

The height of the scaffold is 44.1 m.

2. A boy throws a ball vertically upwards from a seven-metre-high roof.

(a) If, after 2 seconds, he catches the ball on its way down again, with what speed was it thrown?

(b) What is the velocity of the ball when it is caught?

(c) If the boy fails to catch the ball with what speed will it hit the ground?

Taking the upward direction as positive:

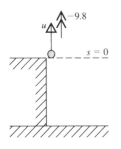

(a) Known: $t = 2$ Required: u

$a = -9.8$

$s = 0$

Using $s = ut + \frac{1}{2}at^2$ gives

$0 = 2u + \frac{1}{2}(-9.8)(4)$

$\therefore \quad u = 9.8$

The ball was thrown with a speed of 9.8 m/s.

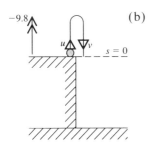

(b) Known: $t = 2$ Required: v

$a = -9.8$

$s = 0$

Using $s = vt - \frac{1}{2}at^2$ gives

$0 = 2v - \frac{1}{2}(-9.8)(4)$

$\therefore \quad v = -9.8$

The ball is *falling* at 9.8 m/s when caught.

(c) The ground is 7 m below the point of projection so the final displacement of the ball is -7 m.

Working from the time when the ball was thrown we have:

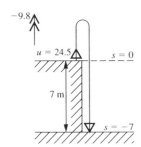

Known: $u = 9.8$ Required: v

$s = -7$

$a = -9.8$

Using $v^2 - u^2 = 2as$ gives

$v^2 - (9.8)^2 = 2(-9.8)(-7)$

$\Rightarrow \quad v^2 = 96.04 + 137.2 = 233.24$

$\therefore \quad v = 15.27\ldots$

The ball hits the ground at 15.3 m/s (3 sf).

Note that in parts (a) and (b) we have shown that a particle thrown upwards with a velocity u, returns to the same level with a velocity $-u$, i.e. with equal speed in the downward direction.

3. **One clay pigeon, A, is fired vertically upwards with a speed of 40 m/s and 1 second later another one, B, is projected from the same point with the same velocity. Taking the acceleration due to gravity as 10 m/s², find**

(a)   the time that elapses before they collide

(b)   the height above the point of projection at which they meet.

When B has been in the air for t seconds, A has been in the air for $(t+1)$ seconds. (In a case like this, using t for the shorter time interval and $(t+1)$ for the longer one, rather than t for the first and $(t-1)$ for the second, avoids the algebraic mistakes that can occur when minus signs are involved.)

Taking the upward direction as positive we have:

	For A		For B
Known:	$u = 40$	Known:	$u = 40$
	$a = -10$		$a = -10$
	time $= (t+1)$		time $= t$
Required:	s	Required:	s

(a)   Using $s = ut + \frac{1}{2}at^2$ gives:

For A $s = 40(t+1) + \frac{1}{2}(-10)(t+1)^2$

For B $s = 40t + \frac{1}{2}(-10)t^2$

$\therefore$ $40(t+1) - 5(t+1)^2 = 40t - 5t^2$

$\Rightarrow$ $40 - 10t - 5 = 0$

$\Rightarrow$ $t = 3.5$

The projectiles collide 3.5 s after B is fired.

(b)   As A and B are at the same height when they collide we can find s for either of them.

Considering B,

$s = 40(3.5) - 5(3.5)^2 = 78.75$

A and B collide 78.8 m above the point of projection (3 sf).

4. A missile is fired vertically upwards with speed **47 m/s**. Find, to **3 significant figures**, the time that elapses before the missile returns to the firing position if it is projected from a point on

(**a**) a seashore where g can be taken as **9.819**

(**b**) a high mountain where g can be taken as **9.783**

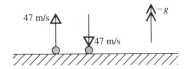

We know that the missile will return with a speed of 47 m/s downwards. Taking the upward direction as positive, we have in either case,

> Known: $u = 47$ Required: t
> $v = -47$
> the value of g

(**a**) Using $v = u + at$ with $a = -9.819$ gives
$$-47 = 47 + (-9.819)t \quad \Rightarrow \quad t = 9.573\ldots$$

The time of flight of the missile is 9.57 s (3 sf)

(**b**) Again using $v = u + at$ but with $a = -9.783$ gives
$$-47 = 47 + (-9.783)t \quad \Rightarrow \quad t = 9.608\ldots$$

The time of flight of the missile is 9.61 s (3 sf).

5. A youth is playing with a ball in a garden surrounded by a wall **2.5 m** high and kicks the ball vertically up from a height of **0.4 m** with a speed of **14 m/s**. For how long is the ball above the height of the wall?
Give answers corrected to **2 significant figures**.

We will measure displacement upward from a point 0.4 m above the ground.

> Known: $u = 14$ Required: t
> $a = -9.8$
> $s = 2.1$
> Using $s = ut + \frac{1}{2}at^2$ gives
> $$2.1 = 14t - 4.9t^2$$
> $\Rightarrow$ $0.3 = 2t - 0.7t^2$
> $\Rightarrow$ $7t^2 - 20t + 3 = 0$

$s = (2.5 - 0.4)$

-9.8

$u = 14$

$s = 0$

0.4

t is found from this quadratic equation by using the formula.

i.e. $t = \dfrac{20 \pm \sqrt{(400 - 84)}}{14} = 2.698$ or 0.159

The ball is at the height of the top of the wall at two different times. Therefore it takes 0.159 seconds to reach the top of the wall when going up, and returns to that height 2.698 seconds from the start. So the ball is above the wall for 2.5 s, corrected to 2 significant figures.

EXERCISE 2b

Unless another instruction is given, take g as 9.8 and give answers corrected to 2 significant figures.

1. A stone dropped from the top of a cliff takes 5 seconds to reach the beach.

   (a) Find the height of the cliff.

   (b) With what velocity would the stone have to be thrown vertically downward from the top of the cliff, to land on the beach after 4 seconds?

2. A particle is projected vertically upward from ground level with a speed of 24.5 m/s. Find

   (a) the greatest height reached

   (b) the time that elapses before the particle returns to the ground.

3. A slate falls from the roof of a high-rise building. Find how far it falls

   (a) in the first second

   (b) in the first two seconds

   (c) during the third second.

4. A ball is thrown vertically upward and is caught at the same height 3 seconds later. Find

   (a) the distance it rose

   (b) the speed with which it was thrown.

5. A brick is dropped down a disused well, 50 m deep.

   (a) For how long does it fall?

   (b) With what speed does it hit the bottom?

6. A boulder slips from the top of a precipice and falls vertically downwards on to a plain 200 m below.

   (a) Find, to 3 significant figures, the speed of the boulder when it hits the plain if the precipice is

       (i) in a polar region where, to four significant figures, the acceleration due to gravity is 9.830 m/s^2

       (ii) in a region near to the equator where the acceleration due to gravity is 9.781 m/s^2 (4 sf)

   (b) Can you find a possible explanation for the difference in the values of g at the two locations (you may need to refer to an encyclopedia).

7. A parachutist is descending vertically at a steady speed of 2 m/s when his watch strap breaks and the watch falls. If the watch hits the ground 3 seconds later at what height was the parachutist when he dropped it?

8. A mine inspector has run into trouble 30 m down an open vertical shaft. To summon help he fires a distress flare straight up the shaft. In order for the flare to be seen, it must reach at least 10 m above the ground level. What is the least speed with which it must be fired?

9. A ball is thrown vertically, with a speed of 7 m/s from a balcony 14 m above the ground. Find how long it takes to reach the ground if it is thrown

(a) downwards   (b) upwards.

Find also the speed with which it reaches the ground in each of these cases.

10. A stone is dropped from the top of a building and at the same instant another stone is thrown vertically upward from the ground below at a speed of 15 m/s. If the stones pass each other after 1.2 seconds find the height of the building.

11. A youth playing in a yard surrounded by a 3 metre-high wall kicks a football vertically upward with initial speed 15 m/s.

(a)   What is the greatest height reached by the ball?

(b)   For how long can it be seen by someone on the other side of the wall?

12. A small ball is released from rest at a point 1.6 m above the floor. When it hits the floor its speed is halved by the impact. How high does it bounce?

13. A stone is dropped from the top of a building to the ground. During the last second of its fall it moves through a distance which is $\frac{1}{5}$ of the height of the building. How high is the building?

14. A competitor is attempting a dive from a springboard that is 6 m above the water. He leaves the springboard with an upward velocity of 7 m/s. Taking the value of g as 10, find the speed at which the diver enters the water and the time for which he is in the air.

15. The defenders of Castle Dracula dropped large rocks from the battlements on to an attacking army. If the height of the battlements above the ground was 35 cubits find, in cubits per second, the speed at which the rocks that missed their targets hit the ground. Take the value of g as 20 cubits per second per second. (A cubit is an ancient measure of length based on a man's forearm and is approximately half a metre.)

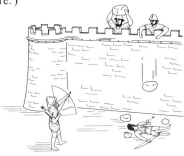

CHAPTER 3

VECTORS

DEFINING A VECTOR

A vector quantity is one for which direction is important as well as magnitude (i.e. size). We have already met, in Chapter 1, some important vector quantities.

Displacement is a distance measured in a particular direction, e.g.
'10 miles due north' is a vector,
whereas '10 miles' is a distance with no specified direction so is not a vector.
A quantity that possesses magnitude only is a *scalar* quantity.
Distance is the magnitude of displacement.

Velocity includes both speed and direction of motion, so
'150 km/h on a bearing of 154°' is a velocity and is a vector,
whereas 'a speed of 150 km/h' is a scalar.
Speed is the magnitude of velocity.

Acceleration is the rate at which velocity is increasing so it follows that acceleration depends on both the speed and direction of motion, i.e. acceleration is a vector.
(Note that there is no different word for the magnitude of acceleration.)

Force is another quantity which plays an important part in the study of mechanics. Clearly if a force pushes an object we need to know both the size of the push and also which way it is acting, i.e. force is a vector.
Force is measured in newtons (N).

VECTOR REPRESENTATION

Any vector can be represented by a section of a line (called a *line segment*). The direction of the line gives the direction of the vector and the length of the line represents the magnitude of the vector.

If the line is labelled AB,
the vector it represents is written $\overrightarrow{AB}$.
The order of the letters indicates the direction of
the vector.
(The vector $\overrightarrow{BA}$ is represented by the same line
but in the opposite direction.)

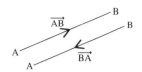

Alternatively a single lower case letter in the middle of the line can be used.
In this case there *must* be an arrow on the line to show the direction of the
vector.
In print the letter is set in bold type, e.g. **a**.
For hand-written work use $\underline{a}$ or $\underset{\sim}{a}$.
The magnitude of **a** is written as a

Related Vectors

Two vectors are *equal* if they have equal magnitudes
and the same direction.

We write **a** = **b**

If the direction of **b** is reversed then **a** and **b** are
equal and opposite.

We write **a** = −**b**

If two vectors have the same direction
but different magnitudes then one can be
expressed as a multiple of the other, e.g.

$$\mathbf{b} = 2\mathbf{a} \quad \text{and} \quad \mathbf{q} = 3\mathbf{p}$$

In general, if **a** and **b** are parallel then

$$\mathbf{a} = k\mathbf{b} \quad \text{where } k \text{ is a constant of proportion and is scalar.}$$

For example, if $\overrightarrow{AB}$ represents a vector **p**,
then $\frac{1}{3}\mathbf{p}$ is represented by $\overrightarrow{AC}$
where $AC = \frac{1}{3}AB$
and $2\mathbf{p}$ is represented by $\overrightarrow{AD}$
where AB is extended so that $AD = 2AB$.

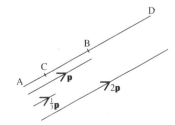

ADDITION OF VECTORS

Consider what happens if a hiker starts from the corner A of a field, walks for 30 m beside the hedge along one side to B and then 40 m along the side perpendicular to the first, to C. The hiker could have reached the same point C by walking directly across the field (assuming this to be allowed!).

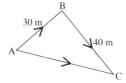

So combining the two displacements $\overrightarrow{AB}$ and $\overrightarrow{BC}$ gives the same final result as the single displacement $\overrightarrow{AC}$.
This is what is meant by *adding vectors* and we can write

$$\overrightarrow{AB} + \overrightarrow{BC} = \overrightarrow{AC}$$

or $\qquad\qquad \mathbf{a} + \mathbf{b} = \mathbf{c}$

$\overrightarrow{AC}$ is called the *resultant* of $\overrightarrow{AB}$ and $\overrightarrow{BC}$.

Note that, in this context,   +   means 'together with' or 'followed by'
          and   =   means 'is equivalent to'.

Note also that, from A we can go to C either directly or via the vectors $\overrightarrow{AB}$ and $\overrightarrow{BC}$. The first point and last point are the same in both cases. Triangle ABC is known as a *triangle of vectors* and when we use it to add vectors we are using the *triangle law*.

Using the Triangle Law

Given two vectors **a** and **b**, represented by line segments $\overrightarrow{OA}$ and $\overrightarrow{OB}$, we can draw diagrams to represent various combinations of **a** and **b**.

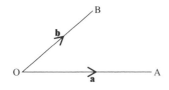

Now **a** + **b** means **a** *followed by* **b**. We can represent this by drawing AC equal and parallel to OB. Then $\overrightarrow{OC}$ represents **a** + **b**, i.e.

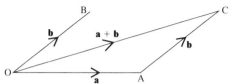

Note that we could equally well have drawn **b** followed by **a**.

To represent **a** − **b** we can draw **a** followed by −**b**.

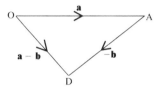

Alternatively we can draw −**b** followed by **a**.

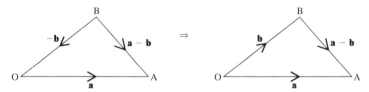

Any number of vectors can be added by the same process, e.g. **a** + **b** + **c** + **d** means **a** followed by **b** followed by **c** followed by **d** and can be represented by $\overrightarrow{OA}$, $\overrightarrow{AB}$, $\overrightarrow{BC}$ and $\overrightarrow{CD}$ as shown.

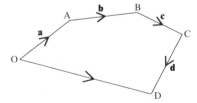

From the diagram we see that $\overrightarrow{OD}$ is equivalent to $\overrightarrow{OA} + \overrightarrow{AB} + \overrightarrow{BC} + \overrightarrow{CD}$

i.e. $\overrightarrow{OD} = \mathbf{a} + \mathbf{b} + \mathbf{c} + \mathbf{d}$

So $\overrightarrow{OD}$ is the resultant of **a**, **b**, **c** and **d**

Note that the arrow we use for marking a resultant vector is larger than those used for the vectors being added.

Examples 3a

1.

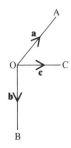

Given the vectors **a**, **b** and **c**, represented by OA, OB and OC as shown, sketch a diagram to illustrate **a** + **b** + **c**.

a + **b** + **c** means **a** followed by **b** followed by **c** so we draw OA followed by a line AD, equal and parallel to OB, followed by a line DE, equal and parallel to OC.

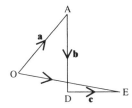

$\overrightarrow{OE}$ represents **a** + **b** + **c**.

2. In a triangle ABC, M is the midpoint of AC. $\overrightarrow{AB} = $ a and $\overrightarrow{BC} = $ b.
Find, in terms of a and b, (a) $\overrightarrow{AC}$ (b) $\overrightarrow{CA}$ (c) $\overrightarrow{AM}$ (d) $\overrightarrow{MB}$.

In the diagram, $\overrightarrow{AC}$ is equivalent to $\overrightarrow{AB}$ followed by $\overrightarrow{BC}$.

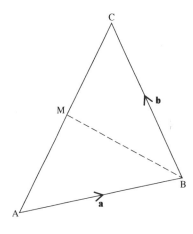

(a) $\overrightarrow{AC} = \overrightarrow{AB} + \overrightarrow{BC}$
$= \mathbf{a} + \mathbf{b}$

(b) $\overrightarrow{CA} = -\overrightarrow{AC}$
$= -(\mathbf{a} + \mathbf{b})$

(c) $\overrightarrow{AM} = \tfrac{1}{2}\overrightarrow{AC}$
$= \tfrac{1}{2}(\mathbf{a} + \mathbf{b})$

(d) In triangle MAB,

$\overrightarrow{MB} = \overrightarrow{MA} + \overrightarrow{AB} = -\overrightarrow{AM} + \overrightarrow{AB}$
$= -\tfrac{1}{2}(\mathbf{a} + \mathbf{b}) + \mathbf{a}$

$\therefore \qquad \overrightarrow{MB} = \tfrac{1}{2}(\mathbf{a} - \mathbf{b})$

3. PQRST is a polygon in which $\overrightarrow{PQ} = p$, $\overrightarrow{QR} = q$, $\overrightarrow{RS} = r$, and $\overrightarrow{ST} = s$.

(a) Find the vector represented by $\overrightarrow{PS}$.

(b) Describe the line that represents $\frac{1}{3}(p + q + r + s)$.

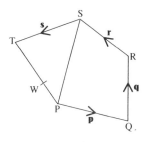

(a) From the diagram we see that $\overrightarrow{PS}$ is equivalent to $\overrightarrow{PQ} + \overrightarrow{QR} + \overrightarrow{RS}$

$\therefore \quad \overrightarrow{PS} = p + q + r$

(b) $p + q + r + s$ is represented by $\overrightarrow{PT}$

$\frac{1}{3}(p + q + r + s)$ is represented by $\frac{1}{3}$ of $\overrightarrow{PT}$

$\therefore \quad \frac{1}{3}(p + q + r + s)$ is represented by $\overrightarrow{PW}$ where $PW = \frac{1}{3}PT$.

4. A, P, Q and R are four points such that $\overrightarrow{AP} = 2a$, $\overrightarrow{AQ} = 3b$ and $\overrightarrow{AR} = 9b - 4a$. Show that P, Q and R are collinear.

Collinear means 'in the same straight line'. For P, Q and R to be collinear, PQ and QR (or PR) must be parallel.

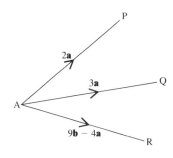

$$\overrightarrow{PQ} = \overrightarrow{PA} + \overrightarrow{AQ} = -2a + 3b$$

$$\overrightarrow{QR} = \overrightarrow{QA} + \overrightarrow{AR} = -3b + (9b - 4a)$$

$$= -4a + 6b$$

Hence $\quad \overrightarrow{QR} = 2\overrightarrow{PQ}$

i.e. QR is parallel to PQ

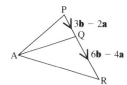

$\therefore$ P, Q and R are collinear.

5. In the triangle OAB, $\overrightarrow{OA} = 3a$, $\overrightarrow{OB} = 3b$, $OQ = \frac{1}{3}OA$ and $OP = \frac{1}{3}OB$. AP and BQ meet at R.

(a) Express $\overrightarrow{PA}$ and $\overrightarrow{QB}$ each in terms of a and b.

(b) Given that $QR = kQB$ and $PR = hPA$, find two expressions for $\overrightarrow{OR}$ in terms of a, b, h and k by considering

(i) $\triangle OQR$ (ii) $\triangle OPR$.

(c) By equating these two expressions find the value of k and hence find the ratio in which R divides QB.

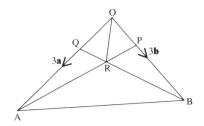

(a) $\overrightarrow{PA} = \overrightarrow{PO} + \overrightarrow{OA} = \frac{1}{3}\overrightarrow{BO} + \overrightarrow{OA} = -\mathbf{b} + 3\mathbf{a}$

$\overrightarrow{QB} = \overrightarrow{QO} + \overrightarrow{OB} = \frac{1}{3}\overrightarrow{AO} + \overrightarrow{OB} = -\mathbf{a} + 3\mathbf{b}$

(b) (i) $\overrightarrow{OR} = \overrightarrow{OQ} + \overrightarrow{QR} = \overrightarrow{OQ} + k\overrightarrow{QB}$
$= \mathbf{a} + k(3\mathbf{b} - \mathbf{a})$

(ii) $\overrightarrow{OR} = \overrightarrow{OP} + \overrightarrow{PR} = \mathbf{b} + h(3\mathbf{a} - \mathbf{b})$

(c) $\mathbf{a} + k(3\mathbf{b} - \mathbf{a}) = \mathbf{b} + h(3\mathbf{a} - \mathbf{b})$

$\Rightarrow (1 - k)\mathbf{a} + 3k\mathbf{b} = 3h\mathbf{a} + (1 - h)\mathbf{b}$

These vectors can be equal only if they are identical, i.e. the coefficients of both **a** and **b** must be equal.

i.e. $1 - k = 3h$ and $3k = 1 - h$

$\therefore \quad 1 - k = 3(1 - 3k) \qquad \Rightarrow \qquad 8k = 2$

Hence $k = \frac{1}{4}$

$\therefore \quad QR = \frac{1}{4}QB$ and $RB = \frac{3}{4}QB$

$\therefore \quad$ R divides QB in the ratio $1:3$.

EXERCISE 3a

1.

Express, in terms of **a** and **b**, the vector represented by
(a) $\overrightarrow{AB}$ (b) $\overrightarrow{BA}$.

2. Given the vectors **p**, **q** and **r** as shown, draw diagrams to illustrate

(a) $\mathbf{p} + \mathbf{r}$ (b) $\mathbf{q} - \mathbf{p}$ (c) $\mathbf{r} - \mathbf{q}$

(d) $\mathbf{p} + \mathbf{q} - \mathbf{r}$ (e) $\mathbf{r} - \mathbf{p} + \mathbf{q}$.

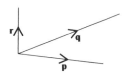

3. 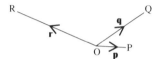 Find the vector represented by

(a) $\overrightarrow{PQ}$ (b) $\overrightarrow{QR}$ (c) $\overrightarrow{PR}$

Give your answers in terms of **p**, **q** and **r**.

4. In the diagram, $\overrightarrow{OA} = \mathbf{a}$, $\overrightarrow{OB} = \mathbf{b}$ and C is the midpoint of AB. Express each of the following vectors in terms of **a** and **b**.

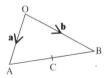

(a) $\overrightarrow{AB}$ (b) $\overrightarrow{AC}$ (c) $\overrightarrow{BC}$

5. In a quadrilateral ABCD, $\overrightarrow{AB} = \mathbf{a}$, $\overrightarrow{BC} = \mathbf{b}$, and $\overrightarrow{CD} = \mathbf{c}$. Express in terms of **a**, **b** and **c**,

(a) $\overrightarrow{AC}$ (b) $\overrightarrow{BD}$ (c) $\overrightarrow{DB}$ (d) $\overrightarrow{DA}$

6. Draw diagrams to illustrate each vector equation.

(a) $\overrightarrow{AB} = 2\overrightarrow{PQ}$ (b) $\overrightarrow{AB} - \overrightarrow{CB} = \overrightarrow{AC}$ (c) $\overrightarrow{AB} + \overrightarrow{BC} = 3\overrightarrow{AD}$

7. In triangle ABC, D bisects BC. Prove that $\overrightarrow{BA} + \overrightarrow{AC} = 2\overrightarrow{DC}$.

8. In a regular hexagon PQRSTU, $\overrightarrow{QR}$ represents a vector **a** and $\overrightarrow{UR}$ represents a vector 2**b**.

Express in terms of **a** and **b** the vectors represented by

(a) $\overrightarrow{PQ}$ (b) $\overrightarrow{ST}$ (c) $\overrightarrow{QT}$

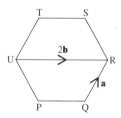

9. In triangle ABC, P and Q are the midpoints of AB and BC respectively. Use vectors to prove the midpoint theorem, i.e. that PQ is parallel to BC and half its length.

10. Given a pentagon ABCDE,

(a) express as a single vector

   (i) $\overrightarrow{AB} + \overrightarrow{BC} + \overrightarrow{CD}$

   (ii) $\overrightarrow{BC} + \overrightarrow{AB}$

   (iii) $\overrightarrow{AB} - \overrightarrow{AE}$

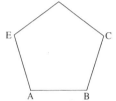

(b) find two ways in which $\overrightarrow{AD}$ can be expressed as a sum or difference of a number of vectors.

11. In the diagram, A, B and C are collinear. Find the value of k.

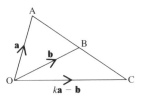

12. In the triangle ABC, $\overrightarrow{AB} = 2\mathbf{a}$, $\overrightarrow{AC} = 2\mathbf{b}$, E and F are the midpoints of AC and AB. BE and CF meet at G.

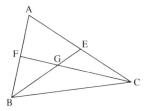

(a) Express $\overrightarrow{BE}$ and $\overrightarrow{CF}$ each in terms of $\mathbf{a}$ and $\mathbf{b}$.

(b) $EG = h\text{EB}$ and $FG = k\text{FC}$. By referring to triangles AFG and AEG, express $\overrightarrow{AG}$ as a sum of vectors in two different ways and hence find the values of h and k.

(c) In what ratio does G divide BE?

RESULTANTS AND COMPONENTS

At this stage in the book the work begins to require some knowledge of trigonometry. Any readers who have not yet acquired this skill should postpone going further with this chapter until they have studied the basics of trigonometry.

When two (or more) vectors are added, the single equivalent vector is called the *resultant vector.* **The vectors that are combined are called** *components.*

Consider the example of a heavy crate being pulled along by two ropes. The unit in which the forces in the ropes are measured is the newton (N).

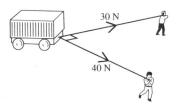

Although the ropes are pulling in different directions, the crate moves in only one direction. This is the direction of the resultant of the tensions (i.e. the pulling forces) in the ropes. By drawing a *triangle of vectors* we can find both the magnitude and the direction of the resultant force. (Remember that we use a larger arrow for the resultant.)

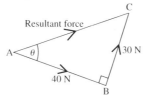

Note that a triangle of vectors *does not necessarily* give the *positions* of the components or the resultant. In this case, for example, each of the components acts on the crate so the equivalent resultant force also acts on the crate.

The magnitude and direction of the resultant force can be found by calculation.

In $\triangle ABC$: $AC^2 = AB^2 + BC^2$ so AC represents 50 N

and $\quad \tan A = \frac{3}{4} \quad (\tan = \frac{\text{opp}}{\text{adj}})$

$\Rightarrow \quad \hat{A} = 37°$ (nearest degree)

Hence the resultant force is of magnitude 50 N acting at an angle of 37° to AB. The calculation is easy because the components are at right angles.

This example can be extended to a general case where two vectors $\overrightarrow{AB}$ and $\overrightarrow{BC}$ are perpendicular. If the magnitudes of $\overrightarrow{AB}$ and $\overrightarrow{BC}$ are p and q, the magnitude of the resultant vector $\overrightarrow{AC}$ is given by $\sqrt{(p^2 + q^2)}$.

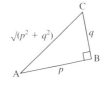

An alternative method is to draw, to scale, AB followed by BC and then join the *starting point* to the *finishing point*. The magnitude and direction of the resultant can then be measured from the drawing.

Scale drawing can be used also to find the resultant of any two or more vectors; this avoids the trigonometric calculations involved when the vectors are not perpendicular to each other.

Examples 3b

1. Find, by calculation, the resultant of two velocities if one is 7 km/h south west and the other is 12 km/h south east.

A sketch is drawn starting with a line representing one velocity followed by a line representing the other one.

AB represents 7 km/h south west.
BC represents 12 km/h south east.
AC represents the resultant velocity.

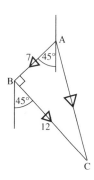

In $\triangle$ABC, $AC^2 = AB^2 + BC^2 = 193$

$\therefore$ $AC = 13.9 \, (3 \text{ sf})$

           $\tan A = BC/AB = \frac{12}{7} = 1.714$

$\therefore$ $A = 60°$ (to the nearest degree)

$\therefore$ the bearing of AC is $225° - 60° = 165°$

The resultant velocity is 13.9 km/h on a bearing of 165°.

2. A light aircraft is flying in still air at 180 km/h on a bearing of 052°. A steady wind suddenly springs up, blowing due south at 70 km/h. Find, by scale drawing, the velocity of the aircraft over the ground.

To find the resultant velocity we add the velocities of the plane and the wind.

Scale: 1 cm to 20 km/h

Measuring from the drawing gives the resultant velocity as

       148 km/h on a bearing of 074°

An alternative method of solution would be to use the cosine rule followed by the sine rule.

3. In a test of strength, a team of four competitors attempts to move a large sack of stones, each one pulling a rope attached to the sack. The magnitude of the force exerted by each competitor is shown in the diagram. Find the resultant force acting on the sack.

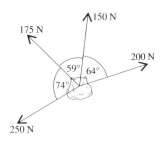

We can find the resultant by drawing lines representing, to scale, the forces taken in order and then joining the first point to the last.

Using a scale of 1 cm to 50 N, the forces are represented by AB, BC, CD and DE. The resultant is then represented by AE.

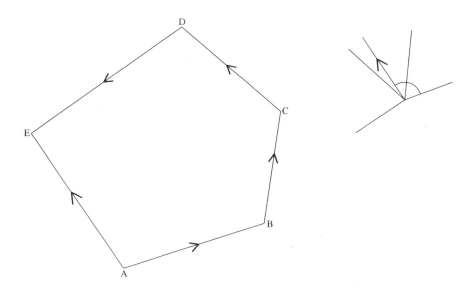

The resultant is a force of 220 N inclined at 108° to the force of 200 N.

If one of two components is given and the resultant is known, the other component can be found by adapting this method.

4. The pilot of a light plane with a speed of 160 km/h when the air is still, has to fly from a base B to an airfield A that is on a bearing of 130° from B. The pilot learns that a wind is blowing at 40 km/h due east.

(a) On what bearing should he set his course in order to fly directly to the airfield?

(b) What is the speed of the plane over the ground?

(c) If the distance AB is 240 km, how long does the flight take?

The resultant velocity of the plane is in the direction $\overrightarrow{BA}$ but we do not know its magnitude. The direction in which the plane must steer is not known but the magnitude is 160 km/h. The velocity of the wind is known in magnitude and direction.

Draw a *sketch* starting with a line BC to represent the wind velocity. From B draw a line on a bearing of 130° to represent the resultant velocity; the length of this line is unknown.
From C a line representing the basic velocity of the plane, i.e. 160 km/h, must be drawn to complete a triangle BCD where D is a point on the resultant velocity. This triangle gives the direction of CD and the length of BD. (To do this accurately we would use a compass and, with centre C, draw to scale an arc of radius 160 to cut the line representing the resultant velocity.)

A scale drawing can now be made based on this sketch.

Because space is restricted here the scale used is 1 cm to 40 km/h, but in practice a larger scale should be used to allow greater accuracy of measurement.

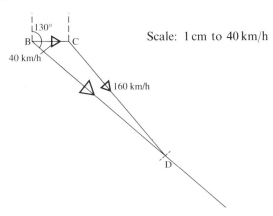

Scale: 1 cm to 40 km/h

(a) The course to be set is 139° (3 sf).

(b) The speed over the ground (BD) is 189 km/h (3 sf).

(c) 240 km at 189 km/h takes 1 hour 16 minutes.

EXERCISE 3b

In questions 1 to 10, find the resultant of the given vectors. Use either
calculation or scale drawing and remember always to give both the magnitude
and the direction of the resultant.

1. A displacement of 12 km south followed by a displacement of 5 km east.

2. A displacement of 5 km east followed by a displacement of 12 km south.
   Is there any difference between your answer to this question and the answer to
   question 1?

3. A velocity of 24 m/s north and a velocity of 7 m/s east.

4. A force of 12 N west and a force of 16 N south.

5. Displacements of 10 m east and 12 m north east.

6. Two velocities, one of 4 m/s south and the other 5 m/s on a bearing 120°.

7. The two forces shown in the diagram (the direction of the
   resultant force can be given as an inclination to either of the
   given forces).

8. 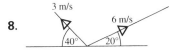   The two velocities shown in the diagram.

9. Displacements of 10 m east, 14 m north and 21 m on a bearing of 260°.

10. A man leaves his home by car and travels 5 km on a road running due east.
    The driver then turns left and travels 2 km due north to a junction where he joins
    a road that goes north west, and drives a further 2 km. Find, from a scale
    drawing, his displacement from home by then.

11. An aircraft, flying with an engine speed of 400 km/h, is set on a course due
    north, in a wind of speed 60 km/h *from* the south west. At what speed and in
    what direction is the aircraft covering the ground?

12. On an orienteering exercise a woman starts from base and walks 500 m on a
    bearing of 138° and then 750 m on a bearing of 080°. What is her displacement
    from base? What bearing should she set to return to base?

In questions 13 and 14, a river running from north to south is flowing at 3 km/h.

13. A girl who can swim at 4 km/h is aiming directly across the river from east
    to west. Find the actual direction of her course across the river and the speed at
    which she passes over the river bed.

14. Another girl, who also swims at 4 km/h, notices that her friend is not moving straight across the river and works out how to go directly across. Draw a sketch of the way she does it and find the speed at which she crosses.

15. A plane leaves an airfield P, flying in a wind of 50 km/h blowing in a direction 048°. The plane arrives at an airfield Q, 400 km due east of P, 2 hours later. By making an accurate scale drawing find the plane's engine speed and the bearing the pilot set for the flight.

16. A fisherman wants to row from one side, P, of the harbour to the other, Q. He can row at 8 mph when the water is still. There is a tide running out of the harbour at 4 mph.

(a) If he steers the boat in the direction PQ, in which direction will he actually move over the sea bed?

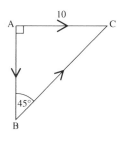

(b) Find the direction in which he must steer in order to cross over directly.

FINDING THE COMPONENTS OF A VECTOR

We have seen that two vectors can be combined into a single resultant vector. Now we will examine the reverse process, i.e. the replacing one vector by an equivalent pair of vectors. This process is called *resolving a vector into components*.

Suppose for example that a vector of magnitude 10 m east is to be replaced by two components, one of them due south and the other north east.

In the diagram, $\overrightarrow{AC}$ represents the given vector.
$\overrightarrow{AB}$ represents the component due south and
$\overrightarrow{BC}$ represents the component north east.
The directions of these components are known but their magnitudes are not.
However, the lengths of the lines AB and BC, and hence the magnitudes of the components, can be found.

$\triangle$ABC is an isosceles right-angled triangle, therefore   AB = AC = 10.
Also Pythagoras' theorem gives $BC^2 = AB^2 + AC^2$ $\Rightarrow$ BC = 14.1   (3 sf)

Therefore the required components are 10 m due south and 14.1 m north east.

Perpendicular Components

A vector can be resolved into an infinite variety of components in different directions but the most useful components, and the easiest to find, are a perpendicular pair.

Consider, for example, a plane taking off at an angle of 30° to the runway at 150 km/h.

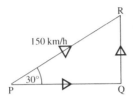

The horizontal and vertical components of the velocity can be found by using trigonometry, i.e.

the horizontal component is given by $PQ = 150 \cos 30° = 130$ (3 sf)

and the vertical component is given by $QR = 150 \sin 30° = 75$.

Calculating the components of a vector plays a very important part in solving mechanics problems so it is important that they can be written down immediately in the form above.

Any reader who, up to now, would first have written down $\dfrac{PQ}{150} = \cos 30°$

should practice going straight to the form $PQ = 150 \cos 30°$; otherwise a great deal of time will be wasted.

Examples 3c

1. A skier ascends at a constant speed of 5 m/s in a chair lift inclined at 27° to the horizontal.

   (a)   Find the horizontal and vertical components of her velocity.

   (b)   What difference is there between the components found in part (a) and the horizontal and vertical components of the velocity of the chair as it returns to base at the same speed?

(a)

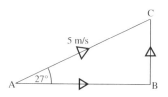

AB = 5 cos 27°

∴ the horizontal component is 4.46 m/s (3 sf).

BC = 5 sin 27°

∴ the vertical component is 2.27 m/s (3 sf).

(b)   As the chair descends, each velocity component has the same magnitude, but is in the opposite direction, from that in part (a)

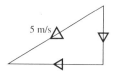

This can be indicated by a minus sign,

i.e. the horizontal component is −4.46 m/s
and the vertical component is −2.27 m/s.

2. A force of 98 N is pressing vertically downward on the inclined face of a wedge. If the angle of inclination of the wedge is 40° find the components of the force parallel to, and perpendicular to, the face of the wedge.

The required components, although not horizontal and vertical in this case, are perpendicular and can again be found from a right-angled triangle.

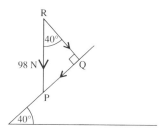

PQ = 98 sin 40° = 62.99...
QR = 98 cos 40° = 75.07...

Therefore, to 3 significant figures,

the component parallel to the face is 63.0 N
and the component perpendicular to the face is 75.1 N.

EXERCISE 3c

1. For each triangle write down an expression for the required side directly as a product of AB and a trig ratio of the given angle.

(a)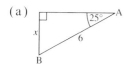

(b)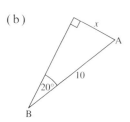

(c)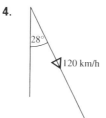

(d)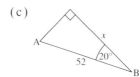

(e)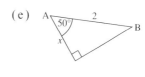

(f)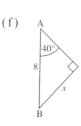

In each question from 2 to 7 find the horizontal and vertical components of the given vector, indicating the direction of each component by an arrow on a diagram.

2.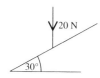

4.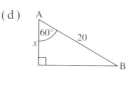

6.

3.

5.

7.

In each question from 8 to 15 find the components parallel to and perpendicular to the inclined line.

8.

9.

10.

11.

12.

13.

14.

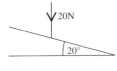

15.

16. A stone is thrown up at an angle of 20° to the vertical with an initial velocity of 35 m/s. What are the initial horizontal and vertical components of the velocity of the stone?

17. A train is travelling at 125 mph on a railway line that runs N 24°E and the direction of a canal is due east. Find the components, parallel and perpendicular to the canal, of the velocity of the train.

18. A boulder is falling vertically downward towards a hillside inclined at 23° to the horizontal. Its velocity just before impact is 176 m/s. Find the components of this velocity, parallel and perpendicular to the surface of the hillside.

19. The diagram shows a rectangular field with dimensions 120 m by 88 m. A boy pulls a truck directly from A to C with a force of 100 N. Find the components of this force parallel and perpendicular to AB.

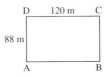

CARTESIAN UNIT VECTOR NOTATION

A vector which has a magnitude of 1 is called a unit vector, irrespective of its direction.

There is, however, a set of special unit vectors each of which has a direction along one of the Cartesian axes of coordinates.

A unit vector in the direction of O*x* is denoted by i, and a unit vector in the direction of O*y* is denoted by j.

A vector $2\mathbf{i} + 5\mathbf{j}$ is made up of

   2 units in the positive direction of the *x*-axis together with

   5 units in the positive direction of the *y*-axis,

i.e. $\overrightarrow{AB}$ represents the vector $2\mathbf{i} + 5\mathbf{j}$

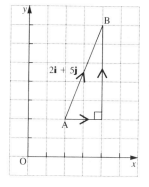

Similarly $2\mathbf{i} - 5\mathbf{j}$ is made up of

   2 units in the positive direction of O*x* together with

   5 units in the negative direction of O*y*

i.e. $\overrightarrow{PQ}$ represents the vector $2\mathbf{i} - 5\mathbf{j}$

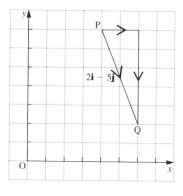

Any vector in the *xy* plane can be given in terms of a multiple of **i** together with a multiple of **j**.

For this reason **i** and **j** are called *Cartesian base vectors.*

OPERATIONS ON CARTESIAN VECTORS

Modulus

In this diagram $\overrightarrow{AB} = 3\mathbf{i} + 4\mathbf{j}$
The length of AB is $\sqrt{(3^2 + 4^2)} = 5$
This is the magnitude, or *modulus*, of $3\mathbf{i} + 4\mathbf{j}$
We denote the modulus of the vector by $|3\mathbf{i} + 4\mathbf{j}|$

i.e. $|3\mathbf{i} + 4\mathbf{j}| = 5$

Further, $\tan A = \frac{4}{3} \Rightarrow A = 53°$ (nearest degree)
so $3\mathbf{i} + 4\mathbf{j}$ is a vector with magnitude 5 units at 53° to AC.
This direction can also be described as being at 53° to the direction of $\mathbf{i}$.

Addition and Subtraction

We know that the resultant, i.e. the sum, of two vectors $\overrightarrow{AB}$ and $\overrightarrow{BC}$, is given
by drawing AB followed by BC and joining A to C.

If $\overrightarrow{AB}$ and $\overrightarrow{BC}$ are given in $\mathbf{i}\,\mathbf{j}$ form, they can be drawn in the xy plane and
their sum 'read' from the graph.

For example, if $\overrightarrow{AB} = 2\mathbf{i} + 5\mathbf{j}$ and $\overrightarrow{BC} = 7\mathbf{i} - 4\mathbf{j}$
$\overrightarrow{AB} + \overrightarrow{BC}$ is seen to be $9\mathbf{i} + \mathbf{j}$ which is

$(2 + 7)\mathbf{i} + (5 - 4)\mathbf{j}$

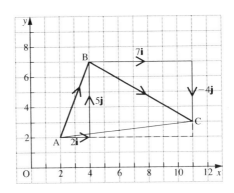

In general,

$(a\mathbf{i} + b\mathbf{j}) + (p\mathbf{i} + q\mathbf{j}) = (a + p)\mathbf{i} + (b + q)\mathbf{j}$

i.e. we add the coefficients of $\mathbf{i}$ and add the coefficients of $\mathbf{j}$.

Now suppose that we want to *subtract* $7\mathbf{i} - 4\mathbf{j}$ from $2\mathbf{i} + 5\mathbf{j}$.

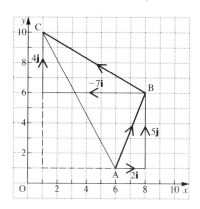

We draw $2\mathbf{i} + 5\mathbf{j}$ ($\overrightarrow{AB}$) followed by a vector equal to $7\mathbf{i} - 4\mathbf{j}$ but in the opposite direction, i.e. $-7\mathbf{i} + 4\mathbf{j}$ ($\overrightarrow{BC}$).

The resultant this time is seen to be $-5\mathbf{i} + 9\mathbf{j}$ which is

$$(2 - 7)\mathbf{i} + (5 - [-4])\mathbf{j}$$

In general,

$$(a\mathbf{i} + b\mathbf{j}) - (p\mathbf{i} + q\mathbf{j}) = (a - p)\mathbf{i} + (b - q)\mathbf{j}$$

This time we subtract the coefficients of $\mathbf{i}$ and of $\mathbf{j}$.

Examples 3d

1. **Taking i as a unit vector due east and j as a unit vector due north, express in the form** $a\mathbf{i} + b\mathbf{j}$ **a vector of magnitude 24 units on a bearing of 220°.**

$$AB = 24\cos 40° = 18.4 \text{ (3 sf)}$$

$$BC = 24\sin 40° = 15.4 \text{ (3 sf)}$$

$$\overrightarrow{AC} = \overrightarrow{AB} + \overrightarrow{BC}$$

$$= -18.4\mathbf{j} + (-15.4\mathbf{i})$$

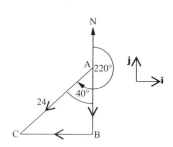

The required vector is $-15.4\mathbf{i} - 18.4\mathbf{j}$

2. **A vector V is in the direction of the vector $12\mathbf{i} - 5\mathbf{j}$ and its magnitude is 39. Find V in the form $a\mathbf{i} + b\mathbf{j}$.**

V is parallel to $12\mathbf{i} - 5\mathbf{j}$

$\therefore$ $\mathbf{V} = k(12\mathbf{i} - 5\mathbf{j}) = 12k\mathbf{i} - 5k\mathbf{j}$

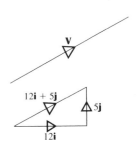

$$|\mathbf{V}| = |12k\mathbf{i} - 5k\mathbf{j}|$$
$$= \sqrt{[(12k)^2 + (-5k)^2]}$$
$$= \sqrt{(169k^2)} = 13k$$

But $|\mathbf{V}| = 39$

$\therefore$ $13k = 39$ $\Rightarrow$ $k = 3$

$\therefore$ $\mathbf{V} = 36\mathbf{i} - 15\mathbf{j}$

EXERCISE 3d

1. Express each given vector in the form $a\mathbf{i} + b\mathbf{j}$

(a)

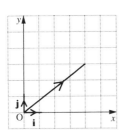

(d)

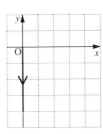

(b)

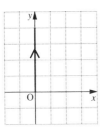

(e)

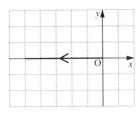

(c)

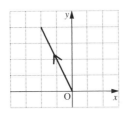

(f)

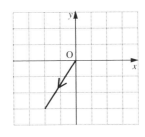

2. Find the modulus of each vector given in question 1.

3. Taking **i** as a unit vector due east and **j** as a unit vector due north, express in the form $a\mathbf{i} + b\mathbf{j}$,

(a) a vector of magnitude 14 units on a bearing 060°

(b) a vector of magnitude 20 units on a bearing 180°

(c) a velocity of 75 units south west

(d) a displacement of 400 units in the direction of $3\mathbf{i} + 4\mathbf{j}$.

In each question from 4 to 6, find (a) $\mathbf{a} + \mathbf{b}$ (b) $\mathbf{a} - \mathbf{b}$ and illustrate the results in a diagram.

4. $\mathbf{a} = 6\mathbf{i} + \mathbf{j}, \quad \mathbf{b} = 3\mathbf{i} - 5\mathbf{j}$

5. $\mathbf{a} = \mathbf{i}, \quad \mathbf{b} = -\mathbf{j}$

6. $\mathbf{a} = -2\mathbf{i} - \mathbf{j}, \quad \mathbf{b} = \mathbf{i} - \mathbf{j}$

In questions 7 to 11, $\mathbf{p} = 4\mathbf{i} - 3\mathbf{j}, \quad \mathbf{q} = -12\mathbf{i} + 5\mathbf{j}, \quad \mathbf{r} = \mathbf{i} - \mathbf{j}$

7. Find (a) $|\mathbf{p}|$ (b) $|\mathbf{q}|$ (c) $|\mathbf{r}|$, giving the answer as a square root.

8. Find a vector in the direction of **p** and with magnitude 30 units.

9. Find (a) the resultant of **q** and **r** (b) $|\mathbf{q} + \mathbf{r}|$

10. Find two vectors, each with a magnitude twice that of **p** and parallel to **p** (remember that parallel vectors can be in the same or opposite directions).

11. Find a vector **v** such that

(a) $|\mathbf{v}| = 35$ and **v** is in the direction of $\mathbf{q} + 5\mathbf{r}$

(b) **v** is parallel to $\mathbf{p} - \mathbf{q}$ and is half the size of $\mathbf{p} - \mathbf{q}$.

CHAPTER 4

FORCE

THE CONCEPT OF FORCE

So far in this book we have been dealing mainly with motion (kinematics). Now we must consider how motion of different kinds is caused (dynamics).

Force is such an everyday quantity that we all have an intuitive idea of what it is. We know that to move a heavy cabinet across a room we push it; to raise a bucket of cement from ground level to roof height, a builder pulls it up. Pushes and pulls are both forces and these simple situations illustrate that force is needed to make an object start to move.

On the other hand, a runaway shopping trolley can be stopped either by getting in front of it and exerting a push, or by holding it from behind and pulling. So a force can also cause a moving object to begin to stop.

Now consider what happens when someone who is holding a stone lets it drop. The stone begins to move downwards. What makes it move? Something must be *pulling it* down. That something is the weight of the stone; *weight is the name we give to the force that attracts each object to the earth, i.e. the force due to gravity.*

A book resting on a horizontal surface, on the other hand, does not fall down, so a force must be preventing it from falling. This is the force exerted by the surface to hold the book up. As the book does not move we deduce that this force (upward) and the weight of the book (downward) must balance.

When an object does not move, we say that the object, and the forces that act on it, are *in equilibrium*.

Conventions

Mechanics is a complex subject and it is impossible to deal with all aspects of the topic at the beginning. So certain simplifications have to be made in order to deal with one idea at a time. You will see that, although some of these approximations are too ideal to be factual, they are reasonably close to reality and make it possible for a student to absorb the principles of mechanics without being hampered by too many details at this stage.

A small object is regarded as existing at a point and is called a *particle*.

An object whose mass is small is considered to be of negligible weight. It is called *light* and its mass is ignored. Such an object may be a particle or it can be a fine string, wire or rod. (Note that 'small' is a comparative term, e.g. a person could be thought of as a 'small object' and treated as a weightless particle in relation to the Forth Bridge, but certainly not in relation to a chair.)

A flat shape whose thickness is small is regarded as being two-dimensional and is called a *lamina*.

Any object attached rigidly to the earth is called *fixed* and is considered to be immoveable, hence forces that act on a fixed object are disregarded.

TYPES OF FORCE

Forces of Attraction

Gravitational attraction is the most important force of this type and almost the only one we shall meet in this book. The gravitational attraction of the earth on an object (also referred to as a body) is called the weight of the object and its acts vertically downward on the object. It is almost always given the symbol W.

The effect of weight can be seen when an object falls and also it can be felt when an object is held, i.e. *the weight of an object acts on it at all times whether the object is moving or not.*

(There are other forces of attraction such as those between a magnet and an iron object.)

Contact Forces

Consider again a book resting in contact with a horizontal surface. The force exerted upward on the book by the surface is a *contact force* called a *normal reaction*. A normal reaction acts in a direction that is perpendicular to the surface of contact and away from that surface. So in this case the *normal reaction* acts vertically upward on the book.

The two forces acting on the book,
i.e. the normal reaction, R, and the weight, W,
are shown in the diagram.

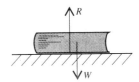

Now suppose that a small push *P* is applied horizontally to the book. If the book and the surface do not have a slippery covering, it is quite likely that the book will not move. Why not? There must be another force, equal and opposite to the push, balancing it. This is a frictional force. Friction can occur only when objects are in rough contact, so it is another example of a contact force.

A frictional force acts on a body *along the surface of contact* and in a direction which *opposes the potential movement* of that body.

This diagram shows all the forces acting on the book.

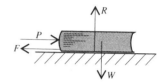

It is very rare for there to be no friction at all between an object and a surface but there can be so little that its effect can be discounted. In this case the contact is said to be *smooth*.

Forces of Attachment

Consider a mass, hanging by a string from a fixed point. The mass does not move so its weight acting downward must be balanced by an upward force.
This force is the *tension* in the string; it is a force of attachment.

Note that a string can never push and it can pull only if it is taut. For a simple attachment, a string cannot be taut at one end and slack at the other, so a taut string exerts an *inward* pull at each end on the object which is attached at that end.

Another way in which objects can be attached to each other is by means of a hinge or pivot. So the force exerted by a hinge is another force of attachment. One difference between it and the tension in a string is that a hinge *can* push as well as pull; another is that the direction of a hinge force is not usually known whereas *the tension in a string acts along the string*.

We may meet other forces occasionally, such as wind force, the driving force of a vehicle etc., but the vast majority of the forces involved at this level belong to the three types described above, i.e. gravitational attraction, contact and attachment.

DRAWING DIAGRAMS

When considering any situation concerning the action of forces on a body the first, and vital, step is to draw a clear, uncomplicated diagram of the forces acting on the object.

Some useful points to remember are:

- Unless the object is light, its weight acts vertically downwards.

- If the body is in contact with another surface, a normal reaction always acts on the body. In addition, unless the contact is smooth, there may be a frictional force.

- If the body is attached to another by a string or hinge, a force acts on the body at the point of attachment.

There is a common misconception that, all the time an object is moving, there *has* to be a force in the direction of motion. This is not true. One of the types of force described earlier in the chapter *may* be acting in the direction of motion but if none of them is, do not fall into the trap of introducing an 'extra' force.

Two or more objects may be linked, or may be in contact with each other. When considering *one* of these objects, make sure that you draw only those forces that act on that separate part and do not include forces which apply to another part of the system. This applies particularly in problems where two objects are attached to each other; if they are connected by a string, one object is affected only by the tension at the end of a string to which it is attached – the tension at the other end acts on the other object.

Finally, do not draw too small a diagram and make the force lines long enough to be seen clearly.

Examples 4a

Each of the following examples describes a situation and shows how a working diagram can be drawn.

1. **A small block is sliding down a smooth inclined plane.**

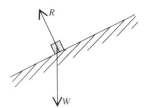

The normal reaction R is perpendicular to the plane which is the . surface of contact.

2. A load is being pulled along a rough horizontal floor by a rope inclined at 50° to the floor.

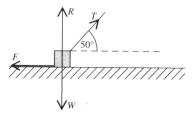

The frictional force acts along the plane in the direction opposite to the motion of the load.

3. A uniform ladder rests with its foot on rough ground and the top against a smooth wall. (The mass of a uniform body is evenly distributed, so the weight of a uniform ladder acts through the midpoint.)

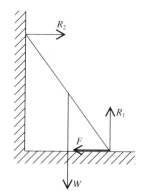

At the foot, the normal reaction is perpendicular to the ground. The frictional force acts along the ground and towards the wall because, *if* the ladder moved, its foot would slip *away* from the wall.

At the top of the ladder the normal reaction is perpendicular to the wall and there is no friction. The weight of the ladder acts through the centre of gravity (the point of balance of the ladder).

4. A uniform beam is hinged at one end to a wall to which it is inclined at 60°. It is held in this position by a horizontal chain attached to the other end.

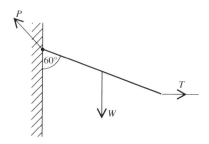

The direction of the force that the hinge exerts on the beam is not known at this stage so we cannot mark it at any specific angle.
(Later on we will see how to determine this angle.)

5. **A truck is attached by a rope to an electric-powered engine which is being driven along a horizontal smooth track.**

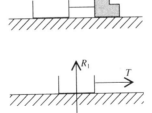

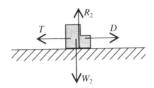

Considering the truck, the only forces acting are the weight of the truck, the vertical normal reaction and the tension in the rope which acts away from the truck (i.e. inward along the rope). Note that the driving force of the engine *does not* act on the truck, it is the tension in the rope that pulls the truck forward.

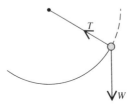

Acting on the engine alone we have the weight, the normal reaction, the driving force and the tension in the rope which acts towards the centre of the rope (i.e. it is a drag on the engine).

6. **A particle is fastened to one end of a light string. The other end of the string is held and the particle is whirled round in a circle in a vertical plane.**

The only forces acting on the particle are its weight and the tension in the string. There is *no force in the direction of motion of the particle.*

EXERCISE 4a

In each question from 1 to 6, copy the diagram and draw the forces acting on the given object in the specified situation. Take any rod, ladder etc. as being uniform.

1.

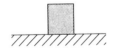

A block at rest on a smooth horizontal surface

2.

A plank resting on two supports, one at each end.

3.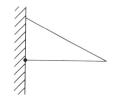

A rod hinged to a wall and held in a horizontal position by a string

5.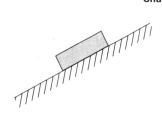

A small block at rest on a rough inclined plane.

4.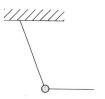

A light string fixed at one end has a particle tied to the other end being pulled aside by a horizontal force.

6.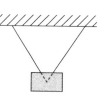

A small picture supported by two cords.

For the remaining questions draw a diagram of the specified object and mark on the diagram all the forces acting on the object.

7. A ladder with its foot on rough ground is leaning against a rough wall.

8. A particle is attached to one end of a light string whose other end is fixed to a point A. The particle is

(a) hanging at rest

(b) rotating in a horizontal circle below A

(c) held at 30° to the vertical by a horizontal force.

9. A stone that has been thrown vertically up into the air when it is

(a) going up   (b) coming down.

10. A rod of length 1 m is hinged at one end to a wall. It is held in a horizontal position by a string joining the point of the rod that is 0.8 m from the wall, to a point on the wall 1 m vertically above A.

11. A shelf AB is supported by two vertical strings, one at each end. A vase is placed on the shelf, a quarter of the way along the shelf from A. Draw separate diagrams to show the forces acting on

(a) the shelf   (b) the vase.

12. Two bricks are placed, one on top of the other, on a horizontal surface. Draw separate diagrams to show the forces acting on

(a) the top brick

(b) the lower brick.

13. A beam is hinged at one end A to a wall and is held horizontal by a rope attached to the other end B and to a point on the wall above A. The rope is at 45° to the wall. A crate hangs from B. Draw separate diagrams to show the forces acting on

(a) the beam

(b) the crate.

14. The diagram shows a rough plank resting on a cylinder and with one end of the plank on rough ground.

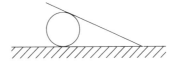

Draw diagrams to show:

(a) the forces acting on the plank,

(b) the forces acting on the cylinder.

15. A person standing on the edge of a flat roof lowers a package over the edge, by a rope, for a colleague to collect. Draw diagrams to show

(a) the forces acting on the package

(b) the forces acting on the person on the roof (use a pinman to represent the person).

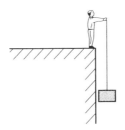

FINDING THE RESULTANT OF COPLANAR FORCES

It is quite common, as we have seen from the examples above, for an object to be under the action of several coplanar forces (i.e. all in one plane) in different directions. To investigate the overall effect of these forces we need to be able to find their resultant.

In Chapter 3 we saw that the magnitude and direction of the resultant of two perpendicular forces was easy to find. For more than two forces, however, scale drawing and measurement was used. As this method gives only a rough result we are now going to consider a way to *calculate* the resultant of any set of coplanar forces.

To do this we choose two perpendicular directions and find the components of all the forces in these directions i.e. we resolve each force in these directions. Components in a chosen direction are positive while those in the opposite direction are negative.

By collecting each set of components we can now replace the original set of forces by an equivalent pair of forces in perpendicular directions.

Note that defining a direction as, say, 'along an inclined plane' is ambiguous because it could mean either up or down the plane. A simple way to clarify the definition is to add an arrow in the correct direction.

Consider, for example, a particle resting on a rough plane inclined to the horizontal at 30°. The forces acting on the particle are shown in the diagram.

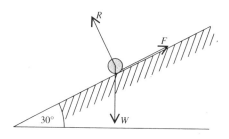

As the normal reaction and the frictional force are perpendicular to each other, it is sensible to resolve each force in these two directions, i.e. along ($\nearrow$) and perpendicular ($\nwarrow$) to the plane, i.e.

	Friction	Reaction	Weight
Component $\nearrow$	$-F$	0	$W \sin 30°$
Component $\nwarrow$	0	R	$-W \cos 30°$

Now we can collect the components of force down and perpendicular to the plane and to indicate these operations we write:

Resolving $\nearrow$ gives $\quad W \sin 30° - F$ [1]

Resolving $\nwarrow$ gives $\quad R - W \cos 30°$ [2]

Calculating the Resultant

If the expression [1] on p. 78 is represented by X and expression [2] by Y, we have

$$X = W \sin 30° - F$$
$$Y = R - W \cos 30°$$

The magnitude of the resultant, R, of X and Y is

$$\sqrt{(X^2 + Y^2)}$$

and R makes an angle α with the plane where

$$\tan \alpha = Y/X$$

In a case where each of the forces is given in the form $a\mathbf{i} + b\mathbf{j}$ the forces are already expressed as components in the directions of $\mathbf{i}$ and $\mathbf{j}$ and it remains only to find X, Y and R.

Examples 4b

1. **A uniform ladder of weight W rests with its top against a rough wall and its foot on rough ground which slopes down from the base of the wall at $10°$ to the horizontal. Resolve, horizontally and vertically, each of the forces acting on the ladder.**

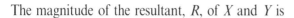

Drawing the components of R_2 and F_2 on separate small diagrams can help.

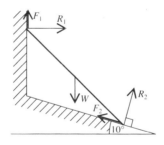

Resolving $\rightarrow$ gives $R_1 - F_2 \cos 10° + R_2 \sin 10°$

Resolving $\uparrow$ gives $F_1 - W + F_2 \sin 10° + R_2 \cos 10°$

2. Find the magnitude of the resultant of the set of forces $3\mathbf{i} + 5\mathbf{j}$, $-7\mathbf{j}$, $-4\mathbf{i} + 11\mathbf{j}$, $5\mathbf{i}$ and $\mathbf{i} + 3\mathbf{j}$ Each force is measured in newtons.
Find the angle between the resultant and the unit vector $\mathbf{i}$.

Let the resultant be $X\mathbf{i} + Y\mathbf{j}$

$$X\mathbf{i} + Y\mathbf{j} = (3\mathbf{i} + 5\mathbf{j}) + (-7\mathbf{j}) + (-4\mathbf{i} + 11\mathbf{j}) + 5\mathbf{i} + (\mathbf{i} + 3\mathbf{j})$$
$$= (3 - 4 + 5 + 1)\mathbf{i} + (5 - 7 + 11 + 3)\mathbf{j}$$
$$= 5\mathbf{i} + 12\mathbf{j}$$

$$|5\mathbf{i} + 12\mathbf{j}| = \sqrt{(25 + 144)} = 13$$
$$\tan \alpha = \tfrac{12}{5} = 2.4$$
$$\alpha = 67° \text{ (nearest degree)}$$

The resultant is 13 N at 67° to $\mathbf{i}$.

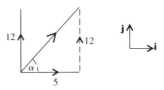

3. Find the resultant of the forces of 4, 6, 2 and 3 newtons shown in the diagram.

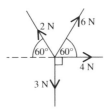

Let the resultant have components X and Y newtons in the directions shown.

Resolving the forces along Ox and Oy we have:

Resolving → $X = 4 + 6 \cos 60° - 2 \cos 60°$ (remember that $\cos 60° = \tfrac{1}{2}$)
$$= 4 + 3 - 1 = 6$$
Resolving ↑ $Y = 6 \sin 60° - 3 + 2 \sin 60°$
$$= 8 \times 0.8660 - 3 = 3.928$$

If the resultant force is R newtons

$$R = \sqrt{(X^2 + Y^2)} = \sqrt{(6^2 + 3.928^2)}$$
$$= 7.17 \quad (3 \text{ sf})$$
and $\tan \alpha = \dfrac{Y}{X} = \dfrac{3.928}{6} = 0.6546\ldots$

⇒ $\alpha = 33°$ (nearest degree)

Therefore the resultant force is 7.17 N at 33° to the force of 4 N.

Sometimes the directions of a group of forces are given with reference to the sides, diagonals etc. of a polygon; the magnitudes of the forces are given separately.

It is important to realise that, in such cases, the forces are *not necessarily* represented by the *lengths* or *positions* of the lines in the polygon.

4. **ABCDEF is a regular hexagon. Four forces act on a particle. The forces are of magnitudes 3 N, 4 N, 2 N and 6 N and they act in the directions of the sides AB, AC, EA and AF respectively.**
Find the magnitude of the resultant force and the angle it makes with AB.

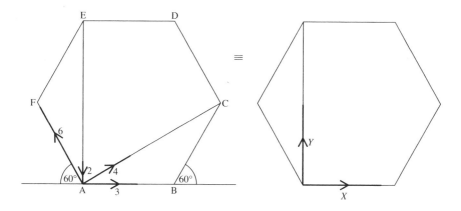

Let the resultant have components X newtons and Y newtons, parallel and perpendicular to **AB** as shown.

Resolving $\rightarrow$ $X = 3 + 4 \cos 30° - 6 \cos 60°$
$= 3 + 3.464 - 3 = 3.464$

Resolving $\uparrow$ $Y = 4 \sin 30° - 2 + 6 \sin 60°$
$= 2 - 2 + 5.196 = 5.196$

If R newtons is the resultant force then, correct to 3 significant figures,

$$R^2 = X^2 + Y^2 = 12 + 27 = 39$$
$$\therefore \quad R = \sqrt{39}$$

and $\quad \tan \alpha = \dfrac{Y}{X} = \dfrac{5.196}{3.464} = 1.5$

The resultant force is $\sqrt{39}$ N at 56° to **AB** (nearest degree).

EXERCISE 4b

In each question from 1 to 6, find the magnitude of the resultant of the given vectors and give the angle between the resultant and the direction of the positive x-axis.

1.

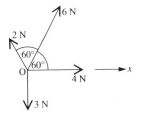

4.

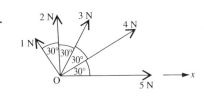

2.

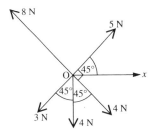

5.

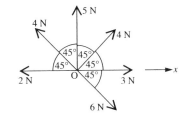

3.

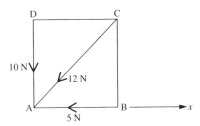

ABCD is a square

6.

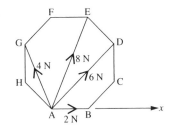

ABCDEFGH is a regular octagon.

In each question from 7 to 10,

(a) illustrate the vectors by a sketch

(b) express, in the form $a\mathbf{i} + b\mathbf{j}$, the resultant of the given vectors

(c) find the magnitude of the resultant and give the angle that the resultant makes with the vector **i**.

7. Four forces, measured in newtons, represented by

$$4\mathbf{i} - 3\mathbf{j}, \quad \mathbf{i} + 6\mathbf{j}, \quad -2\mathbf{i} + 5\mathbf{j}, \quad \text{and} \quad 3\mathbf{i}$$

8. Velocities, in metres per second, represented by

$$4\mathbf{i} - 7\mathbf{j}, \quad -3\mathbf{i} + 8\mathbf{j}, \quad 2\mathbf{i} + 3\mathbf{j}, \quad 8\mathbf{i} \quad \text{and} \quad \mathbf{i} + \mathbf{j}$$

9. Displacements, measured in metres, represented by

$$-6\mathbf{i} + \mathbf{j}, \quad 2\mathbf{i} - 5\mathbf{j}, \quad \mathbf{i} + 4\mathbf{j} \quad \text{and} \quad 3\mathbf{i} + 2\mathbf{j}$$

10. Forces, in newtons, represented by

$$2\mathbf{i} + 2\mathbf{j}, \quad \mathbf{i} - 7\mathbf{j}, \quad -6\mathbf{i} + \mathbf{j}$$

11. ABCD is a rectangle in which AB = 4 m and BC = 3 m. A force of magnitude 3 N acts along AB towards B. Another force of magnitude 4 N acts along AC towards C and a third force, 3 N, acts along AD towards D. Find the magnitude of the resultant of these forces and find the angle the resultant makes with AD.

12. A surveyor starts from a point O and walks 200 m due north. He then turns clockwise through 120° and walks 100 m after which he walks 300 m due west. What is his resultant displacement from O?

13. Three boys are pulling a heavy trolley by means of three ropes. The boy in the middle is exerting a pull of 100 N. The other two boys, whose ropes both make an angle of 30° with the centre rope, are pulling with forces of 80 N and 140 N. What is the resultant pull on the trolley and at what angle is it inclined to the centre rope?

14. Starting from O, a point P traces out consecutive displacement vectors of $2\mathbf{i} + 3\mathbf{j}, \quad -\mathbf{i} + 4\mathbf{j}, \quad 7\mathbf{i} - 5\mathbf{j} \quad \text{and} \quad \mathbf{i} + 3\mathbf{j}$.
What is the final displacement of P from O?

15. A river is flowing due east at a speed of 3 m/s. A boy in a rowing boat, who can row at 5 m/s in still water, starts from a point O on the south bank and steers the boat at right angles to the bank. The boat is also being blown by the wind at 4 m/s south-west. Taking axes Ox and Oy in the directions east and north respectively find the velocity of the boat in the form $p\mathbf{i} + q\mathbf{j}$ and hence find its resultant speed.

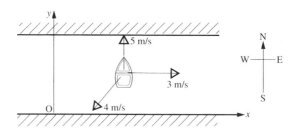

16. A small boat is travelling through the water with an engine speed of 8 km/h. It is being steered due east but there is a current running south at 2 km/h and wind is blowing the boat south-west at 4 km/h. Find the resultant velocity of the boat.

17. Velocities of magnitudes 5 m/s, 7 m/s, 4 m/s and 6 m/s act in the directions north-east, north, south-east and west respectively. Taking **i** and **j** as unit vectors east and north respectively,

(a) draw a sketch showing the separate velocities

(b) find, in the form $a\mathbf{i} + b\mathbf{j}$, the resultant velocity

(c) find the bearing of the resultant velocity

(d) find the resultant speed.

18. ABC is an equilateral triangle and D is the midpoint of BC. Forces of magnitudes 8 N, 6 N and 12 N act along AB, AC and DA respectively (the order of the letters gives the direction of the force). Find the magnitude of the resultant force and the angle between the resultant and DA.

CONSOLIDATION A

SUMMARY

Motion

For motion with constant speed:

$$\text{distance} = \text{speed} \times \text{time}$$

$$\text{average speed} = \frac{\text{total distance}}{\text{total time}}$$

Velocity is the rate of increase of displacement.

Acceleration is the rate of increase of velocity.

Using a displacement–time graph

The velocity at a particular time, t, is given by the gradient of the tangent to the graph at that value of t.

Using a velocity–time graph

The acceleration at a particular time, t, is given by the gradient of the tangent to the graph at that value of t.

The displacement after time t is given by the area under the graph for that time interval.

For motion with constant acceleration

Using the notation

$$u = \text{initial velocity} \quad v = \text{final velocity} \quad a = \text{acceleration}$$
$$t = \text{time interval} \quad \text{and} \quad s = \text{displacement}$$

the equations of motion are:

$$v = u + at$$
$$s = \tfrac{1}{2}(u + v)t$$
$$s = ut + \tfrac{1}{2}at^2 \quad \text{and} \quad s = vt - \tfrac{1}{2}at^2$$
$$v^2 - u^2 = 2as$$

Vectors

A quantity that has both magnitude and direction is a vector.
A vector can be represented by a line segment.

The *magnitude* of the vector is represented by the *length of the line* and the direction of the vector by the direction of the line.

Two *parallel vectors* with the same magnitude are:
  equal if they are in the same direction
  equal and opposite if they are in opposite directions.

If two vectors **a** and **b** are parallel then $\mathbf{a} = k\mathbf{b}$.

When, starting from a point A, lines representing vectors in magnitude and direction are drawn consecutively, the line starting at A that completes the polygon represents the *resultant* vector; all the other vectors are *components*.

The resultant of two perpendicular vectors, **p** *and* **q**
  is of magnitude $\sqrt{(p^2 + q^2)}$
  makes an angle α with the vector **p** where $\tan \alpha = p/q$

The resultant of a set of coplanar vectors can be *calculated* by resolving all forces in two perpendicular directions.

Vectors of unit magnitude in the directions of Ox and Oy are denoted respectively by **i** and **j**.

A vector in the xy plane whose components in the directions Ox and Oy are a and b respectively, can be written $a\mathbf{i} + b\mathbf{j}$.

The magnitude of the vector $a\mathbf{i} + b\mathbf{j}$ is denoted by $|a\mathbf{i} + b\mathbf{j}|$ and is of value $\sqrt{(a^2 + b^2)}$.

Types of Force

Contact forces occur when solid objects are in contact. A pair of equal and opposite forces act, one on each of the objects, perpendicular to the surface of contact.

Forces of attachment act when two objects are connected by, for example, a string or a hinge. Two equal and opposite forces act, one on each of the objects; in the case of a string the forces are an inward pull at each end of the string; the directions of the forces at a hinge are, in general, not known.

The earth exerts a force of attraction on any body outside its surface. This force is the weight of the object. The acceleration produced by the weight of an object is denoted by g; the approximate value of this acceleration at the surface of the earth is 9.8 m/s^2.

MISCELLANEOUS EXERCISE A

Each question from 1 to 4 is followed by several suggested responses. Choose the correct response.

1. If ABCD is a quadrilateral whose sides represent vectors, $\overrightarrow{AB}$ is equivalent to

   A $\overrightarrow{CA} + \overrightarrow{CB}$ **B** $\overrightarrow{CD}$ **C** $\overrightarrow{AC} - \overrightarrow{BC}$

2. $\overrightarrow{AB}$ and $\overrightarrow{PQ}$ are two vectors such that $\overrightarrow{AB} = 2\overrightarrow{PQ}$.

   A AB is parallel to PQ.

   B PQ is twice as long as AB.

   C A, B, P and Q must be collinear.

3. Two forces $\mathbf{F}_1$ and $\mathbf{F}_2$ have a resultant $\mathbf{F}_3$. If $\mathbf{F}_1 = 2\mathbf{i} - 3\mathbf{j}$ and $\mathbf{F}_3 = 5\mathbf{i} + 4\mathbf{j}$ then $\mathbf{F}_2$ is

   A $7\mathbf{i} + \mathbf{j}$ **B** $-3\mathbf{i} - 7\mathbf{j}$ **C** $3\mathbf{i} + 7\mathbf{j}$

4. When a number of particles, all of different weights, are dropped, the acceleration of each particle

   A is constant but different for each particle, depending on its weight.

   B is constant and the same for each particle.

   C increases as the particle falls.

5. Two forces $(3\mathbf{i} + 2\mathbf{j})$ N and $(-5\mathbf{i} + \mathbf{j})$ N act at a point. Find the magnitude of the resultant of these forces and determine the angle which the resultant makes with the unit vector $\mathbf{i}$. (AEB)

6. O is the origin, and the positions of two points A and B are given by

   $$\overrightarrow{OA} = \mathbf{i} + 7\mathbf{j} \quad \text{and} \quad \overrightarrow{OB} = 5\mathbf{i} + 5\mathbf{j}.$$

   Show that the vectors $\overrightarrow{OA}$ and $\overrightarrow{OB}$ have equal magnitudes.

   Points C and D are in positions given by

   $$\overrightarrow{OC} = 2\overrightarrow{OA} \quad \text{and} \quad \overrightarrow{OD} = \overrightarrow{OA} + \overrightarrow{OB}.$$

   Express $\overrightarrow{OC}$ and $\overrightarrow{OD}$ in terms of $\mathbf{i}$ and $\mathbf{j}$, and draw a diagram showing the positions of A, B, C and D.

7. Three forces $(3\mathbf{i} + 5\mathbf{j})$ N, $(4\mathbf{i} + 11\mathbf{j})$ N, $(2\mathbf{i} + \mathbf{j})$ N act at a point. Given that $\mathbf{i}$ and $\mathbf{j}$ are perpendicular unit vectors find

   (a) the resultant of the forces in the form $a\mathbf{i} + b\mathbf{j}$

   (b) the magnitude of this resultant

   (c) the cosine of the angle that the resultant makes with the unit vector $\mathbf{i}$.

   (AEB)

8. The diagonals of the plane quadrilateral ABCD intersect at O, and X, Y are the midpoints of the diagonals AC, BD respectively. Show that

   (a) $\overrightarrow{BA} + \overrightarrow{BC} = 2\overrightarrow{BX}$

   (b) $\overrightarrow{BA} + \overrightarrow{BC} + \overrightarrow{DA} + \overrightarrow{DC} = 4\overrightarrow{YX}$

   (c) $2\overrightarrow{AB} + 2\overrightarrow{BC} + 2\overrightarrow{CA} = 0$

   If $\overrightarrow{OA} + \overrightarrow{OB} + \overrightarrow{OC} + \overrightarrow{OD} = 4\overrightarrow{OM}$, find the location of M. (AEB)

9. A hovercraft travelling horizontally in a straight line starts from rest and accelerates uniformly during the first 6 minutes of its journey when it covers 2 km. Then it moves at constant velocity until it experiences a constant retardation which brings it to rest in a further distance of 4 km.

   (a) Sketch the velocity–time graph and find the maximum velocity, in km/h attained by the hovercraft.

   (b) Determine the time taken during the retardation.

   (c) Given that the total journey time is 42 minutes, determine the distance travelled at constant velocity. (AEB)

10. At time $t = 0$ a particle is projected vertically upwards from a point O with speed 19.6 m/s and, two seconds later, a second particle is projected vertically upwards from O with the same speed. Assuming that the only force acting is that due to gravity, express the heights above O of both particles in terms of t and hence, or otherwise, find the value of t when they collide. Find the speeds of the particles at the instant of collision. (WJEC)

11. A railway train is moving along a straight level track with a speed of 10 m/s when the driver sights a signal which is at green. As soon as the signal is sighted the train starts to accelerate. Given that the acceleration has a constant value of f m/s², show that the distance in metres moved by the train during the nth second after the signal is sighted is

$$\left(10 - \frac{f}{2} + nf\right)$$

   Find the value of f given that the train travels 25 m during the 8th second after the signal is sighted. (NEAB)

12. A train has a maximum speed of 144 km/h which it can achieve at an acceleration of 0.25 m/s². With its brakes fully applied, the train has a deceleration of 0.5 m/s². Two stations are 8 km apart. The train stops at both stations.

   (a) What is the shortest time for the train to travel between these two stations?

   (b) How is your answer to (a) changed if there is a restriction on speed, between the two stations, of 72 km/h? (UCLES)$_s$

13. An airport has a straight level runway of length 3000 metres. During take-off, a jet aircraft, starting from rest, moves with constant acceleration along the runway and reaches its take-off speed of 270 km/h after 40 seconds. Find

(a) the acceleration of the jet during take-off in m/s^2

(b) the fraction of the length of the runway used by the jet during its take-off.

(NEAB)

14. A lift travels vertically upwards from rest at floor A to rest at floor B, which is 20 m above A, in three stages as follows. Firstly, the lift accelerates from rest at A at 2 m/s^2 for 2 s; secondly, it travels at a constant speed; thirdly, it slows down uniformly at 4 m/s^2, coming to rest at B.

Sketch the velocity-time graph for this motion, and show that the journey from A to B takes $6\frac{1}{2}$ seconds. (UCLES)

15. A car is moving along a straight road with uniform acceleration. The car passes a check-point A with a speed of 12 m/s and another check-point C with a speed of 32 m/s. The distance between A and C is 1100 m.

(a) Find the time, in s, taken by the car to move from A to C.

Given that B is the midpoint of AC,

(b) find, in m/s to 1 decimal place, the speed with which the car passes B.

(ULEAC)

16. A train moves from rest at a station, Amesbury, and covers the first 1.8 km of its journey with uniform acceleration. It then travels for 18 km at a uniform speed, and then decelerates uniformly for the final 1.2 km to come to rest at Birchfield. The final journey time is 20 minutes.

(a) Sketch the speed–time graph for this journey.

(b) Calculate, in km/h, the maximum speed attained.

(c) Calculate, in km/h^2, the final deceleration. (ULEAC)

17. A motorist, travelling at 120 km/h on a motorway, passes a police speed check point. The motorist immediately decelerates at a rate of 360 km/h^2.

A police car at the speed check point starts from rest at the instant the motorist passes it, accelerates uniformly to a speed of 130 km/h and then travels at this speed. Given that it overtakes the motorist after 3 minutes and then decelerates,

(a) determine the distance from the speed check to the point where the police car overtakes the motorist

(b) find the time (in minutes) during which the police car is travelling at constant speed. (AEB)

18. Three forces $(\mathbf{i}+\mathbf{j})$ N, $(-5\mathbf{i}+3\mathbf{j})$ N and $k\mathbf{i}$ N, where $\mathbf{i}$ and $\mathbf{j}$ are perpendicular unit vectors, act at a point. Express the resultant of these forces in the form $a\mathbf{i}+b\mathbf{j}$ and find its magnitude in terms of k.
Given that the resultant has magnitude 5 N find the two possible values of k.
Take the larger value of k and find the tangent of the angle between the resultant and the unit vector $\mathbf{i}$. (AEB)

19. A vehicle travelling on a straight horizontal track joining two points A and B accelerates at a constant rate of 0.25 m/s^2 and decelerates at a constant rate of 1 m/s^2. It covers a distance of 2.0 km from A to B by accelerating from rest to a speed of v m/s and travelling at that speed until it starts to decelerate to rest. Express in terms of v the times taken for acceleration and deceleration.

Given that the total time for the journey is 2.5 minutes find a quadratic equation for v and determine v, explaining clearly the reason for your choice of the value of v. (AEB)

20. O is the origin; A and B are points such that $\overrightarrow{OA} = \mathbf{a}$ and $\overrightarrow{OB} = \mathbf{b}$. M is the midpoint of AB. P is a point on OB such that $OP = 2PB$ and Q is a point on OM such that $OQ = \frac{1}{2}QM$. The line PQ is produced to meet OA at R. Express $\overrightarrow{OR}$ in terms of $\mathbf{a}$ and $\mathbf{b}$. (WJEC)

The instruction for answering questions 21 to 24 is:
if the following statement must *always* be true, write T, otherwise write F.

21. If $\mathbf{F}_1 = 2\mathbf{i}+3\mathbf{j}$ and $\mathbf{F}_2 = 2\mathbf{i}-3\mathbf{j}$ then $\mathbf{F}_1$ and $\mathbf{F}_2$ are equal and opposite.

22. A particle of weight W is on a plane inclined at α to the horizontal. The component of the weight parallel to the plane is $W\cos\alpha$.

23. The resultant of $\overrightarrow{AB}$ and $\overrightarrow{BC}$ is $\overrightarrow{CA}$.

24. Velocity is the rate of increase of distance.

25. In $\triangle OPQ$, $\overrightarrow{OP} = 2\mathbf{p}$, $\overrightarrow{OQ} = 2\mathbf{q}$, and M and N are the midpoints of OP and OQ respectively.

(a) By expressing the vectors $\overrightarrow{PQ}$ and $\overrightarrow{MN}$ in terms of $\mathbf{p}$ and $\mathbf{q}$, prove that
$$\overrightarrow{MN} = \tfrac{1}{2}\overrightarrow{PQ}.$$

The lines PN and QM intersect at the point G.

(b) Express, in terms of $\mathbf{p}$ and $\mathbf{q}$, the vectors $\overrightarrow{PN}$ and $\overrightarrow{QM}$.

(c) Given that $\overrightarrow{GN} = \lambda\overrightarrow{PN}$ prove that
$$\overrightarrow{OG} = 2\lambda\mathbf{p}+(1-\lambda)\mathbf{q}.$$

(d) Given that $\overrightarrow{GM} = \mu\overrightarrow{QM}$ find $\overrightarrow{OG}$ in terms of μ, $\mathbf{p}$ and $\mathbf{q}$.

(e) Hence prove that
$$\overrightarrow{OG} = \tfrac{2}{3}(\mathbf{p}+\mathbf{q}).$$
(ULEAC)

26. A particle moving in a straight line with speed u m/s is retarded uniformly for 16 seconds so that its speed is reduced to $\frac{1}{4}u$ m/s. It travels at this reduced constant speed for a further 16 seconds. The particle is then brought to rest by applying a constant retardation for a further 8 seconds. Draw a speed–time graph and hence, or otherwise,

(a) express both retardations in terms of u

(b) show that the total distance travelled over the two periods of retardation is $11u$ m

(c) find u given that the total distance travelled in the 40 seconds in which the speed is reduced from u m/s to zero is 45 m. (AEB)

27. An underground train travels along a straight horizontal track from station A to station B. The train accelerates uniformly from rest at A to a maximum speed of 20 m/s, then travels at this speed for 30 seconds before slowing down uniformly to come to rest at B. The acceleration is f m/s^2, the retardation is $2f$ m/s^2 and the time for the whole journey is 1 minute. Sketch the velocity–time graph for the journey. Calculate

(a) the distance between the stations A and B

(b) the value of f. (NEAB)

28. A car starts from rest at time $t = 0$ seconds and moves with a uniform acceleration of magnitude 2.3 m/s^2 along a straight horizontal road. After T seconds, when its speed is V m/s, it immediately stops accelerating and maintains this steady speed until it hits a brick wall when it comes instantly to rest. The car has then travelled a distance of 776.25 m in 30 s.

(a) Sketch a speed–time graph to illustrate this information.

(b) Write down an expression for V in terms of T.

(c) Show that

$$T^2 - 60T + 675 = 0$$

(ULEAC)

CHAPTER 5

NEWTON'S LAWS OF MOTION

FORCE AND MOTION

Early mathematicians, right up to the Middle Ages, were convinced that whenever a body is moving there must be a force acting on it, i.e. a force is needed to 'make it keep moving'.

We can see, now, that something is wrong with that hypothesis by considering, for example, a puck skimming across the ice rink during an ice hockey match.

The puck is struck with a stick and sent moving (i.e. a force *starts* the motion), but what happens next? The puck continues to move although there is nothing to push it once it has left the hockey stick.

This is just one example where motion exists without a force to cause it.

However, it was not until 1687 that, with the publication of Newton's Laws of Motion, the old hypothesis was discarded and a completely new school of thought established.

NEWTON'S FIRST LAW

This law is the result of one of those brilliant pieces of deduction which from time to time produce an idea so simple that it is difficult to understand why it had eluded thinkers for so long. The law states that:

A body will continue in its state of rest, or of uniform motion in a straight line, unless an external force is applied to it.

This immediately explains the motion of the puck; once struck and set in motion, it continues to move in a straight line until some other force intervenes.

Further deductions can be made from Newton's first law.

- If a body is at rest, or is moving with constant velocity, then there is no resultant force acting on it and any forces that do act must balance exactly, i.e. must be in equilibrium.

- If the speed of a moving object is changing, there must be a resultant force acting on it.

- If the *direction* of motion of a moving object is changing, i.e. it is not moving in a straight line, there must be a resultant force acting on it. (So there is always a force acting on a body that is moving in a curve, even if the speed is constant.)

Newton's first law in effect defines what force is, i.e. force is the quantity that, when acting on a body, changes the velocity of that body.

Now if the velocity of a body changes, there is an acceleration, so we can say:

If a body has an acceleration there is a resultant force acting on it.

If a body has no acceleration there is no resultant force acting on it.

A body has no acceleration when it is at rest, *or when it is moving with constant velocity* so it is clear that no force is needed to keep a body moving with constant velocity.

Examples 5a

1. **A body is at rest under the action of the forces shown, all forces being measured in newtons. Find the values of F and R.**

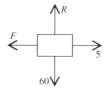

The body is at rest therefore there is no resultant either horizontally or vertically.

Horizontally $\quad 5 - F = 0 \quad \Rightarrow \quad F = 5$

Vertically $\quad R - 60 = 0 \quad \Rightarrow \quad R = 60$

2. **A particle of weight 7 N, hanging at the end of a vertical light string is moving upward with constant velocity. Find the tension in the string.**

The velocity of the particle is constant so it has no acceleration and therefore the resultant force is zero.

Vertically $T - 7 = 0$ $\Rightarrow$ $T = 7$

∴   the tension in the string is 7 N.

3. **The diagram shows the forces that act on a particle. Determine whether the resultant acceleration of the particle is horizontal, vertical or in some other direction, if (a) $P = 5$ (b) $P = 8$.**

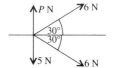

(a)   Horizontally the forces do not balance as there is no force to the left, so there is a horizontal acceleration component.

Vertically the forces balance in pairs, so there is no vertical acceleration.

The resultant acceleration is horizontal.

(b)   As in (a) there is a horizontal acceleration component.

The vertical components of the forces of 6 N balance but the 5 N and 8 N do not, so there is a vertical acceleration component.

The resultant acceleration is the combination of the two components.

The resultant acceleration is neither horizontal nor vertical but is in some other direction.

EXERCISE 5a

In each question from 1 to 6, the diagram shows the forces, all in newtons, acting on an object which is at rest. Find P and/or Q.

1.

3.

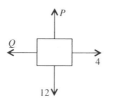

5.

2.

4.

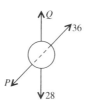

6.

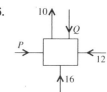

In each question from 7 to 12 determine whether or not the body shown in the diagram has an acceleration. If there is an acceleration state whether it is (i) horizontal (ii) vertical (iii) neither horizontal nor vertical.

7.

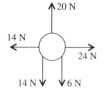

9.

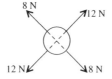

11.

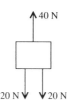

8.

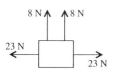

10.

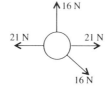

12.

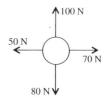

13. The diagram shows a block in rough contact with a horizontal surface. It is being pulled along by a horizontal string.

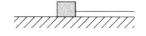

(a) Make a copy of the diagram and on it mark all the forces acting on the block.

(b) What can you say about the tension in the string compared with the frictional force if the block

(i) is accelerating (ii) moves with constant velocity?

14. The forces, in newtons, acting on a body are $-2\mathbf{i} + 6\mathbf{j}$, $5\mathbf{i} - 3\mathbf{j}$, $4\mathbf{j}$ and $-3\mathbf{i} - 3\mathbf{j}$. Determine whether the body is accelerating and, if it is, state in which direction.

15. The diagram shows a plank with one end in rough contact with the ground and resting in smooth contact with a post. The forces acting on the plank are marked on the diagram.

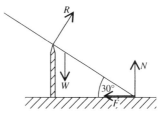

(a) Collect the vertical components of the forces.

(b) Collect the horizontal components of the forces.

(c) Given that the plank is at rest, form equations by giving a numerical value to the expression you found (i) in part (a) (ii) in part (b).

16.

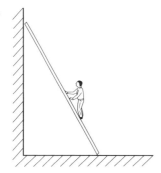

The diagram shows a ladder, with its foot on rough ground, leaning against a smooth wall. The weight of the ladder, W, acts through the midpoint; a man, also of weight W, is standing on the ladder as shown.

(a) Mark all the forces that act on the ladder, on a copy of the diagram (represent the man by a particle).

(b) Write down two expressions involving the forces, that you could use if you were asked to find out whether the ladder is stationary.

17. A wind of strength P newtons is blowing a small boat on a pond. The boat, whose weight is W newtons, is moving with constant velocity. Draw a diagram and mark all the forces acting on the boat, using F for the frictional resistance of the water and R for the supporting force exerted by the water (both in newtons). What is the relationship between

(a) P and F (b) R and W?

NEWTON'S SECOND LAW

This law defines the relationship between force, mass and acceleration. It seems reasonable to accept that

(i) for a body of a particular mass, the bigger the force is, the bigger the acceleration will be

(ii) the larger the mass is, the larger will be the force needed to produce a particular acceleration.

Experimental evidence verifies that the force F is proportional both to the acceleration a and to the mass m.

i.e. $F \propto ma$

or $F = kma$ where k is a constant

Now if $m = 1$ and $a = 1$ then $F = k$, so the amount of force needed to give a mass of 1 kg an acceleration of 1 m/s² is given by k.

If this amount of force is chosen as the unit of force we have $k = 1$ and

$$F = ma$$

The unit of force is called the newton (N) and is defined as the amount of force that gives 1 kg an acceleration of 1 m/s².

Now we know that acceleration and force are both vector quantities and that mass is scalar. We also know that if $\mathbf{p} = k\mathbf{q}$, then $\mathbf{p}$ and $\mathbf{q}$ are parallel vectors.

Therefore, from the equation $\mathbf{F} = m\mathbf{a}$ we see that:

the vectors $\mathbf{F}$ and $\mathbf{a}$ are parallel, i.e. the direction of an acceleration is the same as the direction of the force that produces it.

When more than one force acts on a body, $\mathbf{F}$ represents the resultant force.

If the force is constant the acceleration also is constant and, conversely, if the force varies, so does the acceleration.

If the acceleration is zero, the resultant force is zero – in other words Newton's first law follows from the second.

To sum up:

> **The resultant force acting on a body of constant mass is equal to the mass of the body multiplied by its acceleration.**
>
> $$\mathbf{F(N)} = m(\text{kg}) \times \mathbf{a}(\text{m/s}^2)$$
>
> **The resultant force and the acceleration are in the same direction.**

Examples 5b

1. **A force of 12 N acts on a body of mass 5 kg. Find the acceleration of the body.**

Using $F = ma$ gives $12 = 5a$

$\Rightarrow \qquad a = 2.4$

The acceleration is 2.4 m/s² in the direction of the force.

2. **A set of forces act on a mass of 3 kg and give it an acceleration of 11.4 m/s². Find the magnitude of the resultant of the forces.**

If $m = 3$ and $a = 11.4$ using $F = ma$ gives

$$F = 3 \times 11.4 = 34.2$$

The resultant force is of magnitude 34.2 N.

(The direction of the resultant is not asked for but it is in the direction of the acceleration.)

3. Forces $4\mathbf{i} - 7\mathbf{j}$ and $-\mathbf{i} + 3\mathbf{j}$ act on a particle of mass 2 kg. Given that the forces are measured in newtons, express the acceleration of the particle in the form $a\mathbf{i} + b\mathbf{j}$ and find its magnitude. Find the angle between the direction of the acceleration and the vector $\mathbf{i}$.

The resultant force is $4\mathbf{i} - 7\mathbf{j} + (-\mathbf{i} + 3\mathbf{j}) = 3\mathbf{i} - 4\mathbf{j}$

Using $\mathbf{F} = m\mathbf{a}$ gives $3\mathbf{i} - 4\mathbf{j} = 2\mathbf{a}$

So, measured in m/s², $\mathbf{a} = \frac{1}{2}(3\mathbf{i} - 4\mathbf{j}) = 1.5\mathbf{i} - 2\mathbf{j}$

The magnitude of $\mathbf{a}$ is

$\quad |1.5\mathbf{i} - 2\mathbf{j}|$

$\quad = \sqrt{(1.5^2 + [-2]^2)} = 2.5$

The magnitude of the acceleration is 2.5 m/s².

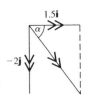

$\tan \alpha = 2/1.5 \quad \Rightarrow \quad \alpha = 53°$ to the nearest degree.

Therefore the direction of the acceleration is at 53° to $\mathbf{i}$ as shown.

4. The diagram shows the forces that act on a particle of mass 5 kg, causing it to move vertically downward with an acceleration a m/s². Find the values of P and a.

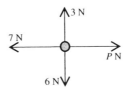

There is no horizontal motion

$\therefore \quad \rightarrow \quad$ gives $P - 7 = 0 \quad \Rightarrow \quad P = 7$

The resultant force vertically downwards is $(6 - 3)$ N $= 3$ N

Using $F = ma \quad \downarrow \quad$ gives

$\quad\quad 3 = 5a \quad \Rightarrow \quad a = \frac{3}{5}$

In some problems we are given some facts about how an object is *being made to move* with constant acceleration and also other information about the *motion* of the object. So we use both $F = ma$ and one of the equations of motion, derived in Chapter 2, which contains a.

5. A force F newtons acts on a particle of mass 3 kg.

(a) If the particle accelerates uniformly from 2 m/s to 8 m/s in 2 seconds, find the value of F.

(b) If $F = 6$ find the displacement of the particle 4 seconds after starting from rest.

(a) For the motion of the particle we have:

$$u = 2, \quad v = 8, \quad t = 2 \text{ and } a \text{ is required.}$$

Using $\quad v = u + at$ gives $\quad 8 = 2 + 2a$

$\Rightarrow \qquad a = 3$

Now using $\quad F = ma$ gives $\quad F = 3 \times 3$

$\Rightarrow \qquad F = 9$

The force is 9 N.

(b) This time we know the force so we use Newton's Law first.

Using $\quad F = ma$ gives $\quad 6 = 3a \qquad \Rightarrow \qquad a = 2$

Now for the motion of the particle:

$$u = 0, \quad a = 2, \quad t = 4 \text{ and } s \text{ is required.}$$

Using $\quad s = ut + \frac{1}{2}at^2$ gives $\quad s = 0 + \frac{1}{2}(2)(4)^2$

$\Rightarrow \qquad s = 16$

The displacement of the particle is 16 m.

EXERCISE 5b

1. A force of 12 N acts on a body of mass 8 kg. What is the acceleration of the body?

2. The acceleration of a particle of mass 2 kg is 14 m/s². What is the resultant force acting on the particle?

3. A force of 420 N acts on a block, causing an acceleration of 10.5 m/s². Assuming that no other force acts on the block, find its mass.

4. A force, measured in newtons, is represented by $5\mathbf{i} - 12\mathbf{j}$. If the force acts on an object of mass 26 kg find, in the form $a\mathbf{i} + b\mathbf{j}$, the acceleration of the object. What is the magnitude of the acceleration?

5. Find, in Cartesian vector form, the force which acts on an object of mass 5 kg, producing an acceleration of $7\mathbf{i} + 2\mathbf{j}$ measured in m/s².

6. In each diagram the forces shown (measured in newtons) cause an object of mass 8 kg to move with the acceleration shown. Find P and/or Q in each case.

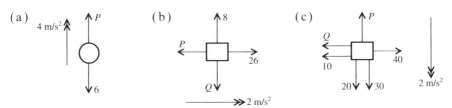

7. Each diagram shows the forces acting on a body of mass 3 kg. Find the magnitude and direction of the acceleration of the body in each case.

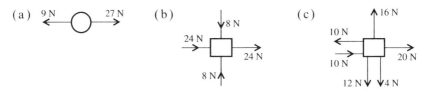

8. In each diagram the mass of the body is m kilograms. Find m and P.

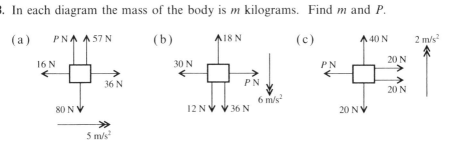

9. A body of mass 3 kg is accelerating vertically downwards at 5 m/s^2 under the action of the forces shown, all measured in newtons. Find the values of P and Q.

10. A body of mass 2 kg accelerates uniformly from rest to 16 m/s in 4 seconds. Find the resultant force acting on the body.

11. A force of 100 N acts on a particle of mass 8 kg. If the particle is initially at rest, find how far it travels in the first 5 seconds of its motion.

12. A block of mass 6 kg is pulled along a smooth horizontal surface by a horizontal string. If the block reaches a speed of 20 m/s in 4 seconds from rest, find the tension in the string.

13. Forces of $2\mathbf{i} + 3\mathbf{j}$ and $-6\mathbf{i} + 7\mathbf{j}$ act on a body of mass 10 kg. Given that the forces are measured in newtons find in the form $a\mathbf{i} + b\mathbf{j}$

(a) the acceleration of the body

(b) the velocity of the body 2 seconds after starting from rest.

14. A constant force of 80 N acts for 7 seconds on a body, initially at rest, giving it a velocity of 35 m/s. Find the mass of the body.

15. A force of 12 N acts on a particle of mass 60 kg causing the velocity of the particle to increase from 3 m/s to 7 m/s. Find the distance that the particle travels during this period.

16. A body of mass 120 kg is moving in a straight line at 8 m/s when a force of 40 N acts in the direction of motion for 18 seconds. What is the speed of the body at the end of this time?

WEIGHT

Consider an object of mass m kg, falling freely under gravity with an acceleration g m/s^2.

We know that the force producing the acceleration is the weight, W newtons, of the object, so using $F = ma$ gives

$$W = mg$$

i.e.   **a body of mass m kilograms has a weight of mg newtons.**

For example, taking the value of g as 9.8,

the weight, W newtons, of a person whose mass is 55 kg is given by

$$W = mg = 55 \times 9.8 = 539$$

therefore the person's weight is 539 N.

If the weight of a rockery stone is 1078 N, its mass, m kg, is given by

$$W = mg \quad \Rightarrow \quad m = \frac{W}{g} = \frac{1078}{9.8} = 110$$

therefore the mass of the stone is 110 kg.

You can get an idea of what a newton is like if you think of the weight of a smallish apple.

Example 5c

A rope with a bucket attached to the end is used to raise water from a well. The mass of the empty bucket is 1.2 kg and it can raise 10 kg of water when full. Taking g as 9.8 find the tension in the rope when

(a) the empty bucket is lowered with an acceleration of 2 m/s²

(b) the full bucket is raised with an acceleration of 0.3 m/s²

(a) The acceleration of the bucket is downward so the resultant force acts downwards.
The weight of the bucket is $1.2g$ N $= 1.2 \times 9.8$ N
The resultant force downward is $(1.2g - T)$ N

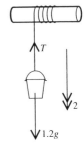

Using $F = ma \downarrow$ gives

$$1.2 \times 9.8 - T = 1.2 \times 2$$
$$T = 1.2 \times 9.8 - 1.2 \times 2$$
$$= 9.36$$

The tension in the rope is 9.36 N.

(b) The acceleration of the full bucket is upward so the resultant force acts upwards.
The weight of the full bucket is 11.2×9.8 N
Resultant force $\uparrow$ is $(T - 11.2g)$ N

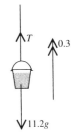

Using $F = ma \uparrow$ gives

$$T - 11.2 \times 9.8 = 11.2 \times 0.3$$
$$T = 11.2 \times 0.3 + 11.2 \times 9.8$$
$$= 113.12$$

The tension in the rope is 113 N (3 sf).

Note that readers who are competent in algebraic factorising may prefer to give the line where T is calculated in part (a) as $T = 1.2 (9.8 - 2) = 9.36$ and the similar line in part (b) as $T = 11.2 (9.8 + 0.3) = 113.12$.

EXERCISE 5c

In this exercise take the value of g on earth as 9.8.

1. (a) Find the weight of a body of mass 5 kg.

   (b) What is the mass of a sack of potatoes of weight 147 N?

   (c) What is the weight of a tennis ball of mass 60 g?

2. On the moon the acceleration due to gravity is 1.2 m/s². What answers to question 1 would a student in a lunar school give?

3. A particle of mass 2 kg, attached to the end of a vertical light string, is being pulled up, with an acceleration of 5.8 m/s², by the string.

Find the tension in the string.

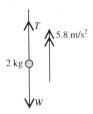

4. A mass of 6 kg is moving vertically at the end of a light string. Find the tension in the string when the mass has an acceleration of

(a) 5 m/s² downwards   (b) 7 m/s² upwards   (c) zero.

5. The tension in a string, which has a particle of mass m kilograms attached to its lower end, is 70 N.

Find the value of m if the particle has

(a) an acceleration of 3 m/s² upwards

(b) an acceleration of 9 m/s² downwards

(c) a constant velocity of 4 m/s upwards

(d) a constant velocity of 4 m/s downwards.

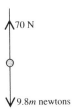

6. A goods lift with a mass of 750 kg can be raised and lowered by a cable. The maximum load it can hold is 1000 kg.

(a) Find the tension in the cable when

(i) raising the fully-loaded lift with an acceleration of $\frac{1}{2}$ m/s²

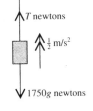

(ii) lowering the empty lift with an acceleration of $\frac{3}{4}$ m/s²

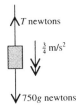

(b) The tension in the cable is 14 700 N when the lift, partly loaded, is being raised at constant speed. Find the mass of the load.

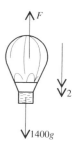

7. A balloon of mass 1400 kg is descending vertically with an acceleration of 2 m/s². Find the upward force being exerted on the balloon by the atmosphere.

8. A block of mass 4 kg is lying on the floor of a lift that is accelerating at 5 m/s². Find the normal reaction exerted on the block by the lift floor if the lift is

(a) going up   (b) going down.

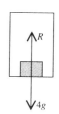

MOTION CAUSED BY SEVERAL FORCES

We have seen that if an object is moving in a particular line, we can analyse the motion by collecting the forces in that direction. However, in the examples so far considered in this chapter the given forces have all been either vertical or horizontal so that collecting them was simple arithmetic. Now we must look at more realistic situations where forces in various directions act on an object.

Examples 5d

1. **A body of mass 3 kg is sliding down a smooth plane inclined at 30° to the horizontal.**

   (a) Find the acceleration of the mass in terms of g.

   (b) Show that the normal reaction exerted by the plane on the mass is given by $\dfrac{3\sqrt{3}g}{2}$.

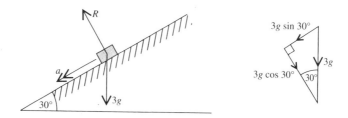

The diagram shows all the forces that act on the body. The body can move in only one way, i.e. down the plane, so the acceleration is marked in that direction. The acceleration is caused only by the components of force that act along the plane so these must be found and collected.

(a)   The resultant force ⟋ is $3g \sin 30°$, i.e. $3g\left(\frac{1}{2}\right)$

   Using $F = ma$ gives $\left(\frac{3}{2}\right)g = 3a$ $\Rightarrow$ $a = \frac{1}{2}g$

(b)   As there is no acceleration perpendicular to the plane, the resultant of the forces in that direction is zero.

   Resolving ⟍ gives $R - 3g \cos 30° = 0$

$$R = 3g\left(\frac{\sqrt{3}}{2 \cdot}\right) = \frac{3\sqrt{3}g}{2}$$

Note that in this example the values of $\sin 30°$ and $\cos 30°$ are used in their exact form (using square roots). Readers who have not yet encountered these values should now learn them as they often occur in problems where exact expressions are required. Taking $\cos 30°$ as 0.8860, is correct only to 4 decimal places and using it can cause difficulty in a problem where exact values are required.

Using Pythagoras' theorem in a sketch of half an equilateral triangle with sides of length 2 units, shows clearly the sine and cosine of both $30°$ and $60°$, as well as the tangent of each angle, and helps in remembering the exact forms.

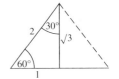

Another angle whose trig. ratios can be given exactly is $45°$. For this one a right-angled isosceles triangle shows the values.

The Form in which Answers should be Given

There is no precise value for g so if a numerical value of g is used in a problem all calculations based on that value are approximations.

Further approximations are made if numerical values of trig ratios are introduced (e.g. $\sin 50° = 0.7660$ correct to 4 dp), so answers based on these approximations should be given only to 2 or 3 significant figures.

For these reasons answers are often given in an exact form, i.e. a quantity which has no exact numerical value is left as a symbol such as g, W, m etc. The trig ratios for angles of $30°$, $45°$ and $60°$ can be expressed in exact surd form and others may be left as, say, $\sin 20°$.

Unless other instructions are given, answers should be presented in exact form. When answers are required to a given degree of accuracy, take the value of g as 9.8 m/s^2 unless a different value is specified.

2. A cyclist is riding up a hill inclined at 20° to the horizontal. His speed at the foot of the hill is 10 m/s but after 30 seconds it has dropped to 4 m/s. The total mass of the cyclist and his machine is 100 kg and there is a wind of strength 15 N down the slope. Find, corrected to 3 significant figures, the constant driving force exerted by the cyclist up the slope.

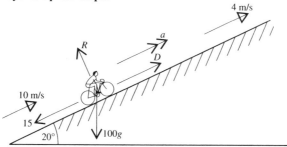

First we will find the acceleration of the cyclist up the slope.

For the motion up the slope:

Known $u = 10$ Required a

             $v = 4$

             $t = 30$

Using $v = u + at$ gives $4 = 10 + 30a$

∴ $a = -\frac{6}{30} = -\frac{1}{5}$

The cyclist has an acceleration of $-\frac{1}{5}$ m/s².

       Resolving   ↗   gives $D - 15 - 100g \sin 20°$

Using $F = ma$ ↗   gives $D - 15 - 100g \sin 20° = 100a$

i.e.   . $D = 15 + 980(0.3420) + 100(-\frac{1}{5})$

                                     $= 330.1\ldots$

The cyclist's driving force is 330 N (3 sf).

EXERCISE 5d

1. The diagram shows a small block of mass 2 kg being pulled up a plane inclined at 30° to the horizontal. The block has an acceleration of 0.5 m/s².

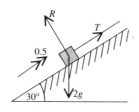

Find an exact expression in terms of g for the tension in the string if

(a) the plane is smooth

(b) the plane is rough and exerts a frictional force of 4 N.

2. A truck is being pulled along a horizontal track by two cables, against resistances totalling 1100 N, with an acceleration of 0.8 m/s². One cable is horizontal and the other is inclined at 40° to the track. The tensions in the cables are shown on the diagram. Taking g as 9.8 find, corrected to 3 significant figures,

(a) the mass of the truck

(b) the vertical force exerted by the track on the truck.

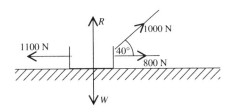

3. Ballast of mass 20 kg is dropped from a balloon that is moving horizontally with a constant speed of 2 m/s.

(a) Mark on a diagram the forces acting on the ballast as it falls.

(b) What is (i) the vertical component (ii) the horizontal component of the acceleration of the ballast?

(c) Without doing any further calculation, sketch *roughly* the path of the ballast as it falls.

4. The diagram shows a small block of mass 5 kg being pulled along a rough horizontal plane by a string inclined at 60° to the plane. There is a frictional force of 18 N.

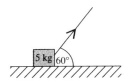

Copy the diagram and on it mark all the forces that act on the block. If the block has an acceleration of 3 m/s², find the tension in the string and show that the normal reaction exerted by the plane on the block can be expressed as $(5g - 33\sqrt{3})$ N.

5.

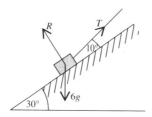

A body of mass 6 kg is sliding down a smooth plane inclined at 30° to the horizontal. Its speed is controlled by a rope inclined at 10° to the plane as shown; the tension in the rope is 10 N. Given that the body starts from rest, find how far down the plane it travels in 5 seconds. Taking g as 9.8, give the answer corrected to 2 significant figures.

NEWTON'S THIRD LAW

The statement of this law is:

Action and Reaction are Equal and Opposite

This means that if a body A exerts a force on a body B then B exerts an equal force in the opposite direction on A. This is true whether the two bodies are in contact or are some distance apart, whether they are moving or are stationary.

Consider, for example, a mass resting in a scale pan. The scale pan is exerting an upward force on the mass and the mass is exerting an equal force downward on the scale pan.

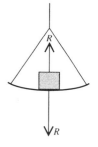

Now consider two particles connected by a taut string. The objects are not in direct contact but exert equal and opposite forces on each other by means of the equal tensions in the string which act inwards at each end.

This is true even if the string passes round a *smooth* body, such as a pulley, which changes the direction of the string.

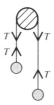

The tensions in the two portions of the string are the same and each portion exerts an inward pull at each end. So in each portion the tension at one end acts on the particle and at the other end the tension acts on the pulley; all these tensions are equal.

(If the string passes round a rough surface the tensions in the different portions of the string are not equal, but the study of this situation is beyond the scope of this book.)

EXERCISE 5e

For each question copy the diagram, making your copy at least twice as big, and mark on it all the forces that are acting on each body (in questions 4 and 6 draw small blocks to represent the people). Use either a different colour or a different type of line (e.g. broken and solid) for the forces that act on separate objects. Ignore forces that act on fixed surfaces – these are indicated in the usual way by hatching.

1.

A load hangs from a beam which is supported at each end.

3.

A mass B hangs by string from another mass A which hangs from a fixed point.

5.

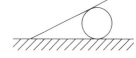

A uniform rod with one end on rough ground rests against an oil drum.

2.

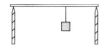

A mass on a table is linked by a string to a mass hanging over the smooth edge.

4.

A passenger is in a lift that is being drawn up by a cable.

6.

A workman is standing on a rung of a uniform ladder that rests on rough ground and a smooth wall.

THE MOTION OF CONNECTED BODIES

In Chapter 4 (p. 70), a number of conventions were introduced. These, although not completely factual, made it possible to apply the principles of mechanics to a simplified situation giving results that were acceptable.

For the work that follows we add further conventions that provide simple methods for solving the mechanics problems that arise in this topic.

For instance we will refer to an inextensible (or inelastic) string, i.e. a string whose length cannot alter, whereas in reality no string is completely inextensible. We shall also be dealing with *smooth* pulleys, i.e. those whose bearings and rim are completely without friction; in practice no pulley is completely smooth.

Having said this, however, solutions based on these conventions can be fairly close to reality.

Consider two particles A and B, of masses m_A and m_B, connected by a light inextensible string passing over a smooth fixed pulley and suppose that $m_A > m_B$.

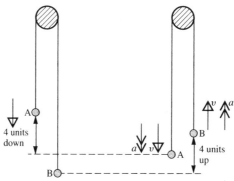

As the particles move, one of them moves upwards in the same way as the other moves downwards, i.e.

    the upward speed of B is equal to the downward speed of A,

    the upward acceleration of B is equal to the downward acceleration of A,

    the distance B moves up is equal to the distance A moves down.

The way in which each particle moves is determined by the forces which act on *that particle alone,* so to analyse the motion we consider the particles separately, i.e.

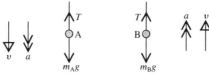

Examples 5f

1. **A light inextensible string passes over a smooth fixed pulley and carries particles of masses 5 kg and 7 kg, one at each end. If the system is moving freely, find in terms of g**

   (a) **the acceleration of each particle**

   (b) **the tension in the string**

   (c) **the force exerted on the pulley by the string.**

   The two particles have the same acceleration, a m/s², and the two parts of the string have the same tension, T newtons.
   For each particle we will find the resultant force in the direction of motion and use it in the equation of motion, $F = ma$, in that direction.

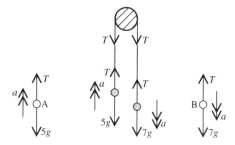

   For A: ↑

   The resultant force   is $T - 5g$

   Using $F = ma$ gives $T - 5g = 5a$ [1]

   For B: ↓

   The resultant force   is $7g - T$

   Using $F = ma$ gives $7g - T = 7a$ [2]

   (a) Adding [1] and [2] gives $2g = 12a$ $\Rightarrow$ $a = \frac{1}{6}g$

   The acceleration of each particle is $\frac{g}{6}$ m/s².

   (b) From [1], $T - 5g = 5\left(\dfrac{g}{6}\right)$ $\Rightarrow$ $T = \dfrac{35g}{6}$

   The tension in the string is $\dfrac{35g}{6}$ N.

   (c) The string exerts a downward pull on each side of the pulley. Therefore the resultant force exerted on the pulley by the string is $2T$ downwards.

   i.e. $\dfrac{35g}{3}$ N downwards.

2. A small block of mass 6 kg rests on a table top and is connected by a light inextensible string that passes over a smooth pulley, fixed on the edge of the table, to another small block of mass 5 kg which is hanging freely. Find, in terms of g, the acceleration of the system and the tension in the string if

(a) the table is smooth

(b) the table is rough and exerts a frictional force of $2g$ N.

We will use the equation of motion for each block in its direction of motion.

(a)

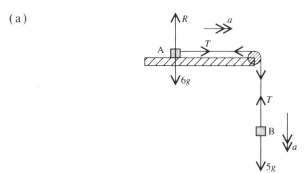

For the block A $\rightarrow$ $T = 6a$ [1]

For the block B $\downarrow$ $5g - T = 5a$ [2]

[1] + [2] $\Rightarrow$ $5g = 11a$ $\Rightarrow$ $a = \frac{5}{11}g$

and $T = 6a = \frac{30}{11}g$

The acceleration is $\frac{5}{11}g$ m/s^2 and the tension is $\frac{30}{11}g$ N.

(b)

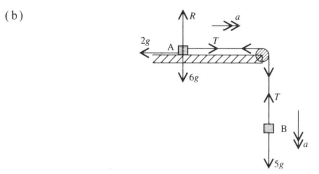

For the block A $\rightarrow$ $T - 2g = 6a$ [3]

For the block B $\downarrow$ $5g - T = 5a$ [4]

[3] + [4] $\Rightarrow$ $3g = 11a$ $\Rightarrow$ $a = \frac{3}{11}g$

and $T = 6a + 2g = \frac{40}{11}g$

The acceleration is $\frac{3}{11}g$ m/s^2 and the tension is $\frac{40}{11}g$ N.

3. A particle P of mass **2m** rests on a rough plane inclined at **30°** to the horizontal. The frictional force is equal to one-half of the normal reaction. P is attached to one end of a light inelastic string which passes over a smooth pulley fixed at the top of the plane and carries a particle Q of mass **3m** hanging freely at the other end. Find in terms of **g**

(a) the normal reaction between P and the plane

(b) the acceleration of P

(c) the force exerted by the string on the pulley.

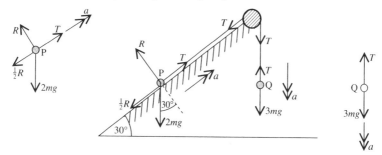

(a) P does not move perpendicular to the plane so the resultant force in this direction is zero.

For P, resolving $\nwarrow$ gives $R - 2mg \cos 30° = 0$

$\Rightarrow$ $R = 2mg(\sqrt{3}/2) = mg\sqrt{3}$

The normal reaction is $mg\sqrt{3}$

(b) For P, resolving $\nearrow$ gives $T - \frac{1}{2}R - 2mg \sin 30° = 2ma$

i.e. $T - \frac{1}{2}mg\sqrt{3} - 2mg(\frac{1}{2}) = 2ma$

$\Rightarrow$ $T - \frac{1}{2}mg\sqrt{3} - mg = 2ma$ [1]

For Q, resolving $\downarrow$ gives $3mg - T = 3ma$ [2]

Add [1] and [2] $mg(2 - \frac{1}{2}\sqrt{3}) = 5ma \Rightarrow 5a = \frac{1}{2}g(4 - \sqrt{3})$

The acceleration is $\frac{1}{10}g(4 - \sqrt{3})$

(c) From [2] $T = 3mg - 3ma$

$= 3mg - \frac{3}{10}mg(4 - \sqrt{3})$

$= \frac{3}{10}mg(6 + \sqrt{3})$

The string exerts two equal tensions on the pulley so the resultant force R on the pulley is midway between these tensions, i.e. it bisects the angle of 60°.

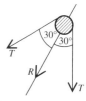

Resolving in this direction gives $R = 2T \cos 30°$

i.e. $R = 2T(\frac{\sqrt{3}}{2}) = T\sqrt{3} = \sqrt{3} \times \frac{3}{10}mg(6 + \sqrt{3})$

$\therefore$ the string exerts a force on the pulley of $\frac{9}{10}mg(2\sqrt{3} + 1)$

Problems in Which the Motion Changes

An aspect of work on pulley systems that has not yet been considered is what happens when a string breaks or goes slack. This situation is covered in the next example.

Examples 5f (continued)

4. **Two particles of masses 1 kg and 3 kg are attached to the ends of a long light inelastic string which passes over a fixed smooth pulley. The system is held with both particles hanging at a height of 2 m above the ground, and is released from rest. In the ensuing motion the heavier particle hits the ground and does not rebound. Find the greatest height reached by the mass of 1 kg.**

Until the 3 kg mass hits the ground, the masses move as a simple connected system.

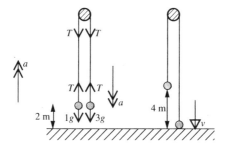

For the 1 kg mass   ↑ $T - 1g = 1a$

For the 3 kg mass   ↓ $3g - T = 3a$

Adding gives $2g = 4a$ $\Rightarrow$ $a = \tfrac{1}{2}g$

For the motion of the 3 kg mass over a distance of 2 m we have $u = 0$, $a = \tfrac{1}{2}g$, $s = 2$ and we want the final speed v, which is the speed of each mass at the moment of impact.

Using $v^2 - u^2 = 2as$ gives

$$v^2 - 0 = 2 \times \tfrac{1}{2}g \times 2 \quad \Rightarrow \quad v = \sqrt{2g}$$

Once the 3 kg mass hits the ground the string becomes slack and no longer exerts any tension on the 1 kg mass. This mass therefore moves on upwards, with an initial speed of $\sqrt{2g}$, under the action of its weight alone, i.e. with an acceleration of $-g$ upwards.

When it reaches its greatest height,
its speed is zero.
For this part of the motion:

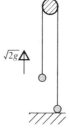
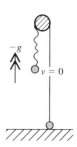

$$u = \sqrt{2g}, \quad a = -g, \quad v = 0$$

Using $v^2 - u^2 = 2as$ gives

$$0 - 2g = 2(-g)s \quad \Rightarrow \quad s = 1$$

The 1 kg mass rises a distance of 1 m above its position when the string went slack. As this point was already 4 m above ground level, the greatest height reached by the 1 kg mass is 5 m above the ground.

EXERCISE 5f

Give the answers in terms of g.

1. Each diagram shows the forces, all in newtons, acting on two particles connected by a light inextensible string which passes over a fixed smooth pulley. In each case find the acceleration of the system and the tension in the string.

(a)
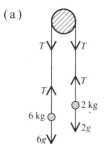

(b)
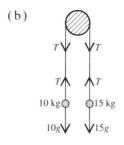

(c)
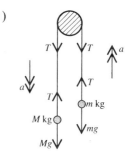

2. Two particles are connected by a light inextensible string which passes over a fixed smooth pulley. Find the acceleration of the system and the tension in the string if the masses of the particles are

(a) 5 kg and 10 kg   (b) 12 kg and 8 kg   (c) 2M and M.

3. Two particles of masses 8 kg and 4 kg hang one at each end of a light inextensible string which passes over a fixed smooth pulley. Find

(a) the acceleration of the system when the particles are released from rest

(b) the distance that each particle moves during the first 5 seconds.

4. A particle of mass 4 kg rests on a smooth plane inclined at 60° to the horizontal. The particle is attached to one end of a light inelastic string which passes over a fixed smooth pulley at the top of the plane and carries a particle of mass 2 kg at the other end. Find

(a) the acceleration of the system   (b) the tension in the string.

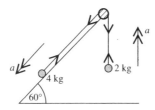

5. A particle of mass 5 kg rests on a smooth horizontal table and is attached to one end of a light inelastic string. The string passes over a fixed smooth pulley at the edge of the table and a particle of mass 3 kg hangs freely at the other end. When the system is released from rest find

(a) the acceleration of the system

(b) the tension in the string.

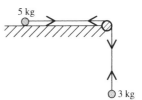

*6. Two particles A and B rest on the smooth inclined faces of a fixed wedge. The particles are connected by a light inextensible string that passes over a fixed smooth pulley at the vertex of the wedge as shown in the diagram. If A and B are each of mass 4 kg find the force exerted by the string on the pulley when the system is moving freely.

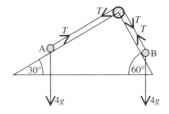

7. Two particles of masses 2 kg and 6 kg are attached one to each end of a long light inextensible string which passes over a fixed smooth pulley. The system is released from rest and the heavier particle hits the ground after 2 seconds. Find the height of this particle above the ground when it was released, and the speed at which it hits the ground.

*8.

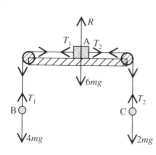

The diagram shows a block A of mass $6m$ resting on a smooth horizontal table. A light inelastic string passes over a fixed smooth pulley at one edge of the table and connects A to a particle B of mass $4m$. Another similar string passes over a smooth pulley fixed at the opposite edge of the table and carries a particle C of mass $2m$. Assuming that all the moving parts are in a vertical plane find

(a) the acceleration of the system
(b) the tension in each string.
(Draw separate diagrams for A, B and C.)

9. Two particles, A of mass 4 kg and B of mass 5 kg, are connected by a light inextensible string passing over a smooth pulley. Initially B is 1 m above a fixed horizontal plane. If the system is released from rest in this position, find

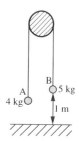

(a) the acceleration of each particle
(b) the speed of each particle when B hits the plane
(c) the further time during which A continues to rise (assuming that it does not reach the pulley).

0. Two particles P and Q are connected by a long, light inextensible string passing over a smooth pulley. The mass of P is m, the mass of Q is $2m$ and the particles are held so that each is 3 m below the pulley. The system is released from rest and after 1 second the string breaks. Find

(a) the speed of each particle at that instant
(b) the further distance that P rises.

1.

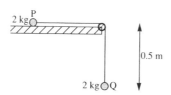

The diagram shows a particle P lying in contact with a smooth table top 1.5 m above the floor. A light inextensible string of length 1 m connects P to another particle Q hanging freely over a small smooth pulley at the edge of the table. The mass of each particle is 2 kg, and P is held at a point distant 0.5 m from the edge of the table. When the system is released from rest find

(a) the speed of each particle when P reaches the edge of the table
(b) the tension in the string.

*12. If, in question 11, P slips over the pulley without any change in its speed, find, for the subsequent motion,

  (a) the acceleration of the system

  (b) the tension in the string.

In questions 13 and 14 use $g = 10$ and give answers corrected to 2 significant figures.

13. A particle A of mass $10m$ lies on a horizontal table and is connected by a light inextensible string to a particle B of mass $8m$. The string passes over a smooth pulley fixed at the edge of the table and B hangs freely. The table is rough and exerts a frictional force of magnitude $2mg$ on A.

  (a) Find the acceleration of the system.

  (b) If the system is released from rest when A is 1.2 m from the pulley find, in terms of g, the speed of the particles when A reaches the pulley.

*14. If in question 13 the string snaps when A is 0.6 m from the edge of the table

  (a) state the way in which B now moves

  (b) find the speed with which A reaches the pulley.

MATHEMATICAL MODELLING

Already in this book we have encountered a variety of problems involving 'real-life' situations such as men climbing ladders, cars pulling caravans and so on. If we had taken into account all the complications of size, irregular shape, non-uniformity etc., a solution would have been very difficult to find. However, so that mathematical techniques based on known physical laws could be applied easily and directly, the effect of certain quantities was ignored or simplified. For example, a man or a car was treated as a particle, a ladder as a uniform rod, air resistance as negligible etc., and in doing this we were 'making assumptions'.

The process is called *mathematical modelling*.

Whenever a problem is to be solved by making a mathematical model, all the necessary assumptions should be stated clearly at the outset and it should be considered whether these assumptions are reasonable in the context of the problem. The results obtained from such a model can only be approximate but they are accurate enough for most purposes.

When a practical problem has been modelled it becomes a simplified mathematical exercise. The working diagram used in the solution no longer need be realistic (and often time-consuming); instead of large objects such as trees, vehicles, planks, crates etc., it can be simply made up of points (particles) and lines (rods). Then forces, velocities, dimensions and so on can be marked much more clearly.

In this type of problem, where the assumptions made are usually some way from the actual situation, it is inappropriate to give answers to more than two significant figures.

Examples 5g

1. **A tractor of mass M is pulling a trailer of mass M_1. The tractor exerts a steady driving force D. Construct a mathematical model stating all assumptions made and hence find, in terms of M, M_1 and D,**

   (a)   the acceleration of the trailer

   (b)   the tension in the tow rope.

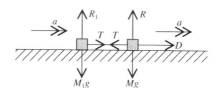

Model the tractor and the trailer each as a small block; assume no resistance to motion and that the tow bar is light and does not stretch.

The tractor and trailer have the same acceleration.

Considering the trailer, the only force acting $\rightarrow$ is T

$F = ma \rightarrow$ gives $T = M_1 a$ [1]

Considering the tractor, the force acting $\rightarrow$ is $D - T$

$F = ma \rightarrow$ gives $D - T = Ma$ [2]

(a)   Add [1] and [2] $D = (M + M_1)a$

   The acceleration of the trailer is $\dfrac{D}{M + M_1}$

(b)   From [1] $T = M_1 \times \dfrac{D}{M + M_1}$

   The tension in the tow rope is $\dfrac{M_1 D}{M + M_1}$

Note that the results are given without a unit as no units are given in the question.

2. A large box of mass 15 kg can be raised and lowered by a crane. The box contains a load of mass 20 kg. Find

(i)   the tension in the cable of the crane

(ii) the force exerted by the load on the bottom of the box when the box is accelerating at 2 m/s² (a) upwards (b) downwards.
(Take $g = 10$ and give answers corrected to 2 significant figures.)

(a) We will model the load as a particle and the box also as a particle when it is considered alone.

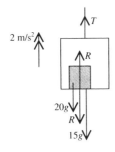

                                                          Forces acting on box   Forces acting on load

Resolving ↑   for the load $R - 20g = 20 \times 2$

                                      ⇒ $R = 240$

                   for the box $T - R - 15g = 15 \times 2$

                                      ⇒ $T = 420$

(i)   The tension is 420 N.   (ii) The force on the box is 240 N.

(b)

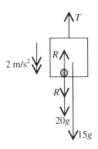

                                                          Forces acting on box   Forces acting on load

Resolving ↓   for the load $20g - R = 20 \times 2$

                                      ⇒ $R = 160$

                   for the box $R + 15g - T = 15 \times 2$

                                      ⇒ $T = 280$

(i)   The tension is 280 N.   (ii) The force on the box is 160 N.

EXERCISE 5g

In each question state any assumptions you make in order to form a mathematical model that can be used to solve the problem.

1. A lift of mass 500 kg carrying a load of 80 kg is drawn up by a cable. The lift first accelerates at $\frac{1}{12}g$ m/s^2 from rest to its maximum speed which is maintained for a time, after which the lift decelerates to rest at $\frac{1}{10}g$ m/s^2. For each of these three stages of motion find

   (a) the tension in the cable

   (b) the force exerted by the load on the floor of the lift.

2. A car of mass 1 tonne is pulling a caravan of mass 800 kg along a level straight road. There is a total resistance to the motion of 450 N; the individual resistances on the car and caravan are in the ratio of their masses. If the combination accelerates uniformly from rest to 20 m/s in $12\frac{1}{2}$ seconds find

   (a) the tension in the tow bar

   (b) the driving force exerted by the car's engine.

3. If in question 2 the tow bar snaps at the instant when the speed reaches 20 m/s and the car continues with the same driving force, find

   (a) the subsequent acceleration of the car

   (b) the deceleration of the caravan

   (c) how long it takes for the caravan to stop.

4. A car of mass 800 kg exerting a driving force of 2.2 kN (i.e. 2200 N) is pulling a trailer tent of mass 300 kg along a level road. If there is no resistance to the motion of either the car or the trailer find the acceleration of the car and the tension in the towbar.

5. A lift of mass 800 kg is operated by a cable as shown in the diagram. A passenger of mass 70 kg is standing in the lift. Find, stating what object you can use to represent the passenger,

   (i) the force exerted by the passenger on the floor of the lift

   (ii) the tension in the cable

   when the lift is accelerating

   (a) upwards   (b) downwards at 0.5 m/s.

   (Draw separate diagrams for the lift and for the passenger.)

CHAPTER 6

FORCES IN EQUILIBRIUM. FRICTION

CONCURRENT FORCES IN EQUILIBRIUM

A body that is at rest, or is moving with constant velocity, is in a state of equilibrium.

The acceleration of a body in equilibrium is zero in any direction therefore the resultant force in any direction is also zero.

The converse of this statement is not necessarily true because, although forces with zero resultant cannot make an object move in a line they can, as we shall see later on, cause an object to *turn,* e.g.

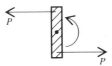

However a set of *concurrent* forces (i.e. all passing through one point) can never cause turning so, as at present we will deal only with concurrent forces, the problem of turning will not arise yet.

We saw in Chapter 3 that the resultant of a set of forces can be found by collecting components in each of two perpendicular directions, giving

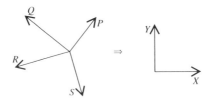

Now if the resultant is zero, the collected components in each direction must individually be zero, i.e. $X = 0$ and $Y = 0$.

Applying this fact to a concurrent system in equilibrium, in which some forces are unknown, provides a method for finding the unknown quantities.

Examples 6a

1. A particle of weight 16 N is attached to one end of a light string whose other end is fixed. The particle is pulled aside by a horizontal force until the string is at 30° to the vertical. Find the magnitudes of the horizontal force and the tension in the string.

Let P newtons and T newtons be the magnitudes of the horizontal force and the tension respectively.

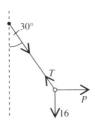

Resolving $\rightarrow$ gives $\quad P - T \sin 30° = 0$

i.e. $\qquad\qquad P - T \times \frac{1}{2} = 0$ $\hspace{2cm}$ [1]

Resolving $\uparrow$ gives $\quad T \cos 30° - 16 = 0$

i.e. $\qquad\qquad T \times \frac{1}{2}\sqrt{3} - 16 = 0$ $\hspace{2cm}$ [2]

From [2] $\qquad\qquad T = \frac{32}{\sqrt{3}} = \frac{32\sqrt{3}}{3}$

From [1] $\qquad\qquad P = \frac{1}{2}T = \frac{16\sqrt{3}}{3}$

Therefore the magnitude of the horizontal force is $\frac{16\sqrt{3}}{3}$ N and the magnitude of the tension is $\frac{32\sqrt{3}}{3}$ N.

2. A load of mass 26 kg is supported in equilibrium by two ropes inclined at 30° and 60° to the horizontal as shown in the diagram. Find in terms of g the tension in each rope.

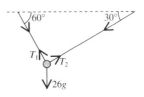

We will assume that the ropes are light and that the load can be treated as a particle.

The tensions in the ropes act in perpendicular directions so we will resolve in these directions.

Let the tensions in the ropes be T_1 and T_2 newtons.

Resolving $\quad\diagdown\quad T_1 - 26g \cos 30° = 0$

$\therefore \qquad\qquad T_1 = 26g\left(\frac{\sqrt{3}}{2}\right) = 13g\sqrt{3}$

Resolving $\quad\diagup\quad T_2 - 26g \sin 30° = 0$

$\therefore \qquad\qquad T_2 = 26g\left(\frac{1}{2}\right) = 13g$

The tensions in the ropes are $\quad 13g\sqrt{3}$ N $\quad$ and $\quad 13g$ N.

3. In each diagram the forces are measured in newtons and are in equilibrium. Find the values of P and Q.

(a)   (b)

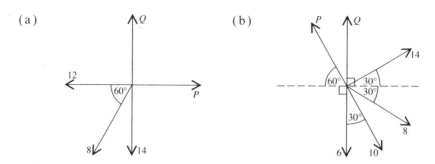

(a)   Resolving horizontally and vertically gives simple equations.

Resolving $\rightarrow$ $P - 12 - 8\cos 60° = 0$

$\Rightarrow$ $P - 12 - 8 \times \frac{1}{2} = 0$

$\therefore$ $P = 16$

Resolving $\uparrow$ $Q - 14 - 8\sin 60° = 0$

$\Rightarrow$ $Q - 14 - 8(\frac{\sqrt{3}}{2}) = 0$

$\therefore$ $Q = 14 + 4\sqrt{3}$

(b)   If we resolve perpendicular to P, the equation we get does not contain P and so gives the value of Q. Similarly, to find P easily we can resolve perpendicular to Q.

Resolving $\perp$ to P $\nearrow$

$14 + Q\cos 60° + 8\cos 60° - 6\cos 60° = 0$

$\Rightarrow$ $14 + \frac{1}{2}Q + \frac{1}{2}(8 - 6) = 0$

$\Rightarrow$ $Q = -30$

Resolving $\perp$ to Q $\rightarrow$

$14\cos 30° + 8\cos 30° + 10\cos 60° - P\cos 60° = 0$

$(14 + 8)(\frac{\sqrt{3}}{2}) + 10(\frac{1}{2}) - \frac{1}{2}P = 0$

$\Rightarrow$ $P = 22\sqrt{3} + 10$

If the forces are given in the form $a\mathbf{i} + b\mathbf{j}$ then, because a and b are the magnitudes of components in the directions of $\mathbf{i}$ and $\mathbf{j}$, the sum of the coefficients of $\mathbf{i}$ is zero and similarly for $\mathbf{j}$.

4. Forces $2\mathbf{i} - 3\mathbf{j},$ $7\mathbf{i} + 4\mathbf{j},$ $-5\mathbf{i} - 9\mathbf{j},$ $P\mathbf{i} + 2\mathbf{j},$ and $\mathbf{i} - Q\mathbf{j},$ are in equilibrium. Find the values of P and Q.

$$(2\mathbf{i} - 3\mathbf{j}) + (7\mathbf{i} + 4\mathbf{j}) + (-5\mathbf{i} - 9\mathbf{j}) + (P\mathbf{i} + 2\mathbf{j}) + (\mathbf{i} - Q\mathbf{j})$$
$$= (5 + P)\mathbf{i} + (-6 - Q)\mathbf{j}$$
$$= 0\mathbf{i} + 0\mathbf{j}$$

$\therefore$ $5 + P = 0$ and $-6 - Q = 0$ $\Rightarrow$ $P = -5$ and $Q = -6$

EXERCISE 6a

In this exercise all forces are measured in newtons.
In questions 1 to 3 the forces shown in the diagram are in equilibrium.
Find the values of P, Q and, where appropriate, the value of θ.

1. (a)
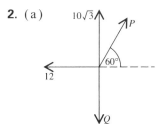

(b)
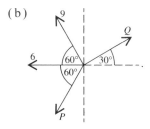

(c)
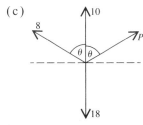

2. (a)
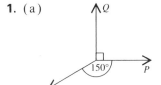

(b)

(c)
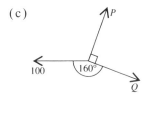

3. (a)
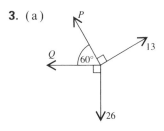

(b)
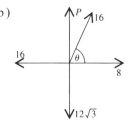

(c)
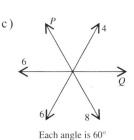

Each angle is 60°

4. A light inextensible string is of length 50 cm.
It is fixed to a wall at one end A and a
particle of mass 4 kg is attached to the other
end B. A horizontal force applied to the
end B holds the particle in equilibrium at
a distance of 30 cm from the wall.
Find, in terms of g, the tension in the string.

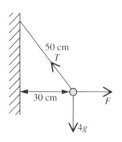

5. A small block of weight 20 N is attached to two
light inelastic strings. The other ends of the strings
are fixed to two fixed points
on the same level, 1 m apart. The lengths of the
strings are 0.6 m and 0.8 m. What is the angle
between the strings? By resolving in the directions
of the strings, find the tension in each string.

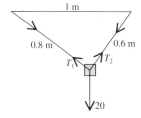

6. A small block of weight W rests on a smooth
plane inclined at $30°$ to the horizontal and is held in
equilibrium by a light string inclined at $30°$ to the
plane. Find, in terms of W, the tension in the string.

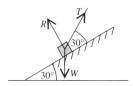

7.

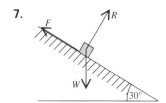

A block of weight W rests in equilibrium on a
rough plane inclined at $30°$ to the horizontal.
Find, in terms of W, the magnitude of the
frictional force.

8. A small block of weight $6W$ rests on a smooth plane inclined at an angle θ to
the horizontal. Find θ if the block is held in equilibrium by

(a) a force $3W$ parallel to the plane   (b) a horizontal force $2W$.

Remember that $\tan \theta = \frac{\sin \theta}{\cos \theta}$.

9. Write down, in the form $a\mathbf{i} + b\mathbf{j}$, each
of the forces shown in the diagram. Given
that the forces are in equilibrium, find P
and Q.

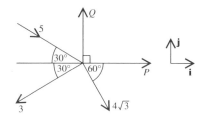

10. Four forces act on a particle keeping it in equilibrium. Find the values of p and q if the forces are

(a) $2\mathbf{i} + 7\mathbf{j}$, $5\mathbf{i} - p\mathbf{j}$, $9\mathbf{i} + 4\mathbf{j}$ and $q\mathbf{i} - 11\mathbf{j}$

(b) $\mathbf{i} - 6\mathbf{j}$, $-8\mathbf{i} + 3\mathbf{j}$, $p\mathbf{i} + q\mathbf{j}$ and $3\mathbf{i} + 10\mathbf{j}$

11. The resultant of the forces $7\mathbf{i} - 2\mathbf{j}$, $-6\mathbf{i} + 5\mathbf{j}$, $3\mathbf{i} + 6\mathbf{j}$ and $a\mathbf{i} + b\mathbf{j}$, is $11\mathbf{i} - 2\mathbf{j}$.

(a) Find the values of a and b.

(b) When a fifth force is added to the given forces, equilibrium is established. Write down in terms of $\mathbf{i}$ and $\mathbf{j}$ the force that is added.

FRICTION

Frictional forces were mentioned briefly in Chapter 4 when we considered a book lying on a table top.

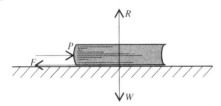

A force P applied to the book does not necessarily move it. That is because if the contact between the book and the table is rough there is frictional resistance to motion.

The frictional force F, acting on the book, acts along the table in a direction opposite to the potential direction of motion. (We know from Newton's Third Law that an equal and opposite frictional force acts on the table but this is ignored as the table is regarded as fixed.)

If P and F are the only forces acting horizontally on the book then, as long as it is stationary, P and F must be equal. (It is obvious that at no stage can we say $F > P$; if this were the case the book would move towards the pushing force!). So the amount of friction is just sufficient to prevent motion.

Now if P gradually increases eventually the book will be *just on the point of moving*. A further increase in the value of P will make the book move.

When the book is on the point of moving, friction is said to be *limiting*; F has reached its maximum value and the book is in *limiting equilibrium*.

Beyond this point the book moves, i.e. $P > F$.

The Coefficient of Friction

For two particular surfaces in rough contact, it can be shown experimentally that the limiting value of the frictional force is a fixed fraction of the normal reaction between the surfaces.

This fraction is called the *coefficient of friction* and it is denoted by the Greek letter μ (pronounced mew, in English), i.e. for limiting friction

$$F = \mu R$$

As this is the maximum value of the frictional force, F can take any value from zero up to μR, i.e.

$$0 \leqslant F \leqslant \mu R$$

Once an object begins to move, the frictional force opposing motion remains at the constant value μR. The *marginal* difference between the value of μ when friction is limiting (the coefficient of static friction) and its value once motion takes place (the coefficient of dynamic friction) is so small that at this level it can be ignored.

The value of μ depends upon the materials of which the *two* surfaces in contact are made – it is *not* a property of *one* surface – so ideally we should always refer to *rough contact* rather than to a rough plane, etc.

However, because wording of the strictly correct definition is a bit lengthy, it is not always used and a phrase such as 'a ladder rests against a rough wall' is often found. This should be taken to mean that there is friction between the ladder and the wall. Similarly 'a block moves on a smooth plane' means that we ignore friction between the block and the plane.

It is interesting to note that rough contact does not necessarily involve surfaces that would ordinarily be described as rough. For instance, a highly polished metal block placed on a highly polished flat metal sheet is extremely difficult to move across the sheet, although each surface on its own would be called smooth. Clearly in the context of mechanics the ordinary meanings of 'rough' and 'smooth' cannot be used; instead

'rough' means that there *is* friction at the contact;
'smooth' means that we ignore friction at the contact.

The properties of friction discussed so far can be summarised to give what are known as the *laws of friction*.

THE LAWS OF FRICTION

- When the surfaces of two objects are in rough contact, and have a tendency to move relative to each other, equal and opposite frictional forces act, one on each of the objects, so as to oppose the potential movement.

- Until it reaches its limiting value, the magnitude of the frictional force F is just sufficient to prevent motion.

- When the limiting value is reached, $F = \mu R$, where R is the normal reaction between the surfaces and μ is the coefficient of friction for those two surfaces.

- For all rough contacts $0 < F \leqslant \mu R$.

- If a contact is smooth $\mu = 0$.

The Angle of Friction

When two objects are in rough contact and friction is limiting, two contact forces act on each object; one is the normal reaction R and the other is the frictional force μR.

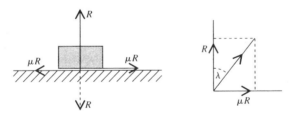

The resultant of R and μR, which is sometimes called the *resultant contact force*, is shown in the diagram.

Its magnitude is $\sqrt{(R^2 + [\mu R]^2)} = R\sqrt{(1 + \mu^2)}$

The angle between this resultant and the normal reaction is called the *angle of friction* and is denoted by λ, where

$$\tan \lambda = \frac{F}{R} = \frac{\mu R}{R}$$

i.e. $\tan \lambda = \mu$

The tangent of the angle of friction is equal to the coefficient of friction

There are occasions when it is convenient to use the resultant contact force.

Examples 6b

1. **A small block of weight 32 N is lying in rough contact on a horizontal plane. A horizontal force of P newtons is applied to the block until it is just about to move the block.**

   (a) If $P = 8$ find the coefficient of friction μ between the block and the plane.

   (b) If $\mu = 0.4$, find the value of P.

   (a)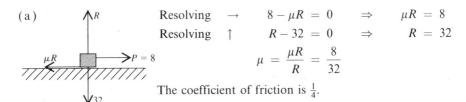

   Resolving $\rightarrow$ $8 - \mu R = 0$ $\Rightarrow$ $\mu R = 8$

   Resolving $\uparrow$ $R - 32 = 0$ $\Rightarrow$ $R = 32$

   $$\mu = \frac{\mu R}{R} = \frac{8}{32}$$

   The coefficient of friction is $\frac{1}{4}$.

   (b)

   Resolving $\rightarrow$ $P - 0.4 \times R = 0$

   Resolving $\uparrow$ again gives $R = 32$

   $\therefore$ $P - 0.4 \times 32 = 0$

   $\Rightarrow$ $P = 12.8$

2. **A small block of weight 24 N rests in rough contact with a horizontal plane. A light string is attached to the block and is inclined at 30° to the plane. The block is just about to slip when the tension in the string is 12 N. Find the coefficient of friction between the block and the plane.**

   Friction is limiting so $F = \mu R$.

   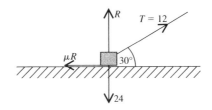

   Resolving $\rightarrow$ $12 \cos 30° - \mu R = 0$

   $\Rightarrow$ $12 \times \sqrt{3}/2 - \mu R = 0$ $\Rightarrow$ $\mu R = 6\sqrt{3}$

   Resolving $\uparrow$ $12 \sin 30° + R - 24 = 0$ $\Rightarrow$ $R = 18$

   $$\frac{\mu R}{R} = \frac{6\sqrt{3}}{18} \quad \Rightarrow \quad \mu = \frac{\sqrt{3}}{3}$$

   The coefficient of friction is $\frac{\sqrt{3}}{3}$ $\left(= \frac{1}{\sqrt{3}} \right)$.

3. A particle of weight 8 N is resting in rough contact with a plane inclined at an angle α to the horizontal where $\tan\alpha = \frac{3}{4}$. The coefficient of friction between the particle and the plane is μ. A horizontal force P newtons is applied to the particle. When $P = 16$ the particle is on the point of slipping up the plane.

(a) Find μ.

(b) Find the value of P such that the particle is just prevented from slipping down the plane.

(a)

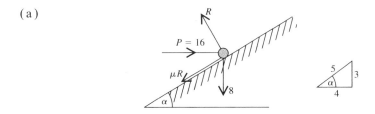

Resolving parallel and perpendicular to the plane involves μ in only one equation.

Resolving $\nearrow$ $16\cos\alpha - \mu R - 8\sin\alpha = 0$

$\Rightarrow$ $16 \times \frac{4}{5} - \mu R - 8 \times \frac{3}{5} = 0$

$\Rightarrow$ $\mu R = 8$

Resolving $\nwarrow$ $R - 16\sin\alpha - 8\cos\alpha = 0$

$\Rightarrow$ $R - 16 \times \frac{3}{5} - 8 \times \frac{4}{5} = 0$

$\Rightarrow$ $R = 16$

$\mu R / R = \frac{8}{16}$ $\Rightarrow$ $\mu = \frac{1}{2}$

(b)

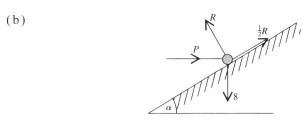

This time resolving horizontally and vertically uses P in only one equation.

Resolving $\uparrow$ $R\cos\alpha + \frac{1}{2}R\sin\alpha - 8 = 0$

$\Rightarrow$ $R = \frac{80}{11}$

Resolving $\rightarrow$ $P + \frac{1}{2}R\cos\alpha - R\sin\alpha = 0$

$\Rightarrow$ $P = \frac{R}{5} = \frac{16}{11}$

i.e. $P = 1\frac{5}{11}$

4. The diagram shows a plane, inclined at an angle α to the horizontal, where $\tan \alpha = \frac{3}{4}$, with a smooth pulley fixed at the top. A light inextensible string passes over the pulley and connects the particle A, which is in rough contact with the plane, to the particle B hanging freely. The mass of B is 3 kg, the mass of A is 2 kg and the coefficient of friction with the plane is 0.2. When the system is released from rest find the acceleration of the particles and the tension in the string. Take the value of g as 10 and give answers corrected to 2 significant figures.

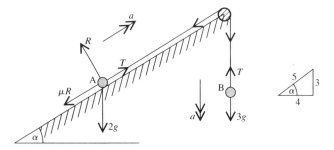

When the system moves, the frictional force has a constant value of $0.2R$.

Using Newton's Law, $F = ma$, in the direction of motion of each particle gives

For B $\downarrow$ $3g - T = 3a$ [1]

For A $\nearrow$ $T - 2g \sin \alpha - 0.2R = 2a$ [2]

           $\nwarrow$ $R - 2g \cos \alpha = 0$ [3]

From [3] $R = 20 \times \frac{4}{5} = 16$

In [2] $T = 2a + 20 \times \frac{3}{5} + 0.2 \times 16 = 2a + 15.2$

In [1] $3a = 30 - (2a + 15.2) \quad \Rightarrow \quad 5a = 14.8$

$\therefore$ $a = 2.96 \quad$ and $\quad T = 21.12$

To 2 sf, the acceleration is 3.0 m/s^2 and the tension is 21 N.

EXERCISE 6b

Give answers that are not exact corrected to 2 or 3 significant figures as appropriate.

In each question from 1 to 7 the particle is of weight 24 N and has rough contact with the specified surface; μ is the coefficient of friction between the particle and the surface.

1. The particle is just about to slip down a plane inclined at 30° to the horizontal. Find the value of μ.

2. The particle is on a horizontal plane and is being pulled by a horizontal string. If it is just on the point of moving when the tension in the string is 8 N, find the value of μ.

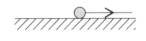

3. The particle is just about to slip up a plane inclined at 30° to the horizontal, when being pushed by a force parallel to the plane. If $\mu = \frac{1}{2}$, find the magnitude of the force.

4. The particle is on a horizontal plane and is being pulled by string inclined at 60° to the horizontal. If it is just on the point of moving when the tension in the string is 16 N, find the value of μ.

5. The particle is supported in limiting equilibrium on a plane inclined at 30° to the horizontal, by a string parallel to the plane. If $\mu = \frac{1}{5}$, find the tension in the string.

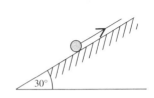

6. The particle is resting on a plane inclined at an angle α to the horizontal, where $\tan \alpha = \frac{4}{3}$. A force of 12 N parallel to the plane is just able to prevent the particle from slipping down the plane. Find the value of μ.

7. The particle is held in limiting equilibrium, on a plane inclined at 30° to the horizontal, by a string inclined at 30° to the plane as shown. Given that the value of μ is $\frac{1}{4}$, find the tension in the string when the particle is on the point of moving

(a)   up the plane

(b)   down the plane.

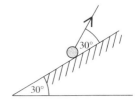

8. A small block of weight W is placed on a plane inclined at an angle θ to the horizontal. The coefficient of friction between the block and the plane is μ.

(a)   When $\theta = 30°$ the block is on the point of slipping. Show that $\mu = \frac{\sqrt{3}}{3}$.

(b)   If $\mu = \frac{1}{5}$ and $\tan \theta = \frac{3}{4}$ find, in terms of W, the magnitude of the horizontal force needed to prevent the block from slipping down the plane.

(c)   If $\mu = \frac{2}{5}$ and $\tan \theta = \sqrt{3}$ find, in terms of W, the magnitude of the horizontal force that will be on the point of making the block slide up the plane.

9. A horizontal force P newtons is applied to a body of weight 80 N, standing in rough contact with a horizontal plane. The coefficient of friction between the body and the plane is $\frac{1}{2}$. What is the magnitude of the frictional force when

(a) $P = 10$ (b) $P = 40$ (c) $P = 50$?

State in each case whether or not the body moves.

10.

 $\tan \alpha = \frac{5}{12}$ $\Rightarrow$

A warehouse porter is trying to push a trolley, of mass 24 kg, up a plane inclined at an angle α to the horizontal, where $\tan \alpha = \frac{5}{12}$. He finds that the trolley is just on the point of moving when the horizontal force he is exerting on the handles reaches 200 N. Stating any assumptions that are necessary, use a suitable model, with $g = 10$, to find the value of the coefficient of friction between the trolley and the plane.

*11. The diagram shows a particle A lying in rough contact with a table. A light inelastic string attached to A, passes over a smooth pulley at the edge of the table and is attached to another particle B hanging freely. The particles are of equal mass M and the coefficient of friction between A and the table is $\frac{2}{5}$. Find in terms of g and M, the acceleration of B and the tension in the string.

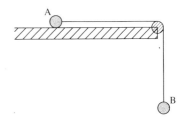

*12.

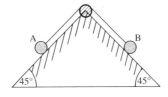

A and B are two particles connected by a light string that passes over a smooth pulley at the top of a wedge as shown in the diagram. The mass of A is m and that of B is $2m$. Contact between each particle and the wedge is rough with a coefficient of friction of $\frac{1}{5}$. When the system is allowed to move, find the acceleration and the tension in the string.

CHAPTER 7

WORK AND POWER

WORK

Anyone pushing a heavy crate across a storeroom floor would be justified in thinking that it was hard work. This is a common concept of work, i.e. making an effort to move an object, and it is reflected in the following definition of *mechanical work*.

When an object moves under the action of a constant force F, the amount of work done by the force is given by:
   the component of F in the direction of motion $\times$ distance moved.

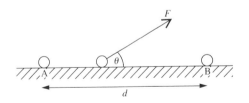

So if a constant force F moves an object from A to B,
the amount of work done by F is $(F \cos \theta) \times (d)$,
i.e. $Fd \cos \theta$

If the force is measured in newtons and the distance in metres, the work done is measured in joules (J).

For example, if a force of 12 N acts on a body and moves it a distance of 3 m in the direction of the force, the amount of work done by the force is 36 J.

Note that work is done *only if the force succeeds in moving the object*.
A force applied to an object that remains at rest, does not do any work.

When several forces act on one body, the work done by each force can be found independently of the others.

Work Done Against a Particular Force

In some circumstances we are more concerned about the work that is needed to *overcome* an opposing force. Consider, for example, an object that is pulled at constant speed from A to B, a distance *s* along a rough surface.

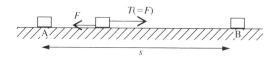

A frictional force of magnitude *F* acts on the object, opposing the motion.

Because there is no acceleration, the force causing the displacement is equal and opposite to the frictional force, i.e. it is of magnitude *F* and therefore the amount of work it does is given by *Fs*.

This work is done to overcome friction and is called *the work done against friction*. So we see that

the work done *against* a force is given by
the magnitude of that force × the distance moved in the opposite direction.

Now consider a body of mass *m*, raised vertically through a height *h*.
The weight, *mg*, acts vertically downward and has to be overcome by an upward force in order to raise the body.

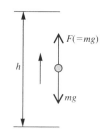

The force needed to raise the body vertically at constant speed is *mg* upwards. The work done by this force is *mg* × *h*. This amount of work is needed to overcome the opposing force of gravity, i.e. the work done against gravity is *mgh*.

Again we see that the work done against gravity is given by

the magnitude of the gravitational force (*mg* downwards)
× the distance moved in the opposite direction (*h* upwards).

It is sometimes convenient to regard work done against a force as being negative work done *by* that force. In the situation above, for example, the work done *by* gravity is $-mgh$.

Work Done by a Moving Vehicle

A variety of forces can act on a moving vehicle, including friction, air resistance, the weight of the vehicle, reaction with the ground etc., but most important is the driving force.

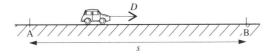

The *work done by a vehicle*   means   the *work done by the driving force*,
i.e. $D \times s$

Examples 7a

Whenever a force is represented on a diagram by a letter, e.g. P, it is understood that the force is P newtons.

1. **A body resting in smooth contact with a horizontal plane, moves 2.6 m along the plane under the action of a force of 20 N. Find the work done by the force if it is applied**

   (a) horizontally

   (b) at 60° to the plane.

   (a)

   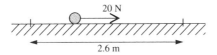

   The whole of the force acts in the direction of motion.

   Work done $= 20 \times 2.6$ J $= 52$ J

   (b)

   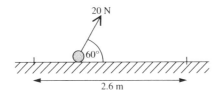

   The component of the force that acts in the direction of motion is $F \cos 60°$

   Work done $= (20 \cos 60° \times 2.6)$ J $= (20 \times \frac{1}{2} \times 2.6)$ J $= 26$ J

2. Sixteen crates, each of mass 250 kg, are raised 3 m by a hoist, to be placed on a platform. Find the work done against gravity by the hoist. State any assumptions you have made.

The work done against gravity in raising one crate $= (250 \times g \times 3)$ J

$= 750g$ J

The work done against gravity in raising 16 crates $= (16 \times 750g)$ J

$= 12\,000g$ J $= 12g$ kJ

(1000 J is 1 kilojoule, i.e. 1 kJ)

Assumptions are that the crates can be treated as particles, that the hoist raises them to exactly 3 m, and that no crates are stacked in successive layers.

3. A small block of mass 2 kg slides, at constant speed, $\frac{4}{5}$ m down the face of a plane inclined at 30° to the horizontal. Contact between the block and the plane is rough. Giving answers in terms of g, find

(a) the work done by
   (i)   the weight of the block
   (ii) the reaction between the block and the plane.

(b) the work done against friction.

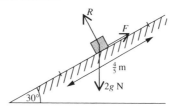

(a)   The direction of motion is down the plane.

   (i)   The component of weight down the plane is $2g \sin 30°$ N $= g$ N
           $\therefore$ the work done by the weight is $g \times \frac{4}{5}$ J $= \frac{4}{5}g$ J
   (ii) The reaction has no component parallel to the plane.
           $\therefore$ no work is done by the reaction.

(b)   The acceleration of the block is zero so the resultant force along the plane is zero.

Resolving $\diagup$ gives $2g \sin 30° - F = 0$

$\therefore$ the frictional force is g N up the plane.

$\therefore$ the work done against friction is $(g \times \frac{4}{5})$ J $= \frac{4}{5}g$ J

4. A tractor climbs, at a steady speed of **5 m/s,** up a slope inclined at an angle α to the horizontal. The mass of the tractor is **1400 kg** and $\sin \alpha = \frac{3}{25}$. By modelling the tractor as a particle find the work done by the tractor against gravity per minute. If the total work done by the tractor in this time is **780 kJ,** find the resistance to motion. (Take g as **10.**)

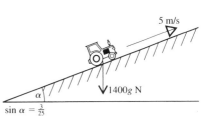

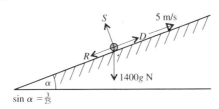

In one minute:

the tractor climbs up the slope a distance $\quad 5 \times 60$ m $\; = \; 300$ m

$\therefore \quad$ the vertical distance climbed is $\qquad 300 \sin \alpha$ m $\; = \; 36$ m

$\therefore \quad$ the work done against gravity is $\quad (1400 \times g \times 36)$ J $\; = \; 504$ kJ

The speed of the tractor is constant so the acceleration is zero.

Resolving $\nearrow$ gives $\quad D - R - mg \sin \alpha \; = \; 0$

$\therefore \qquad\qquad\qquad\qquad D = R + 1400 \times 10 \times \frac{3}{25} \; = \; R + 1680$

The work done by the tractor $=$ the work done by the driving force

$\qquad\qquad\qquad\qquad\qquad\quad = \; (D \times 300)$ J $\; = \; (R + 1680) \times 300$ J

This is known to be 780 kJ

Therefore $\qquad 780\,000 \; = \; 300(R + 1680) \qquad \Rightarrow \qquad R \; = \; 2600 - 1680$

The resistance to motion is 920 N.

EXERCISE 7a

In each question from 1 to 3, a small object moves from A to B under the action of the forces shown in the diagram. Find the work done by each force.

1. AB $= 2$ m

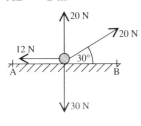

2. AB $= 3$ m

3. AB $= 4$ m

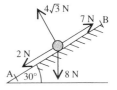

For the rest of this exercise use $g = 10$, state any assumptions that are made in modelling each situation and give answers corrected to 2 significant figures.

In each question from 4 to 8 find the work done against gravity.

4. A block of mass 3 kg is raised vertically through 2.1 m.

5. A workman of mass 87 kg climbs a vertical ladder of length 7 m.

6. Eight crates of beer, each of mass 24 kg, are lifted from the ground on to a shelf that is 1.8 m high.

7. A forklift truck loads a tiger in a cage into the hold of an aircraft. The combined mass of tiger and cage is 340 kg and the floor of the hold is 7.2 m above the ground.

8. A crane lifts a one-tonne block of stone out of a quarry that is 11 m deep.

9.

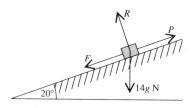

A block of mass 14 kg is pulled a distance of 6 m up a plane inclined at 20° to the horizontal. The contact is rough and the magnitude of the frictional force is 30 N.

Find the work done against (a) friction (b) gravity.

10. There is an average resistance of 480 N to the motion of a train as it travels 5.8 km between two stations. Find the work done against the resistance.

11.

A boy, whose mass is 40 kg, has a sledge of mass 6 kg. He pulls the sledge 36 m up a slope inclined at 30° to the horizontal.

(a) Find the amount of work that the boy has to exert against gravity in order to pull the sledge up.

The boy then sits on the sledge and slides back to the foot of the slope.

(b) Find the work done by gravity during the descent.

12. A wardrobe is lowered by a rope at a steady speed from the balcony of a fifth-floor flat to the ground, 12 m below. Given that the mass of the wardrobe is 37 kg, find the work done by the rope during the descent.

In questions 13 to 16 a box of mass 6 kg is pulled by a rope along a horizontal surface at a constant speed.

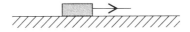

13. The speed is 4 m/s. Find the work done by the rope in 20 seconds if the tension in it is 18 N.

14. The work done by the rope in moving the box 8 m is 200 J. Find the tension in the rope.

15. The coefficient of friction between the box and the surface is $\frac{1}{3}$. If friction is the only resistance to the motion of the box, find the work done by the rope in pulling the box through 5 m.

16. The work done by the rope against friction in pulling the box a distance of 12 m is 180 J. Find the coefficient of friction between the box and the surface.

17.

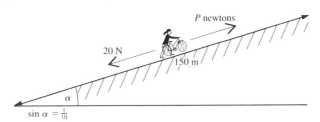

A girl pushes her bicycle 150 m up a hill inclined at an angle α to the horizontal where $\sin \alpha = \frac{1}{10}$. If the combined weight of the girl and her bicycle is 700 N, find the work she does against gravity. If there is an average resistance to motion of 20 N find the total work done by the girl.

18. A crate of mass 40 kg is pulled by a rope at a constant 1.5 m/s down a slope inclined at 15° to the horizontal. Contact is rough and the coefficient of friction is 0.7. Find

(a) the frictional force

(b) the tension in the rope

(c) the work done by the rope per second

(d) the work done by gravity while the crate moves down the slope for 6 seconds.

19.

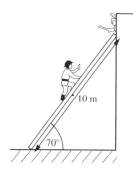

To rescue a woman trapped in a burning flat, a fireman climbs 10 m up the turntable ladder, which is inclined at an angle of 70° to the horizontal. How much work does the fireman, whose mass is 80 kg, do against gravity during his climb? He lifts the woman out through the window and carries her down the ladder to safety. If she has a mass of 50 kg,

(a) write down the magnitude of

    (i)   the force exerted by the·woman on the fireman

    (ii) the force exerted by the fireman on the ladder

(b) the work done by gravity during the descent.

20.

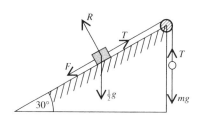

A small smooth pulley is fixed at the top of the rough face of a wedge which is inclined at 30° to the horizontal. A block A of mass $\frac{1}{2}$ kg, lying on the face, is attached to one end of a light inextensible string which passes over the pulley and carries a particle B hanging freely at the other end. The coefficient of friction between the block and the wedge is $\frac{1}{\sqrt{3}}$. If the particle B is moving down with a constant speed, find, in terms of g,

(a) the frictional force acting on the block

(b) the tension in the string

(c) the weight of the particle.

When the particle moves down through 1 m, find

(d) the work done by gravity on the particle

(e) the total work done against gravity and friction by the string attached to the block.

POWER

There are many situations where it is not sufficient to know *how much* work a force can do. It is also important to know *the rate at which the work is being done*. This quantity is known as *power*.

One unit of power is produced when work is done at the rate of 1 joule per second. This unit is called the watt (W).

A machine working at the rate of 1 joule per second has a power of 1 watt. If 1000 joules of work are done per second, the power is 1000 watts or 1 kilowatt (1 kw).

If we know the total work done in a certain time, the *average power* can be found.

For example, a force that does 45 joules of work in 9 seconds is working at an average rate of 5 joules per second,

i.e. the average power of the force is 5 watts.

The Power of a Moving Vehicle

The power of a vehicle is defined as
the rate at which the driving force is working.

A vehicle that has a speed of v m/s is moved v metres in 1 second by the driving force, D newtons.
Therefore the work done in 1 second by the driving force is Dv joules.
i.e. the power of the vehicle is Dv joules/second which is Dv watts.

Therefore, if H watts is the power of a vehicle,

$$H = Dv$$

If the speed of the vehicle is constant, both D and v are constant and therefore the power is constant.
If the speed is not constant, the value of Dv gives the power *at the instant* when the speed is v m/s.

Note that if the vehicle is stationary, its power is zero. This emphasises the difference between the meaning of power in mechanics and the way the word is used in the motor trade.

There is a maximum value of the power a particular vehicle can generate. When the maximum power is used in a given situation, the speed produced is also maximum. In this condition no acceleration is possible so the resultant force acting on the vehicle is zero.

A vehicle can use less power than the maximum available if, for example, a lower speed is desirable or the resistance falls.

When solving problems involving the power of a moving vehicle, it is often helpful to express the driving force in the form $\frac{H}{v}$. Doing this can reduce the length of the solution.

Examples 7b

1. **A bricklayer's mate can carry a hod of bricks up a vertical five-metre ladder in 46 seconds. Find the average power required if the combined mass of the man and his bricks is 92 kg. (Take g as 9.8.) What assumptions have been made?**

The work done against gravity is $92g \times 5$ J $= 4508$ J

This work is done in 46 s

$\therefore$ the work is done at an average rate of $\frac{4508}{46}$ joules per second

i.e. the average power is 98 W.

It is assumed that the man and his bricks can be modelled as a particle which rises exactly 5 m.

2. **On a level track a train has a maximum speed of 50 m/s. The total resistance to motion is 28 kN.**

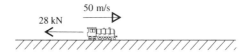

(a)   **Find the maximum power of the engine.**

The resistance is reduced and it is found that the power needed to maintain the same speed as before is 1250 kW.

(b)   **Find the lower resistance.**

(a)   At maximum speed there is no acceleration so the resultant force in the direction of motion is zero. We will model the train as a particle.

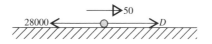

Resolving $\rightarrow$ $D - 28\,000 = 0$ $\Rightarrow$ $D = 28\,000$

To achieve maximum speed, maximum power is needed.

Maximum power $=$ driving force $\times$ maximum velocity

$= 28\,000 \times 50$ W

$= 1400$ kW

(b)

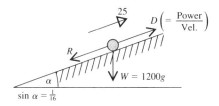

As the velocity is constant the acceleration is zero, so the resultant force is zero.

$$H = Dv \quad \Rightarrow \quad 1\,250\,000 = D_1 \times 50 \quad \Rightarrow \quad D_1 = 25\,000$$

Resolving $\quad \rightarrow \quad 25\,000 - R_1 = 0$

The reduced resistance is 25 kN.

3. When a car of mass 1200 kg is driving up a hill inclined at α to the horizontal, with the engine working at 32 kW, the maximum speed is 25 m/s.
Given that $\sin \alpha = \frac{1}{16}$, and assuming that the car can be modelled as a particle, find the resistance to motion. (Use $g = 10$.)

Speed is maximum so acceleration is zero and the resultant force up the hill is zero.

Driving force = power/velocity.

$$\begin{array}{ll} 25 \nearrow & D \left(= \dfrac{\text{Power}}{\text{Vel.}} \right) \\ R \swarrow & \\ \alpha & W = 1200g \\ \sin \alpha = \frac{1}{16} & \end{array}$$

The driving force is $\dfrac{32\,000}{25}$ N $= 1280$ N

Resolving $\quad \nearrow \quad D - R - W \sin \alpha = 0$

$\Rightarrow \qquad 1280 - R - 1200g \times \frac{1}{16} = 0$

$\therefore \qquad R = 1280 - \dfrac{12\,000}{16} = 530$

The resistance to motion is 530 N.

4. The combined weight of a cyclist and his machine is 850 N. When riding, the resistive forces are proportional to the speed. On a level road, exerting his maximum power of 180 W, the maximum speed attainable is 8 m/s. By treating the cyclist and his machine as a particle, find the maximum speed which can be achieved by working at 30 W down a hill with a gradient of 7%.

(If the gradient of a hill inclined at α to the horizontal is 7%, then $\sin \alpha = 0.07$)

Resistance $\propto$ speed $\quad \Rightarrow \quad R = kv \quad$ where k is a constant of proportion.

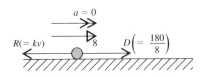 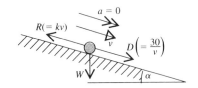

On the level

Resolving $\quad\rightarrow\quad D - R = 0 \quad\Rightarrow\quad \dfrac{180}{8} - 8k = 0$

$\Rightarrow\quad k = \frac{45}{16}$

Downhill

Resolving $\quad\searrow\quad D + W \sin\alpha - R = 0 \quad\Rightarrow\quad \dfrac{30}{v} + 850\times0.07 - \dfrac{45v}{16} = 0$

$\therefore\qquad 2.8125v^2 - 59.5v - 30 = 0$

Solving this quadratic equation by using the formula gives

$$v = \frac{59.5 \pm \sqrt{(59.5^2 + 4\times2.8125\times30)}}{2 \times 2.8125}$$

Only the positive solution has any significance.

Hence $\quad v = 21.64\ldots$

The maximum speed downhill is 21.6 m/s, corrected to 3 significant figures.

EXERCISE 7b

If the speed of a vehicle is given in km/h, the unit must be converted to m/s in order to be consistent with the other units being used. This conversion can be done by using the fact that $1 \text{ km/h} = \frac{5}{18} \text{ m/s}$.
Use $g = 10$, state any assumptions that are made and give answers corrected to 2 significant figures.

In questions 1 to 5 find the average rate at which work is done.

1. A mass of 60 kg is lifted vertically through 4 m in 9 seconds.

2. A mass of 40 kg is lifted vertically at a constant speed of 5 m/s.

3. A cat weighing 24 N climbs up a 3 metre high wall in 2 seconds.

4. A part-time assistant in a supermarket is stacking wine bottles on a shelf.
Each bottle weighs 5 N and is lifted up 1.6 m. The assistant stacks 36 bottles in a minute.

5. An elevator is raising bales of straw, each with a mass of 23 kg, to the floor of a loft that is 2.1 m above ground level. On average, 64 bales are raised per hour.

6. A car driving at a constant speed v, against a constant resistance R, is working at a rate H.

(a) If $v = 25$ m/s and $R = 960$ N, find H.

(b) If $H = 60$ kW and $v = 120$ km/h, find R.

(c) If $R = 1300$ N and $H = 26$ kW, find v.

7. A goods train has a maximum speed of 90 km/h on a level track when the resistive forces amount to a constant 40 kN. Find the maximum power of the engine.

8. A car has a maximum speed of 140 km/h on a level road with the engine working at 54 kW. Find the resistance to motion.

9.

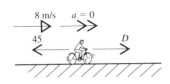 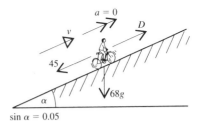

sin α = 0.05

A boy and his bicycle have a combined mass of 68 kg. Working at maximum power the cyclist can achieve a speed of 8 m/s on a level road against resistive forces totalling 45 N. Find the maximum power. Find also the maximum speed up an incline with a 5% gradient.

10. The maximum power that a van of mass 900 kg can exert is 36 kW. If the resistance to the motion of the van is a constant 1500 N, find the maximum speed that the van can reach on a slope of inclination 1 in 15 (i.e. $\sin \alpha = \frac{1}{15}$) when driving (a) up the slope (b) down the slope.

Give one reason why your answers may not reflect the actual maximum speeds of a real van.

11. A car of mass 1100 kg has a maximum power output of 44 kW. The resistive forces are constant at 1400 N. Find the maximum speed of the car

(a) on the level (b) up an incline with gradient 5%

(c) down the same incline when using half the maximum power.

12. A car of mass 950 kg has a maximum power of 30 kW and encounters resistive forces totalling 1050 N.

(a) Find its maximum speed on level ground.

(b) The driver wishes to descend a hill of gradient 1 in 15 without increasing speed. What percentage of the maximum power of the engine should be used?

13. A lorry of mass 2000 kg is subject to a constant resistance. The maximum speed of the lorry down a slope of 1 in 10 is 24 m/s and the maximum speed up the same slope is 12 m/s. Find

(a) the maximum power of the engine

(b) the constant resistance

(c) the maximum speed of the lorry on a level road.

14. The engine of a car of mass 1000 kg is working at the constant rate of 50 kW. The resistance to motion is proportional to the speed.

(a) If the maximum speed on a level road is 40 m/s, find the constant of proportion.

(b) Find the maximum speed up a slope with a gradient of 5%.

(c) When the engine is working at half power, the maximum speed of the car down a slope inclined at an angle α to the horizontal is 35 m/s. Find α to the nearest degree.

ACCELERATING VEHICLES

If a vehicle exerts a driving force that exceeds all the forces opposing motion, there is a resultant force in the direction of motion. As a result, the car has an acceleration which can be found by applying Newton's Law.

Examples 7c

1. A car of mass 1200 kg has a maximum speed of 144 km/h on a level road when there is a resistive force of 56 N. Find the acceleration of the car at the instant when its speed is 81 km/h and the engine is working at maximum power.

Speeds must be expressed in metres per second.

$144 \text{ km/h} = 144 \times \frac{5}{18} \text{ m/s} = 40 \text{ m/s}$; similarly $81 \text{ km/h} = 22.5 \text{ m/s}$

At maximum speed, treating the car as a particle, we have

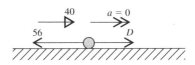

Resolving $\rightarrow$ $D - 56 = 0$ $\Rightarrow$ $D = 56$

$\therefore$ if H watts is the maximum power, $H = Dv = 56 \times 40 = 2240$

The maximum power is 2240 W.

At the lower speed

The resultant force in the direction of motion is $D - R$

Applying $F = ma$ in this direction gives

$$\frac{2240}{22.5} - 56 = 1200a \qquad \Rightarrow \qquad a = \frac{43.56}{1200} = 0.0363$$

Therefore, corrected to 2 significant figures, the acceleration is 0.036 m/s².

2. A motor cyclist whose mass combined with his machine is 240 kg is driving up a road of inclination 1 in 10 at maximum power of 10 kW. When the speed is 25 m/s, the motor bike is accelerating at 0.05 m/s². Taking the value of g as 10,

(a) find the constant resistance to motion.

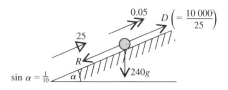

At the top of the hill he picks up a pillion passenger of mass 80 kg and drives on along the road which is now horizontal.

If the resistance is increased by 20%, find

(b) the greatest speed that can be achieved when the engine is working at 70% of the maximum power

(c) the immediate acceleration produced if maximum power is suddenly engaged.

(a) Treating the motor cyclist as a particle, we have

The resultant force up the hill is

$$D - W \sin \alpha - R = \frac{10\,000}{25} - 2400 \times \frac{1}{10} - R$$

Using Newton's Law, $F = ma$, in the direction of motion gives

$$400 - 240 - R = 240 \times 0.05 \qquad \Rightarrow \qquad R = 148$$

The resistance to motion is 148 N.

(b)   The resistance is now 120% of 148 N, i.e. 177.6 N, and the power is 70% of 10 kW, i.e. 7000 W. The motor cyclist, his machine and the pillion passenger are treated as a particle.

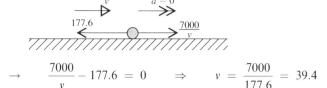

Resolving $\rightarrow$ $\dfrac{7000}{v} - 177.6 = 0$ $\Rightarrow$ $v = \dfrac{7000}{177.6} = 39.4$

The maximum speed is 39.4 m/s.

(c)

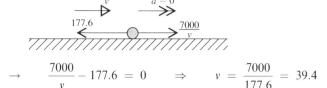

The resultant force in the direction of motion is $D - R$

Therefore using $F = ma$ gives

$$\dfrac{10\,000}{39.4} - 177.6 = 320a \qquad \Rightarrow \qquad a = \dfrac{253.8 - 177.6}{320} = 0.24$$

The immediate acceleration is 0.24 m/s².

Note that this is an instantaneous acceleration; as a result the velocity increases, so reducing the driving force which in turn reduces the acceleration.

EXERCISE 7c

Use $g = 10$, state any assumptions made and give answers corrected to 2 significant figures.

1. A car of mass 1500 kg is being driven up an incline of 1 in 20 against a constant resistance to motion of 1000 N.

   (a)   At the instant when the speed is 20 m/s and the acceleration is 0.1 m/s², find the power being exerted by the engine.

   (b)   If the engine is working at 20 kW, find the acceleration at the instant when the speed is 7.5 m/s.

2. The resistive forces opposing the motion of a car of mass 2000 kg total 5000 N. If the engine is working at 70 kW, find the acceleration at the instant when the speed is 40 km/h.

3. A car of mass 1000 kg has a maximum power of 50 kW. The car is travelling up a hill with a gradient of 8% against resistance to motion of 3000 N. Find the acceleration at the instant when the speed is 30 km/h.

4. A train of mass 400 tonnes is moving down an incline of 1 in 50 using a power output of 50 kW. The resistance to motion is 30 kN. Find the acceleration at the instant when the speed is 20 m/s.

5. The constant resistances to the motion of a car of mass 1200 kg total 960 N.

(a) If the car is driving along a level road and has an acceleration of 0.2 m/s^2 at the instant when the speed is 25 m/s, find the power exerted by the engine.

(b) If the car is moving down a slope of inclination 1 in 15 and working at 40 kW, find the acceleration at the instant when the speed is 25 m/s.

6. The engine of a train of mass 50 000 kg is working at 1800 kW as the train ascends a slope of inclination 1 in n. The train encounters constant resistive forces of 10 kN.

(a) If the maximum speed of the train is 50 m/s, find n.

(b) Find the acceleration at the instant when the speed is 30 m/s.

7. A car of mass 1000 kg encounters resistive forces of 1200 N when ascending a slope with a 10% gradient.

(a) If the engine is working at 30 kW find the maximum speed.

(b) When moving at this speed the driver suddenly increases the power of the engine by 25%. Find the immediate acceleration of the car.

8. A cyclist is riding up an incline of 1 in 20. Working at 2 kW the maximum speed up the incline is 20 km/h.

(a) Find the resistance to motion given that the combined mass of the cyclist and machine is 100 kg.

(b) If there is no change in resistance or power, find the instantaneous acceleration when the cyclist reaches level ground at the top of the slope.

9. A car of mass 800 kg moves against a constant resistance R newtons. The maximum speeds of the car up and down an incline of 1 in 16 are respectively 14 m/s and 42 m/s. If the rate at which the engine is working is H kW, find

(a) the values of R and H

(b) the acceleration at the instant when the speed is 17.5 m/s on level ground.

10. The resistance to the motion of a car of mass 1000 kg is proportional to the square of the speed. With the engine working at 60 kW, the car can drive up an incline of 1 in 20 at a steady speed of 30 m/s. If the car travels down the same slope with the engine working at 40 kW, find the acceleration at the instant when the speed is 20 m/s.

CHAPTER 8

MECHANICAL ENERGY

ENERGY

Anything that has the capacity to do work, possesses energy. This energy can be used up in doing work.

Conversely, in order to give energy to an object, work must be done to it,

i.e. work and energy are interchangeable and so are measured in the same unit, the-joule.

There are various different forms of energy such as light, heat, sound, electrical energy and chemical energy. These can often be converted from one form to another, e.g. electrical energy can be used to give heat or light energy.

In this book however we are concerned primarily with *mechanical energy,* which is the capacity to do work as a result of motion or position.

KINETIC ENERGY (KE)

A body moving with speed v possesses *kinetic energy.* The value of the KE is equal to the amount of work needed to bring that body from rest to the speed v and an expression for its value can be found as follows.

Consider a body of mass m which starts from rest and reaches a speed v after moving through a distance s under the action of a constant force F.

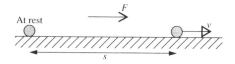

The acceleration, a, is given by

$$v^2 - u^2 = 2as \qquad \Rightarrow \qquad a = \frac{v^2 - 0}{2s}$$

Then Newton's Law, $F = ma$, gives

$$F = \frac{mv^2}{2s} \quad \Rightarrow \quad Fs = \tfrac{1}{2}mv^2$$

Now Fs is the work done in producing the kinetic energy, therefore $\tfrac{1}{2}mv^2$ is the value of the kinetic energy,
i.e. for a body of mass m moving with speed v,

$$\mathbf{KE} = \tfrac{1}{2}mv^2$$

Note that both m and v^2 are always positive quantities showing that KE is always positive and does not depend upon the direction of motion,
i.e. kinetic energy is a scalar quantity.

POTENTIAL ENERGY (PE)

Potential energy is a property of position. If a body is in such a position that, if released, it would begin to move, it possesses PE.

Consider, as an example, a body that is held at a height h above a fixed level. If that body is released it will begin to fall, i.e. it will begin to possess KE. So before it is released it has the *potential* to move, hence the name for energy due to position.

The value of the PE is equal to the work needed to raise the body through a vertical distance h.

The work done in raising a body of mass m is the work done against gravity, i.e. $mg \times h$

$$\mathbf{PE} = mgh$$

If the body falls from rest and reaches a speed v at the bottom then, using $v^2 - u^2 = 2as$ gives $v^2 = 2gh$, i.e. $gh = \tfrac{1}{2}v^2$.

Therefore $mgh = \tfrac{1}{2}mv^2$ confirming that potential energy is converted into kinetic energy.

There is no absolute value for the PE of an object, as the height h is measured from some particular fixed level. If a different level is chosen the PE is changed without the body itself moving. It follows therefore that in every problem the level from which height is measured must be clearly specified. As the PE of an object that is *on* the chosen level is zero, in this book we identify this datum by marking it 'PE = 0'.

Negative Potential Energy

If an object is *below* the datum, the value of h is negative (h is the height *above* the datum), so the object has *negative potential energy.*

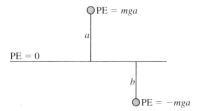

Note that there is another type of potential energy, which readers will meet in Chapter 10. It is called elastic potential energy (EPE) and it is a property of an object attached to a stretched elastic string. However, none of the work in this chapter requires knowledge of this type of mechanical energy.

Examples 8a

1. **A window cleaner of mass 72 kg climbs up a ladder to a second-floor window, 5 m above ground level. Assuming that the window cleaner can be treated as a particle, find his potential energy relative to the ground.**
   He then descends 3 m to clean a first-floor window. Find how much potential energy he has lost. (Use $g = 9.8$.) Is the assumption reasonable?

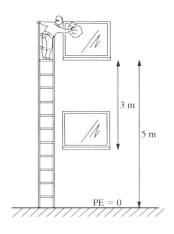

At the second-floor window, A,

$$m = 72$$
$$g = 9.8$$
$$h = 5$$

$\therefore \qquad PE = mgh = 72 \times 5 \times 9.8 \text{ J}$
$$= 3500 \text{ J} \ (2 \text{ sf})$$

In descending to the lower window, B, the reduction in height is 3 m.

Loss in PE $= mg \times (\text{reduction in } h)$
$$= 72 \times 9.8 \times 3 \text{ J}$$
$$= 2100 \text{ J} \ (2 \text{ sf})$$

Alternatively we could find the values of PE at the two windows and subtract.

The height of the point where the weight of the man acts may be quite different from the height of the window, so treating him as a particle that rises *exactly* 5 m is not a reasonable assumption, but gives a rough approximation.

2. **A particle of mass 6 kg has a velocity v metres per second.**
   What is its kinetic energy if (a) $v = 7$ (b) $v = 4i - 3j$?

   (a) $KE = \frac{1}{2}mv^2 = \frac{1}{2} \times 6 \times 7^2 \text{ J} = 147 \text{ J}$

   (b)   The speed of the particle is $|v|$

   When $v = 4i - 3j$, $|v| = \sqrt{(4^2 + [-3]^2)} = 5$

   $KE = (\frac{1}{2} \times 6 \times 5^2) \text{ J} = 75 \text{ J}$

3. **A bird of mass 0.6 kg, flying at 9 m/s, skims over the top of a tree 6.2 metres**
   high. What is the total mechanical energy of the bird as it clears the tree?
   Use $g = 9.8$.

   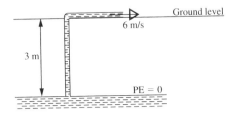

   Treating the bird as a particle,

   KE of the bird is $\frac{1}{2}mv^2$,

   i.e. $\frac{1}{2} \times 0.6 \times 9^2 \text{ J} = 24.3 \text{ J}$

   PE of the bird is mgh

   i.e. $0.6 \times 9.8 \times 6.2 \text{ J} = 36.5 \text{ J}$

   The total mechanical energy is $KE + PE = 60.8 \text{ J}$

4. **Water is being raised by a pump from a storage tank 3 m below ground level and**
   ejected at ground level through a pipe at 6 m/s. If the water is delivered at a rate
   of 420 kg each second, find the total mechanical energy supplied by the pump in one
   second in lifting and ejecting the water. (Take g as 9.8.)

   In 1 second,   PE gained by water $= mgh$

   $= 420 \times 9.8 \times 3 \text{ J} = 12\,348 \text{ J}$

   KE gained by water $= \frac{1}{2}mv^2$

   $= \frac{1}{2} \times 420 \times 6^2 \text{ J} = 7560 \text{ J}$

   The total energy gained by the water is supplied by the pump.

   Total ME supplied by the pump per second $= 19\,908 \text{ J}$.

EXERCISE 8a

Use $g = 9.8$ and give answers corrected to 2 or 3 significant figures as appropriate. If any assumptions are made, state what they are.

1. The potential energy of a particle of mass m kilograms, which is at a height h metres above a given datum, is N joules.

(a) If $m = 4$ and $h = 11$, find N.   (b) If $N = 48$ and $m = 6$, find h.

2. Find, in joules, the kinetic energy of

(a) a block of mass 8 kg moving at 9 m/s

(b) a car of mass 1200 kg travelling at 36 km/h

(c) a bullet of mass 16 g moving at 500 m/s

(d) a body of mass 10 kg with a velocity $\mathbf{v}$ m/s where $\mathbf{v} = 3\mathbf{i} + 2\mathbf{j}$.

3. Find the items missing from the following table.

Mass	Speed	Kinetic energy
7 kg	2 m/s	
14 kg		126 J
	6 m/s	396 J

4.

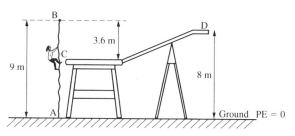

On an assault course a woman of mass 54 kg starts from ground level at A, climbs 9 m up a scramble net, drops from the top B, to a mat C, 3.6 m below B, then runs up a bar to D at a height of 8 m above the ground. Find

(a) her potential energy relative to the ground at   (i) A   (ii) D

(b) the gain in potential energy between A and B

(c) the loss in potential energy between B and C

(d) the gain in potential energy between C and D

(e) Using the answers to parts (b), (c) and (d), find the P.E. at D.

Check that this agrees with your answer to part (a) (ii).

5. (a) Find the gain in kinetic energy when the speed of a body of mass 4 kg increases from 7 m/s to 11 m/s.

(b) Find the kinetic energy lost when the speed of the same body falls from 18 m/s to 5 m/s.

6. A car of mass 900 kg is travelling at 72 km/h.

(a) Find how much kinetic energy is lost if the speed falls to 54 km/h.

(b) If the kinetic energy rises to 281.25 kJ, at what speed is the car travelling?

7. A particle of mass 3 kg is at rest. It begins to move with constant acceleration and five seconds later it has kinetic energy of 150 J. Find

(a) the speed at the end of the 5 seconds

(b) how far the particle has travelled in this time.

8. A pump raises 45 kg of water through a vertical distance of 12 m.

(a) How much potential energy is gained by the water?

The water is then forced through a pipe at 12 m/s.

(b) How much kinetic energy does the water gain?

9. A pump raises water from a depth of 5 m and ejects it through a pipe with a speed of 8 m/s.

(a) If the cross-sectional area of the pipe is 0.06 m^2, find the volume of water discharged per second.

(b) Given that 1 m^3 of water has a mass of 1000 kg, find the mass of water discharged per second.

(c) Find the total mechanical energy gained by the water.

THE PRINCIPLE OF WORK AND ENERGY

Mechanical energy was defined as the capacity of the forces acting on a body to do work. Now the link between energy and work done can be expressed more precisely.

The work done by external forces acting on a body is equal to the change in the mechanical energy of the body.

This is the principle of work and energy.

If the external forces act in a direction that helps promote the motion of the body the mechanical energy increases, whereas opposing external forces cause a decrease in mechanical energy.

The weight of an object is not counted as an external force in this context because work done by weight is Potential Energy and is already accounted for.

In solving problems involving this principle it is wise to use it in the form

Final ME $\sim$ Initial ME = Work Done   ($\sim$ means *the difference between*)

This avoids any confusion in cases where one type of energy increases and the other decreases.

Examples 8b

1. A force acting on a body of mass 6 kg, moving horizontally, causes the speed to increase from 3 m/s to 8 m/s. How much work is done by the force? If the magnitude of the force is 11 N, how far does the body move during this speed change?

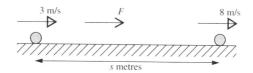

As the body is moving horizontally there is no change in PE so only KE changes.

$$\text{Work done} = \text{Final ME} - \text{Initial ME}$$
$$= \tfrac{1}{2}(6)(8^2) - \tfrac{1}{2}(6)(3^2) \text{ J}$$
$$= 165 \text{ J}$$
$$\text{Work done by force} = Fs$$
$$\therefore \qquad 165 = 11s \qquad \Rightarrow \qquad s = 15$$

The body moves 15 m.

2. A small block of mass 3 kg is moving on a horizontal plane against a constant resistance of R newtons. The speed of the block falls from 12 m/s to 7 m/s as the block moves 5 m. Find the magnitude of the resistance.

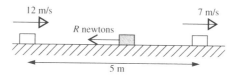

There is no change in PE and, as the speed is reducing, the final KE is less than the initial KE

$$\text{Initial ME} - \text{Final ME} = \tfrac{1}{2}(3)(12^2) - \tfrac{1}{2}(3)(7^2)$$
$$= 216 - 73.5 \text{ J}$$
$$= 142.5 \text{ J}$$
$$\text{Work done by resistance} = \text{Change in ME}$$
$$\therefore \qquad R \times 5 = 142.5 \qquad \Rightarrow \qquad R = 28.5$$

The magnitude of the resistance force is 28.5 N.

3. A stone falls vertically downward through a tank of viscous oil. The speed of the stone as it enters the oil is 2 m/s and at the bottom of the tank it is 3 m/s. Given that the oil is of depth 2.4 m, find the resistance, F newtons, that it exerts on the stone whose mass is 4 kg. Take g as 9.8.

The resistance is an opposing force so the work it does reduces the ME of the stone. Both KE and PE change.

Initial ME $= [\frac{1}{2}(4)(2^2) + (4)(9.8)(2.4)]$ J

$\qquad = 8 + 94.08$ J $= 102.08$ J

Final ME $= \frac{1}{2}(4)(3^2) + 0 = 18$ J

Work done by the resistance $=$ Change in ME

$\therefore F \times 2.4 = 102.08 - 18$

$\Rightarrow \qquad F = 35.03\ldots$

The resistance is 35.0 N (3 sf).

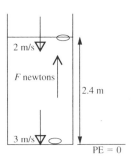

4. **A car of mass 1000 kg drives up a slope of length 750 m and inclination 1 in 25. If resistance forces are negligible, calculate the driving force of the engine if the speed at the foot of the incline is 25 m/s and the speed at the top is 20 m/s. Model the car as a particle and use** $g = 9.8$.

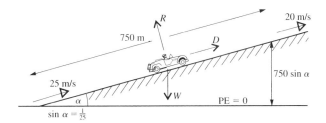

The driving force, D newtons, acts for 750 m

$\therefore$ work done by driving force $= 750D$ J

Both kinetic energy and potential energy change.

$\qquad$ Final KE $= \frac{1}{2}(1000)(20^2)$ J $= 200$ kJ

$\qquad$ Final PE $= (1000)(9.8)(750 \sin \alpha) = 9800(750 \times \frac{1}{25}) = 294$ kJ

$\therefore \qquad$ Final ME $= 494$ kJ

$\qquad$ Initial KE $= \frac{1}{2}(1000)(25^2)$ J $= 312.5$ kJ

$\qquad$ Initial PE $= 0$

$\therefore \qquad$ Initial ME $= 312.5$ kJ

$\qquad$ Work done by driving force $=$ Final ME $-$ Initial ME

$\therefore \qquad\qquad 750 D = 494\,000 - 312\,500$

$\Rightarrow \qquad\qquad\qquad D = 242$

The driving force is 242 N.

EXERCISE 8b

Use the principle of work and energy for each question. Model each large object as a particle, ignore air resistance unless it is specifically mentioned and state any *other* assumptions made. Take *g* as 9.8 and give answers corrected to 2 or 3 sf as appropriate.

1. A mass of 6 kg is pulled by a string across a smooth horizontal plane. As the block moves through a distance of 4.2 m, the speed increases from 2 m/s to 6 m/s. Find the tension in the string.

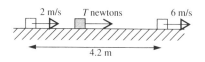

2. A body of mass 8 kg, travelling on a rough horizontal plane at 12 m/s, is brought to rest by friction. Find the work done by the frictional force.

3. A ball of mass 0.4 kg is thrown vertically upwards with a speed of 10 m/s. It comes instantaneously to rest at a height of 3.6 m above the point of projection, P. Find the resistance to its motion. The ball then falls back to P. If the resistance is unchanged, find the speed at P.

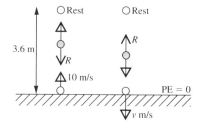

4. A body of mass 0.5 kg is lifted, by vertical force, from rest at a point A to a point B that is 1.7 m vertically above A. If the body has a speed of 3 m/s when it reaches B find the work done by the force. Hence find the magnitude of the force.

5.

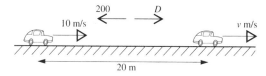

A car of mass 750 kg, is travelling along a level road at 10 m/s against a constant resistance of 200 N. Exerting a driving force of 1200 N, the driver accelerates for 20 m. At what speed will the car then be moving?

6. A block of mass 5 kg lies in contact with a horizontal plane. It is pulled from rest through a distance of 8 m by a horizontal force of 12 N. Find the speed attained if the contact between block and plane is

    (a) smooth   (b) rough, with a coefficient of friction of $\frac{1}{10}$.

7. A bullet of mass 0.02 kg is fired horizontally at a speed of 360 m/s into a fixed block of wood. The bullet is embedded 0.06 m into the block. Find the average resisting force exerted by the wood.

8. A particle of mass 7 kg is pulled by a force of F newtons, 4 metres up a smooth plane inclined at 30° to the horizontal. Find the work done by the pulling force if

(a) the particle is pulled at a constant speed

(b) the speed changes from 1 m/s initially to 2 m/s at the end.

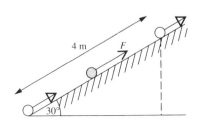

9. A block of mass 3 kg slides down a plane inclined at α to the horizontal where $\sin \alpha = \frac{1}{4}$. The block starts from rest and there is a constant frictional force of 4 N.

(a) How far has the block travelled when its speed reaches 6 m/s?

(b) When the block has moved 4 m down the plane, what is its speed?

10. In raising a body of mass 2 kg from rest vertically upwards by 6 m and giving it a speed of v m/s, a force does work amounting to 800 J. Find v.

11. A body of mass 2 kg falls vertically from rest through a distance of 5 m. If the speed by then is 9 m/s find the air resistance.

In questions 12 to 14, water is pumped at a rate of p m³/s from a tank d m below ground. The water is then delivered at ground level at v m/s, through a pipe whose cross-sectional area is a m². The density of water is 1000 kg/m³.

12. If $d = 8$, $v = 10$ and $a = 0.05$, find

(a) the volume of water discharged per second

(b) the weight of this water

(c) the work done per second by the pump in raising and delivering the water.

13. If $p = 0.2$, $d = 5$ and $a = 0.1$, find

(a) the value of v

(b) the power the pump exerts in moving the water.

14. The work done per second by the pump in raising and ejecting the water is 160 kJ. Given that $p = 6$ and $d = 2$ find v.

15. Two boys are kicking a ball about. It has a mass of 0.5 kg. The ball comes towards one boy at 8 m/s and he passes it back at 12 m/s.

(a) Find the work he does in bringing the ball to instantaneous rest.

(b) Find the work he does in giving it a speed of 12 m/s.

16. A boy makes a slide on level icy ground. The slide is 6 m long. He runs up to the slide and steps onto it at a speed of 5 m/s. He reaches the other end at a speed of 4 m/s. His mass is 45 kg.

(a) Find the loss of kinetic energy.

(b) Find the resistance to motion.

(c) Assuming that air resistance is negligible, find the coefficient of friction.

17. A particle of mass m kg is pushed up a plane inclined at an angle θ to the horizontal. The coefficient of friction between the particle and the plane is μ. The particle has an initial speed u m/s. It travels a distance d m up a line of greatest slope of the plane and the speed then is v m/s.
Find

(a) the work done against the frictional force

(b) the work done against gravity

(c) the total work done.

CONSERVATION OF MECHANICAL ENERGY

We know from the principle of work and energy that the total change in the mechanical energy of a body is equal to the work done on the body.
It follows directly that:

If the total work done by the external forces acting on a body is zero there is no change in the total mechanical energy of the body, i.e. energy is conserved

This is the *principle of conservation of mechanical energy.*

Remember that the weight of a body is not an external force in this context as the work done by the weight is already included as potential energy.

At present we are concerned with only two types of mechanical energy, so a problem can be solved by working out the loss in KE, say, and equating it to the gain in PE. (However, for those readers intending to carry on studying mechanics and who will meet problems which also include the third type of mechanical energy (see p. 154) we recommend the method of equating the *total* mechanical energy in two positions.)

Examples 8c

1. **A particle is projected vertically with speed 8 m/s. Find its speed after it has moved a vertical distance of 2 m (a) upwards (b) downwards. Take g as 9.8 and give answers corrected to 2 significant figures.**

Let the mass of the particle be m kg and the final speed be v m/s.

At A $PE = 0$ and $ME = \frac{1}{2}m \times 8^2$ J

$\therefore$ Total $ME = 32m$ J

(a) At B $PE = mgh = 19.6m$ J

and $KE = \frac{1}{2}mv^2$ J

$\therefore$ Total $ME = (19.6m + \frac{1}{2}v^2m)$ J

Total ME at A $=$ Total ME at B

$\therefore$ $32m = 19.6m + \frac{1}{2}v^2m$

$\Rightarrow$ $v^2 = 24.8$

$\Rightarrow$ $v = 5.0\ (2\text{ sf})$

The speed at B is 5.0 m/s (2 sf).

(b) At C $PE = mg(-2)$ J

and $KE = \frac{1}{2}mv^2$ J

$\therefore$ Total $ME = (\frac{1}{2}mv^2 - 19.6m)$ J

Total ME at A $=$ Total ME at C

$\therefore$ $32m = \frac{1}{2}mv^2 - 19.6m$

$\Rightarrow$ $v^2 = 103.2$

$\Rightarrow$ $v = 10\ (2\text{ sf})$

The speed at C is 10 m/s.

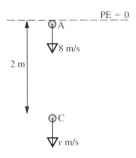

Note that in the example above the value of the mass, m, was not given and was not needed as it cancelled in each conservation of energy equation. This will always be the case as m is a factor of both PE and KE.

Note also that an alternative solution could be given using Newton's Law and the equations of motion with constant acceleration.

2. A bead is threaded on to a circular ring of radius 0.5 m and centre O, which is fixed in a vertical plane. The bead is projected from the lowest point of the ring, A, with a speed of 4 m/s, and first comes to instantaneous rest at a point B. Contact between the ring and the bead is smooth and there is no other resistance to motion. Find the height of B above A. Take g as 9.8 and give answers corrected to 2 significant figures.

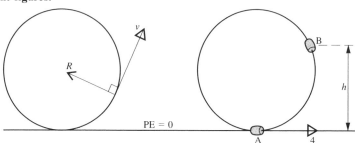

The normal reaction R is always perpendicular to the direction of motion of the bead therefore no work is done by R. No other external force is acting so energy is conserved.

At A $PE = 0$ and $KE = \frac{1}{2}mv^2 = \frac{1}{2}m(4)^2$ J

$\therefore$ Total ME $= (0 + 8m)$ J

At B $PE = mgh = 9.8mh$ J   and $KE = 0$

$\therefore$ Total ME $= (9.8mh + 0)$ J

             Total ME at A $=$ Total ME at B

$\therefore$ $8m = 9.8mh$

$\Rightarrow$ $h = \frac{8}{9.8}$

Corrected to 2 significant figures, B is 0.82 m above A.

3. A small block A of mass $2m$, is lying in smooth contact with a table top. A light inextensible string of length 1 m is attached at one end to A, passes over a smooth pulley at the edge of the table, and carries a block of mass m hanging freely at the other end. Initially A is held at rest, 0.8 m from the edge of the table.
If the system is released, find A's speed when it reaches the edge.
Take g as 9.8 and correct your answers to 2 significant figures.

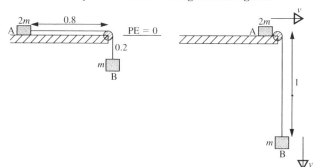

Initially   KE $= 0$,   PE of A $= 0$,   PE of B $= mg(-0.2)$ J

Total ME $= -(0.2)(9.8)m$ J $= -1.96m$ J

When A reaches the edge of the table, the speed of each block is v m/s and B is 1 m below the table top.

When A reaches the edge,

$$\text{KE of A} = \tfrac{1}{2}(2m)v^2 \text{ J} \qquad \text{KE of B} = \tfrac{1}{2}mv^2 \text{ J}$$

$$\text{PE of A} = 0 \qquad\qquad \text{PE of B} = mg(-1) \text{ J}$$

Total ME $= (mv^2 + \tfrac{1}{2}mv^2 - mg)$ J

Initial total ME $=$ Final total ME

$\therefore \qquad -1.96m = mv^2 + \tfrac{1}{2}mv^2 - 9.8m$

i.e. $\qquad 7.84 = \tfrac{3}{2}v^2 \qquad \Rightarrow \qquad v^2 = 5.227 \qquad \Rightarrow \qquad v = 2.3 \,(2 \text{ sf})$

The speed of each block is 2.3 m/s.

Remember that the choice of the datum level for PE is arbitrary. We could just as well have chosen the lowest level reached by B. (You may like to check that the same result would be obtained.)

In Examples 2 and 3 you will notice that the letter m is used both for mass and metre. It is easy to see the difference between the two m's in type because, for mass, m is italic. When hand-writing a solution, however, this cannot be done so it is a good idea to write out the word metre in full.

EXERCISE 8c

For each question use the Conservation of Mechanical Energy. Take g as 9.8 and give answers corrected to 2 or 3 significant figures as appropriate.

1. A particle of mass m kilograms is projected vertically upwards with speed 6 m/s. Find the height it attains before first coming to rest.

2. A stone is thrown vertically downwards with speed 3 m/s. Find its speed after it has fallen 4.5 m.

3. A ball is thrown vertically upwards, from a point A, with speed 8 m/s. Given that A is 1 m above ground, find

   (a) the greatest height reached by the ball

   (b) the speed of the ball when its height is 2.4 m

   (c) the speed of the ball as it hits the ground.

Questions 4 to 6 concern a particle P, moving on a plane inclined to the horizontal at 30°. A and B are two points on the plane. Contact between the body and the plane is smooth.

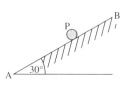

4. P is projected up the plane from A with speed 4 m/s and comes to instantaneous rest at B. Find the distance AB.

5. P is released from rest at B. Find its speed as it passes through A if AB = 2.4 m.

6. P is moving down the plane and passes through B with speed v m/s. If AB = 2.1 m and the particle passes through A with speed 6 m/s, find v.

In questions 7 to 9 a smooth bead is threaded on to a smooth circular wire with centre O and radius a metres. The wire is fixed in a vertical plane.

7. The bead is released from rest at a point level with O. If $a = 0.5$, find the speed of the bead as it passes through the lowest point.

8. The bead is projected from the lowest point on the wire with speed 4.2 m/s. If $a = 0.6$, find the height above O at which the bead first comes to rest.

9. The bead is projected from the lowest point and just reaches the highest point. Given that $a = 0.8$, find the speed of projection.

For questions 10 and 11, give answers in terms of l and g.

10. Two identical particles of mass m are connected by a light inelastic string of length $2l$. One particle, A, rests in smooth contact with a horizontal table and the other particle B hangs freely over the edge of the table. The string is perpendicular to that edge. If A is released from rest when it is at a distance l from the table edge, find its speed when it reaches the edge.

11. Two particles P and Q are connected by a light inextensible string that passes over a smooth pulley. The masses of P and Q are m and $2m$ respectively. The particles are released from rest when each is at a depth $2l$ below the pulley. Find their speed when each has moved a distance l.

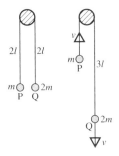

State all the assumptions you make in modelling each problem from 12 to 14.

12. Sue hopes to swing across the stream on a rope, which is attached to an overhanging tree at point A. The bank on the opposite side of the stream is 1.2 m higher than the bank on which she is standing. At what speed must she push off in order just to get there?

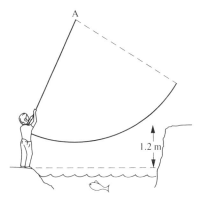

13. Peter Pan 'flies' across the stage on a harness, which slides along a smooth wire AB. End A of the wire is fixed at a height of 6 m and end B at a height of 5.5 m. The lowest point of the wire as he crosses the stage is at a height of 4 m. He starts from rest at A.

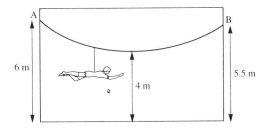

(a) Find his maximum speed.   (b) Find his speed at B.

14.

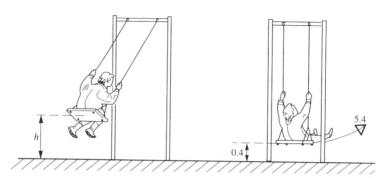

The seat of a swing is 0.4 m above the ground when it is stationary. A girl is swinging so that she passes through the lowest point with speed 5.4 m/s. Find the height of the seat above ground when she first comes to rest.
State what assumptions have been made and comment on their suitability.

15. A particle slides down a smooth plane on a line AB, which is inclined at 40° to the horizontal. At A its velocity down the plane is 3 m/s. The height of A above B is 16 m. Find the velocity of the particle when at B.

CONSOLIDATION B

SUMMARY

Newton's Laws of Motion

1. A body's state of rest or constant velocity is unchanged unless a force acts on it.

2. The acceleration of a body is proportional to the force producing it, i.e. $F = ma$, where the unit of force, called the newton, is the force needed to give a mass of 1 kg an acceleration of 1 m/s^2.
The force and the acceleration are in the same direction.

3. Forces between objects act in equal and opposite pairs.

Weight

Weight is the force of gravity attracting an object to the earth. The acceleration g which it produces is approximately 9.8 m/s^2.

The weight of an object of mass m is mg.

Equilibrium

A body that is at rest or moving with constant velocity is in a state of equilibrium and so is the set of forces acting on that body.

A particle is in equilibrium if the resultant force acting on it is zero.

Friction

Friction exists if two objects are in rough contact and have a tendency to move.

The frictional force F is just large enough to prevent motion, up to a limiting value.

When the limiting value is reached, $F = \mu R$ where R is the normal reaction and μ is the coefficient of friction.

For rough contact $0 \leqslant F \leqslant \mu R$ and for smooth contact $F = 0$.

When friction is limiting, the resultant of μR and R is at an angle λ to R where λ is the angle of friction and $\tan \lambda = \mu$.

Work

When a constant force acts on an object the amount of work done by the force is given by

component of force in the direction of motion
× distance moved by object

The unit of work is the joule (J), which is the amount of work done when a force of 1 newton moves an object through 1 m.

The work done by a vehicle moving at constant speed is

driving force × distance moved

Power

Power is the rate at which work is done and is measured in watts where

1 watt (W) is 1 joule per second.

The power of a moving vehicle is the rate at which the driving force is working and this is given by

driving force × velocity

Energy

Energy is the ability to do work.

Energy and Work are interchangeable so energy is measured in joules.

The Kinetic Energy (KE) of a moving object is given by $\frac{1}{2}mv^2$; it can never be negative.

Potential Energy (PE) is equivalent to work done by gravity and is given by mgh where h is the height of an object above a chosen level.
PE is negative for a body below the chosen datum.

If the total work done by the external forces acting on a body is zero the total mechanical energy of the body remains constant.

Work done by gravity is accounted for as potential energy so weight is not included as an external force.

MISCELLANEOUS EXERCISE B

In this exercise use $g = 9.8$ unless another instruction is given.

1. The diagram shows three coplanar forces of magnitudes 2 N, 3 N and P N all acting at a point O in the directions shown.

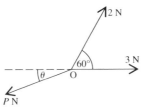

Given that the forces are in equilibrium obtain the numerical values of $P \cos \theta$ and $P \sin \theta$ and hence, or otherwise, find $\tan \theta$ and P.   (AEB)

2. A particle of mass 2 kg is suspended from a point C of a light string, the ends of the string being attached to fixed points A and B. When the particle hangs in equilibrium, AC is horizontal and CB is inclined at 20° to the horizontal.

   Find the magnitude of the tension in the string in

   (a) the section BC   (b) the section AC   (AEB)

3. A particle of mass 2 kg which is free to move along the positive x-axis is at rest at time $t = 0$ s. For the first ten seconds it is acted upon by a force in the positive x direction of magnitude 6 N; for the next 240 seconds no force acts; then the particle is brought to rest by applying a force of magnitude 10 N along the x-axis. Find the time at which the particle comes to rest and the distance travelled up to that time.   (AEB)

4. Two particles A and B are placed on a rough horizontal table at a distance a apart. The coefficient of friction between the table and A is $\frac{1}{2}$, and the coefficient of friction between the table and B is $\frac{3}{4}$.
   The particles are projected simultaneously with velocity u in the direction AB. Given that the particles do not collide, find, in terms of u and g, the distance travelled by each particle before it comes to rest.

   Deduce that $u^2 < 3ga$.   (NEAB)

5. In a race, an athlete of mass 70 kg starts from rest and runs a distance of 100 m along a straight horizontal track. During the first 6 seconds of the race, the net propelling force is horizontal and of magnitude 175 N.
   For the remainder of the race, the net force is a horizontal resistance with a magnitude of F N.

   (a) Find the speed of the athlete after the first 6 seconds.

   (b) Given that the athlete completes the race in 11 s, find the value of F.
                                                                              (AEB)

6. The points A, B and C of a horizontal plane have coordinates (4, 3), (−4, 0) and (4, −3), respectively, these dimensions being in metres. A particle P on the plane is subject to three forces which are directed towards A, B and C.

(a) When P is at the origin the forces directed towards A, B and C have magnitudes 4 N, 2 N and 4 N, respectively, as shown in the diagram.

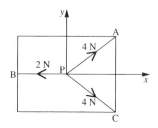

Calculate the magnitude of the resultant of the three forces, and state its direction.
(i) Given that the plane is rough, and that P is in equilibrium at the origin, state the magnitude and direction of the frictional force on P.
(ii) Given that the plane is smooth and that the mass of P is 0.1 kg, calculate the acceleration of P when it is at the origin.

(b) When P is at the point (−4, 3) the force directed towards B is zero, and the forces directed towards A and C have magnitudes 10 N and 14 N, respectively. Calculate the magnitude and direction of the resultant of the two non-zero forces. (UCLES)

7.

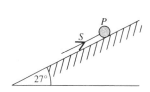

A small parcel P, of mass 1.5 kg, is placed on a rough plane inclined at an angle of 27° to the horizontal. The coefficient of friction between the parcel and the plane is 0.3. A force S, of variable magnitude, is applied to the parcel as shown in the diagram. The line of action of S is parallel to a line of greatest slope of the inclined plane.
Determine, in N to 1 decimal place, the magnitude of S when the parcel P is in limiting equilibrium and on the point of moving

(a) down the plane   (b) up the plane. (ULEAC)

8. A light inextensible string passes over a small fixed smooth pulley. The string carries a particle of mass 0.06 kg at one end and a particle of mass 0.08 kg at the other end. The particles move in a vertical plane, with both hanging parts of the string vertical. Find the magnitude of the acceleration of the particles and the tension in the string. (AEB)

9. A smooth pulley is fixed at a height $3l$ above a
horizontal table and a light inextensible string
hangs over the pulley. A particle of mass m is
attached to one end of the string and a particle
of mass $2m$ is attached to the other end.
The system is held at rest with the particles
hanging at the same level and at a distance l from
the table. The parts of the string not in contact
with the pulley are vertical. The system is then
released from rest. Find, in terms of g and l, the
speed u with which the particle of mass $2m$ strikes
the table.

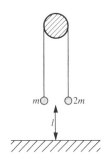

10.

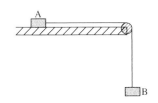

The diagram shows two particles, A of mass $3m$ and B of mass $2m$, connected by
a light inextensible string which passes over a smooth fixed pulley at the edge of
a horizontal table. Initially A is held at rest on the table and B is hanging freely
at a height h above the floor. The particle A is then released and during its
motion along the table experiences a retarding force, due to friction with the
table, of magnitude $\frac{1}{3}mg$. The particle B strikes the floor before A reaches the
edge of the table. Find, in terms of m and g, the tension in the string and the
acceleration of the particles whilst they are both moving. Show that the speed

of A at the instant when B hits the floor is $\sqrt{\left(\dfrac{2gh}{3}\right)}$.

When B hits the floor, A continues moving along the table but eventually comes
to rest before it reaches the edge. Show that the length of the string must be
greater than $4h$.   (NEAB)

11. A long light inextensible string passes over a smooth pulley, and particles of
masses m and $4m$ are fixed to the two ends of the string. The system is released
from rest with the string taut and with each particle at a height 1.2 m from the
floor. (Take g to be 10).

(a) Show that the acceleration of either particle is 6 m/s^2.

(b) Calculate the time taken for the heavier particle to reach the floor, and the
speed on impact.

(c) Assuming the heavier particle does not rebound, calculate the greatest height
above the floor attained by the lighter particle in the subsequent motion.
                                                                 (UCLES)

12.

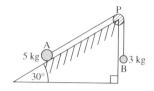

The diagram shows a particle A, of mass 5 kg, resting on a smooth plane which is inclined at 30° to the horizontal. A light inextensible string connects A to a second particle B, of mass 3 kg, which hangs freely. The string passes over a small smooth pulley P fixed at the top of the inclined plane, and the portion AP of the string is parallel to a line of greatest slope of the plane. The system is released from rest with the string taut and the hanging part vertical.

(a) Calculate, in m/s² to 2 decimal places, the acceleration of A.

(b) Calculate, in N to 2 decimal places, the tension in the string.   (AEB)

Each question from 13 to 16 is followed by several suggested responses. Choose which is the correct response.

13. A particle travelling in a horizontal straight line has an acceleration of $+2$ m/s².

A   Its total mechanical energy is constant.
B   Its kinetic energy is constant.
C   Work is being done on the particle.

14. A particle is moving with uniform velocity.

A   The particle is in equilibrium.
B   The particle has a constant acceleration.
C   There is a resultant force acting on the body in the direction of the velocity.

15. Forces represented by $2\mathbf{i} + 5\mathbf{j}$, $\mathbf{i} - 8\mathbf{j}$, and $p\mathbf{i} + q\mathbf{j}$, are in equilibrium, therefore

A $p = 3$ and $q = -3$ C $p = -2$ and $q = 3$
B $p = -3$ and $q = 3$ D $p = 2$ and $q = -40$

16. The potential energy of a body of mass m is mgh where h is

A   the distance from a chosen point
B   the height above the ground
C   the height above a chosen level.

17. A lorry has mass 6 tonnes (6000 kg) and its engine can develop a maximum power of 10 kW. When the speed of the lorry is v m/s the total non-gravitational resistance to motion has magnitude $25v$ N. Find the maximum speed of the lorry when travelling along a straight horizontal road. Find also the maximum speed up a hill which is inclined at an angle α to the horizontal where $\sin \alpha = \frac{1}{100}$, giving your answer to 2 decimal places.   (AEB)

18.

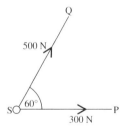

A heavy stone S, resting on rough horizontal ground, is to be moved by pulling horizontally on two ropes, SP and SQ, which are attached to the stone. The tensions in the ropes are 300 N and 500 N respectively, and the angle between the ropes is 60°, as shown in the diagram. Find, either by accurate drawing or by calculation, the resultant force on the stone due to the two ropes, giving its magnitude and the direction that it makes with SP.

The stone is dragged slowly along the ground for a distance of 2 m in the direction of the resultant pull, all forces remaining constant in magnitude and direction. Find the total amount of work done by the forces dragging the stone along the ground.

Assuming that the coefficient of friction between the stone and the ground is 0.5, find the mass of the stone. (Take g to be 10 m/s^2.)   (UCLES)

19. A lorry, travelling at constant speed, moves a distance of 2 km up a hill which is inclined at 10° to the horizontal. Given that the mass of the lorry is 2400 kg, and that the frictional resistance to the motion is of magnitude 800 N, find, in J, to 3 significant figures, the total work done by the engine of the lorry against the resistance and gravity.   (AEB)

20.

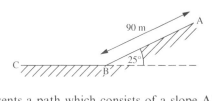

The diagram represents a path which consists of a slope AB, 90 m long, inclined at 25° to the horizontal and a horizontal section BC. A boy on a skate-board starts from rest at A and glides down AB before coming to rest between B and C. The magnitude of the resistive forces opposing the motion are constant throughout the journey. The combined mass of the boy and skate-board is 40 kg and the boy reaches B with a speed of 14 m/s. Calculate, to 3 significant figures,

(a)   the energy lost, in J, by the boy and skate-board in going from A to B

(b)   the magnitude, in N, of the resistive forces

(c)   the distance, in m, the boy travels along BC before coming to rest.
                                                                        (ULEAC)

21. A smooth bead P, of mass 0.15 kg, is threaded on a smooth straight fixed wire which is inclined at 35° to the horizontal. The bead P is released from rest and moves down the wire, starting from a point A. After 2 seconds, P passes through a point B on the wire. Calculate

(a) the distance AB, giving your answer in metres to 1 decimal place,

(b) the kinetic energy of P at B, giving your answer in joules to 1 decimal place.

(AEB)

22. A cyclist working at a rate of 150 watts maintains a steady speed of 27 km/h along a straight horizontal road. Find the resistance to motion, stating your units. (AEB)

The instruction for answering questions 23 to 30 is: if the following statement must *always* be true, write T, otherwise write F giving reasons, where you can, for your conclusion.

23. If a frictional force acts on a body, it is not necessarily of value μR where R is the normal contact force.

24. The angle of friction is the angle between the frictional force and the normal reaction.

25. If a body has a resultant force acting on it the body will accelerate in the direction of the force.

26. A particle is hanging freely attached to a light inextensible string. The string is made to accelerate vertically upward. The tension in the string is greater than the weight of the particle.

27. A car is towing a van at a steady speed. The tension in the tow rope is greater than the resistance to the motion of the van.

28. If a particle has a constant acceleration vector it must be moving in a straight line.

29. As long as no external forces act on a system the kinetic energy must be constant.

30. Some external forces which act on a moving body do not do any work.

31. A lorry of mass 5000 kg is travelling up a slope inclined at an angle α to the horizontal, where $\sin \alpha = \frac{1}{200}$. The engine of the lorry is working at a steady 20 kW and the constant resistances due to friction and to airflow around the lorry amount to 600 N. At the instant when the lorry is moving at a speed of 12 m/s, calculate, in m/s² to 3 significant figures, the acceleration of the lorry.

(ULEAC)

32. A car has an engine of maximum power 15 kW. Calculate the force resisting the motion of the car when it is travelling at its maximum speed of 120 km/h on a level road.

Assuming an unchanged resistance, and taking the mass of the car to be 800 kg, calculate, in m/s², the maximum acceleration of the car when it is travelling at 60 km/h on a level road. (UCLES)

33. The total mass of a train is 400 tonnes. It moves on a straight horizontal track against a constant resistance of 20 kN. Find the rate, in kilowatts, at which the engine is working when it is travelling at a uniform speed of 63 km/h. The tractive force of the engine is now increased to 25 kN and maintained at this value. The resistance remains unchanged. As a result, the speed of the train increases uniformly from 63 km/h to 81 km/h. For this part of the motion, show that the acceleration of the train is $\frac{1}{80}$ m/s² and find

(a) the time taken

(b) the distance travelled. (NEAB)

34. The resistance to the motion of a motor coach is K newtons per tonne, where K is a constant. The motor coach has mass $4\frac{1}{2}$ tonnes. When travelling on a straight horizontal road with the engine working at 39.6 kW, the coach maintains a steady speed of 40 m/s.

(a) Show that $K = 220$.

The motor coach ascends a straight road, which is inclined at an angle α to the horizontal, where $\sin \alpha = 0.3$, with the same power output and against the same constant resisting forces.

(b) Find, in joules to 2 significant figures, the kinetic energy of the motor coach when it is travelling at its maximum speed up the slope. (ULEAC)

35. The resistance to motion of a car of mass 2000 kg is proportional to its speed. With the engine working at 72 kW the car can attain a maximum speed of 12 m/s when travelling up a straight road, which is inclined at an angle α to the horizontal where $\sin \alpha = \frac{5}{49}$.

(a) Show that the resistance to motion is 4000 N at this speed.

(b) Find the greatest speed at which the car could travel *down* this road with the engine working at 72 kW. (AEB)

36. A motor cyclist together with his machine has a mass of 200 kg. He is ascending a straight road inclined at θ to the horizontal where $\sin \theta = \frac{1}{7}$ against a constant resistance of 120 N and the engine is working at 9 kW. Calculate

(a) the acceleration of the motor cyclist when his speed is 20 m/s

(b) the maximum speed which he can attain up the incline.

Later the motor cyclist descends the same hill with a pillion passenger whose mass is 75 kg. Given that the constant resistance is now 165 N and that the engine is switched off, find the time taken and the distance covered as the speed of the motor cyclist increases from 10 m/s to 20 m/s. (WJEC)

37. A car, of mass M kilograms, is pulling a trailer, of mass λM kilograms, along a straight horizontal road. The tow-bar connecting the car and the trailer is horizontal and of negligible mass., The resistive forces acting on the car and trailer are constant and of magnitude 300 N and 200 N respectively. At the instant when the car has an acceleration of magnitude 0.3 m/s^2, the tractive force has magnitude 2000 N.

Show that $\qquad M(\lambda + 1) = 5000.$

Given that the tension in the tow-bar is 500 N at this same instant, find the value of M and the value of λ.   (ULEAC)

38. The total mass of a woman and her bicycle is 80 kg. The woman freewheels down a slope inclined at an angle θ to the horizontal, where $\sin \theta = 0.14$, with constant acceleration of magnitude 0.9 m/s^2.

(a) Prove that the total magnitude of the resistive forces opposing the motion is 37.76 N.

(b) Find the time required for the woman to cover 180 m from rest.

The woman cycles up the same slope at constant speed 6 m/s the resistive forces remaining unchanged.

(c) Find in watts, to 3 significant figures, the power that must be exerted by the woman.

(d) If now the woman suddenly increases her work rate by 240 W, find the magnitude of her acceleration up the slope, to 3 significant figures, at that instant.   (ULEAC)

39. (In this question take g to be 10)

(a) A weightlifter lifts a weight of mass 100 kg from the floor to a height of 2 m above the floor. Calculate the work done on the weight by the weightlifter.
The weightlifter then allows the weight to fall back to the floor. State the loss in potential energy of the weight, and hence calculate the speed of the weight on impact with the floor.

(b) Water flows over a waterfall where there is a vertical drop of 80 m. The water at the top of the waterfall is flowing at a speed of 3 m/s. By considering the potential and kinetic energy of 1 kg of water, or otherwise, find the speed of the water after it has fallen 80 m.

Water flows over the waterfall at a rate of 200 m^3/s and 1 m^3 of water has a mass of 1000 kg. Assuming that 40% of the energy of the water at the bottom of the waterfall can be converted into electricity by suitable generators, calculate the power, in kilowatts, that could be developed.
                                                                (UCLES)

CHAPTER 9

ELASTIC STRINGS AND SPRINGS

ELASTIC STRINGS

When a string can be stretched by forces applied at its ends it is called an *elastic string*.

The *natural length* of an elastic string is its *unstretched length*.

Two forces, one at each end, must be applied to an elastic string in order to stretch it. You may argue that pulling *one* end will cause stretching if the other end is tied to a fixed object such as a wall, but remember that a force acts on the string at the point of attachment, e.g. where it is fastened to the wall.

Clearly the stretching forces must each act outwards; they must also be equal and opposite as otherwise the string would move in position and not just stretch.

An elastic string that has been stretched is *taut*; when it is in a straight line and is of natural length, it is described as 'just taut'.

The difference between the stretched length and the natural length is called the *extension* and is often denoted by *x*.

We know that a taut string exerts an inward pull at each end and, by considering the equilibrium at one end of the string, we see that the tension is equal and opposite to the stretching force there.

HOOKE'S LAW

In the seventeenth century a relationship between the extension of a stretched string and the tension at each end was discovered experimentally by Hooke.

The relationship, known as Hooke's Law, states that, up to a certain point,

**the extension, *x*, in a stretched elastic string
is proportional to the tension, *T*, in the string**

i.e. $T \propto x$ or $T = \dfrac{\lambda x}{a}$

where a is the natural length of the string

and λ is the *modulus of elasticity* of the string.

The Elastic Limit

As the extending forces applied to the string are steadily increased, there comes a time when a further increase suddenly produces an extension much greater than Hooke's Law would suggest. The string has become *overstretched* and will not return to its natural length when it is released; it has gone beyond its *elastic limit*. Subsequently its extension bears no relationship to the tension and, at this level of study, is no longer of any interest to us; in this book we deal only with strings that have not exceeded their elastic limit.

The Modulus of Elasticity

The form in which Hooke's Law is usually used is

$$T = \frac{\lambda x}{a}$$

Considering the dimensions on each side of this formula we see that

on the LHS we have   the dimensions of a force

on the RHS we have   (the dimensions of λ) $\times \dfrac{\text{length}}{\text{length}}$

Therefore λ has the dimensions of force and is measured in newtons.

Further, when the length of an elastic string is doubled, i.e. when $x = a$, then $T = \lambda$ showing that

> **λ is equal to the tension in an elastic string
> whose length is twice the natural length.**

SPRINGS

Hooke's Law applies to springs in a similar way as to elastic strings but there is one important difference – a spring can be compressed as well as stretched.

When stretched, i.e. when it is *in tension*, a spring behaves in exactly the same way as a stretched elastic string, i.e. equal and opposite *tensions* act *inwards* at the ends.

When a spring is compressed, the reduction in its length is called the *compression* and the forces in the spring are an *outward push*, called a *thrust*, at each end.

The spring is said to be *in compression* and it obeys Hooke's Law where T is the thrust and x is the compression.

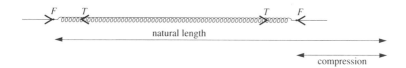

Examples 9a

1. **A light elastic string whose natural length is 0.8 m is stretched to a length of 1.1 m by a force of 12 N as shown. Find the modulus of elasticity of the string.**

Tension = extending force $\quad \Rightarrow \quad T = 12$

Using Hooke's Law gives $\quad\quad T = \lambda\left(\dfrac{0.3}{0.8}\right)$

$\therefore \quad\quad\quad\quad\quad\quad\quad\quad\quad\quad \lambda = 32$

The modulus of elasticity is 32 N.

2. The natural length of a light spring is a and its modulus of elasticity is λ. Find, in terms of a and λ, the length of the spring when

(a) the tension in the spring is $\frac{1}{2}\lambda$

(b) the thrust in the spring is $\frac{1}{2}\lambda$.

When there is a tension in the spring it is extended and when there is a thrust it is compressed.

(a)

$$T = \tfrac{1}{2}\lambda \qquad\qquad\qquad\qquad\qquad\qquad T = \tfrac{1}{2}\lambda$$

Hooke's Law gives $\qquad \frac{1}{2}\lambda = \lambda \times \dfrac{x}{a} \qquad \Rightarrow \qquad x = \frac{1}{2}a$

The extended length is $\qquad a + \frac{1}{2}a = \frac{3}{2}a$

(b)

$$T = \tfrac{1}{2}\lambda \qquad\qquad\qquad T = \tfrac{1}{2}\lambda$$

Again Hooke's Law gives $\qquad x = \frac{1}{2}a \qquad$ but this time it is a compression

The compressed length is $\qquad a - \frac{1}{2}a = \frac{1}{2}a$

---

3. One end of an elastic string, of natural length 1.2 m and modulus of elasticity 20 N, is fixed to a point A on a smooth horizontal surface. A particle of mass 1.2 kg is attached to the other end B and a force acts on the particle, pulling it away from A, until the length of the string has increased to 1.5 m.

(a) Find the tension in the string.

(b) If the force ceases to act on the particle, find the acceleration with which the particle begins to move.

(c) State, with a reason, whether or not the particle continues to move with constant acceleration.

(a)

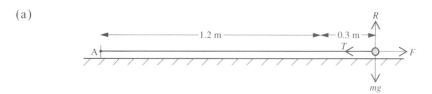

The extension in the string is 0.3 m

Hooke's Law gives $\qquad T = \dfrac{20 \times 0.3}{1.2} = 5$

The tension in the stretched string is 5 N.

(b)

When the force is removed the only horizontal force acting on the particle is the tension in the string and this acts towards A.

Using Newton's Law $F = ma$

gives $5 = 1.2a$ $\Rightarrow$ $a = 4.17$ (3 sf)

The particle begins to move towards A with acceleration $4.17\,\text{m/s}^2$ (3 sf).

(c) As soon as the particle begins to move towards A the string gets shorter and the tension in it reduces, causing a reduction in the acceleration.

Therefore the particle does not move with constant acceleration.

EXERCISE 9a

1. An elastic string of natural length 1 m is stretched to a length of 1.3 m by a force of 3 N. Find its modulus of elasticity.

2. A light elastic string whose modulus of elasticity is 18 N is stretched from its natural length of 1.4 m to a length of 1.8 m. Find the tension in the string.

3. The length of a spring whose modulus of elasticity is 30 N is reduced by 0.4 m when compressed by forces of 20 N. What is the natural length of the spring?

4. A force of 8 N acts outward at each end of a light elastic string of natural length 1.6 m. Find the stretched length of the string if the modulus of elasticity is

   (a) 10 N   (b) 20 N   (c) 40 N.

5. The natural length of a light spring is 0.9 m and its modulus of elasticity is 20 N. Two forces act inwards, one at each end of the spring. Find the compressed length of the spring if the magnitude of each force is F newtons, where

   (a) $F = 10$ (b) $F = 14$ (c) $F = 4$.

6. A force F newtons acts inwards at each end of a light spring and produces a compression of 0.3 m. Find the value of F if the modulus of elasticity of the spring is 12 N and the natural length is

   (a) 0.5 m   (b) 1 m   (c) 2 m.

7. A light elastic string, with modulus of elasticity 22 N, is extended by 0.5 m. Find the natural length of the string if the tension in it is

   (a) 8 N   (b) 10 N   (c) 16 N.

8. One end of an elastic string of natural length 0.5 m is attached to a point A on a smooth horizontal surface. A particle P, of mass 0.4 kg is attached to the other end of the string. This particle is held on the table, at a distance 0.8 m from A, by a person exerting a horizontal force of 6 N.

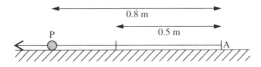

(a) Find the modulus of elasticity of the string.

(b) The person releases the particle. Find its initial acceleration.

(c) Find the tension in the string and the acceleration of P when the distance AP is,

(i) 0.7 m   (ii) 0.6 m   (iii) 0.5 m.

9. An upper body exerciser consists of a spring of length 20 cm with a handle attached at each end. The modulus of elasticity of the spring is 600 N. A boy can stretch the spring to 25 cm in length. Find the force which he is then exerting on each handle.

10. Another model of the upper body exerciser, which uses springs of the same type as in Question 9, is adjustable by inserting extra springs between the handles.

Ben uses two springs and can extend it by 13 cm. Tony uses three springs and can extend it by 9 cm. Who is the stronger?

EQUILIBRIUM PROBLEMS

In Chapter 6 the equilibrium of a particle under the action of a set of coplanar forces was discussed and the following conclusions were reached.

● Two forces that keep a particle in equilibrium must be equal and opposite.

● When three or more forces acting on a particle are in equilibrium, the sum of the force components in any direction is zero.

Using these facts we can now look at some problems in which elastic strings or springs provide one or more of the forces that keep a particle in equilibrium.

Examples 9b

1. A light elastic string, of natural length 1 m and modulus of elasticity 35 N, is fixed at one end and a particle of mass 2 kg is attached to the other end. Find the length of the string when the particle hangs freely in equilibrium.

From the equilibrium of the particle,

$$T = 2g$$

Now using Hooke's Law we have

$$T = \frac{\lambda x}{a} \quad \Rightarrow \quad 2 \times 9.8 = 35 \times x$$

$$\therefore \qquad x = 0.56$$

Therefore the length of the string is 1.56 m.

2. The natural length of a light elastic string AB is 2.4 metres and its modulus of elasticity is 4g newtons. The ends A and B are attached to two points on the same level and 2.4 m apart, and a particle of mass n kg is attached to the midpoint C of the string. When the particle hangs in equilibrium, each half of the string is at 60° to the vertical. Find, corrected to 2 significant figures, the mass of the particle.

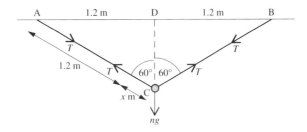

From symmetry the tensions in the two portions of the string are equal.

For the forces acting on the particle:

Resolving vertically

$$2T\cos 60° = ng \quad \Rightarrow \quad T = ng \qquad\qquad [1]$$

Now using Hooke's Law for either portion of the string,

$$T = \frac{\lambda x}{a} \quad \Rightarrow \quad T = 4g \times \frac{x}{1.2} \qquad\qquad [2]$$

From [1] and [2],

$$ng = \frac{4gx}{1.2} \quad \Rightarrow \quad x = 0.3n$$

In △ACD,   AD = AC sin 60°

∴ $AC = \dfrac{1.2}{\sin 60°} = 1.385\ldots$

Also AC is the stretched length of one half of the string

∴   AC = 1.2 + x

⇒ $1.385\ldots = 1.2 + 0.3n$ ⇒ $n = 0.6188\ldots$

The mass of the particle is 0.619 kg   (3 sf).

3. **A light elastic string of natural length 2a is fixed at one end A and carries a particle of mass 3m at the other end. When the particle is hanging freely a horizontal force 3mg is applied to it. When the particle is in equilibrium the string is inclined to the vertical at an angle θ and the extension of the string is a.**

   (a) Find θ

   (b) Show that the modulus of elasticity of the string is 6mg√2 newtons.

(a)

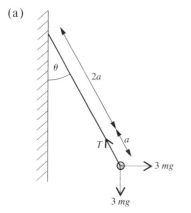

Considering the equilibrium of the particle:

Resolving ← gives

$$T \sin \theta - 3mg = 0 \qquad\qquad [1]$$

Resolving ↑ gives

$$T \cos \theta - 3mg = 0 \qquad\qquad [2]$$

[1] ÷ [2]   gives

$$\frac{T \sin \theta}{T \cos \theta} = \frac{3mg}{3mg} \qquad \Rightarrow \qquad \tan \theta = 1$$

∴ $\theta = 45°$

(b)   The answer is required in surd form so we will express sin 45° in that way.

From [1], $T = \dfrac{3mg}{\sin 45°} = 3mg\sqrt{2}$ (as $\sin 45° = \dfrac{1}{\sqrt{2}}$)

Using Hooke's Law $T = \lambda \dfrac{a}{2a}$ ⇒ $\lambda = 2T = 6mg\sqrt{2}$

The modulus of elasticity is $6mg\sqrt{2}$.

4. **Two identical springs, AC and BC, each of natural length a and modulus of elasticity 2mg, are joined together at C. The ends A and B are attached to two points distant 4a apart vertically; A is above B. A particle of mass m is attached at C. Find the length of BC when the particle rests in equilibrium.**

The extensions in the two springs are not equal as the weight of the particle is helping to stretch the upper spring but not the lower one.

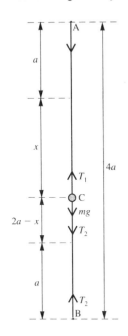

Taking x as the extension in AC we have $AC = a + x$.

Then $CB = 4a - (a + x) = 3a - x$

Therefore the extension in BC is $(3a - x) - a = 2a - x$

The particle is in equilibrium

$$\therefore \qquad T_1 - T_2 - mg = 0 \qquad\qquad\qquad [1]$$

Using Hooke's Law

For AC $\qquad T_1 = \dfrac{\lambda x}{a} = \dfrac{2mgx}{a} \qquad\qquad\qquad [2]$

For CB $\qquad T_2 = \dfrac{\lambda(2a - x)}{a} = \dfrac{2mg(2a - x)}{a} \qquad [3]$

Using [2] and [3] in [1] we have

$$\dfrac{2mgx}{a} - \dfrac{2mg(2a - x)}{a} - mg = 0$$

$$\therefore \qquad \dfrac{4mgx}{a} = 5mg \qquad \Rightarrow \qquad x = \tfrac{5}{4}a$$

$$BC = 3a - x = \tfrac{7}{4}a$$

EXERCISE 9b

In questions 1 to 3 an elastic string, with modulus of elasticity λ N and natural length a m, has one end attached to a fixed point and the other to a particle of mass m kg which hangs in equilibrium.

1. $a = 0.8$, $m = 0.3$. The string is stretched to length 1.3 m. Find λ.

2. $a = 0.7$, $m = 0.6$, $\lambda = 5$. Find the extension of the string.

3. When $\lambda = 24$, the particle causes the string to stretch to three times its natural length. Find m.

In questions 4 to 6 a vertical spring, with modulus of elasticity λ N and natural length a m, has its lower end on the floor. On top of the spring is a light platform on which a particle of mass m kg rests in equilibrium.

4. $a = 0.3$, $m = 2$. The particle is 0.2 m above the floor. Find λ.

5. $a = 0.5$, $m = 1.5$, $\lambda = 40$. Find the height of the particle above the floor.

6. If $\lambda = 35$, the particle compresses the spring to 60% of its natural length. Find m.

7. A spring is fixed at one end. When it hangs vertically, supporting a mass of 2 kg at the free end, its length is 3 m. The mass of 2 kg is then removed and replaced by a particle of unknown mass. The length of the spring is then 2.5 m. If the modulus of elasticity of the spring is 9.8 N, find the mass of the second load.

8. The end A of a light elastic string AB of natural length a and modulus of elasticity $2mg$ is fastened to one end of another light elastic string AC of natural length $2a$ and modulus of elasticity $3mg$. The ends B and C are stretched between two points $6a$ apart, so that BAC is a horizontal line.

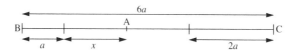

Find the length of AB.

9. A mass of 4 kg rests on a smooth plane inclined at $30°$ to the horizontal. It is held in equilibrium by a light elastic string attached to the mass and to a point on the plane. Find the extension in the string if it is known that a force of 49 N would double the natural length of 1.25 m.

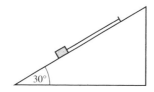

10. Two identical springs AC and BC of natural length a and modulus of elasticity $3mg$ are attached to a particle of mass m at point C. The ends A and B are attached to two points so that A is vertically above B and $AB = 3a$. The particle is in equilibrium between A and B. Find the length of AC.

11. Two springs AB and BC are joined together end to end to form one long spring. The natural lengths of the separate springs are 1.6 m and 1.4 m and their moduli of elasticity are 20 N and 28 N respectively. Find the tension in the combined spring if it is stretched between two points 4 m apart.

12. An elastic spring is fixed at one end. When a force of 4 N is applied to the other end the spring extends by 0.2 m. If the spring hangs vertically supporting a mass of 1 kg at the free end, the spring is of length 2.49 m. Find the natural length and modulus of elasticity of the spring.

13. The natural length of an elastic string is 2.5 m. The ends A and B are attached to two points on the same level and 2 m apart. A particle of mass 0.4 kg is attached to the mid point of the string and it hangs in equilibrium 1 m below the level of AB. Find the modulus of elasticity of the string.

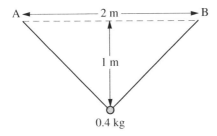

14. A light elastic string AB has natural length 2a and modulus of elasticity $mg\sqrt{3}$. The end A is attached to a fixed point and a particle of mass m is attached to B. The system is held in equilibrium, with B below the level of A and AB inclined at an angle θ to the vertical, by a force P at 90° to AB. In this position AB $= 3a$.

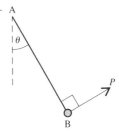

(a) Find θ. (b) Find the force P.

15. ABCD is an elastic string of natural length 3 m and particles of equal mass are attached to the unstretched string at points B and C where AB $=$ BC $=$ CD. The ends A and D are then attached to two points on the same horizontal level and 3 m apart. The particles hang in equilibrium so that the string sections AB and CD are each at 60° to the horizontal.

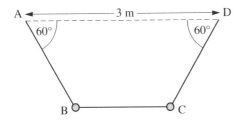

(a) Given that the extension in AB is x m, find an expression for the extension in BC.

(b) Find the value of x.

(c) The modulus of elasticity is 50 N. Find the mass of each particle.

PROBLEMS ABOUT REAL OBJECTS

Some of the problems encountered in earlier chapters involved real objects such as cars, people, cycles etc., which we modelled as particles with the same mass as the object. Other assumptions, such as zero air resistance, were also made in some cases in order to form a model.

In this chapter we have seen how Hooke's Law can be used to solve problems about light elastic strings and springs but in most real situations strings and springs do have some weight. So in order to estimate results in these circumstances we now need to make the further assumption that real strings and springs are light and obey Hooke's Law. The context in which strings and springs occur often includes a weighing machine with a scale pan whose weight is small compared with anything placed in the pan and in this case another assumption is made, i.e. that the pan is light (weightless).

The worked example and the exercise that follow, illustrate the application of this wider range of assumptions needed to form a model.

Example 9c

Raj is designing a machine to weigh heavy crates. A 'mock-up' of one of his designs uses a spring attached to a weighing platform as shown.

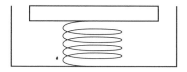

The natural length of the spring is 10 cm. When a crate, known to be of mass 100 kg is placed on the platform, it descends through 1 cm. How far should the platform descend when a mass of 150 kg is placed on it?

We can model this situation by

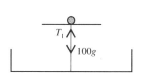

> assuming that the spring obeys Hooke's Law,
> treating the spring and the platform as light,
> treating the package as a particle,

and so produce this simplified diagram.

When the mass of 100 kg is in equilibrium, $T_1 = 100g$

The compression in the spring is 1 cm so Hookes Law gives

$$T_1 = \lambda \times \frac{1}{10} \quad \Rightarrow \quad \lambda = 1000g$$

If the thrust is T_2 when the mass of 150 kg is in equilibrium, $T_2 = 150g$

and Hooke's Law gives $T_2 = \dfrac{\lambda x}{10}$

$\therefore \quad 1500g = 1000gx \quad \Rightarrow \quad x = 1.5$

The estimated distance that a 150 kg mass would descend is 1.5 cm.

EXERCISE 9c

1. In a pinball machine the ball, of mass 30 g, is propelled from a cup, of mass 90 g, which is attached to a spring. The spring has modulus of elasticity 9 N and natural length 15 cm.

   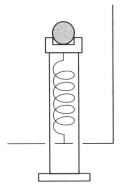

   (a) The mechanism allows the spring to be given a compression of 10 cm. Find the force necessary to do this.

   (b) Assuming that the masses of the spring and the holding mechanism are negligible, find the initial acceleration of the total mass of the cup and ball when the spring is released.

2. The spring for some kitchen scales is of length 10 cm. It is connected to a dial which is intended to measure loads up to 49 N. Design considerations suggest that the compression should not exceed 2.5 cm. By assuming that the scale pan is weightless find the minimum value you would recommend for the modulus of elasticity of the spring.

3. A door latch mechanism contains a spring to return the latch to the closed position. When the latch is in this position the spring is held compressed by means of a peg which prevents it from expanding. The spring has modulus of elasticity 12 N and natural length 8 cm.

   (a) In the closed position the spring exerts a force of 5 N on the peg. Find the compression of the spring.

   (b) In opening the door the spring is compressed by a further 1 cm. Find the force required to hold the latch in the open position.

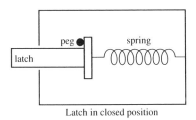

Latch in closed position

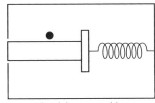

Latch in open position

4. 'Springmakers' Ltd. are testing springs for a new application. They all have the same modulus of elasticity and are of various lengths.

    (a) A spring of natural length 0.3 m is found to extend to 0.35 m when a force of 240 N is applied. Find the modulus of elasticity of the spring under test.

    (b) In the application the spring will be subjected to tensions up to a value of 300 N. It is required that the extension should not exceed 0.05 m. Find the maximum length of spring which could be used.

5. Part of the suspension system on one rear wheel of a car consists of a spring with one end fixed to the body of the car and the other end fixed to the wheel axle. The modulus of elasticity of the spring is 8000 N.

    (a) The car, when empty, puts a load of 2940 N on this spring, which compresses it to a length of 25 cm. Find the natural length of the spring.

    (b) Passengers and luggage increase the load on this spring by 1370 N. What is its compression then?

    (c) The ground clearance of the empty car is 18 cm. Find the value to which this will be reduced (for a stationary car) when the passengers and luggage are in the car.

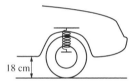

6. A 'baby-bouncer' consists of a safety seat that can be suspended on two identical elastic ropes from the top of a doorway 2 metres high. The seat is designed to rest in equilibrium at a height of 0.5 m above the floor when occupied by a child of mass 7 kg.

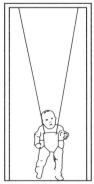

    (a) If the ropes are each 1.1 m long, what should the modulus of elasticity be?

    (b) An older child, of mass 11 kg, climbs into the seat. At what height above the floor will the seat rest in equilibrium?

State all the assumptions you have made in working out your answers.

7. An anti-vibration mount for a
machine contains a spring attached to
a fixed base plate and to a platform,
which can move vertically. A machine
of mass 240 kg has a rectangular base,
and a mount is placed under each
corner of this base. It is required
that, when the machine is not running,
the springs should be compressed to
half their natural length.

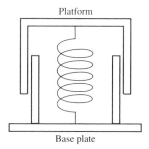

Platform

Base plate

(a) Find the thrust in each spring. State any assumptions you make in doing this.

(b) Find the modulus of elasticity of the springs.

8. The mechanism for a retractable ball-point pen includes a spring with a natural
length of 1.2 cm.

The force needed to bring the pen into use from the retracted position is 1 N and
the spring is then compressed to a length of 0.5 cm.
Find the modulus of elasticity of the spring.

CHAPTER 10

ELASTIC POTENTIAL ENERGY

An object possesses potential energy when the position of the object is such that releasing it from that position results in motion. We are familiar with the potential energy of an object which would fall from rest when released, but there is another type of potential energy.

Consider an elastic string fixed at one end and with a particle P attached to the other end. A force acting on P, away from the fixed end, stretches the string, and when that force is removed the particle begins to move, i.e. a particle attached to a stretched elastic string possesses potential energy because of its position. To distinguish this type of potential energy from gravitational potential energy, it is known as *elastic potential energy* (EPE).

We know that work done to a system causes an equivalent increase in the mechanical energy of the system, so the EPE of a stretched elastic string can be found by calculating the work done by the stretching force.

The Work Done in Stretching an Elastic String

The force that stretches an elastic string is not constant because it is at all times equal to the tension in the string, which, in turn, is directly proportional to the extension. It follows that one way to find the work done is to multiply the *average force* by the total extension produced.

Consider the work done when an elastic string, with a natural length a and modulus of elasticity λ, is stretched from an extension x_1 to an extension x_2.

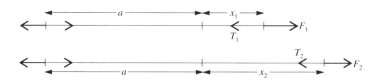

When the extension is x_1 $\qquad F_1 = T_1 = \dfrac{\lambda x_1}{a}$

When the extension is x_2 $\qquad F_2 = T_2 = \dfrac{\lambda x_2}{a}$

Over the period while the extension increases from x_1 to x_2,

$$\text{the average extending force is } \tfrac{1}{2}(T_1 + T_2),$$

$$\text{therefore the work done is given by } \tfrac{1}{2}(T_1 + T_2)(x_2 - x_1)$$

i.e. **work done in stretching an elastic string**
$= $ average tension $\times$ increase in extension

If x_1 is zero, i.e. the string is initially unstretched, then T_1 also is zero and the expression above can be simplified to give $\tfrac{1}{2}Tx$ where T is the final tension. Further, using $T = \dfrac{\lambda x}{a}$, the work done can be expressed as $\dfrac{\lambda x^2}{2a}$,

i.e. **the amount of work needed to stretch an elastic string**

by an extension x is given by $\dfrac{\lambda x^2}{2a}$

An alternative way to find the work done in stretching an elastic string, uses calculus.

When the extension is s, say, the extending force is $\dfrac{\lambda s}{a}$.

Now if the string is further stretched by a small amount δs, the work required is given approximately by $\dfrac{\lambda s}{a}(\delta s)$.

The total work done in stretching the string from its natural length a to a length $(a + s)$ is therefore given approximately by $\displaystyle\sum_0^x \dfrac{\lambda s}{a}\,\delta s$.

Then, as $\delta s \to 0$, $\displaystyle\sum_0^x \dfrac{\lambda s}{a}\,\delta s \;\to\; \dfrac{\lambda}{a}\int_0^x s\,\mathrm{d}s = \dfrac{\lambda x^2}{2a}$

The expressions derived above apply equally well to an extended *or compressed* elastic spring.

Examples 10a

1. The natural length of a light elastic string is 1.2 m and its modulus of elasticity
   is 18 N. Initially the string is just taut and is then stretched until it is 2 m long.
   Find the work done during the extension.

   The initial extension and the initial tension are both zero.

   The work done is given by $\dfrac{\lambda x^2}{2a}$, i.e. $\dfrac{18 \times 0.8^2}{2 \times 1.2}$

   The work done is 4.8 J.

2. The string described in example 1 is being held at a stretched length of 1.6 m when a force
   begins to stretch it further. The force acts until the work it has done amounts to 3 J.

   (a) Find the final extension.

   (b) State an assumption that has been made.

   (a)   The initial extension is 1.6 m; let the final extension be x m.

   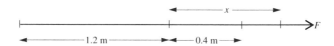

   When extension is 0.4 m, the tension is $\dfrac{18 \times 0.4}{1.2}$ N, i.e. 6 N

   The final tension is given by $\dfrac{18x}{1.2}$

   The average tension is given by $\frac{1}{2}(15x + 6)$

   Work done $=$ average tension $\times$ increase in extension

   $$= \tfrac{1}{2}(15x + 6)(x - 0.4)$$

   $\therefore \qquad 3 = \tfrac{1}{2}(15x^2 - 2.4)$

   $\therefore \qquad 15x^2 = 8.4 \quad \Rightarrow \quad x = 0.7483\ldots$

   The final extension is 0.748 m   (3 sf).

   (b) We have assumed that the string has not exceeded its elastic limit.

3. One end of an elastic spring is fixed to a point A on a horizontal plane. The modulus of
   elasticity of the spring is λ, the natural length is a and the spring is strong enough to
   stand vertically. A particle of mass m is attached to the other end of the spring which is
   held at a distance a vertically above A. If the particle is allowed to descend gently to its
   equilibrium position find how much work is done in compressing the spring.

When the particle is in equilibrium

$$T = mg$$

Using Hooke's Law gives

$$T = \frac{\lambda x}{a}$$

$$\therefore \quad mg = \frac{\lambda x}{a} \quad \Rightarrow \quad x = \frac{mga}{\lambda}$$

The work done in compressing the spring

is given by $\quad \dfrac{\lambda x^2}{2a} \quad$ i.e. $\quad \dfrac{\lambda}{2a}\left(\dfrac{mga}{\lambda}\right)^2$

The work done is $\quad \dfrac{m^2 g^2 a}{2\lambda}$

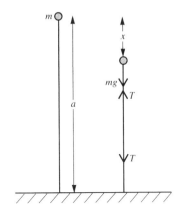

EXERCISE 10a

1. The natural length of an elastic string is 0.3 m and its modulus of elasticity is 8 N. Find the work done in

(a) giving it an extension of 0.2 m

(b) stretching it to a length of 0.6 m

(c) stretching it from a length of 0.4 m, to a length of 0.7 m.

2. The natural length of a spring is 5a and its modulus of elasticity is 2mg. Find the work done in

(a) compressing it to a length 4a

(b) compressing it from length 3a to length 2a

(c) compressing it until the thrust is mg.

3. The modulus of elasticity of a spring is 20 N and its natural length is 1.5 m. Find the work done in

(a) stretching it until the tension is 8 N

(b) stretching it to increase the tension from 8 N to 12 N.

4. An elastic string has a natural length of 2 m. The work done in extending it by 0.5 m is 6 J.

(a) Find the modulus of elasticity.

(b) Find the work done in stretching from its natural length to a length of 3 m.

5. Find the work done in stretching a rubber band round a roll of papers of radius 4 cm if the band when unstretched will just go round a cylinder of radius 2 cm and its modulus of elasticity is 0.5 N. Assume that the rubber band obeys Hooke's law.

6. A light elastic string is fixed at one end to a point A and a particle P attached to the other end hangs in equilibrium at a point B. The natural length of the string is 1.4 m, the modulus of elasticity is 8 N and the mass of the particle is 2 kg.

(a) Find the depth of B below A.

(b) Find the work done in stretching the string from its natural length to the length AB.

7. An upper body exerciser consists of a spring of length 20 cm with a handle attached at each end. The modulus of elasticity of the spring is 600 N. A man can stretch the spring to 25 cm in length.

(a) Find the work he does in performing 15 repetitions of this exercise.

(b) If these repetitions take him 20 seconds, find his power.

8. A jack-in-the-box toy comprises a box of depth 25 cm, with a spring fixed to its base. 'Jack' is a puppet attached to the top of the spring. The natural length of the spring is 30 cm. The height of the puppet is 10 cm and its mass is 0.2 kg.

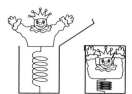

(a) When 'Jack' is in equilibrium with the box open the length of the spring is 25 cm. Find the modulus of elasticity.

(b) Making the assumption that the puppet does not compress, find the work done in compressing the spring so that the box can be closed.

*9. A string of natural length $2a$ and modulus of elasticity λ has its ends attached to fixed points A and B, where $AB = 3a$. Find the work done when the mid-point C of the string is pulled away from the line AB to a position where triangle ABC is equilateral.

ENERGY PROBLEMS

We know that if no external work is being done to or by a system, the total amount of mechanical energy remains constant, and problems involving the conservation of kinetic energy and gravitational potential energy have already been considered.

Now that the elastic potential energy of a stretched elastic string can be found (the work done in stretching an elastic string from its natural length is equal to the elastic potential energy in the string) problems can be tackled where three types of mechanical energy, KE, PE and EPE, arise. As there are now likely to be changes in all three types, equating the *total* mechanical energy in one position with the total in another position is wiser than trying to juggle with which types of energy are decreasing and which increasing.

Examples 10b

1. A particle of mass 2 kg is attached to one end of an elastic string of length 1 m and modulus of elasticity 8 N which is lying on a smooth horizontal plane. The other end of the string is fixed to a point A on the plane and when the string is just taut the particle is at a point B. The particle is pulled away from A until it reaches the point C where AC = 1.5 m and is held in that position.

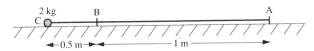

(a) Find the elastic potential energy in the string.

The particle is then released from rest.

(b) Find the velocity of the particle when it passes through the point B.

(c) What is its velocity when it passes through A?

(a) When the particle is at C, the EPE is $\dfrac{\lambda x^2}{2a}$, i.e. $\dfrac{8 \times (0.5)^2}{2 \times 1}$

The EPE in the string is 1 J.

(b) In moving from C to B, there is no change in PE as the motion takes place on a horizontal plane.

At C,   KE = 0   and   EPE = 1 J

At B,   KE = $\frac{1}{2}mv^2 = \left(\frac{1}{2}\right)(2)v^2$ and   EPE = 0

Conservation of ME gives $0 + 1 = v^2 + 0$ $\Rightarrow$ $v = 1$

The velocity of the particle at B is 1 m/s towards A.

(c) When the particle reaches B the string is no longer stretched so there is no EPE in the string.

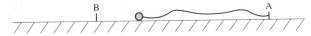

Between B and A there is no change in EPE therefore the KE remains constant.

∴   the velocity at A is 1 m/s in the direction CA.

Note that the speed of the particle remains constant until it reaches D, a point on the opposite side of A where AD = 1 m. Beyond this point the string again begins to stretch and contains EPE.

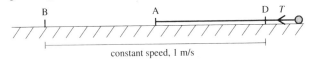

2. A light elastic spring of length 1 m and modulus of elasticity 7 N has one end fixed to a point A. A particle of mass 0.5 kg hangs in equilibrium at a point C vertically below A.

(a) Find the distance AC.

The particle is raised to the point B, between A and C, where AB = 1 m, and is released from rest. Find

(b) the speed of the particle as it passes through C

(c) the distance below B of the point D where the particle first comes to rest.

(a)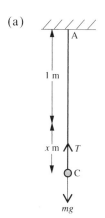

At C $T = mg$

i.e. $T = 0.5 \times 9.8 = 4.9$

Using Hooke's Law gives $T = \dfrac{7x}{1}$

$\therefore$ $7x = 4.9$ $\Rightarrow$ $x = 0.7$

The extension is 0.7 m.

$\therefore$ The distance 'AC is 1.7 m.

(b)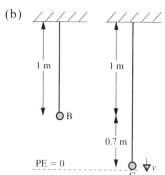

We will use conservation of ME from B to C.

At B $PE = mgh = 0.5 \times 9.8 \times 0.7 = 3.43$

$KE = 0$

$EPE = 0$

At C $PE = 0$

$KE = \frac{1}{2}mv^2 = 0.25v^2$

EPE in the string is given by $\dfrac{\lambda x^2}{2a}$

i.e. $EPE = \dfrac{7 \times 0.7^2}{2 \times 1} = 1.715$

Conservation of ME gives

$$3.43 + 0 + 0 = 0 + 0.25v^2 + 1.715$$

$\therefore$ $v^2 = 6.86$ $\Rightarrow$ $v = 2.619\ldots$

The speed at C is 2.62 m/s (3 sf).

(c)   We could use conservation of ME from C to D but that involves the speed at C which *might* not be correct. By working from B to D we avoid this.

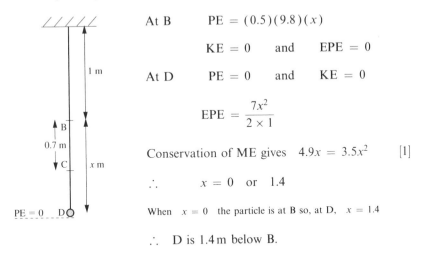

At B $PE = (0.5)(9.8)(x)$

              $KE = 0$ and $EPE = 0$

At D $PE = 0$ and $KE = 0$

              $EPE = \dfrac{7x^2}{2 \times 1}$

Conservation of ME gives $4.9x = 3.5x^2$ [1]

∴ $x = 0$ or 1.4

When $x = 0$ the particle is at B so, at D, $x = 1.4$

∴   D is 1.4 m below B.

Note that the two positions of instantaneous rest, i.e. B and D, are at equal distances from the equilibrium position C.

3. **One end of a light elastic string, with natural length 0.8 m and modulus of elasticity 16 N, is fixed at a point A on a smooth plane inclined at 45° to the horizontal. A particle of mass 1 kg, attached to the other end of the string, rests in equilibrium at a point E on the plane.**

(a) **Find the distance AE.**

The particle is pulled down the plane to a point C, where   AC = 1.6 m,   and is then released from rest

(b) **Find the speed of the particle as it passes through E.**

(a)   Resolving along the plane gives

$T - g \cos 45° = 0$ $\Rightarrow$ $T = 6.929\ldots$

Using Hooke's Law, $T = \dfrac{\lambda x}{a}$,   gives

$6.929\ldots = \dfrac{16x}{0.8}$ $\Rightarrow$ $x = 0.346\ldots$

The length of AE is 1.15 m   (3 sf).

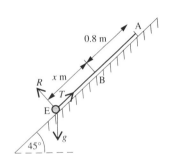

(b)

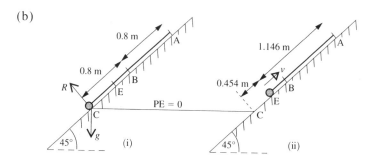

The normal contact force on the particle is perpendicular to the direction of motion, and the plane is smooth, so no work is done.

When the particle is at C, total mechanical energy $= $ EPE

$$= \frac{16(0.8)^2}{2 \times 0.8} \text{J} = 6.4 \text{J}$$

When the particle is at E, total mechanical energy $=$ EPE $+$ PE $+$ KE

$$= \left\{ \frac{16(0.346)^2}{2 \times 0.8} + 1 \times g \times 0.454 \sin 45° + \tfrac{1}{2} \times 1 \times v^2 \right\} \text{J}$$

$$= (0.5v^2 + 4.343\ldots) \text{J}$$

Using conservation of mechanical energy between (i) and (ii) gives

$$0.5v^2 + 4.343\ldots = 6.4 \quad \Rightarrow \quad v = 2.028\ldots$$

The velocity of the particle at E is 2.03 m/s (3 sf).

4. **In a fairground 'test your strength' machine, a spring is fixed at one end and lies, just taut, in a horizontal groove. A metal cylinder is attached to the other end.**

The would-be strong men strike the cylinder so as to compress the spring, and the machine records the compression achieved when the cylinder first comes to rest. The natural length of the spring is 1 m, the modulus of elasticity is 60 N and the mass of the cylinder is 4 kg. If a competitor gives the cylinder an initial speed of 3 m/s, find the recorded compression if

(a) the groove is smooth,

(b) the coefficient of friction between cylinder and groove is 0.2.

State any assumptions you have made in your solution.

At A the only mechanical energy is KE

so total energy at A is $\frac{1}{2}(4)(3^2)$ J, i.e. 18 J

At B the only mechanical energy is EPE

so total energy at B is $\dfrac{60x^2}{(2)(1)}$ J, i.e. $30x^2$ J

These expressions apply to both parts of the question.

(a) No work is done to the system once the cylinder is set moving so we can use conservation of mechanical energy, giving

$$30x^2 = 18 \quad \Rightarrow \quad x = 0.77\ldots$$

The recorded compression is 0.77 m (2 sf).

(b) If the groove is rough, work is done by the frictional force, so energy is not conserved. The frictional force opposes the motion of the cylinder so the work it does causes a loss of energy in the system.

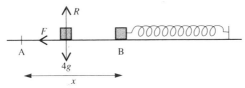

Resolving vertically for the cylinder gives $\qquad R = 4g$

The frictional force, F, is given by μR i.e. $\qquad F = 0.2 \times 4g = 7.84$

Friction acts for a distance x so work done by friction is Fx, i.e. $7.84x$

$$\text{work done} = \text{loss in energy}$$
$$= \text{initial energy} - \text{final energy}$$

i.e. $\qquad 7.84x = 18 - 30x^2 \quad \Rightarrow \quad 30x^2 + 7.84x - 18 = 0$

Solving this quadratic equation by using the formula gives

$$60x = -7.84 \pm \sqrt{(7.84^2 + 2160)} \quad \Rightarrow \quad x = 0.654\ldots$$

The recorded compression is 0.65 m (2 sf).

Assumptions made are:
the cylinder is modelled as a particle; the spring is light

and for part (b)
the groove is uniformly rough; friction between the *spring* and the groove can be neglected.

5. A particle P, of mass **2m**, is fastened to the end of each of two identical elastic strings, each of natural length **3a** and modulus of elasticity **3mg**. The other ends of the strings are fixed to the points A and B that are distant **8a** apart on the same level. When the particle rests in equilibrium at the point D, it is at a distance **3a** below C, the midpoint of AB.

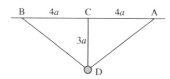

(a) Find the tension in each string.

(b) Measuring gravitational potential energy relative to AB, find the total mechanical energy in the system.

The particle is now raised to C and released from rest in this position.

(c) Find the speed of P as it passes through D.

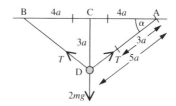

(a) Resolving vertically for T gives $2T \sin \alpha = 2mg$

      $\triangle$ACD is a '3, 4, 5' triangle so   AD $= 5a$

      From $\triangle$ACD, $\sin \alpha = \frac{3}{5}$ $\Rightarrow$ $T = \frac{5}{3}mg$

(b) The extension in each string is $5a - 3a$, i.e. $2a$

      EPE in each string is $3mg \times \dfrac{(2a)^2}{2 \times 3a} = 2amg$

      PE of the particle is $2mg(-3a) = -6mga$

      The particle is at rest so there is no KE

      Therefore the total ME $= 2 \times 2mga - 6mga = -2mga$

(c) When the particle is at C there is no KE or PE

      but the EPE in each string is $3mg \times \dfrac{a^2}{6a} = \frac{1}{2}mga$

      When the particle passes through D

      EPE + PE $= -2mga$ (from (b)) and KE $= \frac{1}{2}(2m)v^2 = mv^2$

      Using conservation of ME from C to D gives

            $2(\frac{1}{2}mga) = -2mga + mv^2$ $\Rightarrow$ $v = \sqrt{3ga}$

      i.e. the speed of the particle as it passes through D is $\sqrt{3ga}$.

EXERCISE 10b

In questions 1 and 2 a particle of mass 2 kg is attached to one end of a light elastic string of natural length a metres and modulus of elasticity λ newtons. The other end of the string is attached to a fixed point A on a smooth horizontal surface.

1. $\lambda = 5$, $a = 0.8$, AB = 1 m and BC = 1 m.

The particle is held at C, with the string stretched and released from rest.

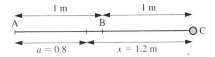

(a) Find the elastic potential energy when the particle is at C.

(b) Find the elastic potential energy when the particle is at B.

(c) Find the kinetic energy when the particle is at B.

(d) Find the speed of the particle as it passes through B.

2. $\lambda = 8$, $a = 0.5$, AB = 0.5 m and BC = 0.7 m.

The particle is projected from point B with velocity 10 m/s in the direction BC.

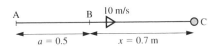

(a) Find the elastic potential energy when the particle is at B.

(b) Find the kinetic energy when the particle is at B.

(c) Find the elastic potential energy when the particle is at C.

(d) Find the speed of the particle as it passes through C.

In questions 3 and 4 a particle of mass 3 kg is attached to one end of a light elastic spring of natural length a metres and modulus of elasticity λ newtons. The other end of the spring is attached to a fixed point A on a horizontal surface. The coefficient of friction between the particle and the surface is μ.

3. $\mu = 0$, $\lambda = 800$, $a = 2$, AC = 2 m.

The particle is projected from C with speed 20 m/s and it has a speed of 10 m/s as it passes through B.

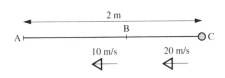

(a) Find the elastic potential energy when the particle is at C.

(b) Find the kinetic energy when it is at C.

(c) Find the kinetic energy when it is at B.

(d) Find the elastic potential energy when it is at B.

(e) Find the distance AB.

4. $\lambda = 150$, $a = 3$, $AB = 2\,m$ and $BC = 3\,m$.

The particle is released from rest at C
and it has a speed of $4\,m/s$ as it passes
through B.

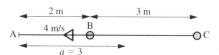

(a) Find the elastic potential energy
when the particle is at C.

(b) Find the elastic potential energy when it is at B.

(c) Find the kinetic energy when it is at B.

(d) Find the work done by friction during the motion from C to B.

(e) Find the value of μ.

5. A particle of mass $2\,kg$ is suspended from a point A by a light elastic spring of
natural length $1\,m$ and modulus of elasticity $80\,N$. The particle is initially held at
a point B, which is $0.6\,m$ vertically below A, with the spring compressed. It is
then released from rest at B. In the subsequent motion the particle is at point C
when the spring has reached its natural length. Take the gravitational potential
energy to be zero at point C.

(a) When the particle is at B find the values of
   (i) the elastic potential energy   (ii) the gravitational potential energy
   (iii) the kinetic energy.

(b) When the particle is at C find the value of the elastic potential energy.

(c) Find the speed of the particle as it passes through C.

6. In this question use $g = 10\,m/s^2$.

One end of a light elastic string of natural length $2\,m$ and modulus of elasticity
$120\,N$ is fixed at a point A; the other end carries a particle of mass $1\,kg$. The
particle is released from rest at the point A and drops vertically. It first comes to
rest at a point B vertically below A and the string then has an extension of x metres.

(a) Write down the kinetic energy of the particle when it is at A and find, in
   terms of x,
   (i) the elastic potential energy
   (ii) the gravitational potential energy relative to B.

(b) Write down the kinetic energy of the particle when it is at B and find the
   elastic potential energy in terms of x.

(c) Find the depth of B below A.

7. A particle of mass $5\,kg$ is suspended from a point A by a light spring of
natural length $0.4\,m$. When in equilibrium the particle is at a point B, vertically
below A, where $AB = 1.8\,m$. The particle is pulled down to point C, also
vertically below A, where $AC = 3\,m$. It is then released from rest at C.

(a) Find the modulus of elasticity.

(b) Find the speed of the particle as it passes through B.

8. A light spring, of natural length $4a$ and modulus of elasticity $2mg$, is fixed at a point A and a particle of mass m is attached to the other end. When the particle is suspended in equilibrium, it is at a point B vertically below A. It is projected vertically downwards from B with speed v and first comes to rest at a point C where AC $= 8a$.

(a) Find the distance AB.   (b) Show that $v = \sqrt{2ga}$.

*9. A ring R, of mass m, can slide freely round a smooth wire in the shape of a circle, of radius a, fixed in a vertical plane. The ring is fastened to one end of a light elastic string of natural length a and modulus of elasticity mg. The other end of the string is attached to the lowest point of the wire. R is held at the highest point of the wire and is slightly disturbed from rest in this position.

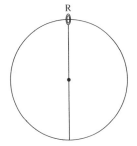
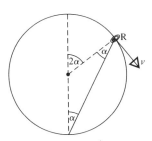

Find the velocity of R when

(a) it is level with the centre of the wire,

(b) the string first goes slack,

(c) the string makes an acute angle 2α with the upward vertical.

*10. In a pin-ball machine the ball, of mass 30 g, is propelled from a cup, of mass 90 g, which is attached to a spring. The spring has modulus of elasticity 9 N and natural length 15 cm. The mechanism allows the spring to be given a compression of 10 cm. It is then released and the ball and cup move over the smooth horizontal surface of the table. The ball leaves the cup when the spring reaches its natural length.

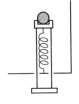

(a) Find the elastic potential energy of the spring when it is fully compressed.

(b) Find the kinetic energy of the cup and ball combined, at the moment when the spring reaches its natural length.

(c) Find the speed with which the ball leaves the cup.

(d) Find the kinetic energy of the cup at the moment when the ball has just left it.

(e) Find the extended length of the spring at the moment when the cup first comes to rest.

11. A particle of mass 2 kg lies on a smooth plane
which is inclined at an angle of 30° to the
horizontal. The particle is attached to one end of
a light elastic string of natural length 1 m and
modulus of elasticity 20 N. The other end of the
string is attached to a point A on the plane.
The particle can rest in equilibrium at a point B
on the plane.

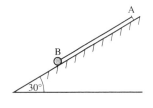

(a) Find the extension of the string when the particle is at B.

If the particle is released from rest at A,

(b) find the speed with which it will pass through B.

(c) find the distance it has travelled down the plane from A when it first comes
to rest.

12. Two points A and B are distant $3a$ apart in a horizontal line. A particle P of
mass m is connected to A by a light inextensible string of length $4a$ and to B by a
light elastic string of length a and modulus of elasticity $\frac{1}{4}mg$. Initially P is held
at a point distant a from B on AB produced, with both strings just taut, as
shown in the diagram. P is released from rest in this position.

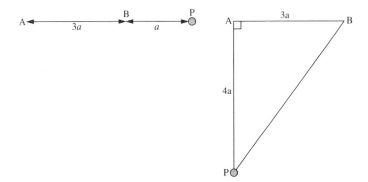

(a) Show that when AP is vertical the speed of P is $2\sqrt{ga}$.

(b) Find the tension in the elastic string at this instant.

*13. A and B are two fixed points on a smooth horizontal plane. $AB = 3.3$ m. A spring,
of natural length 0.9 m and modulus of elasticity 6 N, has one end attached to point A
and the other end to a particle P, of mass 0.03 kg. A second spring, of natural length
1.5 m and modulus of elasticity 8 N has its ends attached to the particle and to point B.

(a) Find the distance AP when the particle is in equilibrium.

(b) The particle is released from rest at point C where $AC = 0.5$ m.
   (i) Find its velocity as it passes through the equilibrium position.
   (ii) Find the length of AP when the particle first comes to rest.

*14. An aircraft of mass 5000 kg lands on the deck of an aircraft-carrier at a speed of 50 m/s relative to the deck. It catches on the mid-point of an elastic cable which lies, just taut, across the deck at 90° to its path. The cable has a natural length of 60 m. The plane is brought to rest in a distance of 40 m by the stretching cable and by a retarding force of magnitude 8000 N in the direction opposite to its motion, which is produced by its engines. Find the modulus of elasticity of the cable.

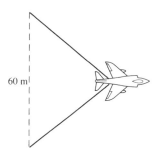

60 m

REFINING MATHEMATICAL MODELS

A mathematical model of a situation is produced by making certain assumptions. These may include assuming that certain objects are small enough to be treated as particles, other subjects are light enough to be treated as weightless, air resistance can be ignored, relevant physical laws are obeyed, and so on.

The model is then used to predict results. If these results, compared with those observed in trials, are not accurate enough to be useful, then the model requires improvement.

One way to achieve this is to include one or more of the quantities previously ignored. Which of these to choose depends upon assessing how reasonable the original assumptions were.

Consider, for example, the loaded weighing machine given below, modelled by a light spring that obeys Hooke's Law, with a particle placed on it as shown.

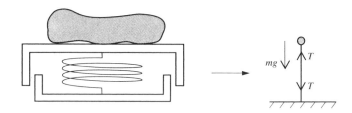

One of the assumptions made in forming this model is that the platform is light. If unreliable results are produced, a refined model could take into account the mass of the platform. Whether or not the refined model produces better estimates can only be assessed by comparing predicted results with observed results.

EXERCISE 10c

1. A climber of mass 60 kg is attached by a rope of length 30 m to a point on a vertical rock-face. When she is 20 m above the point to which the rope is attached she slips and falls. She first comes to rest after falling through a vertical distance of 57 m. Assuming that she is not retarded by contact with the rock-face during her fall, and that the mass of the rope can be neglected, state any further assumptions you would make in order to model this situation.

(a) Use your model to find
(i) the modulus of elasticity of the rope
(ii) her speed at the moment that the rope becomes taut.

(b) What features of the model might lead to unreliable estimates? State any adjustments that could be made.

*2. The Department of Transport in Ruritania, having invested in new trains for its railways, decides to replace all the old buffers in the stations. A technical team, looking into the design of new buffers, decides first of all to produce a mathematical model.

The first model treats the train as a particle, of mass M kilograms, hitting the buffers at a speed of v metres per second and being brought to rest in a distance of d metres. The buffers are modelled as a spring, of natural length a metres, modulus of elasticity λ, fixed at one end and able to take a maximum compression of c metres. The mass of the buffers is taken to be negligible, being small compared to the mass of the train. Using this model there is no loss of energy when the particle collides with the spring. In this first model it is assumed that the retarding force produced by the brakes is negligible.

(a) Assuming that the spring does not become fully compressed, obtain d in terms of M, v, a and λ.

(b) Show that if the train is to be brought to rest before the spring is fully compressed then it is necessary to satisfy the condition $\lambda > \dfrac{M\,av^2}{c^2}$

It is decided to refine the model by introducing a retarding force R, produced by the brakes.

(c) Again assuming that the spring does not become fully compressed, show that now

$$d = \frac{\sqrt{a^2 R^2 + \lambda M\,av^2} - aR}{\lambda}$$

(d) Show that this refined model includes the result obtained in (a) from the original model.

3. A bungee jumping event is to take place from a suspension bridge which is 50 m above a river. The jumpers dive from the bridge attached to an elastic rope. You are required to consider some suitable ropes for this activity. The aim is that the jumper should be brought to rest just above the surface of the water. Carry out the investigation for the case of a man of mass 70 kg, using a model which treats the man as a particle and the rope as light, and assume that the initial velocity is zero. State any other assumptions that you make.

Use the following variables:

natural length of the rope, a metres,

modulus of elasticity of the rope, λ newtons,

extension of the rope when the man reaches the water, x metres,

extension of the rope if the man is suspended in equilibrium on it, e metres,

the speed with which he passes through this equilibrium position, V metres per second (this is the greatest speed he reaches).

(a)   Obtain an expression for λ in terms of a and x.

(b)   Obtain an expression for e in terms of a and λ.

(c)   Obtain an expression for V in terms of a, e and λ.

(d)   Complete the following table of values.

a	10	20	30	40	45
x					
λ					
e					
$2a + e$					
V					

(e)   Can you suggest any factors which might make some of these lengths of rope unsuitable?

(f)   Which assumptions might you consider changing in order to refine the model?

CHAPTER 11

PROJECTILES

THE MOTION OF A PROJECTILE

A particle that is sent moving into the air is called a projectile. At this stage we consider that air resistance is small enough to be ignored so the only force acting on a projectile, once it has been thrown, is its own weight. It follows that the projectile has an acceleration downward of magnitude g, but has constant velocity horizontally.

Therefore the horizontal motion and the vertical motion are analysed separately.

The horizontal and vertical components of velocity and displacement will be involved and a new form of notation is used to denote these quantities.

Consider a particle projected with a velocity V at an angle θ to the horizontal. We take O as the point of projection, Ox as a horizontal axis and Oy as a vertical axis.

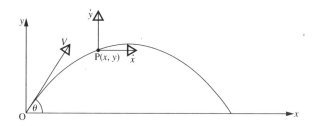

At any time during its flight, the projectile is at a point where:

> the horizontal displacement from O is x
>
> and the vertical displacement from O is y

Now the horizontal velocity is the rate at which x increases and this is denoted by $\dot{x}$. (The dot over the x means 'the rate of increase with respect to time'.)

Similarly the vertical velocity at any time is denoted by $\dot{y}$.

The initial components of velocity are:

$$V \cos \theta \quad \text{in the direction } Ox$$
$$V \sin \theta \quad \text{in the direction } Oy.$$

Consider the horizontal motion, where the velocity is constant.

At any time t seconds after projection,

$$\dot{x} = V \cos \theta \tag{1}$$

$$x = (V \cos \theta) \times t = Vt \cos \theta \tag{2}$$

Consider the vertical motion where there is an acceleration g downwards.

Using $v = u + at$ and $s = ut + \frac{1}{2}at^2$ where $u = V \sin \theta$, gives

$$\dot{y} = V \sin \theta - gt \tag{3}$$
$$y = (V \sin \theta) \times t - \tfrac{1}{2}gt^2 = Vt \sin \theta - \tfrac{1}{2}gt^2 \tag{4}$$

These four equations provide all the information needed to investigate the motion of any projectile.

If, for example, we want the speed v of the particle at a particular time, $\dot{x}$ and $\dot{y}$ can be found from equations [1] and [3], then $v^2 = \dot{x}^2 + \dot{y}^2$.

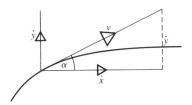

Also, the direction of motion, i.e. the way the particle is moving, is given by the direction of the velocity.

So if the direction of motion makes an angle α with the horizontal, then

$$\tan \alpha = \dot{y} / \dot{x}$$

Examples 11a

1. A particle P is projected from a point O with a speed of 40 m s⁻¹, at 60° to the horizontal. Find, 3 seconds after projection,

   (a) the speed of the particle
   (b) the horizontal and vertical displacements of the particle from O
   (c) the distance of P from O.

   Take g as 10 and give answers corrected to 2 significant figures.

   Note that $\mathrm{m\,s^{-1}}$ is an alternative form for m/s.

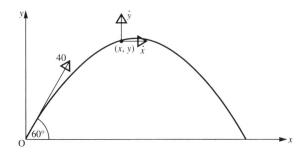

(a)   Use equations [1] and [3] with $V = 40$, $\theta = 60°$ and $t = 3$.

   Horizontally, $\dot{x} = 40 \cos 60° = 20$

   Vertically, $\dot{y} = 40 \sin 60° - gt = 34.64 - 30$

   $\Rightarrow$ $\dot{y} = 4.64$

   If v m s⁻¹ is the speed of the particle after 3 seconds,

   $$v^2 = 4.64^2 + 20^2 \quad \Rightarrow \quad v = 20.53\ldots$$

   The speed of the particle is 21 m s⁻¹ (2 sf)

(b)   Use equations [2] and [4]

   Horizontally, $x = 40 \times 3 \cos 60° = 60$

   Vertically, $y = 40 \times 3 \sin 60° - \tfrac{1}{2}gt^2 = 103.92 - 45$

                   $y = 58.92$

   The displacements of the particle from O are:

   60 m horizontally and 59 m vertically (2 sf)

(c) $OP^2 = x^2 + y^2$

                     $= 60^2 + 58.92^2 = 7072$

   The distance of P from O is 84 m (2 sf)

2. The initial velocity of a particle projected from a point O is $u\mathbf{i} + v\mathbf{j}$ where $\mathbf{i}$ and $\mathbf{j}$ are horizontal and vertical unit vectors respectively.

    (a) Find, in terms of u, v, g and t
        (i) the velocity vector of the particle t seconds after projection
        (ii) the position vector of the particle at this time.

    (b) Given that the position vector of point A is $\dfrac{3uv}{2g}\mathbf{i} + h\mathbf{j}$, find h in terms of v and g.

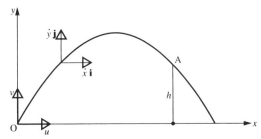

    (a) In the direction Ox, $\dot{x}$ is constant and equal to u $\Rightarrow$ $x = ut$

        In the direction Oy, $\dot{y} = v - gt$ $\Rightarrow$ $y = vt - \frac{1}{2}gt^2$

    $\therefore$ after t seconds

      (i) the velocity of the particle is $u\mathbf{i} + (v - gt)\mathbf{j}$

      (ii) the position of the particle is $ut\mathbf{i} + (vt - \frac{1}{2}gt^2)\mathbf{j}$

    (b) At A, $ut\mathbf{i} = \dfrac{3uv}{2g}\mathbf{i}$ $\Rightarrow$ $t = \dfrac{3v}{2g}$

        Then $h = vt - \frac{1}{2}gt^2 = \dfrac{3v^2}{2g} - \dfrac{g}{2}\left(\dfrac{9v^2}{4g^2}\right) = \dfrac{3v^2}{8g}$

3. A particle P is projected from a point 5 m above the ground. The horizontal and vertical components of the velocity of projection are each 24 m s^{-1}.

    (a) Find the angle of projection.

    (b) Taking g as 10 find the horizontal distance of P from the point of projection when it hits the ground.

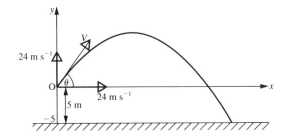

(a)   If the velocity of projection is V then

$$V \cos \theta = 24 \quad \text{and} \quad V \sin \theta = 24$$

$$\therefore \qquad \tan \theta = \frac{V \sin \theta}{V \cos \theta} = \frac{24}{24} = 1$$

$$\Rightarrow \qquad \qquad \theta = 45°$$

(b)   When the particle hits the ground it is 5 m *below* the point of projection so $y = -5$.

At any time t, $\qquad y = 24t - \frac{1}{2}gt^2$

$\therefore \qquad$ when $\qquad y = -5, \quad -5 = 24t - 5t^2 \qquad \Rightarrow \qquad 5t^2 - 24t - 5 = 0$

Hence $\qquad \qquad (5t + 1)(t - 5) = 0 \qquad \Rightarrow \qquad t = -\frac{1}{5} \text{ or } 5$

A negative time has no meaning in this problem, so $\ t = 5$.

At any time t, $\ x = 24t \ $ so, when $\ t = 5, \ x = 120$

When P hits the ground its horizontal distance from O is 120 m.

4. **The top of a tower is 10 m above horizontal ground. A boy fires a stone from a catapult with a velocity of 12 m s^{-1}. Find how far from the foot of the tower the stone hits the ground if**

(a)   **it is fired horizontally**

(b)   **it is fired at 30° below the horizontal.**

Take g as 10 and give answers corrected to 2 significant figures.

(a)

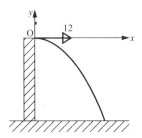

When the initial velocity is horizontal, $\theta = 0$, therefore $\sin \theta = 0$ and $\cos \theta = 1$.

At any time t, $\qquad x = Vt \quad$ and $\quad y = -\frac{1}{2}gt^2$

When the stone hits the ground, $\ y = -10$

$\therefore \qquad -\frac{1}{2}gt^2 = -10 \qquad \Rightarrow \qquad t^2 = \frac{10}{5} = 2$

$\therefore \qquad t = \sqrt{2} \ $ (t cannot be negative)

When $\quad t = \sqrt{2}, \qquad x = Vt = 12\sqrt{2} = 16.97\ldots$

The stone hits the ground 17 m from the foot of the tower (2 sf).

(b)

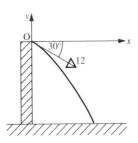

The stone is fired at a downward angle so θ is a negative acute angle.

At any time t, $\quad \cdot \; x = Vt \cos(-30°) = Vt \cos 30°$

and $\qquad\qquad y = Vt \sin(-30°) - \frac{1}{2}gt^2 = -Vt \sin 30° - \frac{1}{2}gt^2$

When the stone hits the ground, $y = -10$

$\therefore \qquad 12t(-\frac{1}{2}) - \frac{1}{2}gt^2 = -10 \qquad \Rightarrow \qquad 5t^2 + 6t - 10 = 0$

$$\Rightarrow \qquad t = \frac{-6 \pm \sqrt{236}}{10}$$

$\therefore \qquad t = 0.936 \quad (t \text{ cannot be negative})$

When $\quad t = 0.936, \quad x = Vt \cos 30° = 12 \times 0.936 \times 0.8660$

$$= 9.726\ldots$$

The stone hits the ground 9.7 m from the foot of the tower (2 sf).

Note that it is possible to solve problems in which θ is negative, without referring to the standard projectile equations, but by starting from first principles taking the downward direction as positive. The reader is given an opportunity to try this approach in question 27 in the next exercise.

5. **A particle P is projected from a point O with velocity V at an angle θ to the horizontal. Find, in terms of V and θ, the time at which P reaches its greatest height, and what the greatest height is.**

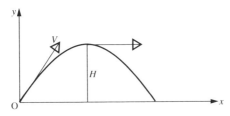

When P is at its greatest height it is momentarily travelling horizontally, i.e. $\dot{y} = 0$.

$$\dot{y} = V \sin \theta - gt$$

$\therefore$ when $\dot{y} = 0$ $\quad V \sin \theta = gt$ $\quad \Rightarrow \quad$ $t = \dfrac{V \sin \theta}{g}$

i.e. the greatest height is reached after a time of $\dfrac{V \sin \theta}{g}$

The height at time t is given by $\quad y = Vt \sin \theta - \frac{1}{2}gt^2$

$\therefore$ for the greatest height H $\qquad H = V\left(\dfrac{V \sin \theta}{g}\right) \sin \theta - \frac{1}{2}g\left(\dfrac{V \sin \theta}{g}\right)^2$

$$= \dfrac{V^2 \sin^2 \theta}{g}\left(1 - \tfrac{1}{2}\right)$$

$\therefore$ the greatest height reached is $\quad \dfrac{V^2 \sin^2 \theta}{2g}$

EXERCISE 11a

Unless another instruction is given, take g as 10 m s^{-2} (i.e. m/s^2) and give answers corrected to 2 significant figures; these approximations are justified as we have already made one approximation in ignoring air resistance.
In questions where vector notation is used, **i** and **j** are horizontal and vertical unit vectors respectively.

In each question from 1 to 5, a particle P is projected from a point O on a horizontal plane with velocity V at an angle θ to the horizontal. All units are based on metres and seconds.

1. Given that $V = 24.5$ and $\theta = 30°$, find the speed of P after

(a) 1 second (b) 2 seconds.

2. If $V = 20$ and $\theta = 60°$, find the height of P above the plane when

(a) $t = 1$ (b) $t = 2$ (c) $t = 3$.

Illustrate your answers with a sketch.

3. Given that $V = 30$ and $\tan \theta = \frac{3}{4}$, find the speed and the coordinates of the position of P after

(a) 1 second (b) 2 seconds.

4. If $V = 10$ and $\theta = 60°$, find the time taken for P to travel a horizontal distance of 5 m and find the height of P at this time.

5. After 4 seconds P hits the plane at a point A. If $\theta = 45°$, find

(a) V (b) the distance OA.

6. A particle is projected with a velocity vector $20\mathbf{i} + 40\mathbf{j}$. Find the velocity vector of the particle after 3 seconds.

7. A particle P is projected from a point O with a velocity of 10 m/s at an angle of 30° to the horizontal. Find the horizontal and vertical displacements of P from O after half a second and hence find the distance from O to P at this time.

8. A stone is thrown from a point O on the top of a cliff of height 60 m, and falls into the sea $2\frac{1}{2}$ seconds later at a point whose displacement from O is $24\mathbf{i} - 60\mathbf{j}$. Find, in the form $a\mathbf{i} + b\mathbf{j}$, the velocity of projection.

9. A particle is projected from a point O with a velocity vector $\mathbf{i} + 2\mathbf{j}$. Find the velocity and the position of the particle in vector form after

(a) t seconds   (b) $1\frac{1}{2}$ seconds.

10. A particle P is projected from a point O with velocity $\mathbf{V}$. Giving answers in terms of g find, in the form $a\mathbf{i} + b\mathbf{j}$, the velocity and position of P after t seconds if

(a) $\mathbf{V} = 5\mathbf{i} + 3\mathbf{j}$

(b) $\mathbf{V} = 4\mathbf{i} - \mathbf{j}$

(c) $\mathbf{V} = 20\,\mathrm{m\,s}^{-1}$ at 60° to the horizontal

(d) $\mathbf{V} = 10\sqrt{5}\,\mathrm{m\,s}^{-1}$ at θ to the horizontal where $\tan\theta = 2$.

11. At fielding practice a cricketer throws the ball with a speed of 26 m/s at an angle α above the horizontal, where $\tan\alpha = \frac{12}{5}$. Find

(a) the times at which the ball is 16 m above the ground

(b) the horizontal distance covered between these times.

12. A stone is thrown downwards from a point A into a quarry that is 25 m deep. Find the initial speed and the direction of projection if, after 2 seconds, the stone lands at the bottom of the quarry at a horizontal distance from A of

(a) 30 m   (b) 20 m.

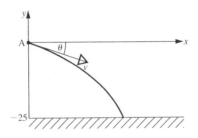

13. A particle is projected with a velocity of $20\sqrt{2}$ m s^{-1} at an angle of 45° to the horizontal. Find, in terms of g, how long it is before the particle reaches its highest point and what the greatest height is.

14. A particle is projected from a point O and after $1\frac{1}{2}$ seconds it passes through a point whose position is represented by the vector $4\mathbf{i} + \mathbf{j}$. Find the initial velocity as a vector.

15. A boy is at the window of his flat and throws a ball to his friend on the ground 24 m below. The ball is thrown with speed 10 m/s at an angle of elevation α, where $\tan \alpha = \frac{5}{12}$.

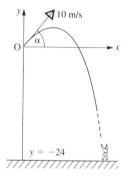

   (a) Find how long it takes for the ball to reach ground level.

   (b) If the friend is standing 8 m from the wall of the flat, will he be able to catch the ball without moving?

16. A particle is projected with a velocity vector $a\mathbf{i} + b\mathbf{j}$. After 5 seconds the velocity vector is $40\mathbf{i} + 60\mathbf{j}$. Find the values of a and b.

17. A particle P is projected from a point O with a velocity of 25 m s^{-1} at an angle of 60° to the horizontal. For how long is P at least 15 m above the level of O?

18. Two seconds after a particle is projected from a point O, its position vector relative to O is given by $12\mathbf{i} + 4\mathbf{j}$. Find in the form $a\mathbf{i} + b\mathbf{j}$

   (a) the velocity of projection   (b) the velocity after 2 seconds.

19. A golf ball is hit by a golfer at 25 m/s towards the green which is 2 m below the level of the tee. The ball is struck at an angle θ to the horizontal where $\tan \theta = \frac{7}{24}$ and lands directly in the hole! What is the horizontal distance from tee to hole?

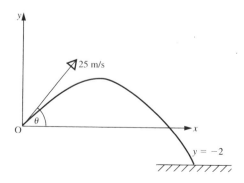

20. In a Highland Games Competition the local strongman, Mac, is hoping to break the record for throwing the heavy hammer. He manages to hurl the hammer with a speed of 25 m s^{-1} at 40° to the horizontal. If the record throw is 60.6 m, does Mac break the record?

21. A stone is thrown from the top of a cliff at an angle α to the horizontal and with a speed of 19.5 m/s. The stone falls into the water 37.5 m from the foot of the cliff. Given that $\tan \alpha = \frac{12}{5}$, find the height of the cliff if α is

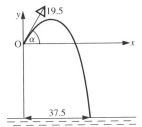

(a) above the horizontal

(b) below the horizontal.

22. Find the angle of projection of a ball, thrown at 20 m s^{-1}, which is at its greatest height when it *just* passes over the top of a tree that is 16 m high.

23. A stone is thrown from a point A, 1 m above the ground, with a velocity of $4\sqrt{2}$ m s^{-1} at 45° to the ground. Taking A as the origin and **i** and **j** as horizontal and vertical unit vectors respectively,

(a) write down the initial velocity vector of the stone,

(b) find the position vector of the stone t seconds after being thrown,

(c) use (b) to find the value of t when the stone hits the ground,

(d) what is the horizontal distance from A of the point where the stone hits the ground?

24. From the top of a fifty-metre high cliff, the angle of depression of a marker buoy is 30°. A student throws a stone from the cliff top in an attempt to hit the buoy but is foolish enough to think that the stone should be thrown at 30° below the horizontal. How far short of the buoy will the stone land if it is thrown with a speed of 15 m s^{-1}?

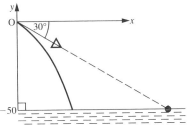

25. The horizontal and vertical components of the initial velocity of a projectile are each of magnitude u. Express in the form $a\mathbf{i} + b\mathbf{j}$

(a) the initial velocity

(b) the velocity of the projectile after (i) t seconds (ii) 2 seconds

(c) the position of the projectile after 2 seconds.

26. An arrow is shot from the top of a building 26 m high. The initial speed of the arrow is 30 m s^{-1}. Find how long the arrow is in the air if it is fired at an angle of 20° to the horizontal

(a) upwards   (b) downwards.

27. A particle P is projected from a point O, with speed V at an angle θ *below* the horizontal.

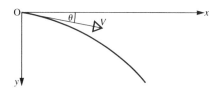

Taking the axes shown in the diagram find, t seconds after projection, an expression in terms of g for

(a) the horizontal velocity component

(b) the vertical velocity component

(c) the horizontal displacement of P from O

(d) the vertical displacement of P from O.

28. A particle P is projected from a point O, with speed 20 m s^{-1} at an angle of 30° below the horizontal. Use the results found in question 27 to find

(a) the speed of P   (b) the distance of P from O,

after 2 seconds.

29.

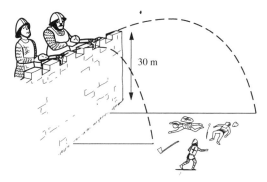

From the battlements of a castle, 30 m above the ground, defenders are catapulting rocks at their attackers below. They all think that the best angle of projection is 20°. The strongest catapulter can fire a rock with a speed of 6 m/s and even the weakest fires at 2 m/s. Find the distances from the foot of the castle walls between which the attackers may be hit.
State any assumptions that have been made.

PROPERTIES OF THE FLIGHT PATH OF A PROJECTILE

The Equation of the Path

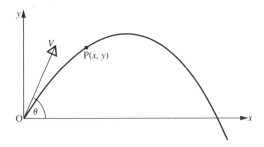

Using x and y axes through the point of projection, we have seen that the coordinates of the position of the projectile at any time can be expressed as

$$x = Vt \cos \theta$$
$$y = Vt \sin \theta - \tfrac{1}{2}gt^2$$

As these two equations give x and y each in terms of the variable t, they are the parametric equations of the path of the projectile (which is also called the trajectory).

From the first equation, $\qquad t = \dfrac{x}{V \cos \theta}$

Substituting in the second equation gives

$$y = \frac{Vx \sin \theta}{V \cos \theta} - \tfrac{1}{2}g\left(\frac{x}{V \cos \theta}\right)^2$$

$\Rightarrow \qquad\qquad\qquad y = x \tan \theta - \dfrac{gx^2}{2V^2 \cos^2 \theta} \qquad\qquad\qquad [\mathbf{1}]$

For any particular projectile, g, V and θ are constants, so y is a quadratic function of x, showing that the path of a projectile is a parabola with a vertical axis of symmetry.

Equation [1] is called *the equation of the path* of the projectile, or the equation of the *trajectory*.
It is very useful in problems where the position of the projectile is involved but the time taken to reach the position is not.

In particular if the angle of projection, θ, has to be found, the equation of the path can be rearranged as follows:

Using $\dfrac{1}{\cos^2 \theta} = \sec^2 \theta = \tan^2 \theta + 1$ gives the equation of the path as

$$y = x \tan \theta - \frac{gx^2}{2V^2}(\tan^2 \theta + 1)$$

If we are given the coordinates of a point through which the projectile passes, and V is known, this becomes a quadratic equation in $\tan \theta$.

The reader should be prepared to use the equation of the trajectory in either of the forms above to solve a problem in which time does not appear.

Note that the equation is valid for the whole of the path, including those cases where the projectile moves on below the level of the point of projection, e.g. a stone thrown from the top of a cliff.

The Greatest Height Reached

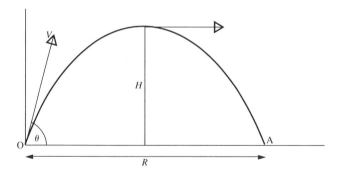

In Examples 11a, we saw in example 5 that, for a projectile given an initial velocity V at an angle θ to the horizontal, the greatest height H reached above the point of projection is given by

$$H = \frac{V^2 \sin^2 \theta}{2g}$$

Although this expression can sometimes be *quoted* in a solution, its derivation may be asked for. So it is important to remember that it is found by using the time taken to reach the point where $\dot{y}$ is zero.

The Range on a Horizontal Plane

The range of a projectile on a horizontal plane is the distance between the point of projection O and the point where the projectile returns to the level of O. The range is usually denoted by R as shown in the diagram opposite.

When the projectile reaches A, $y = 0$,

i.e. $\quad Vt \sin \theta - \tfrac{1}{2}gt^2 = 0 \quad \Rightarrow \quad t = \dfrac{2V \sin \theta}{g}$

Note that this is the time taken for the whole of the journey; it is known as *the time of flight*.

Note also that the time of flight is twice the time taken to reach the highest point on the path.

Now R is the value of x ($= Vt \cos \theta$) at this time,

i.e. $\quad R = V \cos \theta \left(\dfrac{2V \sin \theta}{g} \right) = \dfrac{V^2}{g} \left(2 \sin \theta \cos \theta \right)$

$$\therefore \qquad R = \frac{2V^2 \sin \theta \cos \theta}{g} = \frac{V^2 \sin 2\theta}{g}$$

The second of these forms is best if the problem is simply to find R, but when R is being used in conjunction with other distances, e.g. the greatest height, the first form is usually better.

Note that we are referring above to the range *on a horizontal plane,* i.e. the distance between O and the point where the projectile is again level with O.

If a projectile ends its flight above O (e.g. a ball that lands on a roof), or below O (e.g. a stone thrown from the top of a tower to the ground), the range is the distance between O and the landing point but is *not* given by the expressions above.

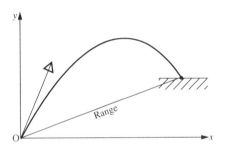

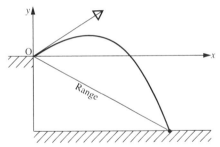

The Maximum Horizontal Range

The formulae derived above for the horizontal range of a projectile show that the value of R depends upon the initial velocity *and* the angle of projection.

Therefore, for a given value of V, the range varies only with θ and the maximum range occurs when $\sin 2\theta$ is maximum.

The greatest value of $\sin 2\theta$ is 1 and this is when $2\theta = 90°$ $\Rightarrow$ $\theta = 45°$

Therefore   **the *maximum* horizontal range, R_{max}, is** $\dfrac{V^2}{g}$

and   **it is achieved when $\theta = 45°$**

Examples 11b

1. **Two seconds after it is projected from a point O on a horizontal plane, a particle P passes through a point represented by $3\mathbf{i} + 5\mathbf{j}$,   where $\mathbf{i}$ and $\mathbf{j}$ are, respectively, vectors of $1\ \text{m s}^{-1}$ horizontally and vertically.   Find**

   (a)   the speed and direction of projection

   (b)   the range of the projectile.

   Take g as 10 and give answers corrected to 2 significant figures.

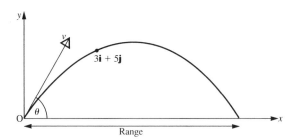

In this problem, although we have the coordinates of a point on the trajectory, we do not use the equation of the path, as time *is* involved.

(a)   At P, $x = 3$ and $y = 5$

   Using $x = Vt \cos \theta$ and $y = Vt \sin \theta - \tfrac{1}{2}gt^2$ when $t = 2$ gives

$$3 = 2V \cos \theta \qquad\qquad [1]$$

   and $5 = 2V \sin \theta - 20$ $\Rightarrow$ $25 = 2V \sin \theta$ $\qquad [2]$

$[2] \div [1]$ $\dfrac{2V \sin \theta}{2V \cos \theta} = \dfrac{25}{3}$ $\Rightarrow$ $\tan \theta = \dfrac{25}{3}$

$\Rightarrow$ $\theta = 83.15°,$ i.e. $83°$ (2 sf)

From [1] $V = \dfrac{3}{2 \cos 83.15°} = 12.576\ldots$

The speed of projection is 13 m s^{-1} (2 sf).

(b)   Range $= \dfrac{2V^2 \sin \theta \cos \theta}{g} = \dfrac{2(12.58)^2 \sin 83.15° \cos 83.15°}{10}$

$= 3.748\ldots$

The range is 3.7 m (2 sf).

2. **A gun is fired with a muzzle speed of 100 m s^{-1} in a tunnel whose roof is 4 m above the point of projection. Find the greatest permissible angle of projection if the bullet is to avoid hitting the roof. Find also the range of the gun with this angle of projection.**

Take g as 10 and give answers corrected to 2 significant figures.

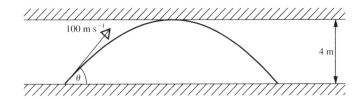

The greatest height of the bullet is 4 m above O.

As we are not asked to derive the expression for the greatest height, it can be quoted.

$H = \dfrac{V^2 \sin^2 \theta}{2g}$ $\Rightarrow$ $4 = \dfrac{10^4 \sin^2 \theta}{20}$

$\therefore$ $\sin \theta = 0.08944$ $\Rightarrow$ $\theta = 5.13°,$ i.e. 5.1° (2 sf)

The range R is given by $R = \dfrac{V^2 \sin 2\theta}{g}$

$\therefore$ $R = (10^4 \sin 10.26°) \div 10 = 180 \text{ m}$ (2 sf)

3. **A missile fired from a point O, with velocity 40 m s⁻¹ at an angle α to the horizontal, passes through a point distant 32 m horizontally and 45 m vertically from O. Show that there are two possible angles of projection and give their values (take g as 10 and give your answers to the nearest degree). Illustrate your answers on a diagram.**

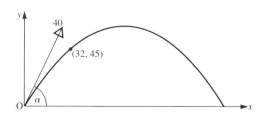

The coordinates of a point on the path of the missile are given and the time when the point is reached does not matter; so we use the equation of the path.

Using $y = x \tan \alpha - \dfrac{gx^2 \, (\tan^2 \alpha + 1)}{2V^2}$

gives $45 = 32 \tan \alpha - \dfrac{10 \times (32)^2 \times (\tan^2 \alpha + 1)}{2 \times 1600}$

Using $\tan \alpha = T$, this equation simplifies to

$$45 = 32T - 3.2(T^2 + 1) = 0$$

$\Rightarrow$ $3.2T^2 - 32T + 48.2 = 0$

This is a quadratic equation for T in which $a = 3.2$, $b = -32$, $c = 48.2$

$\therefore$ $b^2 - 4ac = 407.04$,   which is positive.

So there are two values of T and therefore two different values of α.

Solving the equation by using the formula gives

$$T = \tan \alpha = \frac{32 \pm \sqrt{407.04}}{6.4} = \frac{32 \pm 20.18}{6.4} = 8.153 \quad \text{or} \quad 1.847$$

$\therefore$ the two values of α are $62°$ and $83°$.

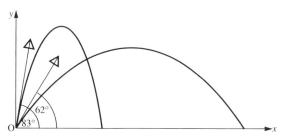

4. A particle is projected from a point O at ground level, at an angle of elevation of 55°. It just clears the top of each of two walls that are 2 m high. If the first of the walls is distant 2 m from O, find

(a) the speed of projection

(b) the distance of the second wall from O.

Take g as 10 and give answers corrected to 2 significant figures.

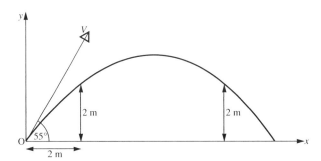

(a)   As the projectile *just* clears the first wall we can take $(2, 2)$ as a point on the path.

Using $y = x \tan \theta - \dfrac{gx^2}{2V^2 \cos^2 \theta}$ gives

$$2 = 2 \tan 55° - \frac{10 \times 4}{2V^2 \cos^2 55°} = 2.856 - \frac{60.79}{V^2}$$

$\Rightarrow \qquad V^2 = 60.79 \div 0.856 = 71.02 \ldots \qquad \Rightarrow \qquad V = 8.427 \ldots$

The speed of projection is 8.4 m s^{-1} (2 sf).

(b)   To find the location of the second wall we need the other value of x for which $y = 2$ so we use the equation of the path again.

When $y = 2$ and $V = 8.426$, the equation of the trajectory becomes

$$2 = 1.4281x - \frac{10 x^2}{2 \times 71.02 \times 0.3290} = 1.428x - 0.2140x^2$$

$\Rightarrow \qquad\qquad x^2 - 6.673x + 9.346 = 0$ \qquad\qquad\qquad [1]

Solving this quadratic equation by formula gives

$$x = \frac{6.673 \pm \sqrt{(6.673^2 - 4 \times 9.346)}}{2} = \frac{6.673 \pm 2.673}{2}$$

$\therefore \qquad x = 2 \text{ or } 4.673$

The second wall is 4.7 m from O (2 sf).

Note that the second root of equation [1] could have been found directly, using the fact that one of the roots is 2 and also that the sum of the roots is $-b/a$. Hence $x + 2 = 6.673 \quad \Rightarrow \quad x = 4.673$.

5. A particle P is projected from a point O on a horizontal plane with speed V at an angle θ to the horizontal.

   (a) State the value of θ for which the range is maximum and give the value of the maximum range.

   (b) Find, in terms of V and g, the ratio of the greatest height to the maximum range.

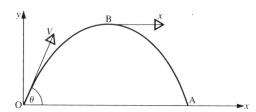

(a) The maximum range is achieved when $\theta = 45°$

$$R_{max} = \frac{V^2}{g}$$

(b) When the particle is at its greatest height, $\dot{y} = 0$

$$\therefore \qquad V \sin 45° - gt = 0 \qquad \Rightarrow \qquad t = \frac{V \sin 45°}{g}$$

The greatest height is the value of y when $t = (V \sin 45°)/g$

i.e. $\qquad y = Vt \sin 45° - \frac{1}{2}gt^2 = \frac{V^2 \sin^2 45°}{g} - \frac{1}{2}g \frac{(V^2 \sin^2 45°)}{g^2}$

$$\therefore \qquad H_{max} = \frac{V^2}{2g} - \frac{V^2}{4g} = \frac{V^2}{4g}$$

The ratio $H_{max} : R_{max}$ is $\dfrac{V^2}{4g} : \dfrac{V^2}{g} = 1 : 4$

EXERCISE 11b

For questions where quantities are given in the form $a\mathbf{i} + b\mathbf{j}$, $\mathbf{i}$ and $\mathbf{j}$ are respectively horizontal and vertical unit vectors, measured in the unit consistent with other quantities in the question.

In questions 1 to 6, a particle P is projected from a point O on a horizontal plane, with speed V and angle of elevation θ. The greatest height reached is H and the range on the plane is R. The maximum range is R_{max}. If you can *remember* the formulae for H and R, you may quote them but if not, derive them as quickly as you can. Do *not* rely on looking up each formula as you need it.

Throughout the exercise take g as 10 and give answers corrected to 2 significant figures unless another instruction is given.

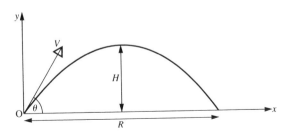

1. $V = 24 \text{ m s}^{-1}$ and $\theta = 30°$, find H and R.

2. $V = 20 \text{ m s}^{-1}$ and $R = 28 \text{ m}$, find θ.

3. $\theta = 45°$ and $H = 10 \text{ m}$, find V and R.

4. $R = 60 \text{ m}$ and $\theta = 60°$, find V and H.

5. $V = 30$ and R has its maximum value, find R_{max}.

6. $H = 16 \text{ m}$ and $V = 40 \text{ m s}^{-1}$, find θ and R.

7. A stone is thrown with speed 26 m s^{-1} at an angle α to the horizontal where $\tan \alpha = \frac{5}{12}$. Find how far it has travelled horizontally when

   (a) it is at the same level as the point of projection
   (b) it is 2 m below the point of projection.

8. A particle is projected from a point O with speed 50 m s^{-1} at an angle of elevation of 40°. Taking Ox as the horizontal axis and Oy vertically upward, *find* (i.e. do not quote) the equation of the flight path of the particle. Hence find the height of the particle when it is distant 20 m horizontally from O.

9. A particle is projected from a point on a horizontal plane with a speed of 12 m/s. Find the angle of projection and the range if the time of flight is half a second.

10. Using a horizontal x axis and a vertical y axis find the equation of the path of a projectile whose initial velocity is

    (a) 5 m s^{-1} at an angle θ above the horizontal where $\tan \theta = \frac{4}{3}$
    (b) 30 m s^{-1} at an angle of 45° below the horizontal.

11. A ball is thrown with a speed of 16 m s^{-1} from a point on a level playground. The angle of projection is α where $\tan \alpha = \frac{3}{4}$. Find the time for which the ball is in the air and the horizontal distance it travels in this time.

12. A stone is thrown from ground level with a speed of 15 m/s, so that when it is travelling horizontally it just passes over a tree of height 3 m. Find the angle of projection.

13. A particle is projected from a point on a horizontal plane with speed $\sqrt{(2gh)}$. Find in terms of h, the range on the plane if the angle of projection is

(a) $30°$ (b) $45°$ (c) $60°$.

14. A missile is fired at 80 m s^{-1} at an angle α to the horizontal. The missile must pass over an obstruction that is 20 m high and 120 m away in the line of flight. Find the smallest permissible value of α.

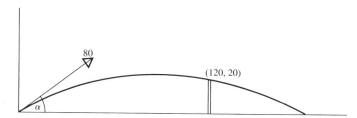

*15. In a children's game played with large pebbles, the aim is to throw a pebble so that it lands within a circle of diameter 0.2 m, drawn on the ground. A player stands at A, a point marked on the ground, which is 3 m from Q as shown. One child can throw the stone with a speed of 5 m/s from a height of 0.6 m above the ground and always projects it at an angle of elevation of at least 35°. If θ is the angle of projection, find the range of possible values of θ that will land the stone in the circle.

Take g as 10 and give angles to the nearest degree.

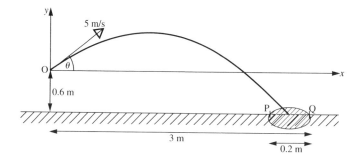

16. A particle is projected from a point O with speed $2\sqrt{gh}$ at an angle α to the horizontal where $\tan \alpha = 2$. Find its height above the point of projection when its horizontal distance from O is

(a) h (b) $2h$ (c) $3h$ (d) $4h$.

Explain the reason for the answer to part (d).

17. A cricketer strikes the ball with speed 36 m s^{-1} at a height of 0.5 m above the pitch. He 'skies' it, hoping to hit it for six. Unfortunately it is caught on the boundary line, 70 m away, at a height of 2.2 m. At what angle was the ball struck?

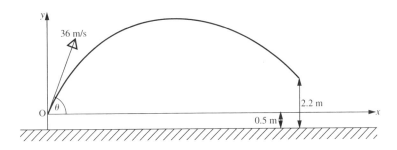

18. The maximum range of a shell fired from a particular gun is 2400 m. At what angle and at what speed is the shell fired?

19. A ball is thrown from O with initial velocity 5**i** + 7**j**. Find the Cartesian equation of its path.

20. The speed of a tennis player's serve is 45 m/s. At the moment when the racquet strikes, the ball is exactly over the base line and at a height of 2.8 m. The height of the net is 0.9 m and the distance from the base line to the point where the ball reaches the net is 12 m. If the ball is served at an angle of 8° below the horizontal, will it clear the net? (Take g as 9.81 and work to 3 sf.)

21. A golf ball is struck with a speed of 26 m s^{-1} at an angle α to the horizontal where tan α = 2. The golfer finds that he has not sent the ball in quite the direction he wanted and directly in the line of flight are two trees. One is 22 m high and 20 m from the tee; the other is 21 m high and 48 m from the tee. Determine whether the ball will clear either or both of the trees.

22. A ball is projected with velocity 10**i** + 20**j** from a point O and moves freely under gravity. The ball strikes a wall, 30 m away from O. If **i** and **j** are vectors of magnitude 1 m s^{-1} horizontally and vertically respectively, find

(a) the time, in seconds taken for the ball to reach the wall

(b) the height above O, in metres, of the point where the ball strikes the wall

(c) the acute angle, to the nearest degree, which the direction of motion of the ball makes with the horizontal at the instant when it strikes the wall.

FURTHER PROBLEMS

Examples 11c

1. A and B are two points $60\sqrt{3}$ m apart on level ground. A particle P is projected from A towards B with speed 45 m s^{-1} at 30° to the horizontal. At the same instant a particle Q is projected from B towards A with speed $15\sqrt{3}$ m s^{-1} at 60° to the horizontal.

(a) Using exact trig ratios prove that P and Q are always at the same height.

(b) Find t when P and Q collide and find in terms of g their height above the ground when they collide.

For P we will use an x-axis in the direction AB and for Q a separate x-axis in the direction BA.

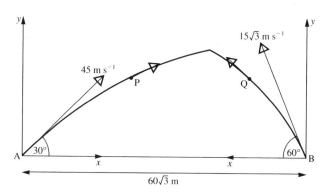

(a) After t seconds, using $y = Vt \sin\theta - \frac{1}{2}gt^2$ gives:

For P $\qquad y_P = 45t \times \frac{1}{2} - \frac{1}{2}gt^2 \qquad = \frac{45}{2}t - \frac{1}{2}gt^2$

For Q $\qquad y_Q = 15\sqrt{3}t \times \frac{\sqrt{3}}{2} - \frac{1}{2}gt^2 = \frac{45}{2}t - \frac{1}{2}gt^2$

i.e. $\qquad$ at any time t, $\quad y_P = y_Q$

$\therefore \qquad$ P and Q are always at the same height.

(b) Using $x = Vt \cos\theta$ gives:

For P $\qquad x_P = 45t \times \frac{\sqrt{3}}{2}$

For Q $\qquad x_Q = 15\sqrt{3}t \times \frac{1}{2} = 15t \times \frac{\sqrt{3}}{2}$

P and Q will collide when $\qquad x_P + x_Q = AB$

i.e. when $\qquad\qquad\qquad 60t \times \frac{\sqrt{3}}{2} = 60\sqrt{3}$

$\Rightarrow \qquad\qquad\qquad\qquad\qquad t = 2$

When $t = 2 \qquad y = \frac{45}{2}t - \frac{1}{2}gt^2 \qquad \Rightarrow \qquad y = 45 - 2g$

P and Q collide $(45 - 2g)$ m above the ground, when $t = 2$.

2. A particle is projected from a point O on a plane inclined at 30° to the horizontal. The velocity of projection is 20 m s^{-1} at 30° to an upward line of greatest slope. If the particle hits the plane at the point A, find, taking g as 10,

(a) the horizontal distance travelled by the particle

(b) the range up the plane.

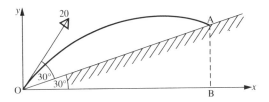

(a) The point A is on the path of the projectile so its coordinates satisfy the equation of the path. A is also on the plane so its coordinates satisfy the equation of the line OA.

The equation of the path of the projectile is

$$y = x \tan 60° - \frac{gx^2}{2(20)^2 \cos^2 60°} \qquad [1]$$

The equation of the line OA is

$$y = x \tan 30° \qquad [2]$$

The coordinates of A can be found by solving these two equations simultaneously.

From [1] and [2]

$$x \tan 30° = x \tan 60° - \frac{10x^2}{800 \cos^2 60°}$$

$$\Rightarrow \qquad \frac{10x}{800 \times 0.25} = 1.7321 - 0.5774$$

$$\Rightarrow \qquad x = 1.1547 \times 20 = 23.094$$

This value of x is also the x coordinate of B.

The particle travels a horizontal distance of 23 m (corrected to 2 sf).

(b)

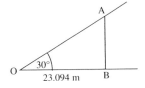

The range up the plane is the distance OA

$$OA = \frac{23.094}{\cos 30°} \text{ m} = 26.6\ldots \text{ m}$$

The range up the plane is 27 m (2 sf).

EXERCISE 11c

The problems in this exercise are varied in type and are generally a little harder than those in the previous exercises. If g is given a numerical value, use 10 and correct answers to 2 significant figures.

1. A particle P is projected from a point O with velocity $12\mathbf{i} + 16\mathbf{j}$. Two seconds later, another particle Q is projected from O and collides with P after another second. Find the initial velocity of Q.

2. For environmental reasons, golfers playing on the village golf course are required to restrict the height of golf balls in flight to 15 m. If a player tees off with an initial speed of 35 m s^{-1} at an angle of projection of 40°, for how long does he contravene the regulations?

3. A particle is projected from a point A on a horizontal plane, with velocity 60 m s^{-1} at 30° to the horizontal. At the same instant a second particle is projected, in the opposite direction with speed 50 m s^{-1}, from a point B on the same plane.

    (a) Given that the particles collide, find the angle of projection of the second particle.

    (b) If the time interval from projection to collision is 1.1 seconds find, to 2 significant figures, the distance between A and B.

4. Two particles A and B are projected at the same instant from the same point O on a horizontal plane. The initial velocity of A is V at 30° to the plane and that of B is $V\sqrt{3}$ at 60° to the plane.

    (a) Show that, as long as both particles are in the air, one of them is vertically above the other.

    (b) Find, in terms of V and g, the distance between the two points where the particles return to the plane.

5. A missile P is projected from a point O with speed 21 m s^{-1} at an angle α to the horizontal. One second later a missile Q is projected from a point 0.3 m below O with initial velocity 31.5 m s^{-1} at an angle β to the horizontal. Given that $\tan\alpha = \frac{4}{3}$ and $\tan\beta = \frac{3}{4}$

    (a) prove that the particles collide

    (b) find the time of the collision

    (c) find the direction in which each missile is moving just before the collision.

6. A particle is projected up a line of greatest slope of a plane inclined at 30° to the horizontal. The initial velocity is 15 m s^{-1} at 30° *to the plane*. Find

    (a) the range up the plane   (b) the time of flight.

7. A particle is projected *down* a line of greatest slope of a plane inclined at 30°
to the horizontal. The initial velocity is 15 m s^{-1} at 60° *to the plane*. Find

   (a) the range down the plane   (b) the time of flight.

8. A particle P is projected up a line of greatest slope of a plane inclined at 30°
to the horizontal. The initial velocity of P is inclined at 15° to the plane. If P
strikes the plane after 3 seconds, find the speed of projection.

*9. The fairway between the tee and the green on the 18th hole at the golf course has
an upward slope of 10%. The maximum speed at which Vic Alder can hit a golf
ball is 50 m s^{-1} and he wants to strike the ball so that it lands as far up the
fairway as possible. Find the angle of projection he should choose. Find also
how far up the fairway the ball will land if this angle is used.

CHAPTER 12

RELATIVE VECTORS

RELATIVE VELOCITY

Most of the time we judge the position or motion of an object with reference to the earth's surface, which is fixed.

Sometimes, however, we 'see' motion that is not relative to the earth. Consider, for example, an observer A sitting in a moving railway carriage, who looks out of the window at B, who is a passenger in another train moving at the same speed on a parallel track. To A, B *appears* to be stationary. Relative to the earth B is moving but, *relative to* A, B is stationary.

Now if A's train is travelling at 70 mph and the speed of B's train is 80 mph, B passes A at 10 mph. The velocities of A and B relative to the earth (often called their true velocities) are 70 mph and 80 mph but, relative to A, the velocity of B is (80−70) mph.

What is happening in both cases is that the observer, in this case A, is disregarding his own velocity and seeing only the *difference* between B's velocity and his own.

Now consider the situation when A gets up from his seat and walks across the carriage to the window opposite, i.e. *relative to the carriage* he has moved at right angles to the railway track. However, as A crossed the carriage, the carriage itself moved forward along the track; so A's true velocity is at an angle to the track as shown in this diagram,

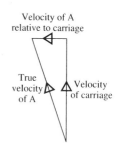

Velocity of A relative to carriage

True velocity of A

Velocity of carriage

i.e.   velocity of A relative to the carriage

= true velocity of A − velocity of carriage

In general if the true velocities of two moving objects, P and Q, (i.e. their velocities relative to the earth) are denoted by v_P and v_Q and the velocity of Q relative to P is denoted by $_Q v_P$ then

$$_Q v_P \ = \ v_Q - v_P$$

Examples 12a

The units used throughout are metres and seconds.

1. **A passenger on a ship, whose velocity is $14i - 22j$, is watching a boat whose velocity is $8i + 5j$. What does the velocity of the boat appear to be to the passenger on the ship? Illustrate your solution with a sketch.**

The velocity of the ship, v_S, is $14i - 22j$

The velocity of the boat, v_B, is $8i + 5j$

The velocity of the boat relative to the ship is $_B v_S$ where

$$_B v_S = v_B - v_S$$
$$= 8i + 5j - (14i - 22j)$$
$$= -6i + 27j$$

The boat appears to have a velocity of $-6i + 27j$

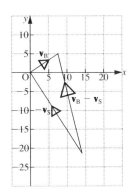

2. **The velocity of a particle P is $-2i + 3j$. Relative to P, another particle Q has a velocity of $6i + 9j$. Find the true velocity of Q and hence find Q's speed.**

$$v_P = -2i + 3j$$

$$_Q v_P = 6i + 9j$$

$$v_Q - v_P = {_Q v_P}$$

$$\therefore \quad v_Q = {_Q v_P} + v_P = 6i + 9j + (-2i + 3j)$$

$$= 4i + 12j$$

Q's velocity is $4i + 12j$

Q's speed is $|4i + 12j|$, i.e. $4\sqrt{10}$ m s^{-1}.

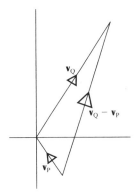

If the velocities of two moving objects are not given in vector form, the relative velocity can be found by drawing and measurement or by trigonometry.

3. **A helicopter is flying at 120 mph on a bearing of 052° above a car travelling east on a straight road at 70 mph. What does the speed and direction of the helicopter appear to be to the driver of the car?**

Taking $\mathbf{v}_h$ and $\mathbf{v}_c$ as the respective velocities of the helicopter and car and $_h\mathbf{v}_c$ as the velocity of the helicopter relative to the car we have

$$_h\mathbf{v}_c = \mathbf{v}_h - \mathbf{v}_c$$

If we draw a line PQ to represent $\mathbf{v}_h$, followed by a line QR representing $-\mathbf{v}_c$ (i.e. $\mathbf{v}_c$ reversed), the line PR that completes the triangle represents $_h\mathbf{v}_c$.

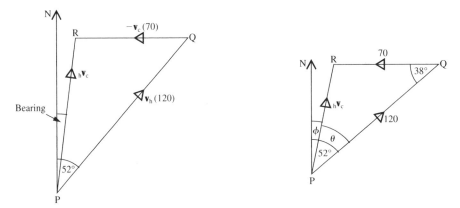

If triangle PQR is drawn to scale, PR can be measured to give the relative speed, while the angle measured clockwise from the north line gives the direction of the relative velocity.

Alternatively, using the cosine rule in $\triangle PQR$ gives

$$PR^2 = 120^2 + 70^2 - 2 \times 120 \times 70 \cos 38° \quad \Rightarrow \quad PR = 77.9..$$

Then the sine rule gives θ

$$\frac{\sin \theta}{70} = \frac{\sin 38°}{77.9} \quad \Rightarrow \quad \theta = 33.6..°$$

Hence $\phi = 18.4°$

So the relative velocity is of magnitude 78 mph (2 sf)
                              on a bearing of 018° (nearest degree).

4. To a cyclist riding due south at $11\,\mathrm{km\,h^{-1}}$, the wind appears to be blowing from the south east at $4\,\mathrm{km\,h^{-1}}$. Find the true wind velocity.

Using $\mathbf{v}_w$ and $\mathbf{v}_c$ for the velocities of the wind and the cyclist, then

$$_w\mathbf{v}_c = \mathbf{v}_w - \mathbf{v}_c$$

where $_w\mathbf{v}_c$ represents the velocity of the wind relative to the cyclist. As it is $\mathbf{v}_w$ that we want to find, we rewrite this equation in the form

$$\mathbf{v}_w = {}_w\mathbf{v}_c + \mathbf{v}_c \quad \text{where} \quad {}_w\mathbf{v}_c = 4\,\mathrm{km\,h^{-1}} \quad \text{and} \quad \mathbf{v}_c = 11\,\mathrm{km\,h^{-1}}$$

Then in a triangle formed by a line representing $_w\mathbf{v}_c$ followed by a line representing $\mathbf{v}_c$, the line that completes the triangle represents $\mathbf{v}_w$.

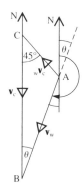

Using the cosine rule in $\triangle ABC$ gives

$$AB = \sqrt{(121 + 16 - 88\sin 45^\circ)} = 8.647\ldots$$

Then the sine rule gives

$$\frac{\sin\theta}{4} = \frac{\sin 45^\circ}{8.647\ldots} \quad \Rightarrow \quad \theta = 19.1\ldots^\circ$$

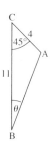

Hence the true wind velocity is $8.6\,\mathrm{km\,h^{-1}}$ (2 sf) on a bearing of 199° (nearest degree).

Alternatively a scale drawing of $\triangle ABC$ can be used.

EXERCISE 12a

In questions where units are not specified, any given vectors are measured in units based on metres, seconds, newtons and kilograms.

1. A car, A, is travelling at 50 mph on a straight road and another car, B, is being driven on the same road at 40 mph. Find the velocity of B relative to A if the cars are travelling

(a) in the same direction   (b) in opposite directions.

2. A horse breaks loose from his groom and gallops away in a straight line at 32 mph. The groom runs directly after the horse at 10 mph.

    (a) What is the velocity of the groom relative to the horse?

    (b) If the horse suddenly stops and begins to gallop at the same speed in the opposite direction, what now is the velocity of the groom relative to the horse?

3. A launch is travelling with velocity $21\mathbf{i} + 16\mathbf{j}$ and the velocity of a pleasure boat is $8\mathbf{i} - 3\mathbf{j}$. Find

    (a) the velocity of the launch relative to the pleasure boat

    (b) the velocity of the pleasure boat relative to the launch.

4. The velocities of a helicopter and a light aircraft are $15\mathbf{i} + 8\mathbf{j}$ and $-3\mathbf{i} + 9\mathbf{j}$. Relative to the light aircraft, what is

    (a) the velocity of the helicopter   (b) the speed of the helicopter?

5. To the pilot of a transport plane flying with velocity $14\mathbf{i} + 11\mathbf{j}$, the velocity of a liner appears to be $-5\mathbf{i} + 2\mathbf{j}$. What is the true velocity of the liner?

6. To a hiker walking due north at 8 km h^{-1} the wind appears to be blowing from the north west at 5 km h^{-1}.

    (a) Taking $\mathbf{i}$ and $\mathbf{j}$ as 1 km h^{-1} east and north respectively, express each of the given velocities in $\mathbf{ij}$ form.

    (b) Find the true wind velocity in the form $a\mathbf{i} + b\mathbf{j}$.

7. The velocity of a particle A is $4\mathbf{i} + 5\mathbf{j}$. Relative to A the velocity of another particle B is $-2\mathbf{i} + \mathbf{j}$ and relative to B the velocity of a third particle C is $3\mathbf{i} - 5\mathbf{j}$. Find

    (a) the true velocity of B

    (b) the true velocity of C

    (c) the velocity of C relative to A.

8. To an observer in a boat moving North East at 20 km h^{-1} an aeroplane appears to be flying due West at 100 km h^{-1}. What is the true course and speed of the aeroplane?

9. A passenger in a train travelling North East at 100 km h^{-1} watches a car moving on a straight road. The car seems to be travelling S 30° W at 125 km h^{-1}. What is the true velocity of the car?

*10. Two aircraft are flying at the same height on straight courses. The first is flying at 400 km h^{-1} due North. The true speed of the second is 350 km h^{-1} and it appears, to the pilot of the first aircraft, to be on a course S 40° W. Find the true course of the second aircraft.

RELATIVE POSITION

The position of a point A can be given, relative to 0, by its coordinates but we can equally well use the vector $\overrightarrow{OA}$. For example, if A is the point $(2, 3)$ then $\overrightarrow{OA}$ is $2\mathbf{i} + 3\mathbf{j}$ and is called the *position vector* of A relative to O.

The position vector of a point can also be given relative to a point other than the origin. Suppose that A and B are points whose position vectors relative to 0 are $20\mathbf{i} + 16\mathbf{j}$ and $15\mathbf{i} + 22\mathbf{j}$ respectively. From the diagram we see that the position (i.e. the displacement) of A *relative to B* is $5\mathbf{i} - 6\mathbf{j}$, i.e. $20\mathbf{i} + 16\mathbf{j} - (15\mathbf{i} + 22\mathbf{j})$.

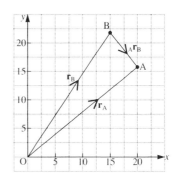

This shows that if we use notation similar to the velocity notation, i.e. $\mathbf{r}_A$, $\mathbf{r}_B$ and $_A\mathbf{r}_B$, we have

$$_A\mathbf{r}_B = \mathbf{r}_A - \mathbf{r}_B$$

EXERCISE 12b

In questions 1 to 4 find, in the form $a\mathbf{i} + b\mathbf{j}$, the displacement of

(a) A relative to B   (b) B relative to A.

1.

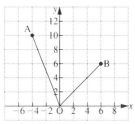

3.

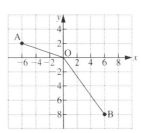

2.

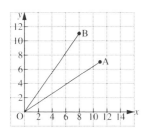

4.

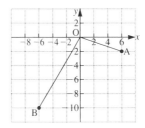

5. Two particles, P and Q, both start moving from the origin with constant velocities $6\mathbf{i} - \mathbf{j}$ and $3\mathbf{i} + 7\mathbf{j}$ respectively.

(a)   Find the position vector of each particle after t seconds.

(b)   What is the displacement of Q relative to P after 3 seconds?

6. A particle P starts from rest at the origin when $t = 0$, and moves along the positive x-axis with an acceleration after t seconds given by $\frac{1}{2}t\mathbf{i}$.

(a)   Find the velocity vector and the position vector of P after t seconds.

A second particle Q starts from the point $(3, 0)$ and moves with constant velocity $3\mathbf{i} + 4\mathbf{j}$.

(b)   Find the velocity vector and the displacement vector of Q after t seconds.

(c)   What is the velocity of Q relative to P when $t = 4$?

(d)   What is the displacement of Q relative to P when $t = 4$?

In the remaining questions, $\mathbf{i}$ and $\mathbf{j}$ are unit vectors east and north respectively. All quantities are measured in units based on kilometres and hours.

7. A helicopter A leaves a heliport and flies with velocity $10\mathbf{i} + 4\mathbf{j}$. At the same time another helicopter B takes off from a field whose position vector relative to the heliport is $36\mathbf{i} + 2\mathbf{j}$. The velocity of B is $-8\mathbf{i} + 3\mathbf{j}$.

(a)   Find, after t seconds,

(i)   the position vector of A relative to the heliport

(ii)   the position vector of B relative to the heliport

(iii) the displacement of B from A.

(b)   Explain what happens when $t = 2$.

8. At 2 p.m. the position vector, relative to a lighthouse, of a ship A is $10\mathbf{i}$ and A's velocity is $12\mathbf{i} + 5\mathbf{j}$. At the same time another ship B, whose velocity is $-3\mathbf{i} + 10\mathbf{j}$, is in a position $20\mathbf{i} - 4\mathbf{j}$ relative to the same lighthouse. Find, after t hours,

(a) the position vector of A relative to the lighthouse

(b) the position vector of B relative to the lighthouse

(c) the position vector of A relative to B

(d) the time when A is due north of B.

9. A particle P starts from the origin O with initial velocity $2\mathbf{i} - \mathbf{j}$ and moves with acceleration $6t\mathbf{j}$. Another particle Q starts from the point $\mathbf{i}$ with initial velocity $\mathbf{i} + \mathbf{j}$ and moves with acceleration $-4\mathbf{i}$. Find

(a) the velocity of P relative to Q at any time t

(b) the speed of P relative to Q when $t = 2$

(c) the displacement of P relative to Q at any time t

(d) the distance between P and Q when $t = 3$.

CLOSEST APPROACH AND INTERCEPTION

Closest Approach

Consider two moving objects A and B, with velocities $\mathbf{v}_A$ and $\mathbf{v}_B$, and suppose that at the time when observation begins, they are at points A_0 and B_0. Now if we consider all the motion relative to B, then B is apparently stationary at the point B_0 and A appears to move from the point A_0 in the direction of $_A\mathbf{v}_B$.

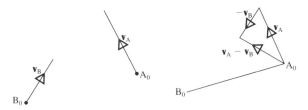

Once the direction of the relative velocity, $\mathbf{v}_A - \mathbf{v}_B$, is found another diagram can be drawn, to a *distance* scale, showing the line B_0A_0 and a line from B_0 in the direction of $\mathbf{v}_A - \mathbf{v}_B$.

The distance between A and B at any time is represented by the distance from A_0 to the corresponding point on the line of relative motion.
In particular, the *shortest distance* between A and B is given by measuring the *perpendicular* distance from A_0 to that line, using the chosen scale.

Examples 12c

1. **An aircraft P is 1200 m due north of another aircraft Q. When observed, both are flying at the same height with constant velocities of 150 m s^{-1} due west and 200 m s^{-1} on a bearing 330° respectively. Find the distance between the aircraft when they are closest together and the time when this occurs.**

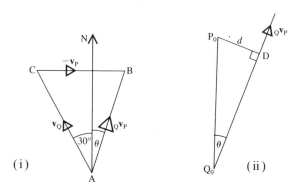

The first sketch shows the triangle giving the velocity of Q relative to P. The second sketch shows the initial distance apart, P_0Q_0, and the direction of motion of Q relative to P (i.e. the *direction* of $_Q\mathbf{v}_P$).

On a scale drawing the first sketch would be drawn to a speed scale, the angle θ measured and used in the second drawing.

From a scale drawing of (i), $\theta = 16°$.

The second sketch would be drawn to a distance scale; measuring d gives the shortest distance between the aircraft.

From a scale drawing of (ii), $d = 330$ m.

The time when P and Q are closest together is found by measuring the distance from Q_0 to D on the distance scale in drawing (ii) (i.e. the distance covered at the relative speed) and dividing it by the relative speed measured on the speed scale in drawing (i).

From scale drawing (ii), $Q_0D = 1150$ m and from (i), $_Q\mathbf{v}_P = 180$ m s^{-1}.

Therefore the time when the aircraft are closest is $\frac{1150}{180}$ seconds, i.e. 6.4 s, after the first observation.

The use of trigonometry gives a more accurate solution so it is always worth looking first to see whether the trigonometry seems to be straightforward. In the example above it requires several steps.

In $\triangle ABC$, the cosine rule gives AB

then the sine rule gives $A\widehat{B}C$ or $C\widehat{A}B$, either of which gives θ

then d is found from $\triangle Q_0 P_0 D$.

Interception

Suppose that one moving object plans to intercept (i.e. meet) another moving object. This means that it must move in a direction such that the shortest distance between them is zero.

Now we know that the shortest distance apart is measured from one end of the initial line joining the objects, to a point on a line drawn from the other end in the direction of their relative velocity. Therefore, for the shortest distance apart to be zero, the direction of the relative velocity must lie along the initial line.

2. **At 1200 hours a destroyer that is 12 nautical miles south west of a cruiser, sets off at 20 knots to intercept the cruiser which is travelling due east at 15 knots. (A knot is 1 nautical mile per hour.)**

    (a) On what bearing should the destroyer travel?

    (b) At what time will interception occur?

Diagram (i) shows the direction at noon of the line joining the destroyer D to the cruiser C. For interception, this direction must also be the direction of the velocity of D relative to C.

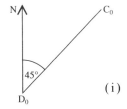

(i)

Diagram (ii) shows the triangle used to find $_D\mathbf{v}_C$ where $_D\mathbf{v}_C = \mathbf{v}_D - \mathbf{v}_C$

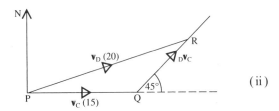

(ii)

We know the magnitude and direction of $\mathbf{v}_C$ and the direction of $_D\mathbf{v}_C$, but only the magnitude of $\mathbf{v}_D$.

So we rearrange the equation as $\mathbf{v}_C + {}_D\mathbf{v}_C = \mathbf{v}_D$ and interpret it by:

drawing a line PQ representing $\mathbf{v}_C$ followed by a line from Q, of unknown length, in the direction of $_D\mathbf{v}_C$. Then the third side PR has to be 20 units long.
The direction of PR gives the bearing of the destroyer.

Scale drawings of these diagrams would give a solution to the problem but this time we will use trigonometry.

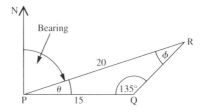

(a)   In $\triangle PQR$, the sine rule gives

$$\frac{\sin \phi}{15} = \frac{\sin 135°}{20} \quad \Rightarrow \quad \sin \phi = 0.5303..$$

$\Rightarrow \qquad \phi = 32.027..°$

$\Rightarrow \qquad \theta = 12.974°..$

$\therefore \qquad$ the destroyer should set a bearing of $77°$ (nearest degree)

(b)   Then using the sine rule again gives

$$\frac{QR}{\sin \theta} = \frac{20}{\sin 135°} \quad \Rightarrow \quad QR = 6.350..$$

The destroyer covers 12 nautical miles at a relative speed of 6.350 knots.

The time taken is therefore $\dfrac{12}{6.350}$ hours,   i.e. 1 hour 53 minutes.

So interception takes place at 1353 hours.

In some problems on interception the initial positions and the velocities are expressed in **ij** vector form. These tend to be less practical than the 'real-life' situations so far examined and the solutions are, in general, simpler.

3. **Two particles P and Q start simultaneously from points P_0 and Q_0, with position vectors $13i + 5j$ and $i + 7j$ respectively, relative to a fixed point O. The particles move with constant velocities represented respectively by $-5i - 2j$ and $i - 3j$. Given that the units are metres and seconds**

   (a) **show that P and Q collide,**

   (b) **find how long after the start the collision occurs,**

   (c) **find the position vector of the point of collision.**

(a) The particles will collide if the direction of the relative velocity is the same as the direction of $Q_0 P_0$.

$$v_Q - v_P = i - 3j - (-5i - 2j) = 6i - j$$

The direction of $Q_0 P_0$ is $\quad 13i + 5j - (i + 7j)$

i.e. $\qquad\qquad\qquad\qquad 12i - 2j \quad$ which is $\quad 2(6i - j)$

Therefore $Q_0 P_0$ is parallel to $v_Q - v_P$ and the particles will collide.

(b) The time of collision is given by $\quad \dfrac{\text{the distance PQ}}{\text{the relative speed}}$

i.e. $\quad t = \dfrac{|12i - 2j|}{|6i - j|} = 2$

Therefore the particles collide 2 s after the start.

(c) After 2 seconds, the position vector of P is given by

$$r_P = 13i + 5j + 2(-5i - 2j) = 3i + j$$

The position vector of Q could equally well be used, as $r_P = r_Q$ at impact.

The particles collide at the point with position vector $\quad 3i + j$

EXERCISE 12c

1. At noon an observer on a ship travelling due east at $20 \, \text{km h}^{-1}$ sees another ship $20 \, \text{km}$ due north which is travelling S $30°$ E at $8 \, \text{km h}^{-1}$. At what time are the ships nearest together?

2. Two aircraft P and Q are flying at the same height at $300 \, \text{km h}^{-1}$ SE and $350 \, \text{km h}^{-1}$ N $60°$ E respectively. If P is initially $10 \, \text{km}$ north of Q, how close do they get to one another?

3. A runaway horse is galloping across a field at $40 \, \text{km h}^{-1}$ on a bearing $020°$. It is already $300 \, \text{m}$ away in a direction due east, from a mounted rider who takes off in pursuit with a speed of $48 \, \text{km h}^{-1}$. In what direction should she ride to catch the runaway?

4. A yacht in distress is $8 \, \text{km}$ from a harbour on a bearing of $220°$ and is drifting S $10°$ E at $4 \, \text{km h}^{-1}$. On what bearing should a lifeboat travel to intercept the yacht if the speed of the lifeboat is $30 \, \text{km h}^{-1}$?

5. Two cyclists are riding one along each of two perpendicular roads which meet at A. At one instant each cyclist is 500 m from A and both are approaching A. If the speed of one cyclist is $8\,\text{m s}^{-1}$ and the shortest distance between the cyclists is 50 m, find the two possible speeds of the second rider.

6. Two aircraft A and B are flying at the same height in directions N 30° E and N 10° W respectively. At the instant when B is 10 km due east of A it is realised that they are on a collision course. If the speed of A is $500\,\text{km h}^{-1}$ find the speed of B.

7. A destroyer moving on a bearing of 030° at $50\,\text{km h}^{-1}$ observes at noon a cruiser travelling due north at $20\,\text{km h}^{-1}$. If the destroyer intercepts the cruiser one hour later find the distance and bearing of the cruiser from the destroyer at noon.

CONSOLIDATION C

SUMMARY

Elastic Strings and Springs

An elastic string is one that can be stretched and will return to its original length.

When it is stretched it is said to be taut.

Its unstretched length is its natural length and in this state it is said to be just taut.

An elastic string is stretched by two equal forces acting outwards, one at each end of the string, which are equal to the tension in the string.

These properties apply also to a stretched elastic spring but a spring, unlike a string, can be compressed. It is then said to be in thrust and exerts an outward push at each end, equal to the compressing force.

Hooke's Law states that the extension, x, in an elastic string or spring is proportional to the tension, T, in the string, i.e.

$$T = \frac{\lambda x}{a}$$

where a is the natural length of the string
and λ is its modulus of elasticity.

The unit for λ is the newton and the value of λ is equal to the force that doubles the length of the string.

A string that has reached its elastic limit no longer obeys Hooke's Law and will not return to its original length if stretched further.

Elastic Potential Energy

A stretched elastic string or a stretched or compressed elastic spring possesses elastic potential energy (EPE) equal to the amount of work needed to provide the extension or compression.

The amount of EPE in an elastic string stretched by an extension x from its natural length a is given by

$$\text{EPE} = \frac{\lambda x^2}{2a}$$

If a string is already stretched by an amount x_1 and is then further stretched until the extension is x_2, the work done in producing the extra extension is given by either of the following expressions:

$$\text{Work done} = (\text{average of initial and final tensions}) \times (x_2 - x_1)$$

$$= \frac{\lambda}{2a}(x_2{}^2 - x_1{}^2)$$

Projectiles

For a particle moving in a vertical xy plane with initial speed V at an angle θ to the horizontal, at any time t:

$$\dot{x} = V\cos\theta \qquad \dot{y} = V\sin\theta - gt$$

$$x = Vt\cos\theta \qquad y = Vt\sin\theta - \tfrac{1}{2}gt^2$$

The equation of the path is

$$y = x\tan\theta - \frac{gx^2}{2V^2\cos^2\theta} \quad \text{or} \quad y = x\tan\theta - \frac{gx^2}{2V^2}(1 + \tan^2\theta)$$

The greatest height occurs when $\dot{y} = 0$

The range on a horizontal plane is $\dfrac{V^2\sin 2\theta}{g}$ and is found by using $y = 0$

The maximum range is $\dfrac{V^2}{g}$ and is given when $\theta = 45°$

Relative Motion

If an object A is at a position vector $\mathbf{r}_A$ with velocity $\mathbf{v}_A$ and another object B is at a position vector $\mathbf{r}_B$ with velocity $\mathbf{v}_B$ then

the displacement of A relative to B is $\mathbf{r}_A - \mathbf{r}_B$
and the velocity of A relative to B is $\mathbf{v}_A - \mathbf{v}_B$

For interception $\mathbf{v}_A - \mathbf{v}_B$ must be parallel to $\mathbf{r}_A - \mathbf{r}_B$

MISCELLANEOUS EXERCISE C

In this exercise use $g = 9.8$ where necessary, unless another instruction is given.

In questions 1 and 2 a problem is set and is followed by a number of suggested responses. Choose the correct response.

1. If the force needed to compress a spring to half of its natural length is T, then the force needed to stretch it to twice its natural length is

   A $\frac{1}{4}T$ **B** $\frac{1}{2}T$ **C** T **D** $2T$

2. If the work done in stretching a spring to twice its natural length is E, then the work done in compressing it to half of its natural length is

   A $\frac{1}{4}E$ **B** E **C** $4E$ **D** $16E$

In question 3 a problem is set and is followed by a number of statements. Decide whether each statement is true (T) or false (F).

3. The diagram shows an elastic string of natural length a and modulus of elasticity $2mg$, fixed at one end to a point A and with a particle P, of mass m, attached to the other end. P is released from rest at a point C where AC $= 2a$.

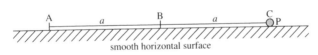

   smooth horizontal surface

   (i) At C the acceleration is $2g$.
   (ii) At C the elastic potential energy is $2mga$.
   (iii) When P reaches B the kinetic energy is zero.
   (iv) When P reaches B the acceleration is zero.
   (v) When P reaches A the velocity is zero.

4. A canal has long straight parallel banks that run north/south. The canal has width $25\,\text{m}$. There is a uniform current of speed $1.5\,\text{m s}^{-1}$ towards the south. A girl wishes to row from a point O on the west bank to a point P on the east bank. The point on the east bank directly opposite to O is E, and P is south of E with $\angle OPE = \theta$ as in the diagram. Her speed in still water is $1\,\text{m s}^{-1}$. Show that she can only row directly to P if $\sin \theta \leqslant \frac{2}{3}$.

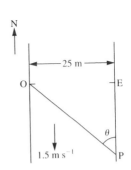

   Given that $\sin \theta = \frac{1}{3}$, find the direction in which she should point the boat in order to get directly to P, as quickly as possible.

   Find the time taken in this case.

   (UCLES)$_s$

5.

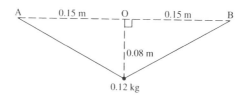

Two light elastic strings each have natural length 0.15 m and modulus 75 N. One end of each string is attached to a particle of mass 0.12 kg, and the other ends are attached to points A and B, 0.3 m apart at the same horizontal level. The particle is initially held at rest 0.08 m vertically below O, the mid-point of AB, as shown in the diagram.

(a) Find the tension in each of the elastic strings.

(b) Show that the energy stored in each of the strings is 0.1 J.

The particle is released from rest in this position, and subsequently passes through O with speed v metres per second.

(c) Use the principle of conservation of energy to calculate v, giving your answer to 2 significant figures.   (ULEAC)

6. In the sport of bungee jumping, a light elastic rope has one end attached to the participant and the other end attached to a fixed support on a high bridge. The participant then steps off the bridge and falls vertically. Given that a particular participant has mass 60 kg, and that the rope has natural length 30 m and modulus of elasticity 588 N, find, in metres to one decimal place, the distance of the participant below the point of support at the first instant of instantaneous rest. (You may assume that the participant does not reach the ground.)   (AEB)$_s$

7. A light elastic string of natural length 0.3 m has one end fixed to a point on a ceiling. To the other end of the string is attached a particle of mass M. When the particle is hanging in equilibrium, the length of the string is 0.4 m.

(a) Determine, in terms of M and g, the modulus of elasticity of the string.

A horizontal force is applied to the particle so that it is held in equilibrium with the string making an angle α with the downward vertical. The length of the string is now 0.45 m.

(b) Find α, to the nearest degree.   (ULEAC)

8. A light elastic string, of natural length 2 m and modulus 39.2 N, has one end attached to a fixed point A. A particle P, of mass 1 kg, is attached to the other end of the string.

(a) Show that, when P hangs in equilibrium, the length of AP is $2\frac{1}{2}$ m.

The particle P is released from rest at A and falls vertically.
Use the work–energy principle to calculate

(b) the speed, to 3 significant figures, of P at a distance $2\frac{1}{2}$ m below A,

(c) the greatest length of the string.   (ULEAC)

9. A particle is projected from a point on horizontal ground. The initial horizontal and vertical components of velocity are 14 m s^{-1} and 21 m s^{-1} respectively. Find

   (a) the times when the particle is moving at $45°$ to the horizontal

   (b) the horizontal and vertical distances from the point of projection to the particle at *each* of these times.   (WJEC)

10. A stone is thrown, at an angle of $30°$ above the horizontal, from the edge of a vertical cliff at a height of 35 m above sea level. If the initial speed of the stone is 7 m s^{-1}, find the time taken for the stone to hit the sea. Find also the horizontal distance from the bottom of the cliff at which the stone enters the sea.   (AEB)

11. This question is followed by several suggested responses. Choose which is the correct response.

   A projectile is projected from a point O on level ground with initial velocity u at $45°$ to the horizontal. When it is about to hit the ground

   A $y = 0$ **B** $x = 0$ **C** it is travelling vertically downwards.

12.

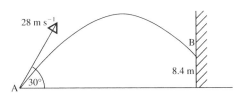

   A golf ball is driven from a point A with a velocity which is of magnitude 28 m s^{-1} and at an angle of elevation of $30°$. The ball moves freely under gravity. On its downward flight, the ball hits a vertical wall, at a point B which is 8.4 m above the level of A, as shown in the diagram. Calculate

   (a) the greatest height achieved by the ball above the level of A,

   (b) the time taken by the ball to reach B from A.

   By using the principle of conservation of energy, or otherwise,

   (c) find the speed, in m s⁻¹ to 1 decimal place, with which the ball strikes the wall.   (ULEAC)

13. The unit vectors **i** and **j** are horizontal and vertically upwards respectively. A particle is projected with velocity $(8\mathbf{i} + 10\mathbf{j}) \text{ m s}^{-1}$ from a point O at the top of a cliff and moves freely under gravity. Six seconds after projection the particle strikes the sea at the point S. Calculate

   (a) the horizontal distance between O and S,

   (b) the vertical distance between O and S to the nearest metre,

   (c) the speed with which the particle strikes the sea, giving your answer in m s⁻¹ to 1 decimal place.   (ULEAC)

14. A yacht is sailing due east at a speed of $2.5\,\mathrm{m\,s^{-1}}$ and a motorboat is moving due
south at a speed of $7\,\mathrm{m\,s^{-1}}$. Find the direction of the velocity of the yacht
relative to the motorboat.   (UCLES)ₛ

15. A mountain rescue team is investigating
whether or not to use a new type of
flexible rope. They take a length of 20 m
of the rope. One end is attached to the top S
of a fixed crane and a harness is attached
to the other end. Kirsty, a member of
the team, whose mass together with that of
the harness is 60 kg, is lowered gently until
she hangs at rest. The stretched rope is
then 21 m long. By modelling the rope as
a light elastic string and Kirsty as a particle,

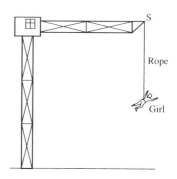

(a) estimate the modulus of elasticity of the rope.

Kirsty climbs back to S and releases herself to fall vertically, strapped in the
harness attached to the rope. She comes to instantaneous rest at the point C at
the end of her descent.

Estimate

(b) the length SC of the stretched rope,

(c) the greatest speed that Kirsty achieves during her descent.

State any further assumptions you have made in modelling Kirsty's descent from
S to C.   (ULEAC)ₛ

16.

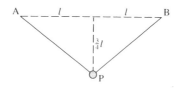

The diagram shows a particle P, of mass $6M$, suspended by two identical light
elastic strings from the points A and B which are fixed and at a horizontal
distance $2l$ apart. Each string has natural length l and P rests in equilibrium at a
vertical distance $\frac{3}{4}l$ below the level of AB. Determine

(a) the tension in either string,

(b) the modulus of elasticity of either string.   (ULEAC)

17. With respect to a fixed origin O, the unit vectors **i** and **j** are directed horizontally and vertically upwards respectively. At time $t = 0$ a ball is projected with velocity $(10\mathbf{i} + 20\mathbf{j})$ m s^{-1} from O and moves freely under gravity. The ball strikes a vertical post, which is 30 m horizontally away from O, at a point P above the horizontal plane through O. Calculate

(a) the time t, in seconds, when the ball strikes the post

(b) the height, in metres to 2 significant figures, of P above the plane

(c) the acute angle, in radians to 2 significant figures, which the velocity of the ball makes with the horizontal at the instant when it strikes the post.

(ULEAC)

18.

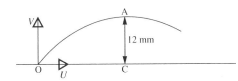

A missile P is projected from a point O on horizontal ground with a velocity whose components are U horizontally and V vertically upwards. At the point A on its flight path the missile is at its maximum height H above the ground. The time taken for the missile to travel from O to A is T. Express both H and T in terms of V and g.

When P has been travelling for a time $\frac{2}{3}T$, a second missile Q is projected vertically upwards with speed V_1 from the point C which is on the ground vertically below A.

If Q subsequently collides with P at A, show that $V_1 = \frac{5}{3}V$.

Given also that $OC = \dfrac{3U^2}{8g}$, find V in terms of U.

Show that, just before P and Q collide, the speed of P is twice that of Q.

(NEAB)

19.

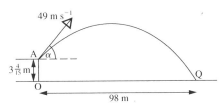

A golf ball is projected with speed 49 m s^{-1} at an angle of elevation α from a point A on the first floor of a golf driving range. Point A is at a height of $3\frac{4}{15}$ m above horizontal ground. The ball first strikes the ground at a point Q which is at a horizontal distance of 98 m from the point A as shown in the diagram.

(a) Show that $6\tan^2\alpha - 30\tan\alpha + 5 = 0$

(b) Hence find, to the nearest degree, the two possible angles of elevation.

(c) Find, to the nearest second, the smallest possible time of direct flight from A to Q.

(ULEAC)

20. The unit vectors **i** and **j** are directed due east and due north respectively. The airport B is due north of airport A. On a particular day the velocity of the wind is $(70\mathbf{i} + 25\mathbf{j})$ km h^{-1}. Relative to the air an aircraft flies with constant speed 250 km h^{-1}. When the aircraft flies directly from A to B determine

(a) its speed, in km h^{-1}, relative to the ground

(b) the direction, to the nearest degree, in which it must head.

After flying from A to B, the aircraft returns directly to A.

(c) Calculate the ratio of the time taken on the outward flight to the time taken on the return flight. (ULEAC)

21. A river with long straight banks is 500 m wide and flows with a constant speed of 3 m s^{-1}. A man rowing a boat at a steady speed of 5 m s^{-1}, relative to the river, sets off from a point A on one bank so as to arrive at the point B directly opposite A on the other bank. Find the time taken to cross the river.

A woman also sets off at A rowing at 5 m s^{-1} relative to the river and crosses in the shortest possible time. Find this time and the distance downstream of B of the point at which she lands. (AEB)

22. While practising her tennis serve, Jenny hits the ball from a height of 2.5 m with a velocity $(25\mathbf{i} - 0.5\mathbf{j})$ m s^{-1}. (**i, j** represent unit vectors in horizontal and vertical directions respectively.)

(a) Find the horizontal distance from the serving point at which the ball lands.

(b) Determine whether the ball would clear a net, which is 1 m high and 12 m from her serving position in the horizontal direction, **i**. (AEB)

23. The two forces $(4\mathbf{i} - 6\mathbf{j})$ N and $(6\mathbf{i} + 2\mathbf{j})$ N act on a particle P, of mass 2 kg.

(a) Show that the acceleration of P is $(5\mathbf{i} - 2\mathbf{j})$ m s^{-2}.

At time $t = 0$, P is at a point with position vector $(5\mathbf{i} + 3\mathbf{j})$ m relative to a fixed origin O and has velocity $(-7\mathbf{i})$ m s^{-1}. Calculate, at time $t = 2$ seconds,

(b) the position vector of P relative to O

(c) the velocity, in metres per second, of P

(d) the kinetic energy, in joules, of P. (ULEAC)

24. Sally can swim in still water at 1.1 m s^{-1}. She swims across a river flowing at 0.7 m s^{-1} between parallel banks 25 m apart. Find the time, in seconds to 1 decimal place, she takes to swim from a given point on one bank to the nearest point on the opposite bank. (ULEAC)

25. A cricket ball was thrown by one player and caught at the same height from which it was thrown by another player, 30 m away. The ball moved freely under gravity. The greatest height reached by the ball above the point from which it was thrown was 10 m.

(a) Show that the vertical component of the initial velocity of the ball was 14 m s^{-1}.

Calculate

(b) the time of flight, in seconds, of the ball,

(c) the speed, in m s^{-1}, with which the ball left the thrower's hand.   (ULEAC)

26. In the dangerous sport of bungee diving an individual attaches one end of an elastic rope to a fixed point on a river bridge. He/she is attached to the other end and jumps over the bridge so as to fall vertically downwards towards the water. The rope should be such that the diver comes to rest just above the surface of the water. In order to find out which particular ropes are suitable experiments are carried out with weights attached to the rope rather than people.
In one experiment it was found that when a weight of mass m was attached to a particular rope of natural length a and dropped from a bridge at a height of $3a$ above the water level then the weight just reached the level of the water.
Show that the modulus of elasticity of the rope is $3mg/2$.

State, with justification, whether the above result is an underestimate or an overestimate of the modulus of elasticity of the rope.

The weight of mass m is then removed and a weight of mass $5m/2$ is then attached to this rope and dropped from the same height so that the weight enters the water.

When the weight emerges from the water its speed has been reduced to zero by the resistance of the water. Show by using conservation of energy or otherwise and assuming that the rope does not slacken, that the subsequent speed v of the weight at height h above the water level is given by

$$v^2 = \frac{gh}{5a}(2a - 3h).$$ (WJEC)$_s$

27. A string of natural length $2a$ and modulus of elasticity λ has its ends fixed to two points, A and B, which are at the same horizontal level and at a distance $2a$ apart. The centre of the string is pulled back to a point C in the same horizontal plane as A and B such that ABC forms an equilateral triangle.

Find the tension in the stretched string and the energy stored in it.

A small mass m is placed inside the stretched string at the point C and the string released. The mass is catapulted through the mid-point of AB.

Neglecting the effects of gravity find the speed of the mass as it leaves the string.
    (MEI)$_p$

28.

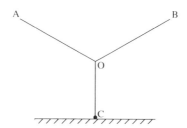

The diagram shows three identical elastic strings OA, OB and OC, each of natural length a and modulus λ. The strings are joined at O, and A and B are fixed to points in a horizontal line at a distance $2a$ apart. String OC is vertical and has a particle of mass m attached to the end C. The system is in equilibrium with the particle resting on a horizontal table. Given that OA and OB are each inclined at $60°$ to the vertical, show that

(a)   the magnitudes of the tensions in the two strings are equal;

(b) $OA = \dfrac{2a}{\sqrt{3}}$;

(c)   the magnitude of the tension in each string is equal to $\dfrac{\lambda(2 - \sqrt{3})}{\sqrt{3}}$.

Find in terms of m, g and λ, an expression for the magnitude of the reaction exerted on the particle by the table.

Deduce that $\lambda \leqslant \dfrac{mg\sqrt{3}}{2 - \sqrt{3}}$.   (AEB)

29. A steel ball B, of mass 0.125 kg, is attached to one end of a light elastic string OB, the end O being attached to the ceiling. The modulus of elasticity of the string is 52.5 N, and the string has natural length 1.5 m. In equilibrium the ball is at E. Show that the depth of E below O is 1.54 m, correct to 3 significant figures.

The ball is released from rest at O, and does not hit the floor. State an assumption necessary for conservation of energy to apply.

Hence find
  (i)   the speed as B passes through E,
  (ii)   the maximum depth of B below O.   (UCLES)ₛ

30. A light elastic string of natural length a and modulus $7mg$ has a particle P of mass m attached to one end. The other end of the string is fixed to the base of a vertical wall. The particle P lies on a rough horizontal surface, and is released from rest at a distance $\frac{4}{3}a$ from the wall. The coefficient of friction between P and the surface is $\frac{1}{4}$.

Use the work-energy principle

(a)   to show that P will hit the wall,

(b)   to find, in terms of a and g, the speed of P when it hits the wall. (ULEAC)

31. The buffer at the end of a railway siding is designed to stop trucks that run into it without any damage being done. The system in the buffer that absorbs the energy of the truck is modelled by an elastic spring which immediately begins to be compressed when the truck runs into the buffer. Once the spring has been compressed, it is prevented from returning to its natural length. Tests show that a truck of mass 10 tonnes moving at $0.5\,\mathrm{m\,s^{-1}}$ is stopped by the buffer when the spring has been compressed by $0.4\,\mathrm{m}$. The maximum compression allowed for in the design of the buffer is $1.2\,\mathrm{m}$. Calculate the maximum compression force that the buffer system is able to withstand. (UCLES)$_\mathrm{s}$

32. Two light springs are joined and stretched between two fixed points A and C which are $2\,\mathrm{m}$ apart as shown in the diagram. The spring AB has natural length $0.5\,\mathrm{m}$ and modulus of elasticity $10\,\mathrm{N}$. The spring BC has natural length $0.6\,\mathrm{m}$ and modulus of elasticity $6\,\mathrm{N}$. The system is in equilibrium.

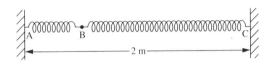

(i) Explain why the tensions in the two springs are the same.

(ii) Find the distance AB and the tension in each spring.

(iii) How much work must be done to stretch the springs from their natural length to connect them as described above?

A small object of mass $0.012\,\mathrm{kg}$ is attached at B and is supported on a smooth horizontal table. A, B and C lie in a straight horizontal line and the mass is released from rest at the mid-point of AC.

(iv) What is the speed of the mass when it passes through the equilibrium position of the system? (MEI)

33. In a charity event, a man is attached to the end of a light elastic rope, the other end of which is secured to a platform on a viaduct. The platform is $120\,\mathrm{m}$ above the ground. The natural length of the rope is $80\,\mathrm{m}$ and its modulus of elasticity is $17\,640\,\mathrm{N}$.

The man drops from the platform and falls without encountering any obstructions. (Air resistance may be neglected.) A 'safe' jump is one in which the man comes instantaneously to rest at least $10\,\mathrm{m}$ above the ground.

Using the principle of conservation of energy and treating the man as a particle,

(a) find, to the nearest kg, the mass of the heaviest man who can make a 'safe' jump,

(b) calculate the speed of a $75\,\mathrm{kg}$ man at a height of $20\,\mathrm{m}$ above the ground. (ULEAC)

CHAPTER 13

IMPULSE AND MOMENTUM

MOMENTUM

The momentum of a body is the product of its mass and its velocity,

i.e. **for a body of mass m, moving with velocity v,**
momentum $= mv$

Because momentum is a scalar multiple of velocity, which is a vector, it follows that momentum also is a vector quantity.

When the velocity of a body is constant and its mass does not change, its momentum is constant.

We know that a force is needed to change the velocity of an object and it follows that a force must act on the object in order to change its momentum. The precise relationship between a force and the change in momentum that it produces can be found by combining Newton's Second Law with the equations of motion with constant acceleration.

Consider a constant force F that acts for a time t on a body of mass m in the direction of its motion, causing the velocity to increase from u to v. As the force is constant, the acceleration a that it produces, is also constant.

Using $F = ma$ and $v = u + at$ gives

$$v = u + t\left(\frac{F}{m}\right)$$

$\Rightarrow \qquad Ft = mv - mu$

So we see that the change in momentum, i.e. final momentum minus initial momentum, is given by the product of the force and the time for which it acts.

IMPULSE

The product of the force and the time for which it acts is called the *impulse* of the force and is denoted by the symbol J,

i.e. $$J = Ft$$

Therefore   **impulse = change in momentum**

i.e. $$J = mv - mu$$

This relationship shows that impulse, too, is a vector quantity. Hence, if a force exerts an impulse on an object in a direction opposite to that of motion, the impulse is negative. It follows that the change in momentum is negative, i.e. the final momentum is less than the initial momentum.

As with all problems involving vector quantities, it is important when dealing with impulse and momentum to define the chosen positive direction.

Although readers will often find that I is used as the symbol for impulse, we prefer to use J. Our reason is that at a further stage in the study of mechanics (beyond the scope of this book) a completely different quantity is represented by I.

Units

The unit of impulse is, as might be expected, the product of a force unit and a time unit so, for a force in newtons acting for a time in seconds, the unit of impulse is the newton second, N s.

(Note that this is *not* newton *per* second.)

Momentum (mass × velocity) can be measured in kilogram metres per second, ($kg\,ms^{-1}$) but usually the impulse unit, N s, is used instead.

Examples 13a

1. **A hammer of mass 0.8 kg is moving at 12 ms^{-1} when it strikes a nail and is brought to rest. What is the magnitude of the impulse exerted on the hammer?**

   All the initial momentum of the hammer is lost when it hits the nail.

   The change in the momentum of the hammer is 0.8×12 N s $= 9.6$ N s

   $\therefore$ the impulse exerted on the hammer is 9.6 N s.

2. A particle of mass 2 kg is moving in a straight line, with a speed of 5 m s^{-1}. A force of 11 N acts on the particle for 6 seconds, in the direction of motion. Find

   (a) the magnitude of the impulse exerted on the particle

   (b) the speed of the particle at the end of this time.

2 kg 5 m s^{-1} 11 N v m s^{-1} + ve

Take the positive direction as being to the right.

(a) The impulse of the force is J N s where

$$J = Ft$$
$$= 11 \times 6$$

The magnitude of the impulse is 66 N s.

(b) Initial momentum is 2×5 N s

   Final momentum is $2v$ N s

   Using $J = mv - mu$ gives

$$66 = 2v - 10 \quad \Rightarrow \quad v = 38$$

The velocity after 6 seconds is 38 m s^{-1}.

3. The velocity of a particle of mass 7 kg, travelling along the x-axis, changes from 13**i** m s^{-1} to 3**i** m s^{-1} in 5 seconds under the action of a constant force. Find the magnitude of the force and state its direction.

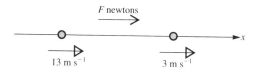

F newtons

13 m s^{-1} 3 m s^{-1}

Let the constant force be **F** N.

Initial momentum is 7×13**i** N s; Final momentum is 7×3**i** N s

The impulse of the force is 5**F** N s

Using $\mathbf{J} = m\mathbf{v} - m\mathbf{u}$ gives

$$5\mathbf{F} = 21\mathbf{i} - 91\mathbf{i} = -70\mathbf{i}$$
$$\Rightarrow \qquad \mathbf{F} = -70\mathbf{i} \div 5 = -14\mathbf{i}$$

The magnitude of the force is 14 N and it acts in the direction $-\mathbf{i}$.

4. A truck of mass **1200 kg** is travelling at **4 m s⁻¹** when it hits a buffer and is brought to rest in **3 seconds**. What is the average force exerted by the buffer?

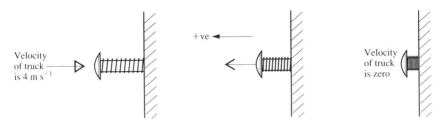

Taking the direction of the force as the positive direction, the initial velocity of the truck is -4 m s^{-1} and the final velocity is zero.

Using $\quad Ft = mv - mu \quad$ gives

$$F \times 3 = 1200 \times 0 - 1200 \times (-4)$$

$$\Rightarrow \qquad F = 1600$$

The average force exerted by the buffer is 1600 N.

5. A particle of mass **3 kg** is moving along a straight line in the direction $\overrightarrow{AB}$ with speed **6 m s⁻¹** when a force is applied to it. After **4 seconds** the particle is moving in the direction $\overrightarrow{BA}$ with speed **2 m s⁻¹**. Find the magnitude and direction of the force.

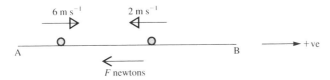

In the direction $\overrightarrow{AB}$

the initial momentum is 3×6 N s, the final momentum is $3 \times (-2)$ N s and the impulse of the force is $(-F) \times 4$ N s

Using $\quad Ft = mv - mu \quad$ gives

$$-4F = -6 - 18 \qquad \Rightarrow \qquad F = 6$$

The force is 6 N acting in the direction $\overrightarrow{BA}$.

Alternatively we could take $\mathbf{i}$ as one unit in the direction $\overrightarrow{AB}$ and use $\mathbf{F}$ as the force vector, leading to

$$4\mathbf{F} = 3 \times (-2\mathbf{i}) - 3 \times 6\mathbf{i} \qquad \Rightarrow \qquad 4\mathbf{F} = -24\mathbf{i}$$

i.e. $\qquad \mathbf{F} = -6\mathbf{i}$

6. A jet of water strikes a wall, at right angles to the jet, with a speed of 20 m s^{-1}. The water does not bounce off the wall. Given that the average force exerted by the wall in stopping the flow is 360 N, find the mass of water being delivered per second.

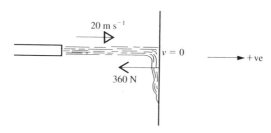

In one second the amount of water that hits the wall is m kilograms.
Therefore the initial momentum of the water is $20m$ N s.
The time taken for m kg of water to be brought to rest is 1 second.

Using
$$Ft = mv - mu \quad \text{gives}$$

$$-360 \times 1 = 0 - 20m$$

$$\Rightarrow \qquad m = -360 \div (-20) = 18$$

The mass of water delivered per second is 18 kg.

Note that the force which the wall exerts on the water is equal and opposite to the force exerted *by* the water on the wall.

EXERCISE 13a

1. Write down the momentum of

(a) a child of mass 40 kg running with a speed of 3 m s^{-1}

(b) a lorry of mass 1200 kg moving at 20 m s^{-1}

(c) a missile of mass 92 kg travelling at 120 m s^{-1}

(d) a train of mass 214 tonnes travelling at 55 m s^{-1}

(e) a bullet of mass 100 g travelling at 40 m s^{-1}.

2. Find the magnitude of the impulse exerted by

(a) a force of 14 N acting for 6 s

(b) a force of 12 tonnes acting for 1 minute

(c) a force that causes an increase in momentum of 88 N s

(d) the weight of a block of mass 20 kg acting for 30 s.

In questions 3 to 6 a force of magnitude F newtons acts in the direction AB on a particle P of mass 2 kg. Initially the velocity of P is u m s^{-1} and t seconds later it is v m s^{-1}, each in the direction AB.

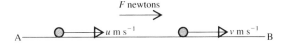

3. If $u = 4$, $v = 7$ and $t = 3$, find F.

4. If $u = 7$, $v = 4$ and $t = 3$, find F.

5. If $u = 5$, $t = 4$ and $F = 10$, find v.

6. If $u = 8$, $t = 5$ and $F = -4$, find v.

7. If $u = 10$, $v = 6$ and $F = -3$, find t.

8. In what time will a force of 12 N reduce the speed of a particle of mass 1.5 kg from 36 m s^{-1} to 12 m s^{-1}?

9. A body of mass 5 kg is moving with a velocity of $10\mathbf{i}$ m s^{-1} when a force $\mathbf{F}$ is applied to it for 4 seconds. Find the velocity at the end of this time if
(a) $\mathbf{F} = 20\mathbf{i}$ (b) $\mathbf{F} = -20\mathbf{i}$.

10. A body of mass 4 kg is moving with speed 7 m s^{-1} when a force is applied to it for 8 seconds. Its speed then is again 7 m s^{-1} but in the opposite direction. Find the magnitude of the force that has caused this change.

11. A dart of mass 40 g hits the dartboard at a speed of 16 m s^{-1}. If the dart comes to rest in the board in 0.02 seconds, find the average force exerted by the board on the dart.

12. A particle of mass 5 kg has a velocity $16\mathbf{i}$ m s^{-1} when a force $-4\mathbf{i}$ N begins to act on it. Find the velocity of the particle when the force has been acting for (a) $\frac{1}{3}$ s   (b) 5 s.
After what times will the *speed* of the particle be 2 m s^{-1}?

13. A high pressure hose is being used to clean the wall of a town hall. The hose delivers a horizontal stream of water which hits the wall at a speed of 20 m s^{-1}. Find the average force exerted on the wall, assuming that the water does not bounce back off the wall, if

(a) 8 kg of water is delivered per second

(b) the cross-sectional area of the hose pipe is 0.5 cm^2.
(Take the density of water as 1000 kg m^{-3})

14. A stationary truck is shunted into a siding by a locomotive that exerts a force of 2600 N on the truck for 12 seconds.

(a) What is the momentum of the truck at the end of this time?

The truck carries on without change of speed until it is brought to rest in 2 seconds when it hits the buffers at the end of the line.

(b) What is the magnitude of the impulse exerted on the truck by the buffers?

(c) What is the average force exerted on the truck by the buffers?

IMPULSIVE FORCES

There are circumstances where a large force acts for a very short time, so that neither the force nor the time for which it acts can easily be evaluated separately, e.g. a cricket bat hitting the ball, a shot being fired, a footballer kicking a ball, etc.

These are examples of *impulsive forces* and in such cases the impulse of the force cannot be calculated using $J = Ft$. The change in momentum caused by the impulse, however, can be used to evaluate the impulse.

Example 13b

A cricket ball of mass 0.2 kg has a speed of 20 m s⁻¹ when the bat strikes it at right angles and reverses the direction of the ball's flight. If the speed of the ball immediately after being struck is 36 m s⁻¹, find the impulse given by the bat to the ball.

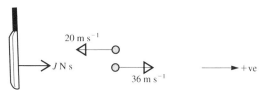

The final momentum of the ball is $0.2 \times 36 \text{ N s} = 7.2 \text{ N s}$
The initial momentum of the ball is $0.2 \times (-20) \text{ N s} = -4 \text{ N s}$
The impulse, J N s, given by the bat is given by

$$J = 7.2 - (-4) = 11.2$$

Therefore the bat exerts an impulse of 11.2 N s on the ball.

EXERCISE 13b

In each question calculate the impulse given.

1. A ball of mass 1.1 kg strikes a wall at right angles with a speed of 6 m s⁻¹ and bounces off at 5 m s⁻¹.

2. The speed of a tennis ball just before it hits the racquet is 38 m s^{-1}. The racquet strikes the ball at right angles, giving it a return speed of 30 m s^{-1}. The mass of the ball is 0.15 kg.

3. A shot of mass 50 g, fired at 250 m s^{-1}, is stopped dead when it hits a steel barrier.

4. A bird of mass 60 g is stunned when it flies at 12 m s^{-1} directly into a window pane.

5. A stone, of weight 24 N, dropped from a high window, hits the ground at 45 m s^{-1} and does not bounce. (Take g as 10.)

COLLISION

Whenever two objects are in contact, they exert *equal and opposite* forces on each other.

Whether they are in contact for a measurable time, or just for a split second, it is clear that each is in contact with the other for the *same time*. Therefore they exert *equal and opposite impulses* on each other.

Provided that neither object is fixed, these equal and opposite impulses produce equal and opposite changes in momentum, so the overall change in momentum of the two objects caused by the collision is zero. Hence, so long as no external force acts on either object, the total momentum of the two objects (which we refer to as *the system*) remains constant.

This property is known as The Principle of Conservation of Linear Momentum and is expressed formally as follows:

If in a specified direction, no external force affects the motion of a system, the total momentum in that direction remains constant.

In some problems involving a collision the two colliding objects bounce and so have individual velocities after impact. Other objects collide and join together at impact, e.g. trucks which become coupled. Such objects are said to *coalesce*.

In either case we are dealing with different velocities before and after impact so it is advisable to draw separate 'before' and 'after' diagrams and to define the chosen positive direction particularly carefully.

JERK IN A STRING

Consider two particles A and B connected by an inextensible string of length *l* and lying on a smooth table. The distance between A and B is less than *l*, so the string is not taut.

Now if A is projected away from B it will move with constant velocity until AB = *l*. At that instant the string jerks tight and suddenly exerts equal and opposite impulsive tensions, *J*, on A and B.

These impulsive tensions cause A and B to experience equal and opposite changes in momentum. Therefore, just as in the case of collision, the total momentum of the system is unchanged by the jerk, i.e. the principle of conservation of linear momentum can be applied.

Now the impulse that acts on B jerks B into motion, while the impulse that acts on A gives A a jerk backwards and, because the string is now taut, A and B begin to move on with equal speeds.

Examples 13c

1. **A particle A of mass 3 kg, travelling at 5 m s^{-1} collides head-on with a particle B with mass 2 kg and travelling at 4 m s^{-1}. If, after impact, B moves in the opposite direction at 2 m s^{-1}, find the velocity of A.**

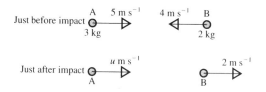

Total momentum before impact is $3 \times 5 + 2 \times (-4)$ N s
Total momentum after impact is $3u + 2 \times 2$ N s

Using conservation of linear momentum gives

$$15 - 8 = 3u + 4$$

$\Rightarrow \qquad u = 1$

The velocity of A is 1 m s^{-1} in the same direction as before.

Note that in this example the direction of motion of A after collision is not given. We guessed that it was to the right and got a positive value for u showing that the guess was correct. Had we thought that A moved with speed u to the left, we would have found that $u = -1$; this also shows that A moves to the right. In other words it does not matter in which direction we mark an unknown velocity; the sign of the answer defines the correct direction.

2. **A three-tonne truck is moving along a track at 8 m s^{-1} towards a five-tonne truck travelling at 5 m s^{-1} on the same track. If the trucks become coupled at impact find the velocity at which they continue to move if they are travelling**

(a) **in the same direction** (b) **in opposite directions.**

(a)

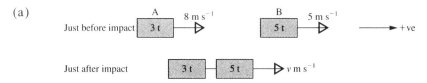

Before impact, the momentum of A is $3000 \times 8 \text{ N s} = 24\,000 \text{ N s}$
the momentum of B is $5000 \times 5 \text{ N s} = 25\,000 \text{ N s}$
the total momentum is $49\,000 \text{ N s}$

After impact, the combined momentum is $8000 \times v \text{ N s}$

Using conservation of linear momentum gives

$$8000v = 49\,000 \quad \Rightarrow \quad v = 6.125$$

The velocity of the coupled trucks is 6.1 m s^{-1} (2 sf).

(b)

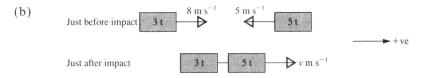

Before impact, the momentum of A is $3000 \times 8 \text{ N s} = 24\,000 \text{ N s}$
the momentum of B is $5000 \times (-5) \text{ N s} = -25\,000 \text{ N s}$
the total momentum is -1000 N s

After impact, the combined momentum is $8000 \times v \text{ N s}$

Using conservation of linear momentum gives

$$8000v = -1000 \quad \Rightarrow \quad v = -0.125$$

The velocity of the coupled trucks is 6.1 m s^{-1} (2 sf) in the direction of motion before impact of the heavier truck.

3. Two particles A and B, joined by a light inextensible string, are lying together on a smooth horizontal plane. The masses of A and B are 1 kg and 1.5 kg respectively. A is projected away from B with a speed of 5 m s^{-1}. Find the speed of each particle after the string jerks taut.

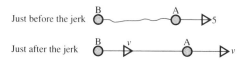

The momentum of the system before the jerk is 1×5 N s

When the string becomes taut, its ends have equal speeds, i.e. A and B have equal speeds.

The momentum of the system after the jerk is $(1 + 1.5) \times v$ N s

Conservation of linear momentum gives

$$5 = 2.5v \qquad \Rightarrow \qquad v = 2$$

Each particle has a speed of 2 m s^{-1} after the string jerks taut.

EXERCISE 13c

Keep your solutions to the questions in this exercise; you will need to refer to them in a later exercise.

In each question from 1 to 5 a body A of mass m_A travelling with velocity u_A, collides directly with a body B of mass m_B moving with velocity u_B. They coalesce at impact. The velocity with which the combined body moves on is v.

1. $m_A = 4$ kg, $u_A = 4$ m s^{-1}, $m_B = 2$ kg, $u_B = 1$ m s^{-1}. Find v.

2. $m_A = 6$ kg, $u_A = 1$ m s^{-1}, $m_B = 2$ kg, $u_B = -3$ m s^{-1}. Find v.

3. $m_A = 9$ kg, $u_A = 5$ m s^{-1}, $m_B = 4$ kg, $v = 3$ m s^{-1}. Find u_B.

4. $m_A = 3$ kg, $u_A = 16$ m s^{-1}, $m_B = 5$ kg, $v = 6$ m s^{-1}. Find u_B.

5. $u_A = 3$ m s^{-1}, $m_B = 6$ kg, $u_B = -5$ m s^{-1}, $v = -1$ m s^{-1}. Find m_A.

6. A bullet of mass 0.1 kg is fired horizontally, at 80 m s^{-1}, into a stationary block of wood that is free to move on a smooth horizontal plane. The wooden block, with the bullet embedded in it, moves off with speed 5 m s^{-1}. Find the mass of the block.

7. A particle A of mass 5 kg travelling with speed 6 m s^{-1}, collides directly with a stationary particle B of mass 10 kg. If A is brought to rest by the impact find the speed with which B begins to move.

8. The masses of two particles, P and Q, are respectively 0.18 kg and 0.1 kg. They are moving directly towards each other at speeds of 4 m s^{-1} and 12 m s^{-1} respectively. After they collide the direction of motion of each particle is reversed and the speed of Q is 6 m s^{-1}. Find P's speed after impact.

9. A sphere P of mass 2 kg is moving at 4 m s^{-1} when it collides with a sphere Q of mass 1 kg moving in the same direction at 3 m s^{-1}. After the impact, both P and Q move on in the same direction as before, P at $u \text{ m s}^{-1}$ and Q at $v \text{ m s}^{-1}$. Given that $7u = 2v$, find u and v.

10. A ball of mass 0.2 kg strikes a wall at right angles with a speed of 8 m s^{-1}. After rebounding it has 25% of its initial kinetic energy.

(a) Find its speed after rebounding.

(b) Find the impulse it exerts on the wall.

(c) If it is contact with the wall for 0.01 s, find the average force it exerts on the wall.

11. Two skaters, a father and son, are standing at rest on the ice. The masses of the father and son are 70 kg and 50 kg respectively. The father then slides a stone over the ice giving it a velocity of 8 m s^{-1}. The son catches it. If the mass of the stone is 5 kg find

(a) the speed with which the father starts to move backwards after releasing the stone,

(b) the common velocity of the son and the stone after he has caught it,

(c) the impulse exerted by the stone on the son.

State any assumptions you need to make to solve this problem.

In this diagram, A and B are particles resting on a smooth table and connected by a slack light inextensible string. Use the diagram for questions 13 and 14.

12. A is of mass 2 kg and B is of mass 4 kg. B is projected with a speed of 10 m s^{-1}. Find the common speed of A and B when the string jerks taut.

13. B is of mass 0.5 kg and is projected with speed 12 m s^{-1}. The jerk when the string becomes taut causes both particles to begin to move with a speed of 4 m s^{-1}. Find the mass of A.

14. Two particles A and B of equal mass m are connected by a light inextensible string of length l. Initially they are held at rest, side by side. A is then released from rest.

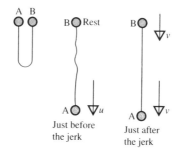

Just before the jerk

Just after the jerk

(a) Find, in terms of l and g, A's speed just as the string is about to jerk taut.

(b) If B is released at this instant find, in terms of l and g, the common speed with which A and B together begin to move.

LOSS IN KINETIC ENERGY

When there is a sudden change in the motion of a system, there is usually a change in the total kinetic energy of the system.

Consider the system given in question 1 in Exercise 13c.

Just before impact   4 kg ▷ 4 m s⁻¹   2 kg ▷ 1 m s⁻¹

Just after impact   6 kg ▷ 3 m s⁻¹

You should have found the common speed after impact to be 3 m s⁻¹

Before impact the KE of A is $\frac{1}{2}(4)(4^2)$ J and that of B is $\frac{1}{2}(2)(1^2)$ J. Therefore the total KE of the system is $(32+1)$ J, i.e. 33 J

After impact the KE of A is $\frac{1}{2}(4)(3^2)$ J and that of B is $\frac{1}{2}(2)(3^2)$ J. Therefore the total KE of the system is $(18+9)$ J, i.e. 27 J

From this we see that a loss in KE of 6 J has been caused by the collision. This can be understood by remembering that when objects collide there is usually a 'bang', i.e. some mechanical energy is converted into sound energy; some may also be converted into heat energy.

FINDING THE IMPULSE

We know that when objects collide, equal and opposite impulses act on the two objects. When we consider the system as a whole, these impulses cancel and do not appear in our calculations.

If the magnitude of the impulses is required we must consider *only one* of the colliding objects. The impulse which acts on one object causes the change in momentum of that object alone.

In the same way the impulsive tension in a jerking string can be found by considering its effect on the particle at one end only.

Looking again at question 1 in Exercise 13c:

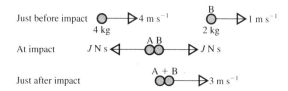

For the body A alone: the initial momentum is $4 \times 4 \, \mathrm{N\,s}$
the final momentum is $4 \times 3 \, \mathrm{N\,s}$

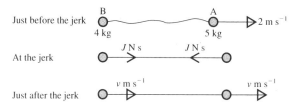

Using   Impulse $=$ Change in momentum   gives

$$-J = 12 - 16 \quad \Rightarrow \quad J = 4$$

The impulse that acts on A is $4 \, \mathrm{N\,s}$.
(The impulse that acts on B is also $4 \, \mathrm{N\,s}$ but in the opposite direction.)

To find the impulse in a jerking string (often called an impulsive tension) a very similar approach is used.
Suppose, for example, that a mass of 5 kg is projected at speed $2 \, \mathrm{m\,s^{-1}}$ away from a mass of 4 kg to which it is attached by a slack inelastic string.

Just before the jerk   B $\circ\!\!-\!\!-\!\!-\!\!-\!\!-\!\!-\!\!-\!\!-\!\!\circ\!\!\!\!\to 2 \, \mathrm{m\,s^{-1}}$
4 kg   5 kg

At the jerk $J \, \mathrm{N\,s}$ $J \, \mathrm{N\,s}$
$\circ\!\!\to$ $\leftarrow\!\!\circ$

Just after the jerk $v \, \mathrm{m\,s^{-1}}$ $v \, \mathrm{m\,s^{-1}}$
$\circ\!\!\to$ $\circ\!\!\to$

First we need the common speed, $v \, \mathrm{m\,s^{-1}}$, after the string jerks tight.

Using conservation of momentum, $5 \times 2 = (5+4)v \quad \Rightarrow \quad v = \frac{10}{9}$

Now for B alone we use   Impulse $=$ Change in momentum   giving

B
O At rest

O $\longrightarrow$ J N s

O $\longrightarrow \triangleright \frac{10}{9}$ m s^{-1}

$J =$ final momentum $-$ initial momentum

$= 4 \times (\frac{10}{9}) - 0$

$\therefore$ the impulse in the string is $4\frac{4}{9}$ N s.

EXERCISE 13d

1. For parts (a), (b), (c), (d), (e) and (f), use your solutions to questions 2 to 5, 7 and 8 in Exercise 13c, to find in each case
   (i) the loss in kinetic energy caused by the impact
   (ii) the impulse exerted on each object.

2. Using your solutions to questions 13 and 14 in Exercise 13c, find in each case the impulse in the string when it jerks tight.

3. A truck A of mass 400 kg, moving at 2 m s^{-1}, runs into a stationary truck B. The two trucks become coupled together and move on with speed 0.8 m s^{-1}. Find

   (a) the mass of truck B

   (b) the impulse exerted on truck B by truck A

   (c) the loss in kinetic energy caused by the collision.

4. The masses of two particles A and B are m and $3m$ respectively. They lie at rest on a smooth horizontal plane and are joined by a light inextensible string.

   (a) A is projected directly away from B with speed $4u$.

   (b) B is projected directly away from A with speed $4u$.

   In each case find

   (i)   the common speed of A and B after the string jerks tight

   (ii) the impulsive tension in the string

   (iii) the loss in kinetic energy caused by the jerk in the string.

5. An empty punt of mass 50 kg is drifting down a river at 2 m s^{-1}. It is approached by a second punt moving at 3 m s^{-1} and, when they are level, a man of mass 70 kg steps across sideways into the empty punt. Find the velocity that he and his new punt have just after he does this

   (a) if he approaches by overtaking the empty punt

   (b) if he approaches from the opposite direction.

6. A sphere of mass 3 kg is dropped on to a horizontal plane from a height of 2 m above the plane. Take g as 10.

(a) Find the speed of the sphere when it hits the plane.

(b) If the sphere does not bounce find the impulse it exerts on the plane.

If the sphere rebounds to a height of 1.4 m, find

(c) the speed at which the sphere rises off the plane

(d) the impulse exerted by the plane on the sphere.

7. A particle A of mass 2 kg lies on the edge of a table of height 1 m. It is connected by a light inelastic string of length 0.65 m to a second particle B of mass 3 kg which is lying on the table 0.25 m from the edge (AB is perpendicular to the edge). If A is pushed gently over the edge find the velocity with which B begins to move. Find also the impulsive tension in the string.

CHAPTER 14

COLLISIONS, LAW OF RESTITUTION

ELASTIC IMPACT

Most of the collisions that have been considered so far have resulted in the objects coalescing at impact. Impacts of this type are called *inelastic*.

If, on the other hand, a bounce occurs at collision, we have an *elastic impact* and the colliding object(s) are said to be *elastic*.

The simplest example of an elastic impact is a *direct* impact, i.e. an impact in which the direction of motion just before impact is parallel to the impulses that act at the instant of collision, e.g.

NEWTON'S LAW OF RESTITUTION

When two particles approach each other, collide directly and then move apart, the speed with which they separate is usually less than the speed at which they approached each other.

Experimental evidence suggests that, for two particular colliding particles, the separation speed is always the same fraction of the approach speed. It was from such evidence that Newton formulated a law known as Newton's *Law of Restitution*, which states that

separation speed = e × approach speed

or

relative speed after impact = e × relative speed before impact

The quantity represented by e is called the *coefficient of restitution* and it is constant for any two particular objects; its value depends upon the materials of which the two objects are made.

For colliding particles, e can take any value from zero to 1.

If the particles coalesce the separation speed is zero, i.e. $e = 0$.

If the relative speeds before and after impact are equal, $e = 1$, and we say that the particles are *perfectly elastic*; we shall see later on that in this case there is no loss in kinetic energy because of the impact.

Note that we have been referring only to particles colliding. A particle is regarded as having no measurable size, so its shape cannot be distorted and the time for which the particles are in contact is infinitesimal.

A larger object however, such as a ball, can undergo distortion on impact and the internal changes in its structure may be such that the basic methods for collision are not completely accurate. It is, however, reasonable to model most balls as elastic particles, as the results are accurate enough for most purposes.

Note also that when we refer to the speeds before and after impact, we mean the speeds *immediately* before and after. If, for instance, a particle falls vertically from a height above a fixed plane, its speed increases as it falls; the speed with which it *approaches the plane* is understood to be the final speed a split second before the collision with the plane.

Similarly the speed of separation from the plane is the initial speed with which the particle begins to rise again.

COLLISION WITH A FIXED OBJECT

Consider first the case of a particle of mass m, moving on a smooth horizontal surface with speed u, towards a fixed block whose face is perpendicular to the direction of motion of the particle. When the particle hits the block an impulse J is exerted on the particle by the block and, if the impact is elastic, the particle bounces off the block in the opposite direction with speed v, say.

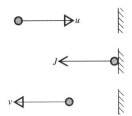

The approach speed is u, and the separation speed is v.

Therefore using the law of restitution gives $v = eu$

Now, taking the direction of J as positive,

using impulse = final momentum − initial momentum

gives $J = mv - (-mu)$

These are the two principles that can be applied to situations of this type in which one of the colliding objects is fixed. The conservation of linear momentum is not valid in such cases for the impulse applied to the particle by the fixed surface is an *external* impulse; hence the momentum of the particle is changed but the momentum of the fixed object is not changed by an equal and opposite amount.

Examples 14a

1. During a game of squash the ball strikes the front wall at right angles, at a speed of $20\,\mathrm{m\,s^{-1}}$. At the moment of impact the ball is travelling horizontally. The coefficient of restitution between the ball and the wall is 0.9. By modelling the squash ball as an elastic particle, find

   (a) the speed with which the ball bounces off the wall
   (b) the impulse exerted on the ball by the wall, given that the mass of the ball is $0.085\,\mathrm{kg}$.

   (a) Using the law of restitution, $v = 0.9 \times 20 = 18$

   The ball leaves the wall at $18\,\mathrm{m\,s^{-1}}$.

   (b) Using impulse $=$ increase in momentum in the direction of J,

   $$J = mv - m(-20) = 0.085(18 + 20)$$

   The impulse exerted is $3.23\,\mathrm{N\,s}$.

2. A ball of mass $0.15\,\mathrm{kg}$ is dropped from a height of $2.5\,\mathrm{m}$ above horizontal ground. After bouncing it rises to a height of $1.6\,\mathrm{m}$.
   Stating any assumptions you need to make, find

   (a) the coefficient of restitution between the ball and the ground
   (b) the ball's loss in mechanical energy caused by the impact with the ground and suggest a possible explanation for this loss.

   Assumptions are: the ball can be regarded as an elastic particle; there is no air resistance or wind; the ball rises vertically after impact with the ground.

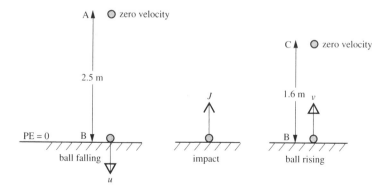

First find the speed of the ball at impact.

Using conservation of mechanical energy from A to B

$$mgh + 0 = 0 + \tfrac{1}{2} mu^2 \quad \Rightarrow \quad u^2 = 2 \times 9.8 \times 2.5 \quad \Rightarrow \quad u = 7$$

$\therefore$ the speed of the ball just before impact is $7\,\text{m s}^{-1}$.

Now find the speed of the ball immediately after impact.

Using conservation of mechanical energy from B to C

$$\tfrac{1}{2} mv^2 = mg(1.6) \quad \Rightarrow \quad v^2 = 31.36 \quad \Rightarrow \quad v = 5.6$$

$\therefore$ just after impact the speed of the ball is $5.6\,\text{m s}^{-1}$.

(a) Using the law of restitution,

$$5.6 = 7e \quad \Rightarrow \quad e = 0.8$$

(b) The loss in mechanical energy due to the impact is the difference in kinetic energy immediately before and after impact.

$$\text{Loss in mechanical energy} = \tfrac{1}{2}(0.15)(7^2) - \tfrac{1}{2}(0.15)(5.6)^2 \text{ joules}$$
$$= 1.32\,\text{J} \quad (3\text{ sf})$$

This loss may have been converted into sound energy at impact.

Kinetic energy can also be converted into heat at impact.

EXERCISE 14a

In questions 1 to 4 a particle moves directly towards a fixed plane surface, strikes it and rebounds. The coefficient of restitution is e.

1. If $e = \tfrac{1}{3}$, find

   (a) the speed after impact

   (b) the impulse exerted by the particle on the plane.

2. If $e = \tfrac{2}{3}$, find

   (a) the speed after impact

   (b) the loss of kinetic energy.

3. Given that $e = 0$, find

   (a) the speed after impact

   (b) the impulse exerted by the particle on the plane

   (c) the loss of kinetic energy.

4. Given that $e = 1$, find

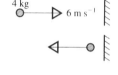

   (a) the speed after impact

   (b) the impulse exerted by the particle on the plane

   (c) the loss of kinetic energy.

5. A small ball, of mass 0.02 kg, is dropped from a height of 0.4 m onto horizontal ground. The impact causes the ball to lose 75% of its mechanical energy. Find

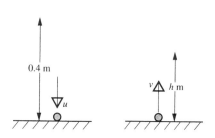

   (a) its speed just before impact

   (b) its speed just after impact

   (c) the coefficient of restitution

   (d) the height to which it rebounds.

6. A sphere of mass 2 kg falls from rest from a height of 10 m above an elastic horizontal plane. Find the height to which the sphere will rise again after its first bounce, if the coefficient of restitution is $\frac{1}{2}$. Find also the impulse exerted by the sphere on the plane.

In each question from 7 to 9 a particle, travelling horizontally with velocity **u**, strikes a vertical wall at right angles and rebounds with velocity **v**. The coefficient of restitution is e. Draw a diagram to illustrate each question before solving it.

7. (a) $\mathbf{u} = 8\mathbf{i}$, $e = 0.75$. Find **v**.   (b) $\mathbf{u} = -5\mathbf{j}$, $\mathbf{v} = 2\mathbf{j}$. Find e.

8. (a) $\mathbf{v} = 4\mathbf{i}$, $e = 0.2$. Find **u**.   (b) $\mathbf{u} = -6\mathbf{i} + 10\mathbf{j}$, $e = 0.5$. Find **v**.

9. (a) $\mathbf{u} = 6\mathbf{i}$, $\mathbf{v} = 0$. Find e.   (b) $\mathbf{u} = 6\mathbf{i} - 8\mathbf{j}$, $\mathbf{v} = -3\mathbf{i} + 4\mathbf{j}$. Find e.

*10. A small sphere is dropped onto a horizontal plane from a height of 20 m. The coefficient of restitution between the sphere and the plane is $\frac{1}{2}$. (Use $g = 10$.)

   (a) Find the height to which the particle rises after each of the first, second and third impacts.

   (b) Show that the heights found in (a) are in geometric progression.

   (c) Find the total distance travelled by the sphere up to the fourth impact.

*11. An ice-hockey player is skating up the rink at a constant speed of $4\,\mathrm{m\,s^{-1}}$. When he is 10 m from the end he strikes the puck, giving it a speed of $20\,\mathrm{m\,s^{-1}}$. The puck hits the end boards and rebounds directly towards him so that he receives it back one second after hitting it. Stating any simplifying assumptions which you make, construct and use a mathematical model to find the coefficient of restitution between the puck and the end boards.

2. A light inextensible string AB, of length a, has the end A fixed to a vertical wall. The end B is attached to a particle which is drawn away from the wall until the string makes an angle of 60° with the wall. The particle is then released from rest. If the value of e is $\frac{3}{4}$, find

  (a) the velocity of the particle just before impact with the wall

  (b) the velocity of the particle just after impact

  (c) the vertical distance through which the particle travels before it next comes to instantaneous rest.

3. Some manufacturers are making a table game in which a small disc is projected by a spring mechanism to slide over a horizontal surface, hit a vertical back-board and rebound into areas marked with scores. The mass of a disc is $0.1\,\text{kg}$. The spring mechanism allows the disc to hit the back-board with speeds between $0.5\,\text{m s}^{-1}$ and $3\,\text{m s}^{-1}$. The coefficient of friction between a disc and the surface is 0.2. Dimensions are shown on the diagram. Model the game by treating the disc as a particle and assume that it strikes the back-board at right-angles.

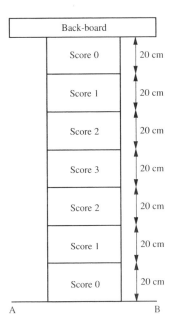

  (a) In choosing material for the back-board it is required that the coefficient of restitution, e, should be such that the disc cannot rebound past the line AB. Find the maximum possible value of e for this requirement to be satisfied.

  (b) Given that $e = 0.75$, find the range of speeds with which the disc can hit the back-board when the score is 3.

COLLISION OF TWO OBJECTS BOTH FREE TO MOVE

The principle of conservation of linear momentum can be applied to the direct impact of particles both of which are free to move, as this is a case where the pair of impulses at impact cause equal and opposite changes in momentum and so have no overall effect on the total momentum of the system.

For this situation therefore, in addition to the law of restitution, we can also use conservation of momentum.

Because we will now be dealing with more than one moving particle, it is particularly important to choose a positive direction when dealing with the momentum of the system.

Examples 14b

1. A particle A, of mass 0.1 kg, is moving with velocity 2 m s^{-1} directly towards another particle B, of mass 0.2 kg, which is at rest. Both particles are on a smooth horizontal table. If the coefficient of restitution between A and B is 0.8, find the velocity (i.e. the speed and direction of motion) of each particle after their collision.

We know that B moves in the +ve direction after impact, as B is at rest when an impulse acts on it in that direction. However, we are not sure which way A will move after impact. We have marked its velocity as +ve and the sign we get for u will determine whether or not this is the correct direction.

Using conservation of momentum $\rightarrow$

$$0.1 \times 2 = 0.1u + 0.2v$$

$\Rightarrow \qquad u + 2v = 2$ [1]

Using the law of restitution

$$v - u = 0.8 \times 2$$ [2]

$[1] + [2]$ gives $\qquad 3v = 3.6 \qquad \Rightarrow \qquad v = 1.2$

Then from [1] $\qquad u = 2 - 2.4 = -0.4$

The minus sign shows that A is, in fact, moving in the −ve direction after impact.

$\therefore$ after impact, the velocity of B is 1.2 m s^{-1} in the +ve direction

and $\qquad$ the velocity of A is 0.4 m s^{-1} in the −ve direction.

Note that it is easier to combine the two equations if equation [1] is arranged with u and v on the LHS.

At the moment of impact two equal and opposite impulses act, one on each of the particles. So if the magnitude of each impulse is required, we must consider the change in momentum of *only one* of the particles.

In the example above, for example, we would find the value of J by considering the motion of B (easier than A as B has no initial momentum),

i.e. $\xrightarrow{+} \qquad J = 0.2v - 0$

2. **A particle P of mass 1 kg, moving with speed $4\,m\,s^{-1}$, collides directly with another particle Q of mass 2 kg moving in the opposite direction with speed $2\,m\,s^{-1}$. The coefficient of restitution for these particles is 0.5. Find**

(a) **the velocity of each particle just after the collision**

(b) **the magnitude of the impulses acting on impact**

(c) **the loss in kinetic energy caused by the collision.**

(a)

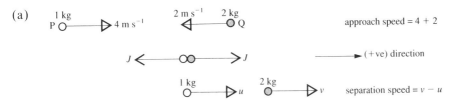

Using conservation of momentum $\rightarrow$

$$1 \times 4 - 2 \times 2 = 1 \times u + 2 \times v$$

$\Rightarrow \qquad u + 2v = 0$ \hfill [1]

Using the law of restitution

$$v - u = 0.5(4+2)$$ \hfill [2]

$[1] + [2]$ gives $\qquad 3v = 3 \qquad \Rightarrow \qquad v = 1 \quad$ and $\quad u = -2$

i.e. after impact P's speed is $2\,m\,s^{-1}$ and Q's speed is $1\,m\,s^{-1}$.

The direction of motion of both particles is reversed by the impact.

(b) Now consider the impulse that acts on Q and the change in momentum it produces.

Impulse $=$ final momentum $-$ initial momentum

$\Rightarrow \qquad J = 2 \times 1 - 2 \times (-2) = 6$

The magnitude of each impulse is $6\,N\,s$.

(c) KE before impact $= [\frac{1}{2} \times 1 \times (4)^2 + \frac{1}{2} \times 2 \times (2)^2]\,J = 12\,J$

KE after impact $= [\frac{1}{2} \times 1 \times (2)^2 + \frac{1}{2} \times 2 \times (1)^2]\,J = 3\,J$

$\therefore \quad$ the loss in KE due to the impact is $9\,J$

3. A sphere of mass $1.5\,\text{kg}$ and speed $2\,\text{m s}^{-1}$ collides head-on with an identical sphere of mass $0.5\,\text{kg}$ and speed $1\,\text{m s}^{-1}$. As a result of the impact the direction of motion of the lighter sphere is reversed and its speed becomes $3.5\,\text{m s}^{-1}$. Find

(a) the speed of the heavier sphere after impact

(b) the coefficient of restitution between the spheres

(c) the loss in kinetic energy due to the impact.

(d) What can you deduce from the answers to (b) and (c)?

(e) State an assumption that has been made.

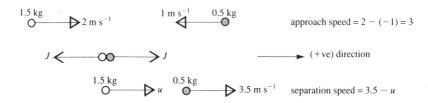

(a) Using conservation of momentum $\rightarrow$

$$1.5 \times 2 - 0.5 \times 1 = 1.5u + 0.5 \times 3.5 \qquad \Rightarrow \qquad u = 0.5$$

The speed of the heavier sphere is $0.5\,\text{m s}^{-1}$.

(b) Using the law of restitution, $3.5 - 0.5 = 3e$

$\Rightarrow \qquad\qquad e = 1$

(c) KE before impact $= \frac{1}{2}(1.5)(2)^2 + \frac{1}{2}(0.5)(1)^2 = 3.25$

KE after impact $= \frac{1}{2}(1.5)(0.5)^2 + \frac{1}{2}(0.5)(3.5)^2 = 3.25$

$\therefore$ there is no loss in energy because of the impact.

(d) From (b) we see that in this problem the spheres are perfectly elastic and, from (c), that no kinetic energy is lost. We deduce that, in general, a perfectly elastic impact causes no loss in kinetic energy. It follows that there is no 'bang' when a perfectly elastic collision occurs.

(e) It is assumed that the spheres can be modelled as particles and do not distort on collision. This requires that the spheres are identical in size. If they were not, the impulses might not act in the direction of motion and so would affect the momentum in a perpendicular direction also.

EXERCISE 14b

In each question from 1 to 4, the diagram shows the velocities of two particles, moving on a smooth horizontal surface, just before and just after they collide.

1. The coefficient of restitution is $\frac{1}{3}$.

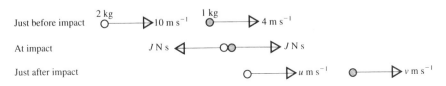

Find the values of u, v, and J.

2.

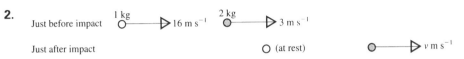

Find   (a)   the value of v (b)   the coefficient of restitution.

Find also the kinetic energy lost at impact.

3. The coefficient of restitution is $\frac{1}{2}$.

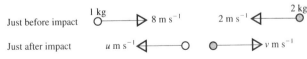

Find   (a) u (b) v (c)   the kinetic energy lost at impact.

4.

Just before impact   3 kg ○——▷1 m s⁻¹   5 m s⁻¹◁——○ 2 kg

Just after impact u m s⁻¹◁——○   ○——▷1 m s⁻¹

Find   (a)   the coefficient of restitution   (b)   the kinetic energy lost at impact.

5. A sphere A of mass 0.1 kg is moving with speed $5\,\mathrm{m\,s^{-1}}$ when it collides directly with a stationary sphere B. If A is brought to rest by the impact and $e = \frac{1}{2}$, find the mass of B, its speed just after impact and the magnitude of the instantaneous impulses.

6. When two spheres of equal mass collide directly at speeds of $4\,\mathrm{m\,s^{-1}}$ and $8\,\mathrm{m\,s^{-1}}$ in opposite directions, half the kinetic energy is lost upon impact.
Prove that $e = \frac{2}{3}$.

7. A car of mass 1000 kg is waiting at rest at traffic lights and its driver has neglected to put the handbrake on. Another car, of mass 800 kg, approaches from behind and, braking too late, is travelling at 25 km h^{-1} when it hits the stationary car, giving it an initial speed of 20 km h^{-1}. Stating any assumptions which you make, estimate the coefficient of restitution between the cars.

8. A charged particle collides directly with a nucleus which is initially at rest. The collision is perfectly elastic and the mass of the nucleus is 1840 times the mass of the particle. What percentage of the particle's initial kinetic energy is transferred to the nucleus?

9. Particles A and B both have mass m and are moving in the same direction along a line, A with speed $3u$ and B with speed u. They collide and after the impact they move in the same direction, A with speed u and B with speed ku. The coefficient of restitution is e.

Just before impact   A $\overset{m}{\bigcirc}\!\!-\!\!\longrightarrow 3u$ B $\overset{m}{\bigcirc}\!\!-\!\!\longrightarrow u$

Just after impact   A $\overset{m}{\bigcirc}\!\!-\!\!\longrightarrow u$ B $\overset{m}{\bigcirc}\!\!-\!\!\longrightarrow ku$

(a) Show that $e = \dfrac{k-1}{2}$.

(b) Deduce that $1 \leqslant k \leqslant 3$.

(c) Find the loss of kinetic energy in terms of m, k and u.

10. A boy's ball lands on a pond covered with thin ice. He attempts to push it to the other side by throwing stones at it. The mass of the ball is 0.4 kg and the stones he selects all have a mass near 0.1 kg. The coefficient of restitution between the stones and the ball is $\frac{1}{2}$ and he throws the stones so that they hit the ball with a horizontal speed of 10 m s^{-1}.
Find the speed of the ball after it has been struck by

(a) the first stone

(b) the second stone.

State any assumptions that have been made.

11. A spacecraft, with the final stage of its rocket still attached, is travelling at 16 000 m s^{-1} when the rocket stage is detached by exploding a charge between it and the spacecraft. The mass of the spacecraft is 4.5×10^4 kg and the mass of the rocket stage is 8×10^4 kg.
The explosion gives an impulse of 2.25×10^6 N s to each part.
Find the magnitudes and directions of the velocities of the two parts after separation.

2. A sphere of mass 0.2 kg, moving at $10 \, \text{m s}^{-1}$, collides directly with another sphere, of mass 0.5 kg moving in the same direction at $8 \, \text{m s}^{-1}$. Their speeds after the collision are u and v respectively and the coefficient of restitution is e.

(a) Find expressions for u and v in terms of e.

(b) Show that $\frac{50}{7} \leqslant u \leqslant \frac{60}{7}$ and find similar inequalities for v.

3. A particle A, of mass m moving with a speed u collides directly with a particle B, of mass km, which is initially at rest. The direction of motion of A is reversed by the impact. The coefficient of restitution is $\frac{1}{3}$.

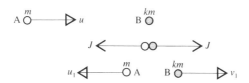

Giving answers in terms of m, k and u,

(a) find the speed of A after the impact.

(b) deduce that $k > 3$

(c) find the speed of B after the impact.

(d) find the impulse exerted by A on B.

4. A small sphere A, of mass m, moving with speed ku (where $k > 1$) collides directly with a small sphere B, of mass $3m$, which is moving with speed u in the same direction. The coefficient of restitution is $\frac{1}{2}$. Find in terms of k and u, an expression for the velocity of A after the impact and hence show that if the direction of motion of A is unchanged by the impact then $k < 9$.

5. Two particles P and Q, of masses m and $3m$ respectively, are connected by a light elastic string of natural length l and modulus of elasticity $3mg$. They are held at rest on a smooth horizontal plane, with the string stretched to a length $5l$, and released from rest.

(a) By considering momentum, show that if V_P and V_Q are their speeds towards each other at any time between being released and colliding, then $V_P = kV_Q$, where k is a constant, and state the value of k.

(b) By considering energy find, in terms of g and l, their speeds just before collision.

Given that they adhere to each other after the collision

(c) find their common velocity just after the impact

(d) find, in terms of m, g and l, the impulse each exerts on the other at impact.

MULTIPLE IMPACTS

Sometimes a collision between two objects leads to further collisions either with another moveable object or with a fixed surface. In such cases each individual impact can be dealt with by the methods already described. It is best to solve one collision completely before starting on the next one, and to begin again with a new set of diagrams. The positive direction can be chosen afresh for each collision – it need not be the same throughout.

Examples 14c

1. **An unfortunate snooker player miscues when he strikes the cue ball, giving it a speed of $8\,\mathrm{m\,s^{-1}}$, so that it hits the brown ball directly and sets it moving towards, and perpendicular to, the cushion. The direction of motion of the cue ball is unaltered by the impact. After hitting the cushion the brown ball collides head-on with the cue ball.**

   (a) Given that the coefficient of restitution between the balls is 0.9 and between the ball and the cushion is 0.8, find the speed of the cue ball after its second collision.

   (b) Several assumptions have to be made. Name any three of them.

   (a) **1st impact** (between the balls)

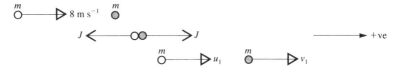

Using conservation of momentum $\rightarrow$

$$8m = mu_1 + mv_1 \qquad \Rightarrow \qquad 8 = u_1 + v_1$$

Using the law of restitution

$$0.9 \times 8 = v_1 - u_1 \qquad \Rightarrow \qquad 7.2 = v_1 - u_1$$

Adding gives $2v_1 = 15.2 \qquad \Rightarrow \qquad v_1 = 7.6$ and $u_1 = 0.4$

   2nd impact (with cushion)

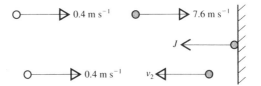

This is an external impact so momentum is not conserved.

Using the law of restitution,

$$0.8 \times 7.6 = v_2 \qquad \Rightarrow \qquad v_2 = 6.08$$

3rd impact (between the balls)

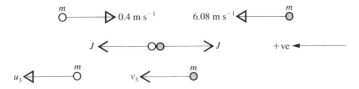

Using conservation of momentum ←

$$6.08m - 0.4m = mu_3 + mv_3 \qquad \Rightarrow \qquad 5.68 = u_3 + v_3$$

Using the law of restitution

$$0.9 \times 6.48 = u_3 - v_3 \qquad \Rightarrow \qquad 5.832 = u_3 - v_3$$

Adding gives $\quad 2u_3 = 11.512 \qquad \Rightarrow \qquad u_3 = 5.76 \quad (3\ \text{sf})$

The speed of the cue ball is $5.76\ \text{m s}^{-1}$ (3 sf).

(b) The cue ball is not given any spin; the table and the cushion are smooth; the balls are identical in size (so that the impulses are horizontal). Also, as the masses of the balls are not given, we must assume them to be equal.

Note that we used u for the speed of one ball and v for the other ball at every impact, then a suffix denotes the impact that has just taken place, e.g. u_1 is the speed of the cue ball after the first impact (u_2 does not appear as the cue ball was not involved in the second collision). This notation is particularly helpful when the number of impacts increases.

2. **A ball A of mass 0.3 kg moves on a smooth horizontal plane towards a ball B, of mass 0.2 kg which is at rest, and strikes it directly with speed $1\ \text{m s}^{-1}$. Ball B then moves with constant speed towards a third ball C, of mass 0.1 kg, which is lying at rest on B's line of motion. If the coefficient of restitution between A and B is 0.5 and that between B and C is 0.7, find the speed of each ball when no more collisions can take place between them.**

As several objects and several impacts are involved, we will denote the speeds of A, B and C by u, v and w, and use the suffixes 1, 2, 3... to denote the number of the collision that has just taken place. It is sometimes helpful to use u_0, etc. for the speeds before any impact has occurred.

1st collision (between A and B)

Conservation of momentum →

$$0.3 \times 1 = 0.3u_1 + 0.2v_1 \qquad \Rightarrow \qquad 3 = 3u_1 + 2v_1 \qquad\qquad [1]$$

Law of restitution

$$0.5 \times 1 = v_1 - u_1 \qquad \Rightarrow \qquad 0.5 = v_1 - u_1 \qquad\qquad [2]$$

$[1] + 3 \times [2]$ gives $4.5 = 5v_1 \qquad \Rightarrow \qquad v_1 = 0.9 \text{ and } u_1 = 0.4$

2nd collision (between B and C)

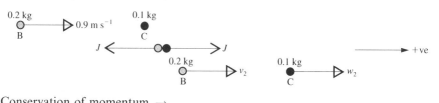

Conservation of momentum $\rightarrow$

$$0.2 \times 0.9 = 0.2v_2 + 0.1w_2 \qquad \Rightarrow \qquad 1.8 = 2v_2 + w_2 \qquad\qquad [3]$$

Law of restitution

$$0.7 \times 0.9 = w_2 - v_2 \qquad \Rightarrow \qquad 0.63 = w_2 - v_2 \qquad\qquad [4]$$

$[3] - [4]$ gives $1.17 = 3v_2 \qquad \Rightarrow \qquad v_2 = 0.39 \text{ and } w_2 = 1.02$

After the second impact the velocities of the balls are:

A $\longrightarrow$ 0.4 m s^{-1} B $\longrightarrow$ 0.39 m s^{-1} C $\longrightarrow$ 1.02 m s^{-1}

A's speed is greater than B's so A will catch up with B and collide again.

3rd collision (between A and B)

Conservation of momentum $\rightarrow$

$$0.3 \times 0.4 + 0.2 \times 0.39 = 0.3u_3 + 0.2v_3 \quad \Rightarrow \quad 0.198 = 0.3u_3 + 0.2v_3 \qquad [5]$$

Law of restitution

$$0.5 \times 0.01 = v_3 - u_3 \qquad\qquad [6]$$

$0.3 \times [6] + [5]$ gives $1.995 = 5v_3 \qquad \Rightarrow \qquad v_3 = 0.399 \text{ and } u_3 = 0.394$

The velocities now are:

A $\longrightarrow$ 0.394 m s^{-1} B $\longrightarrow$ 0.399 m s^{-1} C $\longrightarrow$ 1.02 m s^{-1}

B's speed is greater than A's but less than C's so no further impacts can occur.

The final speeds are: A: 0.394 m s^{-1} B: 0.399 m s^{-1} C: 1.02 m s^{-1}

EXERCISE 14c

In questions 1 to 3, two particles, A and B, lie on a smooth horizontal plane in a line that is perpendicular to a vertical wall. Initially B is at rest when A strikes it directly. B then goes on to strike the wall and rebounds.

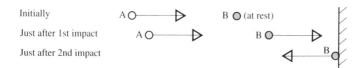

1. The mass of A is 1 kg and it strikes B with speed $4\,\mathrm{m\,s^{-1}}$. B's mass is 8 kg.

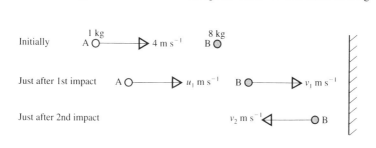

Given that the coefficient of restitution at each impact is $\frac{1}{2}$, find u_1, v_1 and v_2. Explain why there is no further collision.

2. The masses of A and B are 2 kg and 1 kg respectively. A strikes B with speed $9\,\mathrm{m\,s^{-1}}$. The coefficient of restitution between A and B is 1, and that between B and the wall is $\frac{1}{2}$.

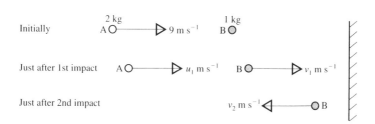

(a) Find the values of u_1 and v_1.

(b) Show that there is a second collision between A and B and find their velocities after it. Is there a further collision?

3. Particle A, of mass 1 kg, strikes B, also of mass 1 kg, at $9\,\mathrm{m\,s}^{-1}$. The coefficient of restitution between A and B is $\frac{1}{3}$, and that between B and the wall is $\frac{1}{2}$.

Just before the first impact   A $\bigcirc\!\!\longrightarrow\!\!\triangleright 9\,\mathrm{m\,s}^{-1}$ B $\bigcirc$ (at rest)

Find the velocities of A and B after each impact until no further collisions can occur.

In questions 4 to 7 three small smooth spheres, all with the same radius, lie in a straight line on a smooth horizontal plane. A is projected directly towards B which is at rest and after that collision B goes on to collide directly with C, also at rest.

4. The masses of A, B and C are 1 kg, 2 kg and 1 kg respectively. The coefficient of restitution between A and B is 1 and that between B and C is $\frac{1}{2}$.

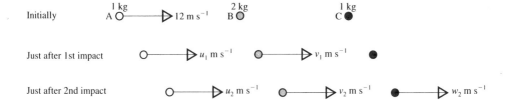

Initially   A $\bigcirc\!\!\longrightarrow\!\!\triangleright 12\,\mathrm{m\,s}^{-1}$ B $\bigcirc$ C $\bullet$

Just after 1st impact $\bigcirc\!\!\longrightarrow\!\!\triangleright u_1\,\mathrm{m\,s}^{-1}$ $\bigcirc\!\!\longrightarrow\!\!\triangleright v_1\,\mathrm{m\,s}^{-1}$ $\bullet$

Just after 2nd impact $\bigcirc\!\!\longrightarrow\!\!\triangleright u_2\,\mathrm{m\,s}^{-1}$ $\bigcirc\!\!\longrightarrow\!\!\triangleright v_2\,\mathrm{m\,s}^{-1}$ $\bullet\!\!\longrightarrow\!\!\triangleright w_2\,\mathrm{m\,s}^{-1}$

Find the velocities of each sphere after each possible impact.

5. The situation just before the first collision is shown in the diagram.

     2 kg   1 kg   2 kg

A $\bigcirc\!\!\longrightarrow\!\!\triangleright 8\,\mathrm{m\,s}^{-1}$ B $\bigcirc$ (at rest)   C $\bullet$ (at rest)

At each impact the coefficient of restitution is $\frac{1}{2}$. Find the velocities of A, B and C after each of the collisions that can occur.

6. The masses of A, B and C are $m\,\mathrm{kg}$, $m\,\mathrm{kg}$ and $km\,\mathrm{kg}$ respectively and A's initial speed is $u\,\mathrm{m\,s}^{-1}$. At each impact the coefficient of restitution is 1.

                                $m\,\mathrm{kg}$ $m\,\mathrm{kg}$ $km\,\mathrm{kg}$

Just before the first impact   A $\bigcirc\!\!\longrightarrow\!\!\triangleright u\,\mathrm{m\,s}^{-1}$ B $\bigcirc$ (at rest)   C $\bullet$ (at rest)

(a) Find, in terms of u, the velocities of A and B after the first collision.

(b) Find, in terms of u and k, the velocities of B and C after the second collision.

(c) State, with reasons, whether there will be further collisions if
     (i) $k > 1$ (ii) $k < 1$.

7. A sphere A, of mass m_1, and velocity u, collides with a stationary sphere B of mass m_2. If sphere A is brought to rest by the collision, find

   (a) the velocity of B after impact

   (b) the coefficient of restitution.

   Sphere B now collides with a stationary sphere C and is brought to rest. Assuming the same coefficient of restitution between B and C as between A and B, find the mass of sphere C.

8. A ball game at a fair consists of a long groove in which there are two heavy balls, B with mass 4 kg and C of mass 3 kg. A competitor bowls another ball A, of mass 3 kg, along the groove to strike B which, in turn, collides with C. The coefficient of restitution between A and B is $\frac{2}{3}$ and that between B and C is $\frac{1}{2}$. At the end of the groove there is an end-stop.

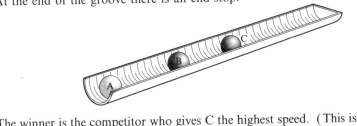

   The winner is the competitor who gives C the highest speed. (This is determined by measuring the time between the second impact and C's collision with the end-stop.)

   By modelling the balls as particles and the groove as smooth, find the speed given to C by a competitor who bowls A off at $10\,\mathrm{m\,s^{-1}}$.

*9. A cunning competitor at the game described in question 10, checks that, to beat the current best score, he must give C a speed of $7\,\mathrm{m\,s^{-1}}$. Find the speed at which he should project A to achieve this.

*10. Two beads, P of mass 0.01 kg and Q of mass 0.02 kg, are threaded on to a smooth wire which is in the form of a circle in a horizontal plane. Initially Q is at rest and P strikes it with speed $0.75\,\mathrm{m\,s^{-1}}$. After this impact they move in opposite directions round the circle. The coefficient of restitution is e.

   (a) Find their speeds after this first impact, in terms of e.

   (b) Given that they collide again after Q has rotated through $270°$ from its initial position, find the value of e and hence the values of the speeds after the first impact.

   (c) Find the velocities after the second impact.

*11. A small sphere is dropped on to a horizontal plane from a height h. The coefficient of restitution between the sphere and the plane is e.

   (a) Find, in terms of h and e, the height to which the particle rises after each of the first, second and third impacts.

   (b) Show that the heights found in (a) are in geometric progression.

   (c) Deduce the total distance travelled by the sphere before it comes to rest.

OBLIQUE IMPACT

All the collisions so far considered have been *direct*, i.e. the impulse that acts on a sphere at impact is in line with the direction of motion of that sphere both before and after the impact.

We now consider a situation where the impact is not direct but *oblique*.

Collision with a Fixed Object

Suppose that when a sphere travelling on a horizontal surface collides with a vertical wall, the direction of its velocity makes an angle θ with the wall.

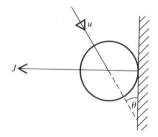

The impulse exerted on the sphere is perpendicular to the wall and causes a change in the momentum of the sphere in that direction; it does not however affect the momentum parallel to the wall.

So if the approach velocity of the sphere is resolved into components parallel and perpendicular to the wall, one of these components is changed by the impact and the other is unchanged.

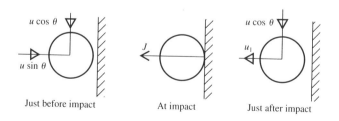

Just before impact   At impact   Just after impact

It follows that the direction, as well as the magnitude, of the velocity of the sphere can be changed by the impact.

Examples 14d

1. A smooth steel ball of mass 0.5 kg is moving in a straight line on a horizontal surface. It collides with a vertical plane when the velocity of the ball is $4\,\text{m s}^{-1}$ at $30°$ to the plane.

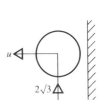

Given that the coefficient of restitution between the ball and the plane is $\frac{1}{2}$, find

(a) the speed of the ball just after impact,

(b) the angle between the plane and the direction of the ball's velocity just after impact,

(c) the magnitude of the impulse that acts on the ball at impact.

Resolve the velocity of the ball parallel and perpendicular to the plane. The speed parallel to the plane is unchanged by the impact; this property is marked on the diagram.

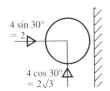

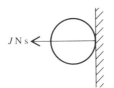

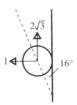

Just before impact   At impact   Just after impact

(a) Using the law of restitution gives

$$u = \tfrac{1}{2} \times 2 = 1$$

The speed, $V\,\text{m s}^{-1}$, of the ball just after impact is given by

$$V = \sqrt{(1^2 + \{2\sqrt{3}\}^2)} = \sqrt{13}$$

∴   the speed is $\sqrt{13}\,\text{m s}^{-1}$

(b) The direction of the velocity after impact is at an angle α to the plane where

$$\tan \alpha = \frac{u}{2\sqrt{3}} = \frac{1}{2\sqrt{3}}$$

$\Rightarrow$ $\alpha = 16°$ (nearest degree)

(c) Using impulse $=$ change in momentum gives

$$J = 0.5u - 0.5(-2) \Rightarrow J = 1.5$$

∴   the impulse is $1.5\,\text{N s}$

Oblique Collision Between Two Moving Objects

If two spheres, of equal radii, are free to move on a horizontal surface and collide when their velocities are not in the same straight line, the two impulses that act on impact are perpendicular to the common tangent of the spheres and so lie on the line joining the centres of the spheres.

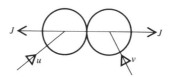

Therefore, for each sphere there is a change in momentum (and hence speed) along this line of centres but not perpendicular to it.

The unchanged components can be incorporated in a working diagram; then along the line of centres the calculation is exactly the same as for direct impact, i.e. conservation of momentum and the law of restitution can be applied in this direction.

Note that the line of centres is horizontal because the spheres have equal radii.

2. **Two smooth spheres, A of mass m and B of mass $3m$, are free to move on a horizontal table. A is projected towards B, which is at rest, and strikes B with speed $2u$. On impact the line through their centres makes an angle of 60° with the velocity of A before impact.**

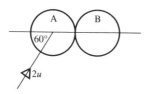

If the coefficient of restitution between the spheres is $\frac{1}{3}$ find, in terms of u, the magnitude and direction of the velocity of each sphere just after impact.

We will resolve the velocity of A along and perpendicular to the line of centres. B is struck along the line of centres so after impact it moves in this direction.

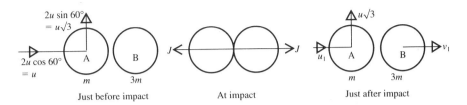

| Just before impact | At impact | Just after impact |

Conservation of moment $\longrightarrow$ gives $mu + 0 = mu_1 + 3mv_1$

$$\Rightarrow \qquad u = u_1 + 3v_1 \qquad [1]$$

Law of restitution gives $\qquad \frac{1}{3}u = v_1 - u_1 \qquad [2]$

$[1] + [2]$ $\qquad \frac{4}{3}u = 4v_1$

$\therefore \qquad v_1 = \frac{1}{3}u \qquad$ and $\qquad u_1 = 0$

For B, the velocity is of magnitude $\frac{1}{3}u$ along the line of centres.

For A, the velocity is of magnitude $u\sqrt{3}$ perpendicular to the line of centres.

3. **Two smooth spheres, A and B, with equal radii, lie on a horizontal plane. The mass of B is twice that of A. The spheres are projected towards each other and they collide when the line joining their centres is in the direction of the unit vector i. The velocity vectors of A and B just before impact are represented by the vectors $2\mathbf{i} + \mathbf{j}$ and $\mathbf{i} - \mathbf{j}$ respectively. If the coefficient of restitution is $\frac{1}{2}$, find their velocity vectors just after impact. Find also the kinetic energy lost due to the collision.**

Let the mass of A be m so that the mass of B is $2m$.

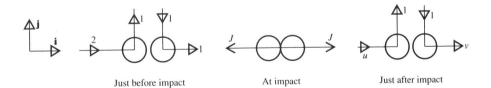

| Just before impact | At impact | Just after impact |

Conservation of momentum $\longrightarrow$ gives

$$2m + 2m = mu + 2mv \qquad \Rightarrow \qquad 4 = u + 2v \qquad [1]$$

Law of restitution gives

$$\tfrac{1}{2}(2 - 1) = v - u \qquad \Rightarrow \qquad \tfrac{1}{2} = v - u \qquad [2]$$

[1] + [2] gives $\qquad \frac{9}{2} = 3v$

$\therefore \qquad v = \frac{3}{2} \qquad$ and $\qquad u = 1$

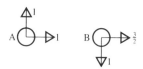

After impact the velocity vector of A is $\quad \mathbf{i} + \mathbf{j}$

and the velocity vector of B is $\quad \frac{3}{2}\mathbf{i} - \mathbf{j}$

The velocity components in the direction of $\mathbf{j}$ are unchanged by the impact so, to find the loss in KE, only the components in the $\mathbf{i}$ direction need be considered.

Initial KE $= \frac{1}{2}m(2)^2 + \frac{1}{2}(2m)(1)^2 = 3m$

Final KE $= \frac{1}{2}m(1)^2 + \frac{1}{2}(2m)(\frac{3}{2})^2 = \frac{11}{4}m$

$\therefore \qquad$ Loss in KE is $\quad \frac{1}{4}m$

EXERCISE 14d

1. A smooth sphere is projected along horizontal ground and collides obliquely with a vertical wall. It hits the wall when moving at $3\,\mathrm{m\,s^{-1}}$ at an angle of $30°$ to the wall. Find the velocity of the sphere just after impact with the wall if
   (a) $e = \frac{1}{2}$, (b) $e = 1$, (c) $e = 0$.

2. A smooth sphere travelling on horizontal ground impinges obliquely on a vertical wall and rebounds at right angles to its original direction of motion. If the sphere is moving at $60°$ to the wall before impact, find the value of e.

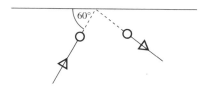

3. Two smooth spheres A and B of equal radius and mass are moving on a horizontal table with velocity vectors $\mathbf{i} + 2\mathbf{j}$, $-3\mathbf{i} + \mathbf{j}$ respectively and collide when the line joining their centres is parallel to $\mathbf{i}$. Find the velocity vectors of A and B after the impact if
   (a) $e = \frac{1}{2}$, (b) $e = 1$, (c) the collision is inelastic.

4. Two smooth spheres A and B of equal radius and mass lie on a horizontal surface. B is at rest and A is projected towards B with velocity vector $4\mathbf{i} + 3\mathbf{j}$ and they collide when their line of centres is parallel to the vector $\mathbf{i}$. If B moves off with speed 3 units, find the value of e and the velocity vector of A after impact.

5. A red ball is stationary on a rectangular billiard table OABC. It is then struck by a white ball of equal mass and equal radius with velocity $u(-2\mathbf{i} + 11\mathbf{j})$ where $\mathbf{i}$ and $\mathbf{j}$ are unit vectors along OA and OC respectively. After impact the red and white balls have velocities parallel to the vectors $-3\mathbf{i} + 4\mathbf{j}$, $2\mathbf{i} + 4\mathbf{j}$ respectively. Prove that the coefficient of restitution between the two balls is $\frac{1}{2}$.

6.

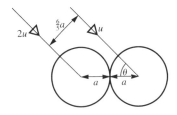

Two uniform smooth spheres, each of mass m and radius a, collide when moving on a horizontal plane. Just before impact the spheres are moving with speeds $2u$ and u as shown in the diagram, their centres moving in parallel lines which are at a distance $\frac{6}{5}a$ apart. The coefficient of restitution between the spheres is $\frac{1}{2}$.

(a) Find the angle θ

(b) Find the speeds of the spheres after impact

(c) Show that the angle between their paths is then approximately $27°$.

7. A ball is thrown down towards a smooth horizontal plane. It strikes the plane when it is travelling at $4\sqrt{2}\,\mathrm{m\,s^{-1}}$ at $45°$ to the plane. If the coefficient of restitution between ball and plane is $\frac{1}{2}$, find

(a) the height above the plane to which the ball rises after impact (use $g = 10$),

(b) the time for which the ball is in the air before it strikes the plane again,

(c) the distance between the first and second impacts with the plane.

*8. A smooth sphere of mass m sliding on a horizontal plane collides obliquely with a sphere of mass $2m$ and of equal radius at rest on the plane. At the moment of impact the velocity u of the moving sphere makes an angle α with the line of centres, and after impact the speed of the heavier sphere is $(\frac{2}{5})u \cos \alpha$. Find the coefficient of restitution between the spheres.

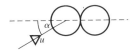

9. A gas molecule having a velocity $300\,\mathbf{i}\,\mathrm{m\,s^{-1}}$ collides with another molecule of the same mass which is initially at rest. After the collision the first molecule has velocity $225\,\mathbf{i} + 130\,\mathbf{j}\,\mathrm{m\,s^{-1}}$. Find, in the form $a\mathbf{i} + b\mathbf{j}$, the velocity of the second molecule after the collision.

CHAPTER 15

MOTION IN A HORIZONTAL CIRCLE

ANGULAR VELOCITY

Consider a particle that moves round the circumference of a circle with centre O. If the particle moves from a point P on the circumference to an adjacent point Q and the angle POQ is θ radians, then the rate at which θ is increasing is $d\theta/dt$. This is the *angular speed* of the particle and is often denoted by the symbol ω.

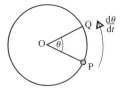

If we also state the direction of rotation (clockwise or anticlockwise), then we are giving the *angular velocity* of the particle.

We use a positive sign to denote the anticlockwise direction of rotation and a negative sign for clockwise rotation.

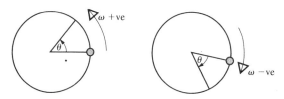

For example, the seconds hand of a clock rotates through 1 revolution in 1 minute so we can say:

its angular speed is $1\,\text{rev}\,\text{min}^{-1}$

and   its angular velocity is $1\,\text{rev}\,\text{min}^{-1}$ clockwise,   or $-1\,\text{rev}\,\text{min}^{-1}$.

Angular velocity can be measured in revolutions per second, but is more usually given in radians per second. Either of these units can be converted to the other by using

$$1 \text{ revolution} = 2\pi \text{ radians.}$$

Constant Angular Velocity

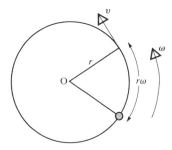

A particle that describes a circle at constant speed covers equal arcs in equal times so its angular velocity is also constant.

Suppose that the radius of the circle is r, the speed round the circumference is $v\,\mathrm{m\,s^{-1}}$ and the angular velocity is $\omega\,\mathrm{rad\,s^{-1}}$. In 1 second the particle travels round an arc of length v and this arc length is also given by $r\omega$, i.e.

$$v = r\omega$$

Examples 15a

1. **Express**   **(a)   3 revolutions per minute in radians per second**

   (b)   0.005 radians per second in revolutions per hour.

(a) $3\,\text{rev/minute} = \dfrac{3}{60}\,\text{rev/second}$

$\qquad\qquad = \dfrac{1}{20} \times 2\pi\,\mathrm{rad\,s^{-1}} = \dfrac{\pi}{10}\,\mathrm{rad\,s^{-1}}$

(b) $0.005\,\mathrm{rad\,s^{-1}} = 0.005 \times 3600\,\text{rad/hour} = 18\,\text{rad/hour}$

$\qquad\qquad\qquad = 18 \div 2\pi\,\text{rev/hour}$

$\qquad\qquad\qquad = \dfrac{9}{\pi}\,\text{rev/hour}$

2. **A point on the circumference of a disc is rotating at a constant speed of $3\,\mathrm{m\,s^{-1}}$. If the radius of the disc is $0.24\,\mathrm{m}$ find, in $\mathrm{rad\,s^{-1}}$, the rate at which the disc is rotating.**

Using $v = r\omega$,   with $v = 3$ and $r = 0.24$ gives

$\qquad 3 = 0.24\omega \qquad \Rightarrow \qquad \omega = 12.5$

$\therefore$ the disc rotates at $12.5\,\mathrm{rad\,s^{-1}}$.

EXERCISE 15a

1. Express (a) 0.2 radians per minute in revolutions per hour
   (b) 100 revolutions per minute in radians per second.

2. Find the angular velocity, in radians per second, of the minute hand of a clock.

3. Find the angular speed of the earth about its axis
   (a) in revolutions per minute (b) in radians per second.

4. A disc is rotating about its centre with angular velocity ω rad s^{-1}. Point P is on the disc at a distance of d metres from the centre and has speed v metres per second.
   (a) $\omega = 6$, $d = 0.2$; find v.
   (b) $v = 5$, $d = 0.4$; find ω.
   (c) $v = 10$, $\omega = 2.5$; find d.

5. A fairground Big Wheel carriage is 8 m from the centre of the wheel, which is rotating at 10 revolutions per minute. Find the speed of the carriage.

6. Find the speed, in km h^{-1}, of a point on the equator of the earth, assuming the equator to be a circle of radius 6400 km.

7. A DIY power drill has a top rotational speed of 2500 revolutions per minute. A drill bit of diameter 3 mm has been inserted. Find the speed of the cutting edge of this drill.

8. A playground roundabout has a diameter of 3 m. A man puts his child onto it at a distance of 1 m from the centre and runs round pushing the edge of the roundabout with a speed of 2.4 m s^{-1}.
   (a) Find the angular speed in rad s^{-1}.
   (b) Find the speed with which the child is moving.

9. At a well a bucket is attached to a thin rope which is wound round a cylinder. The bucket is raised and lowered by turning the cylinder. The cylinder has a radius of 10 cm.

   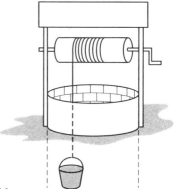

   (a) If the cylinder is turned at 1 revolution per second, at what speed, in m s^{-1}, will the bucket ascend?

   (b) The well owners decide to change the cylinder so that when it is turned at 1 rev/second, the bucket will ascend at 1 m s^{-1}. What should be the radius of the new cylinder?

   State any assumptions you make in solving this problem.

ACCELERATION

If the velocity of a moving object is changing, that object has an acceleration. As velocity is a vector, it can change in magnitude or in direction or both. It is easy to accept that changing *speed* involves acceleration but it not so easy to see that, for example, a car going round a corner at *constant* speed is accelerating because its direction is changing.

A change in speed is caused by a force that acts *in the direction* of motion of the object to which it is applied.

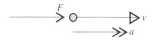

A force of this type cannot produce a change in the direction of the velocity so, if no other force is acting, the object continues to move in a straight line. The acceleration produced is a change in speed, i.e. a change in the magnitude of the velocity.

This type of acceleration was covered in detail in Chapter 2.

A force that is perpendicular to the direction of motion of an object will push or pull the object off its previous line of motion but cannot alter the speed.

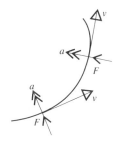

The acceleration in this case is a change in the *direction* of the velocity and the object therefore moves in a curve of some sort; the actual curve described depends upon the particular force acting.

A force that is neither parallel nor perpendicular to the direction of motion of an object has a component in each of these directions. Therefore it causes a change both in the speed and in the direction of motion of the object and the object moves with varying speed on a curved path.

MOTION IN A CIRCLE WITH CONSTANT SPEED

The direction of motion of a particle moving in a circle is constantly changing, so there must be a force acting perpendicular to the direction of motion of the particle at any instant.

If the particle is moving with constant speed there is no force acting in the direction of motion, i.e. no tangential force.

At any point on its path the particle is moving in the direction of the tangent at that point. A force that is perpendicular to this direction acts along the radius at that point. Further, because it is moving the particle from the tangent on to the circumference, the force must act *inwards* along the radius.

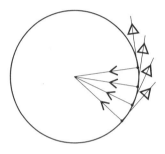

**The force that produces circular motion with constant speed
is at any instant acting radially inwards on the particle,
producing a radial acceleration.**

The Magnitude of the Radial Acceleration

Consider a particle moving at constant speed v, round a circle of centre O and radius r. Suppose that, in a time δt, the particle moves from a point P to a nearby point Q, through a small angle $\delta\theta$ measured in radians. For reasons of clarity, $\delta\theta$ is not drawn as a very small angle in the diagram below.

The length of the arc PQ is $r\delta\theta$ and, as this arc is covered in time δt at speed v, its length is also $v\delta t$.

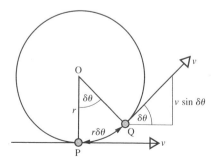

Therefore $\qquad r\delta\theta = v\delta t$

$\Rightarrow \qquad\qquad \dfrac{v}{r} = \dfrac{\delta\theta}{\delta t}$

When the particle is at P it has no velocity component along PO, but when it reaches Q the velocity component parallel to PO is $v \sin \delta\theta$.

Therefore the average acceleration from P to Q, in the direction of PO, is given by

$$\frac{v \sin \delta\theta}{\delta t}$$

The closer Q is to P, the nearer this expression is to the actual radial acceleration at P and, at the same time, both $\delta\theta$ and δt approach zero. When this happens the value of $\dfrac{v \sin \delta\theta}{\delta t}$ becomes indeterminate.

However, for any angle α measured in radians, it can be shown that

$$\text{as} \quad \alpha \to 0, \quad \sin \alpha \to \alpha$$

A formal proof of this property is not given at this stage but you will probably find it quite convincing to compare the values of small angles and their sine ratios using a calculator.

Therefore as $\quad \delta\theta \to 0 \qquad \sin \delta\theta \to \delta\theta$

$$\therefore \qquad \frac{v \sin \delta\theta}{\delta t} \quad \to \quad v \frac{\delta\theta}{\delta t} \quad \to \quad v \frac{d\theta}{dt}$$

But $\dfrac{d\theta}{dt}$ is the rate at which θ increases with respect to time, i.e. $\dfrac{d\theta}{dt}$ is the angular velocity, ω.

Hence the radial acceleration of the particle is $v\omega$.

Then using $v = r\omega$, this acceleration can be expressed as either $r\omega^2$ or $\dfrac{v^2}{r}$.

To sum up:

**The acceleration of a particle travelling in a circle of radius r,
at constant speed v (or constant angular speed ω),
is directed *towards* the centre of the circle
and is of magnitude $r\omega^2$ or $\dfrac{v^2}{r}$.**

It follows that

**a particle can describe a circle with constant speed
only when under the action of a force of constant magnitude
directed *towards the centre* of the circle.**

Examples 15b

1. A particle P is travelling round a circle of radius 0.8 m at a constant speed of $2 \, \text{m s}^{-1}$. Find the acceleration of the particle, giving its magnitude and direction.

The acceleration is given by $\dfrac{v^2}{r}$, where $v = 2$ and $r = 0.8$

The magnitude of the acceleration is $\dfrac{2^2}{0.8} \, \text{m s}^{-2}$, i.e. $5 \, \text{m s}^{-2}$,
and it is directed towards the centre of the circle.

2. A firework consists of a strip of wood, pivoted at its centre, with a rocket attached to each end. The length of the strip is 0.9 m. After the rockets are lit the firework rotates at $12 \, \text{rev s}^{-1}$. Use a suitable model to estimate the radial acceleration of each rocket.

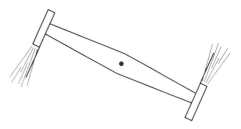

Model each rocket as a particle at a distance 0.45 m from the pivot.

The angular speed is $12 \, \text{rev s}^{-1}$, i.e. $12 \times 2\pi \, \text{rad s}^{-1}$

The acceleration of one particle is given by $r\omega^2$

When $r = 0.45$ and $\omega = 24\pi$, $r\omega^2 = (0.45)(24\pi)^2 = 2560$ (3 sf)

An estimate of the radial acceleration of each rocket is $2560 \, \text{m s}^{-2}$ towards the pivot.

EXERCISE 15b

In questions 1 to 5 a particle is travelling round a circle, centre O, radius r metres, at a constant speed of $v \, \text{m s}^{-1}$ and angular velocity $\omega \, \text{rad s}^{-1}$. Its acceleration is of magnitude $a \, \text{m s}^{-2}$.

1. $v = 16$, $r = 5$; find a.

2. $\omega = 12$, $r = 3$; find a.

3. In what direction is the acceleration

   (a) when the particle has reached a point P on the circle

   (b) when the particle has reached a point Q on the circle.

4. $a = 75$, $r = 12$; find v.

5. $a = 500$, $\omega = 6$; find r.

6. Taking the earth to be a sphere of radius 6400 km, calculate the acceleration, in $m\,s^{-2}$, due to the earth's rotation of

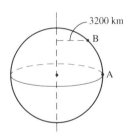

(a)   a person A who is standing on the equator

(b)   a person B who is at a point which is 3200 km from the earth's axis of rotation.

Show on a diagram the direction of the acceleration in each case.

7. An aircraft waiting to land at a busy airport is circling at a constant height, at $500\,km\,h^{-1}$. The passengers experience an acceleration of $0.4g\,m\,s^{-2}$. Find the radius of the circle.

8. A machine is designed to test astronauts in conditions of great acceleration. The astronaut is strapped into a chair which is then moved round in a horizontal circle of radius 5 m. If he can withstand accelerations up to $9g\,m\,s^{-2}$, what is the maximum permissible angular velocity?

9. Passengers on a fairground ride are whirled round in a horizontal circle of radius 6 m, experiencing a radial acceleration of $15\,m\,s^{-2}$. At what speed are they moving?

10. A quarter-scale railway line is to be laid for a model train. It has been decided that the children who ride on the train should not be subjected to a radial acceleration greater than $8\,m\,s^{-2}$. By treating the bends in the line as arcs of circles, find the speed limit you would recommend if the radius of the sharpest bend is 10 m.

MOTION IN A HORIZONTAL CIRCLE

We know that when a circle is described at constant speed, there is an acceleration of constant magnitude v^2/r or $r\omega^2$ towards the centre of the circle and no acceleration tangentially. There must therefore be

(a)   no tangential force acting

(b)   a force of constant magnitude acting towards the centre.

These conditions can be achieved when the circle is in a horizontal plane because the weight of the particle, being a vertical force, has no component in the direction of motion of the particle. So it is possible for the circular motion to be performed at a constant speed. (In a vertical plane, on the other hand, as a particle moves in a circular path the tangential component of the weight varies, causing the speed of the particle to vary.)

There are many ways in which the necessary force towards the centre can be provided, e.g. by a rotating string with one end fixed at the centre or by friction with the road surface as a car turns round a bend. Some of the possibilities are illustrated in the following examples.

Examples 15c

1. One end of a light inelastic string of length a is fixed to a point O on a horizontal plane. A particle P of mass m, attached to the other end of the string, is given a blow which sets it moving in a circle on the plane, with constant angular velocity ω.

   (a) Find the tension in the string and the force exerted on P by the table.

   (b) Explain an assumption that has been made in the question about a certain possible force.

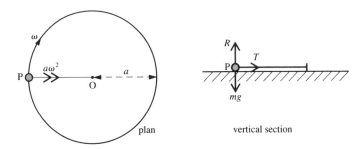

plan   vertical section

Vertically there is no acceleration; horizontally there is an acceleration of $a\omega^2$ towards O.

   (a) Resolving ↑ $R - mg = 0$ [1]

   Using Newton's Law → $T = ma\omega^2$ [2]

   The tension is $ma\omega^2$ and the reaction exerted by the table is mg.

   (b) As P travels with constant speed it is assumed that there is no tangential force, i.e. that there is no friction between the particle and the plane.

2. A small block A, of mass m kg, lies on a horizontal disc which is rotating about its centre B at $3\,\text{rad s}^{-1}$ and A is $0.8\,\text{m}$ from B. If the block does not move relative to the disc, find the least possible value of μ, the coefficient of friction, between the block and the disc.

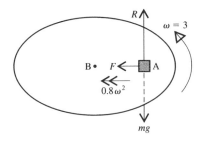

The frictional force F acing on A is towards the centre of the circle because A has to be given a central acceleration in order to travel in a circle; there is no friction tangentially because A has no tendency to move in that direction. Also, as the block is small we treat it as a particle.

Resolving ↑ $\qquad\qquad\quad R - mg = 0 \qquad\qquad \Rightarrow \qquad R = mg$

Using Newton's Law ← $\qquad\quad F = mr\omega^2 \qquad\quad \Rightarrow \qquad F = m(0.8)(3)^2 = 7.2m$

Now $\quad F \leqslant \mu R \qquad \Rightarrow \qquad 7.2m\omega^2 \leqslant \mu mg$

i.e. $\qquad\qquad\qquad\qquad\qquad\qquad \mu \geqslant \dfrac{7.2}{g}$

$\therefore \quad$ the least value of μ is $\dfrac{7.2}{9.8} \quad$ i.e. $0.735 \quad$ (3 sf).

3. **Katie is hoping to prepare her pony, Ben, for showing in a ring. As part of the training programme she is holding one end of an extensible rope and the other end is fastened to Ben's bridle. The rope has an unstretched length of 8.6 m and a modulus of elasticity 740 N.**

   (a) Ben is very obediently trotting at a steady speed of $2.5\,\mathrm{m\,s}^{-1}$ in a circle of radius 11 m.
   Find   (i)   the tension in the rope   (ii) Ben's mass.

   (b) The rope ceases to obey Hooke's Law if the extension exceeds 3 m.
   Find the greatest speed that Ben should be asked to achieve.

We will model Ben as a particle, the rope as a light elastic string, Katie as a point at the centre of the circle and assume that Ben's hooves do not create any friction radially.
As these assumptions are fairly rough, answers corrected to 3 sf are not appropriate.

(a)

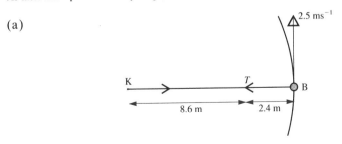

(i) The rope is extended by 2.4 m therefore Hooke's Law, $\quad T = \dfrac{\lambda x}{a} \quad$ gives

$$T = \frac{740 \times 2.4}{8.6} = 206.5 \ldots$$

The tension in the rope is 210 N (2 sf).

(ii) Ben's acceleration towards K is given by $\dfrac{v^2}{r}$, so it is $\dfrac{6.25}{11} \mathrm{m\,s}^{-2}$.

Now using Newton's Law, $\quad T = ma, \quad$ gives

$$206.5 \ldots = m \times \frac{6.25}{11} \qquad \Rightarrow \qquad m = 363.4 \ldots$$

Ben's mass is 360 kg (2 sf).

(b)

When the extension is 3 m, the tension becomes $\dfrac{740 \times 3}{8.6}\,\text{N}$

Then using $T = \dfrac{mv^2}{r}$ gives $v^2 = \dfrac{Tr}{m} = \dfrac{740 \times 3 \times (8.6 + 3)}{8.6 \times 363}$

$\Rightarrow$ $v = 2.87\ldots$

It is dangerous to give the speed *corrected* to 2 sf, as that increases the result slightly. So instead we will *truncate* the calculated figure to 2 sf.

Ben should not exceed a speed of $2.8\,\text{m}\,\text{s}^{-1}$.

EXERCISE 15c

In questions 1 and 2, one end A of an inelastic string AB is fixed to a point on a smooth table. A particle P is attached to the other end B, and moves on the table in a horizontal circle with centre A.

1. The particle is of mass 1.5 kg and its speed is $4\,\text{m}\,\text{s}^{-1}$. If the length of the string is 2.4 m, find the tension in the string.

2. The mass of the particle is 8 kg and it is moving with speed $5\,\text{m}\,\text{s}^{-1}$. Find the length of the string given that the tension in it is 12 N.

   Questions 3 and 4 are about a situation similar to that described above, except that the string AB is elastic.

3. The elastic string AB has a natural length of 2.5 m and its modulus of elasticity is 40 N. The mass of P is 5 kg. If the string is extended by 0.5 m, find the speed of P.

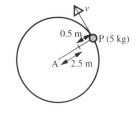

4. When P has a mass of 1.5 kg, and is moving with a speed of $6\,\text{m}\,\text{s}^{-1}$, the string is extended by 0.4 m. Given that the natural length of the string is 2 m, find its modulus of elasticity.

5. A circular tray of radius 0.2 m has a smooth vertical rim round the edge. The tray is fixed on a horizontal table and a small ball of mass 0.1 kg is set moving round the inside of the rim of the tray with speed $4\,\text{ms}^{-1}$. Calculate the horizontal force exerted on the ball by the rim of the tray.

6. A particle of mass 0.4 kg is attached to one end of a light inextensible string of length 0.6 m. The other end is fixed to a point A on a smooth horizontal table. The particle is set moving in a circular path.

(a) If the speed of the particle is $8 \, \text{m s}^{-1}$ calculate the tension in the string and the reaction with the table.

(b) If the string snaps when the tension in it exceeds 50 N, find the greatest angular velocity at which the particle can travel.

The remaining questions in this exercise involve choosing a model. This should be clearly described and all assumptions mentioned.

7. An aircraft is flying at $700 \, \text{km h}^{-1}$ in a horizontal circle of radius 2 km.

(a) Find the horizontal and vertical components of the thrust exerted by the seat, on a passenger of mass 60 kg.

(b) Find the magnitude and direction of the resultant thrust.

8. A satellite, of mass m kilograms, is orbiting the earth on a circular path at a height of 100 km above the surface. At this height the acceleration due to gravity is $9.5 \, \text{m s}^{-2}$. Take the radius of the earth as 6400 km. For the satellite, find

(a) the force exerted on it towards the centre of the earth

(b) its speed in m s^{-1}

(c) its angular speed

(d) the time it takes to perform one orbit.

THE CONICAL PENDULUM

Consider a light inelastic string fixed at one end A and carrying a particle hanging freely at the other end. If the particle, which is not resting on a surface of any sort, is set moving in a horizontal circle, the plane of that circle will be below the level of A. As the particle and the string rotate, they trace out the surface of a cone and the system is known as a *conical pendulum*.

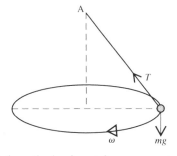

In this situation the tension in the string does two jobs – the horizontal component provides the central force needed to keep the particle rotating and the vertical component balances the weight of the particle. It follows that the string can never be horizontal, as the tension in it *must* have a vertical component.

There is a similar situation when a particle moves on the inner surface of a smooth sphere, in a horizontal circle below the level of the centre of the sphere.

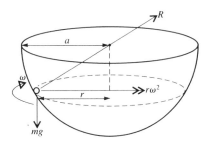

The vertical component of the normal reaction R (which acts through the centre of the sphere) balances the weight of the particle, while the horizontal component of R acts on the particle towards the centre of the circle being described.

Examples 15d

1. A particle P, of mass m, is attached to one end of a light inextensible string of length a and describes a horizontal circle, centre O, with constant angular speed ω. The other end of the string is fixed to a point Q and, as P rotates, the string makes an angle θ with the vertical. Show that

   (a) the tension in the string always exceeds the weight of the particle,

   (b) the depth of O below Q is independent of the length of the string.

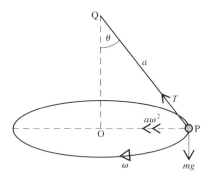

The string cannot be vertical therefore $\theta > 0$.

Resolving ↑ $T\cos\theta - mg = 0$ [1]

Using Newton's Law ← $T\sin\theta = m(a\sin\theta)\omega^2$

⇒ $T = ma\omega^2$ [2]

(a) From [1], $\qquad T = \dfrac{mg}{\cos\theta}$ and $\cos\theta < 1$

$\qquad\qquad\qquad$ ∴ $\qquad\qquad T > mg$

(b) In $\triangle POQ$, $\qquad\qquad QO = a\cos\theta$ $\qquad\qquad$ [3]

$\qquad$ From [1] and [3] $\qquad QO = \dfrac{amg}{T}$

$\qquad$ Then from [2] $\qquad QO = \dfrac{amg}{ma\omega^2} = \dfrac{g}{\omega^2}$

i.e. the depth of O below Q is independent of the length of the string.

2. **An elastic string, of natural length l and modulus of elasticity $2mg$, has one end fixed to a point A and has a particle P, of mass m, attached to the other end B. P is travelling in a horizontal circle with angular speed ω. The string reaches its elastic limit when the tension is $3mg$. Find the angular speed when this state is reached.**

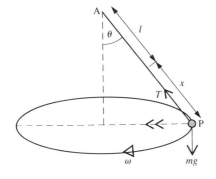

$$T = 3mg \qquad\qquad [1]$$

Resolving ↑ $\qquad\qquad T\cos\theta - mg = 0 \qquad\qquad$ [2]

Newton's Law ← $\qquad\qquad T\sin\theta = m\omega^2(l+x)\sin\theta \qquad\qquad$ [3]

Hooke's Law, $\quad T = \dfrac{\lambda x}{l} \;\Rightarrow\; T = \dfrac{2mgx}{l} \qquad\qquad$ [4]

From [1] and [4] $\qquad 3mg = \dfrac{2mgx}{l} \qquad\Rightarrow\qquad x = \tfrac{3}{2}l$

From [1] and [3] $\qquad 3mg = m\omega^2(l+x) \qquad\Rightarrow\qquad 3g = \omega^2\left(\tfrac{5}{2}l\right)$

Therefore $\quad \omega = \sqrt{\dfrac{6g}{5l}}$.

Note that one of the equations formed was not used. This happens quite often but it is not easy to spot at the outset. So, unless you are *sure* that an equation will not be needed, the best policy is to apply all the relevant principles and then see which equations are useful.

3. A particle P is moving on the inner surface of a smooth hemispherical bowl with centre O and radius $2a$. The particle is describing a horizontal circle, centre C, with angular speed $\sqrt{\dfrac{g}{a}}$.

Find

(a) the magnitude of the force exerted on P by the surface of the bowl

(b) the depth of C below O.

(a) For the forces acting on P,

Resolving ↑ gives

$$R \cos \theta - mg = 0 \qquad [1]$$

Newton's Law → gives

$$R \sin \theta = mr\omega^2$$
$$= m(2a \sin \theta)\left(\frac{g}{a}\right)$$
$$\Rightarrow \qquad R = 2mg \qquad [2]$$

The force exerted on P is $2mg$.

(b) In △OPC, $OC = 2a \cos \theta$

From [1] and [2], $\cos = \frac{1}{2}$

∴ $OC = a$

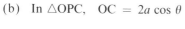

The depth of C below O is a.

4. A smooth ring R, of mass m, is threaded on to a light inextensible string of length 1.4 m. The ends of the string are fixed to two points A and B, distant 1 m apart in a vertical line. When the ring is set rotating in a horizontal circle with angular speed ω radians per second, the distance of the ring from the upper fixed point, A, is 0.8 m.

(a) Show that ARB is a right angle and hence write down the values of sin θ and cos θ where θ is ABR.

(b) Find the radius, r m, of the horizontal circle.

(c) Find the value of ω, corrected to 2 significant figures.

(a)   BR $= (1.4 - 0.8)$ m $= 0.6$ m

   ∴   triangle ARB is a '3, 4, 5' triangle
        with a right angle at R.

   From triangle ABR, $\sin \theta = 0.8$
   and $\cos \theta = 0.6$.

(b)   From triangle ARS, $r = 0.8 \cos \theta$

   The radius of the circle is 0.48 m.

(c)   Using Newton's Law towards the centre of
        the circle gives

$$T \cos \theta + T \sin \theta = m \times 0.48\omega^2$$

$\Rightarrow$ $1.4T = 0.48m\omega^2$ [1]

Vertically the ring is in equilibrium.

Resolving ↑ gives $T \sin \theta - T \cos \theta - mg = 0$

$\Rightarrow$ $0.2T = mg$ [2]

[1] ÷ [2] gives $\dfrac{1.4T}{0.2T} = \dfrac{0.48m\omega^2}{mg}$

$\Rightarrow$ $\omega^2 = \dfrac{7g}{0.48} = 142.9\ldots$

∴ $\omega = 1.2$ (2 sf)

Note.   In this problem the ring is smooth and is *threaded* on to the string;
therefore the tension is the same on both sides of the ring.

If, instead, the ring is *fastened* to a point on the string, the tensions in the two
portions of the string are, in general, different.

EXERCISE 15d

Questions 1 to 3 are about a conical pendulum which consists of an inextensible
string AB with a particle P attached at B. Point A is fixed and B moves in a
circle in a horizontal plane.

1. The length of the string AB is 1.5 m and
   the mass of the particle P is 3 kg. Given
   that P is rotating in a circle with an
   angular speed of 8 rad s^{-1} find

   (a)   the tension in the string

   (b)   the angle between the string and
         the vertical.

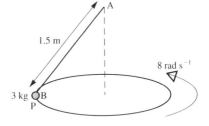

2. The mass of the particle P is 2 kg and P
   is rotating in a circle of radius 0.3 m.
   Given that the string is inclined at 25° to
   the vertical find

   (a) the tension in the string

   (b) the angular speed of P.

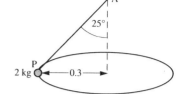

3. The particle P is rotating in a circle with an angular speed of 5 rad s⁻¹. Find the
   depth below A of the plane of this circle.

   Questions 4 and 5 are about a situation similar to that described for questions 1
   to 3, except that the string AB is elastic.

4. The mass of the particle P is 2 kg. The
   elastic string AB has a natural length of 0.4 m
   and its modulus of elasticity is 12 N.
   Given that when P rotates in a circle the
   extension of the string is 0.1 m find

   (a) the tension in the string

   (b) the angular speed of P.

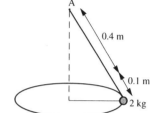

5. The mass of the particle P is 0.5 kg and P is rotating in a circle with an angular
   speed of 2 rad s⁻¹. The modulus of elasticity of the string AB is 3 N.
   Given that the string is inclined at 60° to the vertical find

   (a) the tension   (b) the extended length   (c) the natural length.

6. A particle of mass m, attached to the end A of a light inextensible string
   describes a horizontal circle on a smooth horizontal plane with angular speed ω.
   The string is of length $2l$ and the other end B is fixed,

   (a) to a point on the plane

   (b) to a point which is at a height l above the plane.

   Find, in each case, the tension in the string and the reaction between the particle
   and the plane, giving your answers in terms of m, l, g and ω.

7. One end of a light inextensible string of
   length $3a$ is attached to a fixed point A,
   and the other end to a point B which is
   at a distance $2a$ vertically below A.
   A small bead, P, of mass m, is fastened
   to the midpoint of the string and moves in
   a horizontal circle with speed $\sqrt{5ga}$.

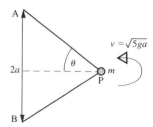

   (a) (i) Find the sine of the angle θ between the string and the horizontal.

        (ii) Express the radius of the circle in terms of θ.

   (b) Find the tensions in the two halves of the string.

8. Suppose that, in question 7, the bead is threaded on to the string (not fastened) and moves in a horizontal circle with centre R. The string is taut and $BR = \frac{5}{6} a$. Find the speed of P.

9. A boy has a ball of mass 0.15 kg on an elastic string of natural length 30 cm. He whirls the ball in a horizontal circle of radius 25 cm with the string at 50° to the vertical. Find

 (a) the tension in the string

 (b) the speed of the ball

 (c) the extended length of the string

 (d) the modulus of elasticity of the string.

10. A powered model aircraft of mass 0.25 kg is attached by a wire of length 10 m to a point A on the ground. The plane flies in a horizontal circle, of radius 5 m, centred above the point A. It is rotating at 12 revolutions per minute. The motion through the air produces a vertical lift force, P newtons, on the plane. Find

 (a) the angle between the wire and the ground

 (b) the tension in the wire

 (c) the lift force P.

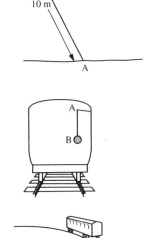

11. Two boys decide to add interest to a train journey by doing an experiment with a pendulum, which they hang from the luggage rack. When the train has constant velocity the pendulum hangs vertically, as shown in the diagram. When the train goes round a bend at $v\,\mathrm{m\,s^{-1}}$, the pendulum deviates through an angle θ from this position.

 Consider the bend to be an arc of a circle. Show on a diagram the position of the pendulum when the train is going round the bend shown.

 The boys can calculate θ by measuring the distance of the pendulum bob, B, from the side of the train. They intend to estimate the radius, r metres, of each bend from a map and use r and θ to calculate v.

 (a) Obtain a formula for v in terms of r and θ.

 (b) On one bend they observe that $\theta = 5°$ and estimate r as 800 m. Find, in $\mathrm{km\,h^{-1}}$, the value they obtain for the speed of the train.

 (c) If the train accelerates on a straight section of track, describe how the pendulum behaves.

12. A small bead B of mass 0.2 kg is rotating in a horizontal circle on the inner surface of a smooth hemisphere of radius 0.3 m and centre O. If the centre of the horizontal circle is 0.1 m below O, find

   (a)   the magnitude of the force exerted by the bowl on the bead,

   (b)   the speed of the bead.

PRACTICAL PROBLEMS

It is not every day that most people see a particle at the end of a string rotating in a horizontal circle; a much more common sight is a vehicle turning round a corner at a roughly constant speed.

By modelling the vehicle as a particle, the speed as constant, the path taken as part of a circle and the road surface as level, we can find an approximation for the frictional force needed between the tyres and the road surface.

We have seen that it is possible to find estimates for the values of quantities in many practical problems by forming a mathematical model. Whether or not these estimates are reasonable depends upon the validity of the assumptions used to produce the model. In many cases the validity of the assumptions can be tested only by experiment. In other cases, common sense is often a good judge of whether a particular assumption is reasonable. For example, modelling a lorry as a particle is sensible if it is going round a wide bend but not if the bend is a tight one (watch an articulated lorry turn a corner and you will see that different wheels follow different paths round the bend). Whatever the case it is important to describe the model carefully, giving all the assumptions that are made.

Example 15e

A car of mass 600 kg turns left at constant speed 4 m s^{-1}, moving on a circular path of radius 4.7 m.

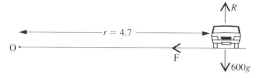

   (a)   **Find the least value of the coefficient of friction between the road and the tyres if the bend is to be taken without skidding.**

   (b)   **State, with comments on their validity, any assumptions that you have made in your solution.**

(a) As the car moves with constant speed while turning the corner, there is no overall force acting in the direction of motion. There is a frictional force on the tyres towards the centre of the circular path.

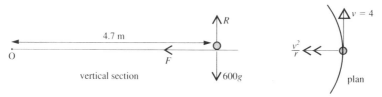

Resolving ↑ $\quad R - 600g = 0 \quad\Rightarrow\quad R = 600g$ [1]

Newton's Law ←, $\quad F = \dfrac{mv^2}{r} \quad\Rightarrow\quad F = \dfrac{(600)(4^2)}{4.7}$ [2]

Using $\quad F \leqslant \mu R \quad$ gives

$$\frac{600 \times 16}{4.7} \leqslant 600 \times 9.8 \times \mu \quad\Rightarrow\quad \mu \geqslant 0.35 \quad (2\text{ sf})$$

The minimum value of μ is 0.35 (2 sf).

(b) Assumptions made are:

The car is small enough to be treated as a particle – this is a substantial assumption as the radius of the corner might not be much greater than the length of the car.

The road surface is level – not unreasonable but unlikely to be true as roads are usually cambered.

No other forces, such as air resistance, affect the motion – although this can never be *true*, it is not unreasonable because its effect on low-speed motion is small.

Note that the frictional forces between the tyres of a road vehicle and the road surface are a complex issue which has been greatly simplified in the example above.

The driving force of the vehicle causes the wheels to rotate so as to drive the car forward but this is possible only if there is a frictional force that prevents the tyres from spinning on the spot, i.e. there *is* some friction along the line of motion.

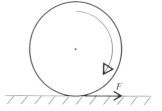

What we can be sure of however is that, at constant speed, the *resultant* force in the direction of motion is zero.

EXERCISE 15e

In this exercise use a mathematical model to solve the problems. Describe the model, state the assumptions that are made and comment on their validity.

1. A space station is planned in the shape of a wheel. The astronauts are to live and work in the space corresponding to the position of a tyre. The station has an overall diameter of 50 m. To provide a simulation of gravity the station will be made to rotate about its centre C. Find the angular velocity necessary to produce an acceleration of 9.8 m s^{-2} at a point 25 m from the centre, e.g. point P on the diagram.

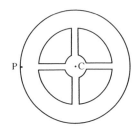

2. An object of mass 0.3 kg is placed on a horizontal turntable, which is rotating at a constant rate, at a distance of 0.2 m from the axis. The coefficient of friction between the object and the turntable is 0.4. Find the greatest possible value for the angular velocity if the object is not to slip outward from its position.

3. A car of mass 800 kg is travelling round a bend of radius 150 m.

(a) Find the frictional force on the tyres when the speed is 30 m s^{-1}.

(b) The coefficient of friction is 0.7. Find the maximum speed at which the car can go without skidding outwards.

4. A van is carrying a parcel of mass 5 kg. When the van goes round a corner the parcel, provided that it does not slip, follows a path which is to be treated as an arc of a circle.

(a) If the radius of the arc on which the parcel moves is 3.8 m and the van is cornering at 30 km h^{-1}, find the frictional force on the parcel, assuming that it does not slip.

(b) In fact parcels have been sliding about and it is decided to provide a rougher surface inside the van. Find the least coefficient of friction for which the parcel described above will not slip.

(c) A surface coating is provided for which the coefficient of friction is 2. If the driver takes a bend at a speed of 50 km h^{-1}, with the parcel moving on an arc of radius 10.5 m, will the parcel now slip?

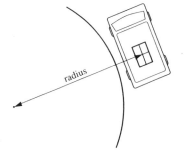

5. A ride at a funfair consists of cars which are made to move in a horizontal circle, of radius 4 m, at a rate of 0.3 revolutions per second. A girl of mass 40 kg is riding in a car. Find the horizontal and vertical components of the force exerted by the car on the girl.

6. On a fairground ride, people stand against the sides of a cylinder, of radius r metres, which then starts to rotate about its axis which is vertical. When a suitable angular velocity ω rad s^{-1} is reached and maintained constant, the floor descends and the people are held in place by the friction between them and the wall. The coefficient of friction is μ. A man of mass m kilograms takes the ride.

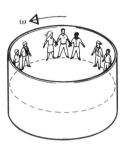

(a) Find the normal reaction force from the wall on the man.

(b) Find the frictional force necessary to prevent him from sliding down the wall.

(c) Find the least angular speed at which the floor can be lowered.

(d) Evaluate this angular speed for the case $r = 2.5$, $\mu = 0.4$, and find the speed with which the man is then moving.

7. Aaron, Beth, Carol and Dipak are skating. They hold hands, in this order, with their arms outstretched. They are of similar size. Their average span from left hand to right hand is 150 cm and each of their masses is approximately 50 kg. Aaron stays on a spot and the others skate round him in circles, with the same angular velocity and staying in a straight line along a radius.

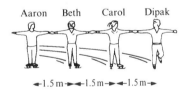

(a) If the angular velocity is ω rad s^{-1}, find the speeds of Beth, Carol and Dipak in terms of ω.

(b) By modelling the skaters as particles and assuming that they are not using their skates to provide any force towards the centre, find in terms of ω,

    (i) the force T_1 which Carol exerts on Dipak

    (ii) the force T_2 which Beth exerts on Carol

    (iii) the force T_3 which Aaron exerts on Beth.

8. A girl is swinging on a rope, of length 5 m, attached to a swivel on top of a pole. It takes her 4 s to complete a circle around the pole. Her mass is 40 kg. Find

(a) the tension in the rope

(b) the angle between the rope and the pole

(c) the radius of the circle.

BANKED TRACKS

When racing cars, cycles, motorbikes, etc. are rounding the bends of a track at high speeds, the available frictional force alone is unlikely to be sufficient to prevent the car from slipping sideways on the track so a further central force has to be provided. This is done by banking the track, i.e. by raising the outside of the curve above the level of the inside. This has the effect of 'tipping' the normal contact force that the ground exerts on the vehicle, away from the vertical so that it has a horizontal component. This component then forms part of the resultant force towards the centre. At the same time, however, only part of the frictional force (the horizontal component) now acts along the radius.

Consider a car of mass m kg, racing on a track that has curved bends with a radius of r metres. The curved sections are banked at an angle θ (i.e. at θ to the horizontal).

If the car travels round the bends at a speed of v metres per second, such that a frictional force acts down the slope, the following relationships can be formed.

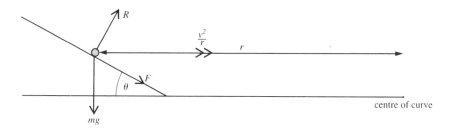

Resolving $\uparrow$ $\qquad R \cos \theta - F \sin \theta - mg = 0$

Newton's Law $\rightarrow$ $\qquad R \sin \theta + F \cos \theta = mv^2/r$

Also $\qquad\qquad\qquad F \leqslant \mu r$

These equations provide the means of solving various problems, e.g. the maximum possible speed for a given angle of banking can be found.

In particular the angle at which the track should be banked for a specified design speed (i.e. the speed at which there is no tendency to slip and therefore no frictional force is needed), can be estimated.

Note that any such values can only be estimates, as a number of assumptions have been made, e.g. the car is modelled as a particle;
the difference between the outer and inner radius is ignored;
constant speed is assumed round the curve.

When a car moves at the design speed it has no tendency to slip sideways on the track so there is no friction up or down the banked track.

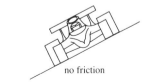

no friction

If the car's speed is higher than the design speed it tends to slip *up* the track and friction therefore acts *down* the track, i.e. it has a component *towards* the centre of the curve.

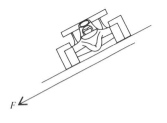

If the car's speed is *lower* than the design speed it tends to slip *down*, and friction therefore acts *up* the track, i.e. it has a component *away from* the centre of the curve.

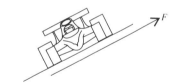

Bends on a railway track are dealt with in a similar way, the outer rail being raised above the level of the inner rail. The difference is that no friction is involved in this case, as movement up or down the banked track is prevented by the lateral force exerted by the rails on the flanges of the wheels.

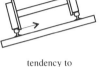

tendency to
move inwards

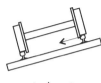

tendency to
move out

In the case of trains moving round banked curved tracks it is much more reasonable to assume constant speed and to ignore the difference in radii as either radius is very large.

Examples 15f

1. A curved section of a race track, where the radius is 120 m, is banked at 40°. By modelling a car that drives round the track as a particle, show that the design speed for this section, $V\,\text{m s}^{-1}$, is independent of the mass of the car and find its value.

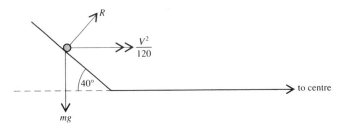

If there is no tendency to slip there is no lateral frictional force.

Resolving ↑ $\qquad R \cos 40° - mg = 0$ $\hfill [1]$

Newton's Law → $\qquad R \sin 40° = m\dfrac{V^2}{120}$ $\hfill [2]$

Hence $\qquad \dfrac{R \sin 40°}{R \cos 40°} = \dfrac{mV^2}{120} \div mg = \dfrac{V^2}{120g}$

∴ $\qquad V^2 = 120g \times \tan 40°$ which is independent of m.

The design speed is $31\,\text{m s}^{-1}$ (2 sf) i.e. $110\,\text{km h}^{-1}$ (2 sf)

2. A railway line is to be laid round a circular arc of radius 500 metres. It is expected that trains will travel over this section of the track at a speed of 45 kilometres per hour. Find

   (a) the force exerted by the outer rail on the flanges of the wheels if the track is level and the mass of the train is 35 tonnes.

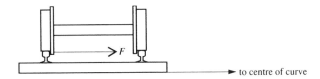

   (b) the height at which the outer rail should be raised above the inner rail to ensure that there is no pressure on the wheel flanges at the expected speed, given that the gauge of the track is 1.5 m (i.e. the rails are 1.5 m apart).

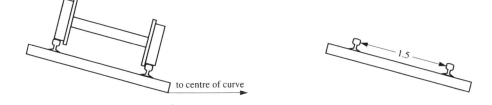

(a) The force acting on the train towards the centre of the curve is provided by the inward pressure on the wheels of the outer rail. The inner rail does not exert any force on the wheels.

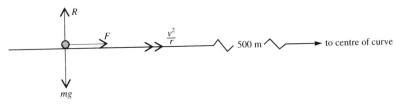

As we are not interested in any vertical forces, we will not resolve vertically.

$$45 \text{ km h}^{-1} = \frac{45\,000}{60 \times 60} \text{ m s}^{-1} = 12.5 \text{ m s}^{-1}$$

Newton's Law → $\quad F = 35\,000 \times \dfrac{(12.5)^2}{500} = 10\,937.5$

The force from the outer rail is $\quad 11 \text{ kN} \quad (2 \text{ sf})$

(b) Each rail exerts a normal reaction on the train and these are probably different, but we will take R as the resultant normal reaction.

We will take m kg as the mass of the train, h metres as the difference in height of the two rails, and α as the angle at which the banked track is inclined to the horizontal.

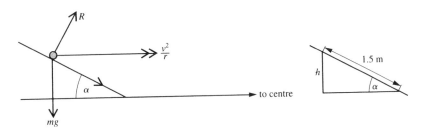

Resolving ↑ $\quad R \cos \alpha - 9.8m = 0$

Newton's Law → $\quad R \sin \alpha = \dfrac{m(12.5)^2}{500}$

Hence $\quad \dfrac{R \sin \alpha}{R \cos \alpha} = \dfrac{(12.5)^2 m}{500 \times 9.8m} \quad \Rightarrow \quad \tan \alpha = 0.03188\dots$

∴ $\quad \alpha = 1.8° \quad (2 \text{ sf})$

Then $\quad h = 1.5 \sin \alpha = 0.0478\dots$

The outer rail should be 48 mm above the inner rail $\quad (2 \text{ sf})$.

3. **If the car in Example 1 is of mass 840 kg and drives round the curved section of the track at 36 m s⁻¹, find the magnitude and direction of the lateral frictional force exerted by the track on the car.**

36 m s⁻¹ is greater than the design speed of 31.4 m s⁻¹ so friction acts down the track.

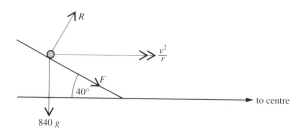

Resolving ↑ $R \cos 40° - F \sin 40° - 840g = 0$

i.e. $R \cos 40° - F \sin 40° = 8232$ [1]

Newton's Law → $R \sin 40° + F \cos 40° = 840 \times \dfrac{36^2}{120}$

$$= 9072 \qquad [2]$$

[2] × cos 40° − [1] × sin 40°   gives

$$F \cos^2 40° + F \sin^2 40° = 9072 \cos 40° - 8232 \sin 40°$$

∴ $F(\cos^2 40° + \sin^2 40°) = 1658$

But $\cos^2 40° + \sin^2 40° = 1$

∴ $F = 1700$ (2 sf)

The frictional force is 1.7 kN acting down the banked track.

EXERCISE 15f

In this exercise give numerical answers corrected to 2 significant figures.

1. A train of mass 50 tonnes travels at 18 m s⁻¹ round a bend which is an arc of a circle of radius 1.5 km. The track is horizontal.

   (a)   Find the force exerted on the side of a rail.

   (b)   On which rail does this act?

2. A locomotive is travelling at 80 km h⁻¹, on a horizontal track, round a bend which is an arc of a circle. The locomotive has a mass of 10 tonnes. The lateral force exerted on a rail by the wheel flanges is 8000 N.

   (a)   On which rail is this force acting?

   (b)   Find the radius of the bend.

3. A road banked at $10°$ goes round a bend of radius $70\,m$. At what speed can a car travel round the bend without tending to side-slip?

4. On a level section of a race track a car can just go round a bend of radius $80\,m$ at a speed of $20\,m\,s^{-1}$ without skidding.

(a) Find the coefficient of friction.

On a section of the track that is banked at an angle θ to the horizontal a speed of $30\,m\,s^{-1}$ can just be reached without skidding, the coefficient of friction being the same in both cases. Taking the value of g as 10,

(b) show that $\dfrac{\cos\theta + 2\sin\theta}{2\cos\theta - \sin\theta} = \dfrac{9}{8}$

(c) find θ to the nearest degree.

5. A circular race track is banked at $45°$ and has a radius of $200\,m$. At what speed does a car have no tendency to side-slip? If the coefficient of friction between the wheels and the track is $\frac{1}{2}$, find the maximum speed at which the car can travel round the track without skidding.

6. An engine of mass $80\,000\,kg$ travels at $40\,km\,h^{-1}$ round a bend of radius $1200\,m$. If the track is level, calculate the lateral thrust on the outer rail. At what height above the inner rail should the outer rail be raised to eliminate lateral thrust at this speed if the distance between the rails is $1.4\,m$?

7. A race track has a circular bend of radius $50\,m$ and is banked at $40°$ to the horizontal. If the coefficient of friction between the car wheels and the track is $\frac{3}{5}$, find within what speed limits a car can travel round the bend without slipping either inwards or outwards.

8. A bend of a race track is banked at $45°$. The coefficient of friction between the wheels of a car and this track is $\frac{1}{2}$. The maximum speed at which the car can go round the bend without skidding is V.

(a) Find the radius of bend in terms of V.

(b) Find the design speed in terms of V.

9. The 'wall of death' at a fairground is in the form of the curved surface of a cylinder, of internal diameter $8\,m$, with its axis vertical. A motorcyclist rides round the wall on a path which is a horizontal circle. The coefficient of friction between wall and tyres is 0.9. Find the minimum speed he must maintain to stay on his circular path without slipping down the wall. State any assumptions you make in modelling this problem.

*10. The force which keeps an aircraft in the air is lift produced by the flow of air over the wings. The direction of this force is perpendicular to a line joining the wing-tips.

Aircraft in level flight   Aircraft in banked flight

(a) Explain why the aircraft must be banked to make it fly on a horizontal circular path.

(b) At what angle must it be banked in order to make it fly on such a path, of radius 2000 m, at $150\,\mathrm{m\,s^{-1}}$.

State any assumptions you make in modelling this problem.

*11. A bend of a race track, of radius r, is banked at $45°$. The coefficient of friction between the wheels of a car and the track is μ. The maximum speed at which the car can go round the bend without skidding is V_1 and the minimum speed at which it can go round without skidding is V_2.

(a) Show that $\dfrac{V_1^2}{rg} = \dfrac{1+\mu}{1-\mu}$.

(b) Obtain a similar expression involving V_2.

(c) Use your results from (a) and (b) to show that $\dfrac{V_1^2\,V_2^2}{r^2\,g^2} = 1$ and hence find r in terms of V_1 and V_2.

(d) Find the design speed in terms of V_1 and V_2.

CHAPTER 16

MOTION IN A VERTICAL CIRCLE

MOTION ON A CURVE WITH VARIABLE SPEED

We saw in Chapter 15 that when a particle describes a curved path at constant speed, the particle has no acceleration in the direction of motion, i.e. no acceleration in the direction of the tangent to the curve at any instant. There is, however, an acceleration perpendicular to the direction of motion which is a measure of the rate of change of the direction of the velocity.

If we now consider motion on a curve when the speed is *not* constant it is clear that

(a) again there is an acceleration component perpendicular to the tangential direction,

(b) there is also an acceleration component *in* the direction of motion which is a measure of the rate of change of the magnitude of the velocity, and which can be expressed as $\mathrm{d}v/\mathrm{d}t$ (or $\dot{v}$).

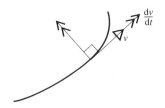

In order to produce these two acceleration components, the force acting on the particle must also have two components that are in the directions of the acceleration components.

This is the situation for a particle describing a curved path in a vertical plane; the weight always acts vertically downwards so, as the direction of motion changes, the weight can be resolved into two components, one along, and one perpendicular to, the direction of motion.

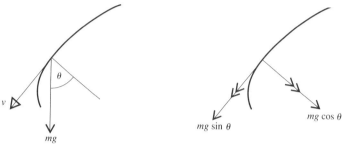

MOTION ON A CIRCLE IN A VERTICAL PLANE

When the path is a circle we know that the component of acceleration towards the centre is v^2/r or $r\omega^2$.

There are various ways by which an object can be made to travel in a vertical circular path. In some cases an object is controlled by a machine which involves technical knowledge beyond the scope of this book. In other cases the particle, once it is set moving, moves under the action of its own weight. It is situations of this type that we deal with here.

Motion Restricted to a Circular Path

Consider a small bead P, threaded on to a circular wire, radius a and centre O, that is fixed in a vertical plane. We will model the bead as a particle and the wire as being friction-free. Suppose that the bead has been set moving round the wire so that it passes through the lowest point A of the wire with speed u and that the speed of the bead is v when it reaches a point B on the wire, where angle AOB is θ.

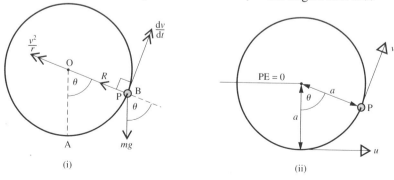

Diagram (i) shows the forces acting on the bead, and the acceleration components produced, as the bead passes through B.

Newtons Law, $F = ma$, can be applied towards O and along the tangent at B.

Using Newton's Law at B gives

$$\diagdown \quad R - mg \cos \theta = \frac{mv^2}{a} \tag{1}$$

$$\diagup \quad -mg \sin \theta = m \frac{dv}{dt} \tag{2}$$

The only external force that acts on the bead, other than its weight, is the normal reaction R. Now R is always perpendicular to the direction of motion so it does no work. Therefore conservation of mechanical energy can be applied and diagram (ii) is useful here.

Taking the PE to be zero at the level of the centre O, we have:

Total ME at A is $\quad \frac{1}{2} mu^2 - mga$

Total ME at B is $\quad \frac{1}{2} mv^2 - mga \cos \theta$

Using conservation of mechanical energy from A to B gives

$$\frac{1}{2} mu^2 - mga = \frac{1}{2} mv^2 - mga \cos \theta \tag{3}$$

These equations provide a solution to most problems, in fact equations [1] and [3] are very often all that are required.

Note that the level of A could have been chosen as the PE zero level, but measuring heights relative to the level of the centre is often more straightforward.

Note also that the bead can move on the wire in one of two basic ways: it can perform complete circles or it can oscillate through an arc. In either case the path is at all times on a circle because the bead is physically prevented from leaving the wire.

Note also that the normal reaction acting on the bead can be either towards the centre of the circle or away from it, e.g.

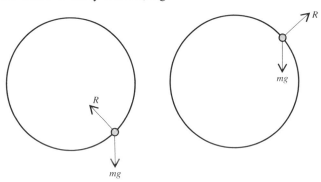

Another example of motion restricted to a circular path is that of a particle attached to one end of a rod that is rotating about the other end, which is pivoted at a fixed point. In this case it is the force in the rod which, together with the weight, causes the circular motion. This force, like the normal reaction in the case above, can act on the particle inwards (tension) or outwards (thrust).

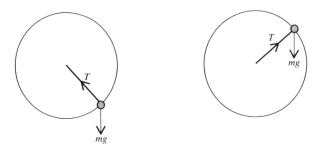

Example 16a

One end of a light rod of length a metres is pivoted at a fixed point O and a particle of mass m kg is attached to the other end. When the rod is hanging at rest, the particle is given a blow that makes it begin to move with velocity V metres per second.

Give answers in terms of a, m and g.

(a)   Find the value of V if the rod first comes to rest when horizontal.

(b)   Show that, for the particle to perform complete circles, $V \geqslant 2\sqrt{ga}$.

(c)   When $V = 2\sqrt{ga}$, find the force in the rod when the particle is at the highest point, and say whether it is a tension or a thrust.

(d)   Given that $V = \sqrt{3ga}$, find the height above O of the particle when the tension in the rod is zero.

The force acting on the particle towards the centre is the tension in the rod. It is always perpendicular to the direction of motion so does no work.

The diagrams show the particle at the lowest point A, and the velocities, forces and accelerations of the particle at a general point B where angle AOB is θ.

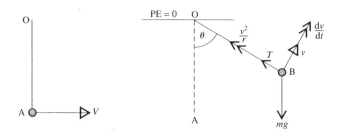

Using Newton's Law at B towards the centre gives

$$T - mg \cos \theta = \frac{mv^2}{a} \qquad [1]$$

Using conservation of mechanical energy from A to B gives

$$\tfrac{1}{2}mV^2 - mga = \tfrac{1}{2}mv^2 - mga \cos \theta \qquad [2]$$

(a) The rod first comes to rest when horizontal,

  i.e. $v = 0$ when $\theta = 90°$

  Hence from [2] $\tfrac{1}{2}mV^2 - mga = 0$ $\Rightarrow$ $V^2 = 2ga$

  The value of V is $\sqrt{2ga}$.

(b) For the particle to describe complete circles it must pass through the highest point with a positive velocity.

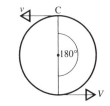

  i.e. $v > 0$ when $\theta = 180°$

  From [2] $v^2 = V^2 - 2ga + 2ga \cos \theta$

  $\therefore$ $V^2 - 2ga + 2ga(-1) > 0$ $\Rightarrow$ $V^2 > 4ga$

  For complete circles, $V > 2\sqrt{ga}$.

(c) If $V = 2\sqrt{ga}$,   the particle *just* reaches the highest point C and the speed of the particle there is zero, i.e. $\frac{v^2}{a}$ is zero.

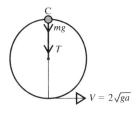

  Applying Newton's Law at C gives

$$T + mg = \frac{mv^2}{a} = 0 \quad \Rightarrow \quad T = -mg$$

As T is negative the force in the rod acts outwards. Therefore the force in the rod is a thrust of magnitude mg.

(d) When $T = 0$,

  from [1] $v^2 = -ga \cos \theta$

  $\therefore$ from [2] $\tfrac{1}{2}V^2 - ga = -\tfrac{1}{2}ga \cos \theta - ga \cos \theta$

  $\Rightarrow$ $V^2 - 2ga = -3ga \cos \theta$

  As $V = \sqrt{3ga}$, $ga = -3ga \cos \theta$ $\Rightarrow$ $\cos \theta = -\tfrac{1}{3}$

  The height of the particle above O is $a \cos(180° - \theta)$

  i.e. the particle is $\tfrac{1}{3}a$ above O.

EXERCISE 16a

Questions 1 to 4 are about a bead P, of mass 1 kg, threaded on to a smooth circular wire, of radius 0.2 m, that is fixed in a vertical plane.

1. The bead is projected from the lowest point of the wire with a speed of $3\,\mathrm{m\,s^{-1}}$. Find the speed of P, $v\,\mathrm{m\,s^{-1}}$, when

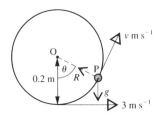

(a) $\theta = 60°$ (b) $\theta = 90°$ (c) $\theta = 180°$

2. For each part of question 1, find the value of R where R newtons is the normal reaction of the wire on the bead. State in each case whether R is acting towards or away from the centre.

3. P is slightly displaced from rest at the highest point. Find v when

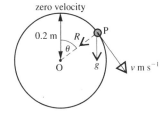

(a) $\theta = 60°$ (b) $\theta = 90°$

(c) $\theta = 180°$ (d) $\theta = 360°$

4. P is projected from the lowest point at $2.5\,\mathrm{m\,s^{-1}}$. Find the angle through which OP has rotated when the reaction between the wire and the bead is zero.

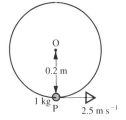

5. A particle P is fastened to one end of a light rod of length 1.4 m. The other end of the rod is smoothly pivoted to a fixed point O. The rod is released from rest when OP is horizontal. Find the speed of P, $v\,\mathrm{m\,s^{-1}}$, when OP has rotated through   (a) 60°   (b) 90°.

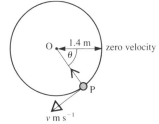

6. If in question 5 the mass of P is 1 kg, find for each angle the force T newtons that the rod exerts on the particle, stating whether the force is a tension or a thrust.

7. A smooth circular wire of radius 3 m is fixed in
a vertical plane. A bead P of mass 1.5 kg is
threaded on to the wire. P is projected at
5 m s⁻¹ from the highest point on the wire.
When the bead has rotated through 90°, find

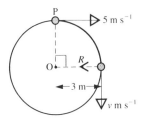

(a) the normal contact force exerted by the
wire on the bead,

(b) the force exerted on the bead in the
direction of the tangent.

8. Referring to the situation described in question 7, when the bead has rotated
through 90°, find the radial and tangential components of the acceleration of P.

9. A light rod of length 3 m is smoothly pivoted at one end to a fixed point O. A
particle P of mass 1.5 kg is attached to the other end of the rod. The particle is
held so that OP is horizontal and is then projected downwards with speed 6 m s⁻¹.
Find the angle through which the rod has rotated when

(a) the force exerted by the rod on the particle first becomes zero,

(b) the speed of P first becomes zero.

In each case find the height of P above O at that instant.

10. If in question 9 the rod is initially held at rest with P vertically above O and is
then slightly displaced, find the radial and tangential components of the
acceleration of P when OP has rotated through

(a) 30°   (b) 90°   (c) 150°.

11. One end of a light rod of length a, is smoothly pivoted to a fixed point and a
particle of mass m is attached to the other end. When the rod is hanging
vertically downwards, the particle is set moving with a speed u. Find the values
of u for which the rod will rotate through a complete revolution.

*12. A particle of mass m is attached to the end A of a light rod AB of length l, free
to rotate in a vertical plane about the end B. The rod is held with A vertically
above B and the particle is projected from this position with horizontal velocity u.
When the particle at A is vertically below B it collides with a stationary particle
of mass $2m$ and coalesces with it. If the rod goes on to perform complete circles
find the range of possible values of u.

*13. Two beads A and B of masses m and $2m$ respectively are free to slide in a vertical
plane round a smooth circular wire of radius a and centre O. The bead A is at
rest at the lowest point C of the wire while B is released from rest at a point on
the same level as O. If the coefficient of restitution between the beads is $\frac{1}{2}$, find
the height above C to which each particle rises after impact.

MOTION NOT RESTRICTED TO A CIRCULAR PATH

So far we have considered situations where all the motion took place on a circular path, whether complete circles or oscillations were performed. There are other cases however in which an object begins by moving on a circle but then moves on a different path. Consider, for example, a particle fastened to one end of an inelastic string whose other end is fixed.

If the particle is set moving in a vertical plane it may, as in the cases considered earlier, rotate in complete circles

or it may oscillate through an arc that is less than a semicircle.

But there is now a third possibility. *If the string goes slack* during the motion, the particle will 'fall inside' the circular path and travel for a time under the action *only* of its weight.

The reason for the third case is that the string, unlike a light rod, cannot exert a thrust; it can only pull. So motion in a circle at the end of a string can take place only as long as the string is taut, i.e. as long as the tension, T, is greater than, or equal to, zero. At the instant when the particle is *about* to leave its circular path, $T = 0$.

If the particle is to describe complete circles we must ensure that $T \geqslant 0$ when $\theta = 180°$. (It is no longer sufficient to say that $v > 0$ when $\theta = 180°$, because this is true even if the particle has left the circular path.)

A particle set moving on the inside of a circular surface gives rise to a similar situation. The particle moves on a circle as long as it is in contact with the surface, i.e. as long as there is a positive inward contact force, R. If the particle loses contact with the surface its path falls inside the circle as shown above. At the point on the surface where contact is lost, $R = 0$.

If there is to be no loss of contact, i.e. complete circles are to be performed, then $R \geqslant 0$ at the top.

Examples 16b

1. A particle of mass m is attached to one end of a light inelastic string of length a whose other end is fixed to a point O. When the particle is hanging at rest it is given a horizontal blow which causes it to begin to move in a vertical plane with initial speed V. Find the ranges of values of V for which the particle at no time leaves a circular path.

The particle will move in a circular path provided that the string does not go slack.

One situation in which the string cannot go slack is when the particle oscillates through no more than 180°.

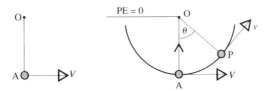

The tension is always perpendicular to the direction of motion so it does no work.

Using conservation of mechanical energy from A to P,

$$\tfrac{1}{2}mV^2 - mga = \tfrac{1}{2}mv^2 - mga \cos \theta \qquad [1]$$

For oscillations through not more than 180°,

$$v = 0 \quad \text{when} \quad \theta \leqslant 90°, \quad \text{i.e. when} \quad \cos \theta \geqslant 0$$

Hence, from [1] $\quad mga \cos \theta = mga - \tfrac{1}{2}mV^2 \geqslant 0$

$$\Rightarrow \qquad V \leqslant \sqrt{2ga}$$

The other situation in which the string does not go slack is when the particle describes complete circles. For this to happen the tension must be greater than or equal to zero when $\theta = 180°$.

Using $\theta = 180°$ in [1] gives $\tfrac{1}{2}mV^2 - mga = \tfrac{1}{2}mv^2 - mga(-1)$

$$\Rightarrow \qquad v^2 = V^2 - 4ga$$

Using Newton's Law towards the centre at the highest point gives

$$T + mg = \frac{mv^2}{a} \qquad \Rightarrow \qquad T = \frac{m}{a}(V^2 - 4ga) - mg$$

When $\theta = 180°, T \geqslant 0 \quad \Rightarrow \quad \dfrac{m}{a}(V^2 - 4ga) - mg \geqslant 0$

$$\Rightarrow \qquad V \geqslant \sqrt{5ga}$$

∴   The particle will not leave the circular path if $V \leqslant \sqrt{2ga}$ or $V \geqslant \sqrt{5ga}$.

2. A smooth cylinder of radius 0.3 m is fixed, with its axis horizontal, on a horizontal plane. A small bead of mass 1 kg is placed at the highest point A on the outside of the cylinder and is just displaced from rest. The bead subsequently loses contact with the surface of the cylinder at a point P.

(a) Find the height of P above the axis of the cylinder.

(b) State, to the nearest degree, the angle between the horizontal and the direction of motion of the bead when it leaves the cylinder.

(c) Find the speed and the tangential acceleration of the bead at this instant.

(d) Describe the subsequent motion of the bead and sketch its path.

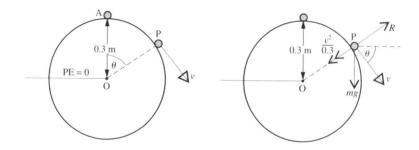

(a)   The normal reaction R acts outward.

The bead will leave the surface when $R = 0$.

Conservation of ME from A to P gives

$$0 + (1)(9.8)(0.3) = \tfrac{1}{2}(1)v^2 + (1)(9.8)(0.3 \cos \theta)$$

$\Rightarrow$ $v^2 = 5.88(1 - \cos \theta)$ [1]

Using Newton's Law at P gives

$$(1)(9.8)\cos \theta - R = (1)\left(\frac{v^2}{0.3}\right)$$ [2]

$$= 19.6(1 - \cos \theta)$$ using [1]

When $R = 0$, $29.4 \cos \theta = 19.6$ $\Rightarrow$ $\cos \theta = \tfrac{2}{3}$

The height of the bead above the centre is then $0.3 \cos \theta$

The bead leaves the surface of the cylinder when it is at the point which is 0.2 m above the centre.

(b)   From (a) $\cos \theta = \frac{2}{3}$ $\Rightarrow$ $\theta = 48°$ (to the nearest degree).

Therefore the direction of motion of the bead when it leaves the surface of the cylinder is at $48°$ to the horizontal.

(c)   From [1], the speed of the bead when it leaves the surface is given by

$$v^2 = 5.88 \left(1 - \tfrac{2}{3}\right) \qquad \Rightarrow \qquad v = 1.4$$

The speed of the bead is $1.4\,\mathrm{m\,s^{-1}}$.

Using Newton's Law along the tangent gives

$$mg \sin \theta = m \frac{dv}{dt}$$

$$\Rightarrow \qquad \frac{dv}{dt} = \frac{g\sqrt{5}}{3} = 7.304\ldots$$

The tangential acceleration is $7.30\,\mathrm{m\,s^{-2}}$ (3 sf).

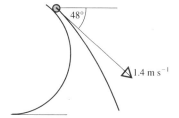

(d)   The bead now moves under the action of its weight only, therefore it travels as a projectile with initial speed $1.4\,\mathrm{m\,s^{-2}}$, at $48°$ to the horizontal.

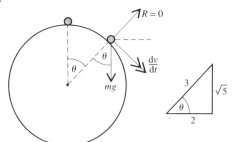

3.   **The diagram shows a section of a toy racing car track. It consists of a slope of length l metres, inclined at 30° to the horizontal, which levels off and then curves upward and round in a complete circle of radius 25 cm, in which the car 'loops the loop'.**

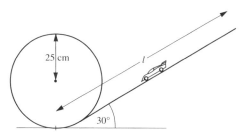

(a)   **Making an assumption, which you should state, use energy considerations to find, in terms of l and g, the velocity V of the car at the foot of the incline, given that it starts from rest at the top.**

(b)   **The car of mass m kg is intended to travel round the inside of the circular loop without losing contact with the track. By defining an appropriate model, show that the value of l must be at least 1.25.**

(a)

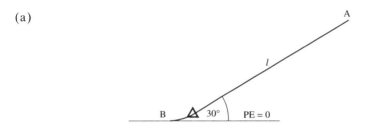

Assuming that there is no resistance to the motion of the car, we can use conservation of mechanical energy from A to B.

$$mgl \sin 30° + 0 = 0 + \tfrac{1}{2}mV^2$$
$$\Rightarrow \qquad\qquad V = \sqrt{gl}$$

(b) Model the car as a particle and the circular track as smooth. Assume that the velocity of the car at lowest point on the circular loop is equal to that at the foot of the incline.

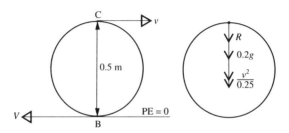

The normal reaction R does no work.

Using conservation of ME from B to C gives

$$\tfrac{1}{2}mV^2 + 0 = \tfrac{1}{2}mv^2 + m(9.8)(0.5)$$
$$\Rightarrow \qquad\qquad \tfrac{1}{2}gl = \tfrac{1}{2}v^2 + 4.9$$
$$\Rightarrow \qquad\qquad v^2 = 9.8(l-1) \qquad\qquad\qquad [1]$$

If the car loses contact with the track, it will be when the contact force disappears, so the condition we want is that $R \geqslant 0$ at the highest point.

Using Newton's Law towards the centre at C gives

$$R + m(9.8) = m\left(\frac{v^2}{0.25}\right)$$

$R \geqslant 0 \qquad \Rightarrow \qquad v^2 \geqslant 2.45$

$\therefore$ from [1] $9.8(l-1) \geqslant 2.45$

$\Rightarrow \qquad\qquad\qquad\qquad l \geqslant 1.25$

The least value of l is 1.25

EXERCISE 16b

Questions 1 to 3 are about a particle P fastened to one end of a light inextensible string whose other end is fixed at a point O.

1. When P, whose mass is 0.2 kg, is hanging vertically below O, it is given a horizontal speed of 5 m s⁻¹. If the length of the string is 0.5 m, find the speed of P when OP has rotated through 60°.

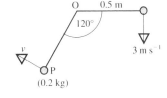

2. If, in question 1, OP is initially horizontal and P is then given a speed 3 m s⁻¹ vertically downwards find

   (a) P's speed when OP has rotated through 120°,

   (b) the tension in the string at that instant.

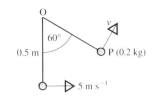

3. From a position where the string, of length 1 m, is horizontal and taut, P is given an initial speed u m s⁻¹ vertically downwards.

   (a) Find the range of values of u for which P will describe complete circles.

   (b) Describe the motion of P if $u = 0$.

In questions 4 to 6 a particle P is set moving in a vertical plane on the inside smooth surface of a cylinder.

4. P is at the lowest point on the surface when it is given a horizontal speed of 6 m s⁻¹. The mass of the particle is 0.5 kg and the radius of the cylinder is 1 m. After P has rotated through 70°, find its speed and the normal reaction exerted by the surface on P.

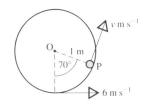

5. P is projected horizontally from the lowest point of the surface, whose radius is a, and after rotating through 90° its speed is $\sqrt{7ga}$. Find the initial speed in terms of a and g.

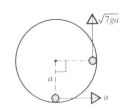

6. The particle, of mass m, is projected vertically upwards from a point level with the centre of the cylinder, with speed $\sqrt{7ga}$. If the radius of the cylinder is a, find

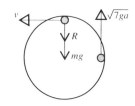

(a) the speed when P reaches the highest point on the surface,

(b) the normal reaction at that point.

7. A particle P is hanging at rest at the end of a light inextensible string of length a m. The other end of the string is fixed at a point O. P is then projected horizontally with speed V m s^{-1}. When OP has rotated through 120°, the string becomes slack (i.e. the tension is zero). Giving answers in terms of a and g find,

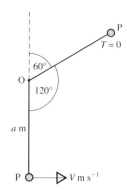

(a) the speed of P at this instant

(b) the value of V.

8. A particle is held in contact with the smooth inner surface of a cylinder of radius 0.08 m, at the point A as shown. It is then given a speed of 4 m s^{-1} vertically downwards and rotates until, at point B, the normal reaction becomes zero. Find the height of B above A and the speed of the particle at B.

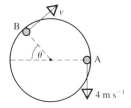

9. An aircraft is looping the loop on a path which is a vertical circle of radius 400 m. Find the minimum speed at the top of the loop for which the pilot would remain in contact with the seat without wearing a seat belt.

Questions 10 and 11 are about a particle P, set moving in a vertical plane on the smooth outer surface of a fixed sphere.

10. The mass of P is 1 kg and the radius of the sphere is 2 m. P is projected horizontally with speed u m s^{-1} from the highest point on the sphere and loses contact with the surface when it has descended a vertical distance 0.5 m. Find the value of u.

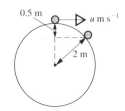

11. The sphere, whose radius is 2 m, is fixed on a horizontal plane and P is just displaced from rest at the highest point. Find

(a)   the height of P above the plane when contact with the sphere is lost,

(b)   P's speed at this instant,

(c)   P's speed when it reaches the plane (use an energy method).

12. A child whirls a basket containing an apple, of mass m kilograms, in a vertical circle of radius 0.6 m. The speed of the apple at the highest point is v metres per second.

(a)   Find, in terms of m and v, the force exerted by the basket on the apple when the apple is at the highest point.

(b)   Find the least value v can have if the apple is not to lose contact with the basket.

⋆13. One very cold winter's day the dome of St Paul's Cathedral becomes icy. A pigeon tries to land gently on the dome but immediately it comes to rest on the surface it finds itself sliding down. Surprised, but interested in this new experience, it sits tight until it loses contact and then flies away. The dome is to be modelled as a hemisphere, of radius 15 m, and its icy surface as frictionless. The pigeon's initial position on the roof is at an angular displacement of 10° from the highest point of the hemisphere.

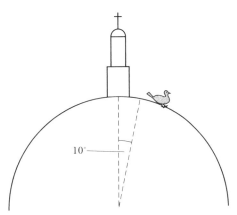

(a)   Find the distance it slides over the roof before losing contact.

(b)   Find its speed just before it starts to fly.

HARDER PROBLEMS

There are many different types of situation involving motion on a circular path in a vertical plane, some requiring ideas that have not been used so far in this chapter. The examples that follow give an indication of the variety of questions that you might meet and the next exercise includes more of them.

Examples 16c

1. A magnet of mass $2m$ kg is attached to one end of a string of length a m. The other end of the string is fixed at a point A. The magnet is held, with the string taut, at a point B level with A, and is released from rest from that position. When the magnet is at the lowest point of its motion it picks up a stationary iron block of mass m kg. Find the height to which the combined mass rises.

First find the speed of the magnet just before it picks up the iron block.

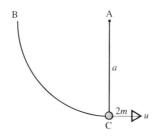

Using conservation of mechanical energy from B to C gives

$$2mga = \tfrac{1}{2}(2m)u^2 \qquad \Rightarrow \qquad u = \sqrt{(2ga)}$$

Just before impact

Just after impact

Using conservation of momentum at impact gives

$$2mu + 0 = 3mv \qquad \Rightarrow \qquad v = \tfrac{2}{3}u = \tfrac{2}{3}\sqrt{2ga}$$

Now we will consider energy again as the mass $3m$ rises to an unknown height h m above C.

Using conservation of mechanical energy
from C to D gives

$$\tfrac{1}{2}(3m)v^2 = 3mgh$$

$$\Rightarrow \qquad h = \frac{v^2}{2g} = \left(\frac{1}{2g}\right)\left(\frac{8ga}{9}\right)$$

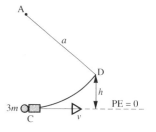

The combined mass rises to a height of $\tfrac{4}{9}a$.

In most of the problems in this chapter an object has been travelling on a vertical circle under the action of its own weight, and therefore with varying speed. There are situations however in which a vertical circular path is described at constant speed, e.g. the motion of a 'Big Wheel' once all the passengers are aboard. The next example illustrates the special features of this case.

2. A thin metal spoke, of length 1.2 m, carries a small
load of mass 0.7 kg at one end and is pivoted at the
other end to a fixed point O. The spoke is being
rotated manually in a vertical plane at a constant
angular speed of 1 revolution per second.

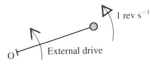

   1 rev s^{-1}

External drive

  (a) Explain why the conservation of mechanical energy cannot be used.

  (b) What is the acceleration of the load towards O when the spoke is
     (i) horizontal   **(ii)** vertically above O   **(iii)** vertically below O.

  (c) Find the force exerted by the spoke on the load in each of the positions defined
     in part (b), stating whether the spoke is in tension or compression.

  (a) External work is being done by whatever is producing the manual rotation so
     as to maintain a constant angular speed. This causes a change in mechanical
     energy, so the total mechanical energy of the system is not constant.

     (Note that the external drive must be a torque (turning effect) and not a linear force)

  (b) The angular speed of the spoke and load is 1 rev s^{-1}, i.e. 2π rad s^{-1}.

     The angular speed of the load, and the radius of the circle, are the same at every point so $r\omega^2$,
     the central acceleration, is also the same at every point.

    In each of the given positions

       the acceleration towards O is $1.2 \times (2\pi)^2$ m s^{-2}

                 i.e.   47 m s^{-2} (2 sf)

(c)

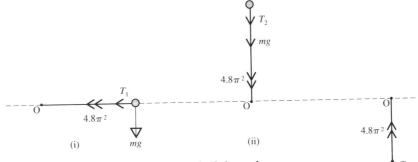

  We will apply Newton's Law towards O in each case.

   (i) $T_1 = (0.7)(4.8\pi^2) = 3.36\pi^2 = 33.16\ldots$

      $\therefore$ the tension is 33 N (2 sf).

  (ii) $T_2 + 0.7g = 3.36\pi^2$ $\Rightarrow$ $T_2 = 26.3\ldots$

      $\therefore$ the tension is 26 N (2 sf).

 (iii) $T_3 - 0.7g = 3.36\pi^2$ $\Rightarrow$ $T_3 = 40.02\ldots$

      $\therefore$ the tension is 40 N (2 sf).

3. Paul is playing with a conker of mass 30 g, fastened to the end of a string 20 cm long. He holds the end of the string in one hand A and holds the conker, C, in the other hand so that the string is taut with AC horizontal. When Paul releases the conker, he intends to watch and see whether it goes right round to be level with his stationary hand again but, after he has let the conker go, one of his friends pushes a stick in the way so that, just as the string becomes vertical, the middle of the string hits the stick at right angles. The conker continues to rotate about the stick as centre.

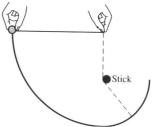

Make a mathematical model for this situation, stating the assumptions made, and use it to

(a) find the speed of the conker just before the string hits the stick and the tension in the string at this instant

(b) explain why the speed of the conker does not change when the string hits the stick

(c) find the tension in the string immediately after it hits the stick

(d) determine whether or not the conker will describe a complete circle about the stick as centre.

Model the conker as a particle, the string as light and inextensible, the stick as having negligible diameter, the end of the string in Paul's hand is perfectly stationary, no resistance to motion, no wind.

(a)

C $\circ$ ———— 0.2 m ———— A $\bullet$

(i)

A

$\bullet$ stick
$\dfrac{v^2}{0.2}$
T
C $\circ\!\!\!\rightarrow\triangleright v$
mg

(ii)

S
$\bullet$ stick
$\dfrac{v^2}{0.1}$
0.1 m $\quad T_1$
C $\circ\!\!\!\rightarrow\triangleright v$
mg

(iii)

Using conservation of mechanical energy from (i) to (ii)

$$0.03 \times 9.8 \times 0.2 = \left(\tfrac{1}{2}\right)(0.03)v^2 \qquad \Rightarrow \qquad v^2 = 3.92$$

$$\Rightarrow \qquad v = 1.98 \quad (3\ \text{sf})$$

Using $F = ma$ in (ii) gives

$$T - mg = \frac{mv^2}{r}$$

$$\Rightarrow \qquad T = (0.03)\left(\frac{3.92}{0.2} + 9.8\right)$$

The tension is 0.88 N (2 sf)

(b) When the string strikes the stick, the radius of the circle being described changes suddenly causing a change in the central acceleration and hence in the tension. However, this instantaneous change in tension is perpendicular to the direction of motion of the conker so the speed of the conker is unchanged.

(c) The radius of the circle being described is now 0.1 m so the tension changes.

$$T_1 - mg = \frac{mv^2}{r}$$

$$\Rightarrow \qquad T_1 = (0.03)\left(9.8 + \frac{3.92}{0.1}\right)$$

The new tension is 1.5 N (2 sf).

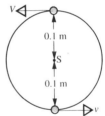

(d) The conker will describe a complete circle about the stick provided that the conker passes through the highest point of that circle, i.e. after 180° rotation the string is still taut.

Supposing that the conker can describe a complete circle then, at the highest point,

conservation of ME would give $\tfrac{1}{2}mv^2 - mg(0.1) = \tfrac{1}{2}mV^2 + mg(0.1)$ [1]

and using $F = ma$ towards S would give $T + mg = \dfrac{mV^2}{0.1}$ [2]

From [1], $V^2 = v^2 - 0.4g = 3.92 - 3.92 \qquad \Rightarrow \qquad V = 0$

Hence from [2] $T = -mg$ which is not possible as tension in a string cannot be negative,

i.e. the string is *not* taut after rotation of 180°.

So we were wrong in supposing that the conker can describe a complete circle. The conker does not perform complete circles about the stick as centre.

Alternatively we could find the angle of rotation θ for which the tension is zero and show that θ is less than 180°.

EXERCISE 16c

1. One end of a light rod of length 3 m, is smoothly pivoted to a fixed point O and a particle P of mass 1.5 kg is attached to the other end. When the rod is hanging at rest, vertically downwards, P is given a horizontal blow of impulse 18 N s as shown in the diagram.

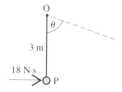

   (a) Find its initial velocity.

   (b) Show that P moves round complete circles.

   (c) Find the force in the rod, T N, when

   (i) $\theta = 160°$ (ii) $\theta = 170°$,

   stating in each case whether it is a tension or a thrust.

   (d) Find the value of θ when $T = 0$.

2. One end of a light rod of length 0.4 m is pivoted at a fixed point A so that it can swing freely in a vertical plane. A particle B of mass 1 kg is attached to the other end of the rod. The rod is released from rest with AB horizontal. When B is vertically below A it collides with a particle C, of mass 0.5 kg, which is approaching it horizontally at $2\,\text{m s}^{-1}$ and the two particles coalesce. Find

   (a) the velocity of B and the tension in the rod just before impact

   (b) the common velocity just after impact

   (c) the tension in the rod just after impact

   (d) the height above the lowest point to which the combined particles now swing.

3. A car, of mass 800 kg, is driven over a hump-backed bridge at a constant speed of $14\,\text{m s}^{-1}$. The upper part of the bridge can be modelled as an arc of a circle, of radius 25 m, this arc subtending an angle of $30°$ at the centre of the circle.

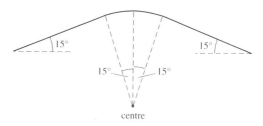

   The gradient of the road approaching the arc is equal to tan 15°. Find

   (a) the reaction on the car when it is at the highest point of the bridge

   (b) the reaction on the car when it is at the end of the curved section of the bridge

   (c) the reaction on the car when it is on the straight section

   (d) the greatest constant speed at which a car may be driven over this bridge without losing contact with the road.

4. A particle of mass 1.5 kg is lying at the lowest point of the inner surface of a hollow sphere of radius 0.5 m when it is given a horizontal impulse. Find the magnitude of the impulse

(a) if the particle subsequently describes complete vertical circles,

(b) if the particle loses contact with the sphere after rotating through 120°.

5. One end of a light inextensible string AB of length l is fixed at A and a particle of mass m is attached at B. B is held a distance l vertically above A and is projected horizontally from this position with speed $\sqrt{2gl}$. When AB is horizontal, a point C on the string strikes a fixed smooth peg so that the radial acceleration of the particle is instantaneously doubled. Express the length of CB in terms of l.

The particle continues to describe vertical circles about C as centre. Compare the greatest and least tensions in the string during this motion.

6. A pair of trapeze artists are performing their act. There is a catcher and a flier. The catcher, of mass 75 kg swings on an arc of radius 7 m and his speed is zero when the trapeze ropes are at 70° to the vertical.
When he reaches the lowest point of his path he catches the flier, who has a mass of 55 kg and is approaching him with a horizontal velocity of 2 m s^{-1}.
Stating all assumptions which you make, find

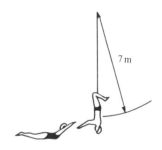

(a) the speed of the catcher just before they connect,

(b) their common velocity just after connecting,

(c) the force with which the catcher must grip the trapeze just after connecting,

(d) the angle the trapeze ropes will make with the vertical when they first come to instantaneous rest.

7. A ballistic pendulum is an instrument used to measure the speed of a bullet. In this problem such a pendulum consists of a bob, of mass 5 kg, suspended by a thin light rod from a fixed point. The bob swings on an arc of a circle of radius 0.6 m against a scale which measures its rotation from the vertical. While the pendulum is hanging vertically at rest a bullet, of mass 0.01 kg is fired horizontally into it. The bob with the bullet embedded in it is observed to swing to a maximum deflection of 24° from the vertical.

(a) Find the common velocity of the bob and bullet just after impact.

(b) Find the tension in the rod just after impact.

(c) Find the velocity of the bullet before hitting the bob.

8.

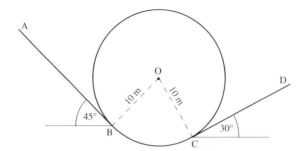

A boy is using his sledge on a snow-covered hillside, which is to be modelled as frictionless. The total mass of the boy and sledge is 50 kg. He starts from rest at point A and slides down the slope AB, which is inclined at 45° to the horizontal and is 20 m long. He then goes into a hollow BC, which is to be treated as an· arc of a circle, of radius 10 m, to which AB is a tangent. He comes out of this hollow, at point C, on to slope CD, which is also a tangent to the arc BC and is inclined at 30° to the horizontal. Find

(a) the normal reaction and his acceleration while on section AB

(b) his speed on reaching point B

(c) the normal reaction just after entering arc BC

(d) his speed on reaching point C

(e) the change in normal reaction as he passes point C.

9. A woman of mass 60 kg is riding on the Big Wheel at a fairground. She is moving on a circle of radius 8 m at a constant angular speed of 5 revolutions per minute. When she has an angular displacement θ from the lowest point, the seat exerts a force on her with components R towards the centre and S along the tangent. Find, in terms of π and θ

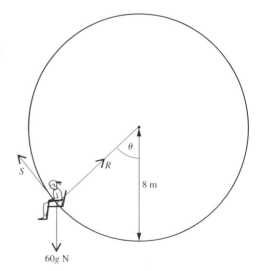

(a) her speed in $m\,s^{-1}$

(b) the components of her acceleration
    (i)   towards the centre
    (ii) along the tangent

(c) R and S

(d) Evaluate, correct to two significant figures, the magnitude of the resultant of R and S when   (i) $\theta = 0°$ (ii) $\theta = 180°$,   and state the direction of this resultant.

10. A light aircraft is looping the loop on a path which is a vertical circle of radius 300 m. At a particular point on its path its speed is $60\,\mathrm{m\,s^{-1}}$ and this speed is increasing at the rate $9\,\mathrm{m\,s^{-2}}$. Find

  (a) its acceleration towards the centre of the circle

  (b) the magnitude of its resultant acceleration.

11. A building is being demolished by swinging a large metal ball at the walls. The ball has a mass of 500 kg and it is attached by a chain to a fixed point. It swings on an arc of a circle of radius 6 m. A cable draws the ball back to a position where the chain is inclined at 30° to the vertical and the ball is then released. It strikes a wall when the chain reaches the vertical position and the impact brings it to rest.

  (a) Find the speed of the ball when it strikes the wall.

  (b) Find the impulse exerted on the wall.

  (c) Find the tension in the chain
    (i) just before the ball strikes the wall.
    (ii) just after the ball strikes the wall.

12. A particle lies at the lowest point on the inside surface of a smooth cylinder, of radius 0.9 m, which is fixed with its axis horizontal. It is then given a horizontal velocity so that it starts to move in a vertical circle. It leaves the surface on reaching point A, after rotating through 120° from the lowest point.

  (a) Find its velocity on reaching point A.

  (b) Find its initial velocity.

  (c) Find the horizontal and vertical components of its velocity at the moment it loss contact.

  (d) By using the vertical component, find the greatest height reached above point A.

  (e) Sketch the path traced out by the particle after leaving the surface.

*13. A particle lies on the outer surface of a smooth sphere, of radius 1 m, which is standing on horizontal ground. It is projected from the highest point with a velocity of $2.5\,\mathrm{m\,s^{-1}}$.

  (a) Find the angle the particle has rotated through when it loses contact with the surface.

  (b) Find its height above the ground at the moment it loses contact.

  (c) Find its velocity at the moment it loses contact.

  (d) Find the horizontal and vertical components of its velocity at the moment it loses contact.

  (e) Consider its motion as a projectile after leaving the surface and find
    (i) the vertical component of its velocity on reaching the ground
    (ii) the time taken to reach the ground
    (iii) the horizontal distance it has travelled in this time.

CONSOLIDATION D

SUMMARY

Motion in a Horizontal Circle

The angular velocity of a particle describing a circle is represented by $d\theta/dt$ or ω, and is measured in radians per second or revolutions per second.

A particle travelling in a circle of radius r metres, with a constant angular velocity ω rad s^{-1}, has a speed round the circumference of $r\omega$ m s^{-1}.

When a particle describes a circle of radius r at a constant speed v (or constant angular velocity ω):

● the acceleration is directed towards the centre of the circle and is of magnitude $r\omega^2$ or v^2/r,
● a force of magnitude $mr\omega^2$ or mv^2/r must act towards the centre.

A string fixed at one end and carrying at the other end a particle performing horizontal circles, is known as a conical pendulum.

The force that enables a vehicle to travel on a horizontal circular path can be provided by friction, or by pressure from rail flanges, or by banking the track on which it moves.

Motion in a Vertical Circle

A particle travelling round a circular path in a vertical plane has

● an acceleration component towards the centre, of $r\omega^2$ where ω is not usually constant,

● a tangential acceleration component of $r\left(\dfrac{d^2\theta}{dt^2}\right)$

If the particle cannot leave the circular path (e.g. a bead on a circular wire) it can describe either complete circles or an arc of any size. For the particle to perform complete circles there must be a positive velocity at the highest point of the circle. Finding the condition for this to apply requires using only the conservation of mechanical energy (provided that no external work is being done).

If there is nothing physical to prevent the particle from leaving the circle, e.g. if it is rotating at the end of a string which can go slack, the particle may describe complete circles or it may oscillate through an arc that is less than a semicircle, or it may leave the circular path and then travel as a projectile.

The condition for complete circles to be described is found by checking that the force, other than the weight, acting along the radius towards the centre (e.g. the tension in the string) does not become zero before the highest point of the circle is reached, (e.g. that $T \geqslant 0$ at the highest point of the circle).

Momentum and Impulse

The momentum of a body of mass m and velocity v is mv.

The impulse of a force F acting for a time t is Ft.

Impulse $=$ Change in momentum

Impulse and momentum are measured in the same unit, the newton second, $N\,s$.

At the instant of a collision or a jerk, an instantaneous impulse occurs. The value of an instantaneous impulse can be found only from the change in momentum it produces.

Elastic Impact

If a collision results in a bounce, the impact is elastic.

Newton's Law of Restitution states that, for two particular colliding particles, the ratio of their relative speed after impact to their relative speed before impact is constant, i.e.

separation speed $= e \times$ approach speed

where e is the coefficient of restitution between the two particles.

In general $\qquad\qquad 0 \leqslant e \leqslant 1$

If $e = 0$ the impact is inelastic and the particles do not bounce.

If $e = 1$ the impact is said to be perfectly elastic and no loss in KE is caused by the collision.

At any impact an impulse acts on each colliding object; the magnitude of the impulse is found by considering the change in momentum of *one* object only.

When both of the colliding objects are free to move, the total momentum in any specified direction is unchanged by the impact.

MISCELLANEOUS EXERCISE D

In this exercise use $g = 9.8$ unless another instruction is given.

1. A brick of mass 3 kg falls from rest at a vertical height of 8 m above firm horizontal ground. It does not rebound.
Calculate the impulse of the force exerted on the ground by the brick.   (AEB)

2. A body of mass 2 kg, moving along a straight line with speed 5 m s^{-1}, collides with a body of mass 1 kg moving in the same direction along the same straight line with speed 2 m s^{-1}. On collision the bodies adhere and move on together. Calculate

(a) their common speed immediately after the collision

(b) the kinetic energy lost during the collision.   (NEAB)

3. A cricket ball, of mass 0.14 kg, is moving horizontally with speed 27 m s^{-1} when it hits a vertically held cricket bat. The ball rebounds horizontally with speed 15 m s^{-1}.
Calculate the magnitude of the impulse, in N s, of the force exerted by the ball on the bat.   (ULEAC)

4. A particle P, of mass 1 kg, is connected to a light inextensible string 0.5 m long. The other end of the string is tied to a fixed point O on a smooth horizontal plane. P moves on the plane in a horizontal circle, centre O, with uniform speed. Given that the string will break when the tension exceeds 8 N, show that P can rotate at 38 revolutions per minute without breaking the string.   (ULEAC)

5.

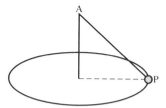

The conical pendulum, shown in the diagram, consists of a light inextensible string which has one end attached to a fixed point A. A particle P, of mass m, is attached to the other end of the string. The particle P moves with constant speed completing 2 orbits of its circular path every second and the tension in the string is $2mg$.

Find, to the nearest cm,

(a) the radius of the circular path of P,

(b) the length of the string.   (ULEAC)

6. Two light strings, AB and BC, are each attached at B to a particle of mass m. The string AB is elastic, of natural length $2a$ and modulus $3mg$. The string BC is inextensible and of length $3a$. The ends A and C are fixed with C vertically below A and AC $= 5a$.

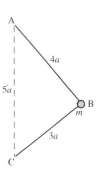

The particle moves with constant speed in a horizontal circle, with both strings taut and AB $= 4a$, as shown in the figure.

(a)   Find the tension in the string AB.

(b)   Find the tension in the string BC.

(c)   Show that the speed of the particle is

$$\sqrt{\left(\frac{44}{5}\, ga\right)}.$$

(ULEAC)

7. A particle of mass 1.8 grams is attached to a fixed point A by a string 1 metre long and describes a horizontal circle below A. Given that the breaking tension of the string is 3 newtons, find the greatest possible number of revolutions per second. (SMP)$_s$

8. A car undergoing trials is moving on a horizontal surface around a circular bend of radius 50 m at a steady speed of 14 m s^{-1}. Calculate the least value of the coefficient of friction between the tyres of the car and the surface. Find the angle to the horizontal at which this bend should be banked in order that the car can move in a horizontal circle of radius 50 m around it at 14 m s^{-1} without any tendency to side-slip.

Another section of the test area is circular and is banked at 30° to the horizontal. The coefficient of friction between the tyres of the car and the surface of this test area is 0.6. Calculate the greatest speed at which the car can move in a horizontal circle of radius 70 m around this banked test area.
(Take the acceleration due to gravity to be 10 m s^{-2}.) (AEB)

9. A child of mass 30 kg keeps herself amused by swinging on a 5 m rope attached to an overhanging tree. She is holding on to the lower end of the rope and 'swinging' in a horizontal circle of radius 3 m.

(a)   Draw a diagram to show the forces acting on the girl.

(b)   Find the tension in the rope.

(c)   Show that the time she takes to complete a circle is approximately 4 seconds.

(d)   State any assumptions that you have made about the rope.

(e)   The girl's older brother then swings, on his own, on the rope in a horizontal circle of the same radius. Show that the tension in the rope is now $5mg/4$ where m is his mass.

Find the time that it takes for him to complete one circle. (AEB)

10. A boy whirls a conker of mass 25 g in a horizontal circle at a constant angular speed. The length of the string supporting the conker is 40 cm. Assume that the boy's hand is at rest.

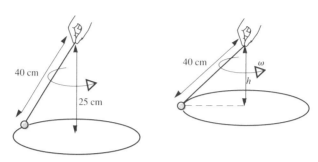

(a) Draw a force diagram showing the forces acting *on the conker*.

(b) Find the tension in the string when the conker is 25 cm below the point of support.

(c) Find the magnitude of the acceleration of the conker.

(d) Find the speed of the conker in m s^{-1}.

(e) The angular speed ω rad s^{-1} of the conker is changed and then becomes constant again. Derive an expression relating ω to the depth, h m, of the conker below the point of support.

(f) Explain why the conker can never be whirled in a horizontal circle level with the point of support. (UODLE)$_s$

In questions 11 to 15 a problem is set and is followed by a number of suggested responses. Choose the correct response

For questions 11 and 12 use this diagram of a bead of mass m, threaded on to a smooth circular wire, of radius a, that is fixed in a vertical plane.

The bead is projected from the lowest point of the wire with speed $\sqrt{6ga}$.

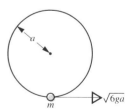

11. The normal reaction of the wire on the bead at the lowest point is

A $6mg$ **B** $7mg$ **C** $5mg$ **D** mg

12. The bead reaches the highest point of the wire with a speed

A 0 **B** $\sqrt{ga}$ **C** $\sqrt{2ga}$ **D** $\sqrt{4ga}$

13. A bead is threaded onto a circular wire fixed in a vertical plane. The bead travels freely round the wire. The acceleration of the bead is

A   towards the centre and constant

B   along the tangent and variable

C   made up of two components one radial and one tangential

D   away from the centre and variable.

14. A smooth hollow cylinder of radius a is fixed with its axis horizontal. A particle of mass m is projected from the lowest point of the inner surface with speed u. If $u = \sqrt{2ga}$ the particle will

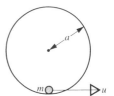

A   oscillate through $90°$

B   perform complete circles

C   leave the cylinder

D   oscillate through $180°$.

15. A particle of mass m, travelling in a vertical circle at the end of an inelastic string of length l, will describe complete circles provided that

A   the kinetic energy at the lowest point exceeds $2mgl$

B   the string never goes slack

C   the potential energy at the highest point is greater than $2mgl$

D   the tension in the string is constant.

16. A particle P of mass m lies inside a fixed smooth hollow sphere with centre O and internal radius a. When P is at rest at the lowest point A of the sphere, it is given a horizontal impulse of magnitude mu.

The particle P loses contact with the inner surface of the sphere at the point B, where $\angle AOB = 120°$.

(a)   Show that $u^2 = \frac{7}{2}ga$.

(b)   Find the greatest height above B reached by P.   (ULEAC)$_s$

17. Two particles A and B, of masses m and $2m$ respectively, are attached to the ends of a light inextensible string which passes over a smooth fixed pulley. The particles are released from rest with the parts of the string on each side of the pulley hanging vertically. When particle B has moved a distance h it receives an impulse which brings it momentarily to rest. Find, in terms of m, g and h, the magnitude of this impulse.   (AEB)

18. A quarry railway truck, of mass 1500 kg, is moving at a speed of 6 m s^{-1} along a horizontal track. It collides with a stationary empty truck, of mass 500 kg. The two trucks immediately couple together and move on together. Calculate

(a) the speed, in m s^{-1} of the pair of trucks immediately after the collision

(b) the total loss of kinetic energy, in J, due to the collision

(c) the magnitude of the impulse, in N s, on the stationary truck due to the collision.
(ULEAC)

19. A small sphere R, of mass 0.08 kg, moving with speed 1.5 m s^{-1}, collides directly with another small sphere S, of mass 0.12 kg, moving in the same direction with speed 1 m s^{-1}. Immediately after the collision R and S continue to move in the same direction with speeds U m s^{-1} and V m s^{-1} respectively. Given that $U:V = 21:26$,

(a) show that $V = 1.3$,

(b) find the magnitude of the impulse, in N s, received by R as a result of the collision.
(ULEAC)

20. A child of mass M kg sits on one of the seats of a 'rotating swing', and moves in a horizontal circle of radius 10 m with constant speed, completing one circuit every 5 s. Each seat has mass m kg. Find the angle between the single chain supporting the seat and the vertical.

Give a reason why the chains are all at the same angle to the vertical, irrespective of the mass of the occupant.

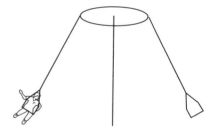

(UCLES)$_s$

21. In this question a situation is described and is followed by several statements. State, with reasons where possible, whether each of the statements is true (T) or false (F).

A sphere A, of mass m, is moving with speed $2u$. It collides directly with another sphere B, also of mass m, which is initially at rest. The coefficient of restitution is $\frac{1}{2}$.

(i) After impact A's speed is zero.

(ii) There is no loss in kinetic energy at impact.

(iii) The impulse that A exerts on B is twice the impulse that B exerts on A.

(iv) After impact B's speed is greater than A's.

In questions 22 to 25 a problem is set and is followed by a number of suggested responses. Choose the correct response.

22. A sphere A of mass $2m$ collides directly with a sphere B of mass m. Before impact each sphere is moving with speed u and A is brought to rest by the collision.

After impact the speed of B is

A $\frac{1}{2}u$ **B** u **C** $2u$ **D** $-u$

23. Two smooth objects, with a coefficient of restitution e, collide directly and bounce as shown

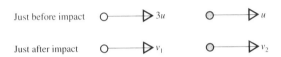

Newton's law of restitution gives

A $e \times 4u = v_2 + v_1$

B $e \times 2u = v_1 - v_2$

C $e \times 2u = v_2 - v_1$

D   it cannot be applied as the masses are not known.

24. Two masses collide and coalesce as shown in the diagram. What is the speed V of the combined mass just after impact?

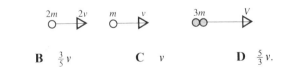

A $3v$ **B** $\frac{3}{5}v$ **C** v **D** $\frac{5}{3}v$.

25. A particle of mass 2 kg moving with speed 4 m s^{-1} is given a blow which changes the speed to 1 m s^{-1} without deflecting the particle from a straight line. The impulse of the blow is

A   10 N s

B   6 N s

C   we do not know whether it is 10 N s or 6 N s.

26. Two small smooth spheres A and B of mass $2m$ and $5m$ respectively, moving along Ox, collide. The velocity of A immediately before collision is $5u$ in the positive x direction, and immediately after collision the velocities of A and B in the positive x direction are $2u$ and $4u$ respectively. Determine

(i) the velocity of B immediately before collision,

(ii) the magnitude of the impulse on B,

(iii) the value of the coefficient of restitution.   (WJEC)$_s$

27. A particle A of mass m, moving with speed u on a smooth horizontal surface, collides directly with a stationary particle B of mass $3m$.

The coefficient of restitution between A and B is e. The direction of motion of A is reversed by the collision.

(a) Show that the speed of B after the collision is $\frac{1}{4}u(1+e)$.

(b) Find the speed of A after the collision.

Subsequently, B hits a wall fixed at right angles to the direction of motion of A and B.

The coefficient of restitution between B and the wall is $\frac{1}{2}$. After B rebounds from the wall, there is another collision between A and B.

(c) Show that $\frac{1}{3} < e < \frac{3}{5}$.

(d) In the case $e = \frac{1}{2}$, find the magnitude of the impulse exerted on B by the wall.   (ULEAC)

28. A small rubber ball is held at height h above a smooth level floor and released from rest at time $t = 0$. If the coefficient of restitution between the ball and the floor is e, show that after the first bounce the ball rises to a height h_1 where $h_1 = e^2h$.

The ball continues to bounce until it comes to rest. Show that the total distance travelled by the ball from initial release to rest is $\dfrac{1+e^2}{1-e^2} h$.

Find

(i) the time when the ball first hits the floor,

(ii) the time between the first and second impacts of the ball on the floor.

Show that the ball comes to rest when

$$t = \frac{1+e}{1-e}\sqrt{\left(\frac{2h}{g}\right)}.$$

                                                                (OCSEB)

29. A stone, of mass 1.5 kg, tied to one end of a light inextensible string, is describing circles in a fixed vertical plane, the other end of the string being fixed. Given that the maximum speed of the stone is twice the minimum speed, prove that when the string is horizontal the magnitude of the tension in the string is 49 N. (AEB)

30.

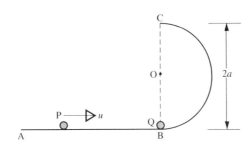

In a mechanics experiment, small marbles P and Q, of the same size but having masses $2m$ and m respectively, slide along a smooth groove in a rail.

The rail, which is fixed in a vertical plane, is in the form of a horizontal straight length AB joined to a semicircular arc BC of radius a and centre O, as shown in the diagram.

Marble P is projected with speed u along AB towards marble Q, which is initially at rest at the point B. There is a direct collision between P and Q, immediately after which the speed of Q is $\frac{5}{4}u$.

(a) Find the speed of P immediately after the collision.

(b) Show that, for Q to remain in contact with the groove on the complete arc BC,

$$u \geqslant \sqrt{\left(\frac{16}{5}ga\right)}.$$ (ULEAC)

31. A ball of mass m is travelling in a vertical plane round the inside of a circular track of radius 1 metre, as shown in the diagram.

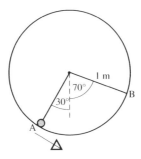

(a) What was its speed at the point A if it just reaches B before rolling back again?

(b) What must its minimum speed be at A if it is to complete the circle with out leaving the track? (SMP)$_s$

32.

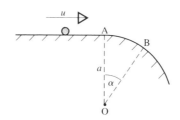

A toboggan of mass M, moving with speed u on smooth level ground, suddenly encounters a smooth downward slope, as shown in the diagram. The motion takes place in a vertical plane, and the downward slope is in the shape of an arc of a circle, radius a and centre O. The points A and B on the arc are such that OA is vertical and $\angle AOB = \alpha$, where $\cos \alpha > \frac{2}{3}$.

(a) Show that, for the toboggan to remain in contact with the slope on arc AB,

$$u^2 \leqslant ag(3 \cos \alpha - 2).$$

Given that $\cos \alpha = \frac{5}{6}$, and that the toboggan loses contact with the slope at the point B,

(b) find the sudden decrease, at the point A, in the magnitude of the force exerted by the ground on the toboggan. (ULEAC)

33. One section of a 'Loop the Loop' ride at an Adventure Park takes passengers round a vertical circle.

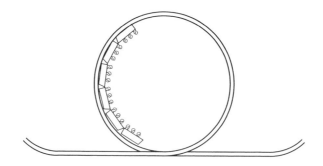

The situation is modelled by considering a small bead P of mass m threaded on a fixed smooth circular wire. The circular wire has centre O and radius a, and its plane is vertical. The bead is projected from the lowest point of the wire with speed $\sqrt{ag}$. When OP makes an angle θ with the downward vertical, find

(a) the speed of the bead.

(b) the reaction of the wire on the bead. (AEB)

34. A ball is dropped from a height of 10 m on to level ground. When it bounces, it leaves the ground with speed 10.5 m s^{-1} travelling vertically upwards. The mass of the ball is 0.1 kg. Air resistance may be neglected.

  (i)   Find the speed with which the ball first hits the ground.

  (ii) Show that the coefficient of restitution between the ball and the ground is $\frac{3}{4}$.

  (iii) Find the impulse on the ball when it first hits the ground.

  (iv) Find the energy lost by the ball on its first bounce.

  (v)   Explain why the ball leaves the ground after the second bounce with a speed of $14 \times \left(\frac{3}{4}\right)^2$ m s^{-1}.

This model for the behaviour of the ball breaks down when the speed of the ball at impact is less than 0.1 m s^{-1}.

  (vi) On which impact does the model break down?   (MEI)

35. The diagram shows the shape of a 'slide' for a children's playground. The section DE is straight and BCD is a circular arc of radius 5 m. C is the highest point of the arc and CD subtends an angle of 30° at the centre.

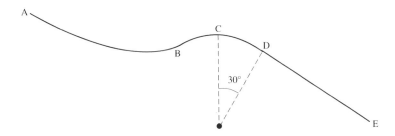

For safety reasons, children should not be sliding so fast that they lose contact with the slide at any point. Neglecting any resistances to motion, find

  (i)   the child's speed as it passes through C, given that it is on the point of losing contact at C,

  (ii) the child's speed as it passes through D, given that it is on the point of losing contact as it reaches D.

Find the greatest possible height of the starting point A of the slide above the level of D, if a child starting rom rest at A is not to lose contact with the slide at any point.

Explain briefly whether taking resistances into account would lead to a larger or smaller value for the greatest 'safe height' above D.   (UCLES)$_s$

CHAPTER 17

THREE-FORCE EQUILIBRIUM
MOMENT COUPLES

COPLANAR FORCES IN EQUILIBRIUM

When any number of forces acting on a body are in equilibrium they cause no change of any sort in the motion of the body, i.e.
(a)   the resultant force is zero
(b)   the set of forces has no turning effect.

In Chapter 6 we considered a *particle* in equilibrium under the action of any number of forces.   A particle is regarded as a mass at a *point* so the forces acting on it all pass through that point, i.e. they are concurrent and therefore cannot have any turning effect.
So, as we saw in that chapter, for a particle to be in equilibrium we need only to ensure that the resultant force in each of two directions is zero.

In certain circumstances there are alternative methods for dealing with the equilibrium of a particle and we now take a look at two special cases.

Two Forces in Equilibrium

The resultant of the two forces is zero so they must be of equal magnitude and act in opposite directions.
But if the forces act along parallel lines they have a turning effect.

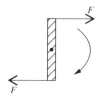

So, for equilibrium, the forces must act in the same straight line.

**Two forces in equilibrium must be equal and opposite
and act in the same straight line.**

THREE FORCES IN EQUILIBRIUM

Consider first the resultant of two non-collinear forces P and Q. We know that when lines representing P and Q are drawn to scale, one after the other, the line joining the starting point to the end point represents the resultant in magnitude and direction; the actual position of the resultant however is through the point of intersection of P and Q.

Now if a force R is added to P and Q so that the three forces are in equilibrium, R must cancel out the effect of the resultant of P and Q. Hence R is equal and opposite to this resultant and passes through the point of intersection of P and Q.

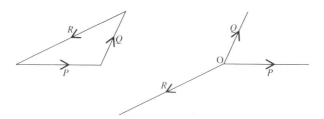

Therefore

**three forces in equilibrium must be concurrent and
can be represented *in magnitude and direction*
by the sides of a triangle taken in order.**

This triangle is known as a *triangle of forces* and it can be used to solve a problem if, in the diagram, there *already is* a triangle whose sides are parallel to the forces acting. Such a triangle is similar to the triangle of forces so the lengths of its sides are proportional to the magnitudes of the corresponding forces.

Do not expect that there always will be a suitable triangle in the diagram. If you cannot spot one, the method used in Chapter 6, of collecting the components in two perpendicular directions and equating to zero in each case, can also be used.

In Chapter 18 we will see how the 'triangle of forces' method can be applied to a rigid body that is in equilibrium under the action of three coplanar forces, (remembering that those forces must be concurrent) but first we will apply it to forces acting on a particle.

Example 17a

A string of length 1 m is fixed at one end to a point A on a wall; the other end is attached to a particle of weight 12 N. The particle is pulled aside by a horizontal force F newtons until it is 0.6 m from the wall. Find the tension in the string and the value of F.

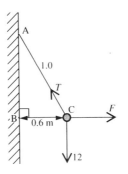

The forces acting on the particle are

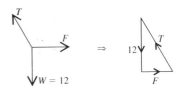

In $\triangle ABC$,   AB is in the direction of the weight

                  BC is in the direction of the force

                  CA is in the direction of the tension

$\therefore$ $\triangle ABC$ is similar to the triangle of forces.

$$\therefore \quad \frac{12}{AB} = \frac{F}{BC} = \frac{T}{CA}$$

$\triangle ABC$ is a 3,4,5 triangle, so $AB = 0.8$

$$\therefore \quad \frac{12}{0.8} = \frac{F}{0.6} = \frac{T}{1}$$

$$\Rightarrow \qquad F = \frac{7.2}{0.8} = 9 \quad \text{and} \quad T = 15$$

The tension in the string is 15 N and the value of F is 9.

EXERCISE 17a

In each question a particle of weight 10 N is attached to one end A of a light
inextensible string. The other end of the string is attached to a fixed point B.
The particle is held in the given position by the force shown in the diagram.
Copy the diagram and mark the forces acting on the particle. Identify a suitable
triangle of forces and hence find the magnitudes of the force and the tension in
the string. (You may need to extend one or more of the force lines).

1.

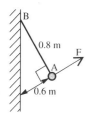

5.

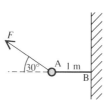

2.

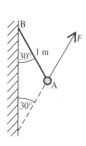

6.

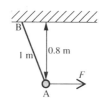

3.

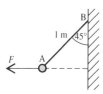

7.

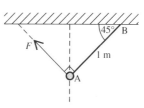

4.

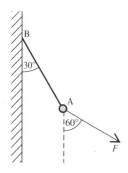

8.

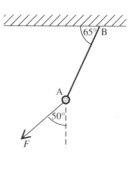

LAMI'S THEOREM

The triangle of forces is useful only when we know some of the lengths involved in the problem. There is an alternative method however, which is applicable when the *angles between the forces* are known rather than any lengths. This method is based on the sine rule.

Consider again three concurrent forces P, Q and R that are in equilibrium, and the corresponding triangle of forces ABC in which BC represents P, CA represents Q and AB represents R.
The angles α, β and γ between the forces are equal to the exterior angles of $\triangle$ABC as shown.

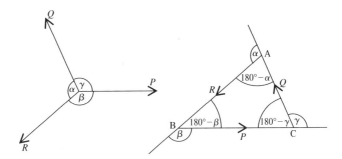

Using the sine rule in $\triangle$ABC gives

$$\frac{P}{\sin(180° - \alpha)} = \frac{Q}{\sin(180° - \beta)} = \frac{R}{\sin(180° - \gamma)}$$

Then, as $\sin(180° - \alpha) = \sin \alpha$, we have

$$\frac{P}{\sin \alpha} = \frac{Q}{\sin \beta} = \frac{R}{\sin \gamma}$$

This relationship is known as *Lami's Theorem* and can be expressed in words as follows.

> **If three forces are in equilibrium they must be concurrent
> and each force is proportional to
> the sine of the angle between the other two forces.**

Lami's Theorem can give a neat solution to many three-force problems *in which all the angles are known*.

Example 17b

A particle of weight 16 N is attached to one end of a light string whose other end is fixed. The particle is pulled aside by a horizontal force until the string is at 30° to the vertical. Find the magnitudes of the horizontal force and the tension in the string.

Let P newtons and T newtons be the magnitudes of the horizontal force and the tension respectively.

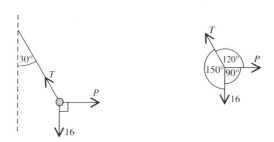

Using Lami's Theorem gives
$$\frac{T}{\sin 90°} = \frac{P}{\sin 150°} = \frac{16}{\sin 120°}$$

Hence
$$\frac{T}{1} = \frac{P}{\frac{1}{2}} = \frac{16}{\frac{\sqrt{3}}{2}}$$

$\Rightarrow$ $\qquad T = \dfrac{32}{\sqrt{3}} = \dfrac{32\sqrt{3}}{3}$ and $P = \dfrac{16}{\sqrt{3}} = \dfrac{16\sqrt{3}}{3}$

Therefore the magnitude of the horizontal force is $\frac{16\sqrt{3}}{3}$ N and the magnitude of the tension is $\frac{32\sqrt{3}}{3}$ N.

A solution to the example above, using the method of resolving in two perpendicular directions, was given on p. 123. We recommend the reader to compare the two methods to see which they prefer.

The two solutions are included to emphasise that there is no right or wrong approach – simply a choice.

EXERCISE 17b

In each question use Lami's Theorem to find the values of P and Q. Give answers corrected to 3 significant figures.

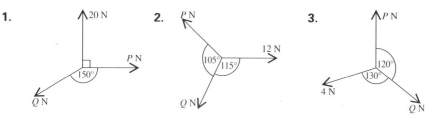

4. 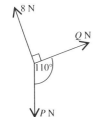   **5.** 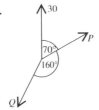   **6.**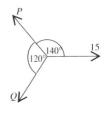

Choice of Method

There are pros and cons for each of the methods we now have for solving three-force equilibrium problems. Here are some points to consider.

- If two of the three forces are perpendicular it is easy to resolve in these two directions and equate the collected components to zero.

- If the given information includes all relevant lengths, and a *suitable triangle is present in the diagram,* using the triangle of forces is quick and easy.

- If all the relevant angles are given, Lami's Theorem at once gives a separate equation for each unknown force.

There is no *best method* for everyone, as individual preference varies, so just remember that *all* the methods work.

EXERCISE 17c

Answer each question by consciously choosing the method *you* think best and then using it.

1. 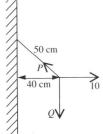   **2.** 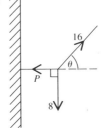   **3.**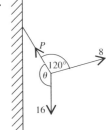

Find P and Q.   Find P and θ.   Find P and θ.

NON-CONCURRENT FORCES

So far we have dealt only with the equilibrium of a particle, i.e. a situation in which the forces are bound to be concurrent; concurrent forces cannot cause rotation so up to now we have not had to worry about turning effect.

When forces act on a rigid body of significant size however, there is no longer any physical reason why these forces should be concurrent. It therefore follows that they are capable of producing rotation and we must now take this into account when considering the equilibrium of a set of forces acting on a rigid body.

First we must take a look at how to measure turning effect.

Consider a rod pivoted at its midpoint P. If it is perfectly uniform, the rod can hang in a horizontal position.

When a downward force F is applied at one end A the rod rotates clockwise as shown. The force has not made the rod move bodily downwards, it has caused the rod to *turn about the pivot.*

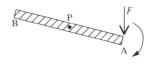

It can be shown experimentally that an additional force $2F$, applied downwards halfway along PB, will maintain the rod in its original position. Each force exerts a turning effect on the rod and together they restore the balance of the rod.

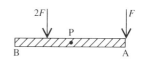

The two forces applied to the rod are not equal however, so clearly the turning effect of a force does not depend entirely on its magnitude.

The other factor is the distance from the pivot of the point of application of the force; a smaller force, further from the pivot can balance a larger force nearer to the pivot.

Experiments show that

the turning effect of a force is given by
magnitude of force × perpendicular distance from pivot

To give a full description of the turning effect of a force we must also give the sense of rotation, i.e. clockwise or anticlockwise.

THE MOMENT OF A FORCE

The turning effect of a force is called the *moment* of the force (or sometimes *torque*).

Not all objects rotate about a pivot, they may turn about a hinge or a fulcrum etc. The general name *axis of rotation* applies to all cases. This name emphasises the fact that rotation does not take place about a point but about a line. The line (axis) is perpendicular to the plane in which the forces act.
So in the examples used above we should really say that the rod rotates about 'a horizontal axis through P'. However it is common practice to refer to *rotation about a point*: it is then taken for granted that the axis of rotation passes through that point and is perpendicular to the plane of the force system.

The Unit of Moment

The magnitude of the moment of a force F, acting at a perpendicular distance d from the axis of rotation, is given by $F \times d$

The unit in which it is measured is the newton metre, Nm.

It may appear at first sight that this unit could apply in another context, as the work done by a force in moving a particle through a linear distance is also the product of force and distance, suggesting the newton metre as the unit. However, as we always use the joule as the unit of work there is no confusion over the Nm which is used exclusively for moments.

The Sign of a Moment

Earlier, when we were collecting components of forces, we chose a *positive direction;* components in that direction had a $+$ sign while components in the opposite direction took a negative sign.

In the same way, we choose a positive sense of rotation when dealing with a system of moments. If, for example, we decide to make anticlockwise the positive sense, an anticlockwise moment has a $+$ sign while a clockwise moment has a $-$ sign. *The resultant moment of a number of forces is then the algebraic sum of the separate moments.*

The positive sense does not always have to be anticlockwise; an individual choice can be made for each problem.

Zero Moment

When a force passes through the axis of rotation, its distance from that axis is zero. Therefore the moment of the force about that axis is zero.

Determining the Sense of Rotation

Most people looking at a diagram can see immediately the sense of rotation that a particular force would cause. From experience however we know that there are a few who have a 'blind spot' here. There is a simple ploy for any readers who have this problem:

> stick a pin into the point on the diagram about which turning will take place and pull the page (gently!) in the direction of the force. You will then *see* the rotation happening.

Examples 17d

1.

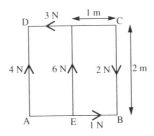

ABCD is a square lamina subjected to the forces shown in the diagram. Find the clockwise moment of each of the forces about an axis through

(a) B   (b) A   (c) E

(a) Magnitude of force	1 N	2 N	3 N	4 N	6 N
⊥ distance from B	0	0	2 m	2 m	1 m
clockwise moment about B	0	0	−6 N m	8 N m	6 N m

(b) Magnitude of force	1 N	2 N	3 N	4 N	6 N
⊥ distance from A	0	2 m	2 m	0	1 m
clockwise moment about A	0	4 N m	−6 N m	0	−6 N m

(c) Magnitude of force	1 N	2 N	3 N	4 N	6 N
⊥ distance from E	0	1 m	2 m	1 m	0
clockwise moment about E	0	2 N m	−6 N m	4 N m	0

2. **The diagram shows a rod AB, free to rotate about the end A. Taking the anticlockwise sense as positive, find the moment about A of each of the forces acting on the rod. Hence find the resultant (total) moment of the forces about A.**

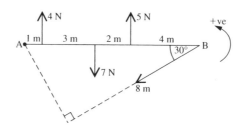

The moment of the 4 N force is $(4 \times 1)\,\mathrm{N\,m} \;=\;\quad 4\,\mathrm{N\,m}$
The moment of the 7 N force is $-(7 \times 4)\,\mathrm{N\,m} \;=\; -28\,\mathrm{N\,m}$
The moment of the 5 N force is $(5 \times 6)\,\mathrm{N\,m} \;=\;\; 30\,\mathrm{N\,m}$

The perpendicular distance from A to the force of 8 N is AB sin 30°, i.e. 10 sin 30°.

The moment of the 8 N force is $-(8 \times 10 \sin 30°)\,\mathrm{N\,m} \;=\; -40\,\mathrm{N\,m}$

The resultant moment is $4 + (-28) + 30 + (-40)\,\mathrm{N\,m}$ i.e. $-34\,\mathrm{N\,m}$

The resultant moment about A is 34 N m clockwise.

When the collected moments of a number of forces are required about an axis through A, say, we say we are *taking moments about A*. This is denoted by the symbol A↺; the sense of the curved arrow indicates the positive sense of rotation, so A↺ means *taking anticlockwise moments about A*.

3. **A force P, represented by $4\mathbf{i} + 2\mathbf{j}$, acts through the point whose position is given by the vector $6\mathbf{i} + \mathbf{j}$, and a second force Q, represented by $\mathbf{i} - 3\mathbf{j}$, acts through the point with position vector $2\mathbf{i}$. Given that the units are newtons and metres, find the magnitude and sense of the resultant moment of P and Q about O.**

For each force the components in the **i** and **j** directions are marked on the diagram so that the moment of each component can be found;   P is   and Q is

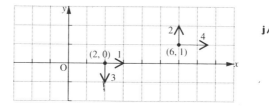

O↺ gives $4 \times 1 \; - \; 2 \times 6 \; + \; 1 \times 0 \; + \; 3 \times 2 \; = \; -2$

The resultant moment about O is 2 N m anticlockwise.

EXERCISE 17d

1. Find the magnitude and sense of the moment of the given force about the point O.

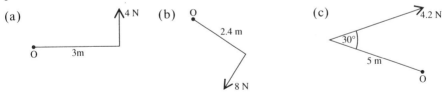

(a)   (b)   (c)

(d) A force, in newtons, represented by $5\mathbf{i} - 7\mathbf{j}$ and acting through the point with position vector $\mathbf{i} + 3\mathbf{j}$.

2. Find, in magnitude and sense, the resultant moment of the given forces about the point A.

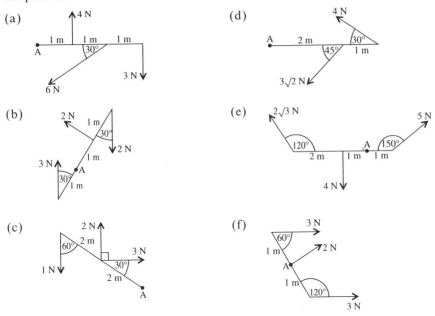

(a)   (d)

(b)   (e)

(c)   (f)

(g) Forces, in newtons, represented by $2\mathbf{i} - 7\mathbf{j}$ and $-3\mathbf{i} + 4\mathbf{j}$, acting through the points with position vectors $6\mathbf{i} + \mathbf{j}$ and $4\mathbf{j}$ respectively, where $\mathbf{i}$ and $\mathbf{j}$ are unit vectors and A is the origin.

3.

ABCD is a square of side 1 m. Find the magnitude and sense of the resultant moment of the given forces about   (a) A   (b) D.

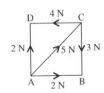

4.

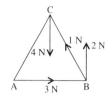

The diagram shows an equilateral triangle of side 2 m. Find the resultant anticlockwise moment of the forces shown about

(a) A   (b) C.

5. (a) Find the resultant clockwise moment about a horizontal axis through A, of the forces acting on the beam AB shown in the diagram.

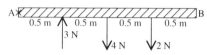

(b) When a force F newtons is applied at B, perpendicular to the beam, the resultant moment is zero. Find the value of F.

6.

A force represented by $7\mathbf{i} + 4\mathbf{j}$ acts through the point with position vector $\mathbf{i} - \mathbf{j}$. The units are newtons and metres. Find the anticlockwise moment of the force about an axis through the point with position vector $2\mathbf{i} + \mathbf{j}$.

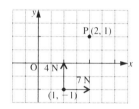

7. Four forces, measured in newtons, are represented by $4\mathbf{i} + 2\mathbf{j}$, $5\mathbf{i}$, $\mathbf{i} - 6\mathbf{j}$, and $-3\mathbf{j}$. They act respectively through points with position vectors measured in metres and represented by $\mathbf{i} - \mathbf{j}$, $\mathbf{i} + \mathbf{j}$, $4\mathbf{i}$ and $3\mathbf{j}$.
Find the magnitude and sense of their resultant moment about

(a) the origin O   (b) the point $(1, 1)$.

COUPLES

Consider two forces, of equal magnitude P, that act in opposite directions along parallel lines distant d apart.

Resolving ↑ gives $P - P = 0$
So the linear resultant of the pair of forces is zero.

But taking moments about A shows that the turning effect of the pair of forces is Pd clockwise, i.e. the resultant moment is not zero.

Therefore the effect of two equal and opposite parallel forces is to produce pure rotation.

Such a pair of forces is called a *couple*.

When a couple acts on a body there is no change in the linear motion of the body but there is a change in its rotation.

Constant Moment of a Couple

Consider again the couple comprising two equal and opposite parallel forces of magnitude P and distant d apart.

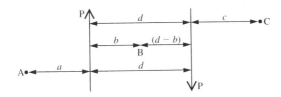

We will take moments about three different axes, through A, B and C

A⤴   clockwise moment $= P(a+d) - Pa = Pd$
B⤴   clockwise moment $= P(d-b) + Pb = Pd$
C⤴   clockwise moment $= P(c+d) - Pc = Pd$

As A, B and C represent any points in the plane of the couple, we see that

**the moment of a couple is the same
about any axis perpendicular to its plane**

The magnitude of the moment, or torque, of a couple is often called simply the magnitude of the couple.

The Characteristics of a Couple

- The linear resultant of a couple is zero.
- The moment of a couple is *not* zero and has the same magnitude regardless of the position of the chosen axis.

A *set* of coplanar forces that satisfies these two conditions is said to
reduce to a couple.

Examples 17e

1. Show that the following forces reduce to a couple and find its magnitude.

$F_1 = -4i + 3j$ acting through the point $2i - j$

$F_2 = 6i - 7j$ acting through the point $-3i + j$

$F_3 = -2i + 4j$ acting through the point $4j$

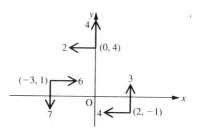

The resultant force is $Xi + Yj$

where $X = -4 + 6 - 2 = 0$

and $Y = 3 - 7 + 4 = 0$

The resultant force is therefore zero.

The resultant turning effect about O (taking anticlockwise as positive)
is $(3 \times 2 - 4 \times 1) + (7 \times 3 - 6 \times 1) + (2 \times 4) = 25$

Since the resultant force is zero and the resultant turning effect is not zero, the forces reduce to a couple of magnitude 25 units anticlockwise.

2. ABCD is a square of side a. Forces of magnitudes 1, 2, 3, P and Q units act along $\overrightarrow{AB}, \overrightarrow{BC}, \overrightarrow{CD}, \overrightarrow{DA}$ and $\overrightarrow{AC}$ respectively. Find the values of P and Q if the set of five forces reduces to a couple.

If the forces reduce to a couple the linear resultant must be zero.

Resolving $\rightarrow$ gives $1 + Q \cos 45° - 3 = 0 \quad \Rightarrow \quad Q = 2\sqrt{2}$

Resolving $\uparrow$ gives $2 + Q \sin 45° - P = 0 \quad \Rightarrow \quad P = 4$

The linear resultant of the given forces is zero if $P = 4$ and $Q = 2\sqrt{2}$

This is not sufficient to ensure that the forces reduce to a couple; we must also show that their turning effect is *not* zero. By choosing an axis through A the moments of forces P, Q and 1 are not involved.

A$\circlearrowleft$ gives $2a + 3a$ which is not zero.

Therefore the forces reduce to a couple if $P = 4$ and $Q = 2\sqrt{2}$

The magnitude of the couple is $5a$ anticlockwise.

3. Forces acting along the sides $\overrightarrow{AB}$, $\overrightarrow{BC}$ and $\overrightarrow{CA}$ of a triangle ABC have magnitudes proportional to the lengths of those sides. Prove that the three forces reduce to a couple.

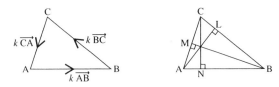

In this problem we shall use the constant moment property of a couple.

Taking moments about axes through A, B and C we have

A⤴ Torque $= |k \overrightarrow{BC}| \times AL = k \times BC \times AL = 2k \times$ area of triangle ABC

B⤴ Torque $= |k \overrightarrow{CA}| \times BM = 2k \times$ area of triangle ABC

C⤴ Torque $= |k \overrightarrow{AB}| \times CN = 2k \times$ area of triangle ABC

Since about each of three axes the turning effect of the given forces is the same, the forces reduce to a couple.

Note that finding equal moments about axes through *two* points is not proof of a couple as there could be a resultant force parallel to the line through those points. The moment about a third axis, through a *non-collinear* point, is needed.

EXERCISE 17e

In questions 1 and 2, ABCD is a square of side $2a$. Forces of magnitudes F and $2F$ act along $\overrightarrow{AB}$ and $\overrightarrow{BC}$ respectively.
Give answers in terms of F and/or a where appropriate.

1. Find the magnitudes of two forces which, acting along AC and AD, combine with the two given forces to form a couple and find the magnitude of the couple.

2. Find the magnitude and direction of a force acting through D which, together with the given forces, form a couple.

3. Forces $\mathbf{i} + 3\mathbf{j}$, $-2\mathbf{i} - \mathbf{j}$, $\mathbf{i} - 2\mathbf{j}$ act through the points with position vectors $2\mathbf{i} + 5\mathbf{j}$, $4\mathbf{j}$, $-\mathbf{i} + \mathbf{j}$ respectively. Prove that this system of forces is equivalent to a couple, and calculate the moment of this couple.

4. Four forces are represented by $\mathbf{i} - 4\mathbf{j}$, $3\mathbf{i} + 6\mathbf{j}$, $-9\mathbf{i} + \mathbf{j}$ and $5\mathbf{i} - 3\mathbf{j}$, and their points of application are given by $3\mathbf{i} - \mathbf{j}$, $2\mathbf{i} + 2\mathbf{j}$, $-\mathbf{i} - \mathbf{j}$ and $-3\mathbf{i} + 4\mathbf{j}$ respectively.

   (a) Show that the forces reduce to a couple and find its magnitude.

   (b) If the first force is moved to the point $-3\mathbf{i} - 3\mathbf{j}$ show that the system is now in equilibrium.

5. Forces of magnitudes $3F$, $4F$, $2F$, F act along the sides $\overrightarrow{AB}$, $\overrightarrow{BC}$, $\overrightarrow{CD}$, $\overrightarrow{DA}$ of a rectangle ABCD in which AB = $4a$ and AD = $3a$. If two more forces pF and qF act along $\overrightarrow{AC}$ and $\overrightarrow{BC}$ respectively, find the values of p and q for which the six forces reduce to a couple.

Explain why it is impossible to find values of p and q for which the system is in equilibrium.

6. Forces 1, 2, 3, 4, P, Q act along the sides of a regular hexagon taken in order. Find values for P and Q for which the six forces reduce to a couple.

EQUILIBRIUM UNDER PARALLEL FORCES

Consider an object that is in equilibrium under the action of the set of parallel forces shown in the diagram.

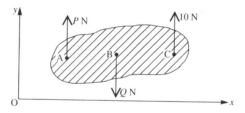

There are no components of force in the direction Ox, so there can be no change in motion in that direction. Therefore, since the object is in equilibrium, we know that

the resultant force in the direction Oy is zero
the resultant moment about A (or any other point) is zero.

As there are only two unknown forces, we need only two equations to find them. So we can

either   collect the parallel forces and take moments about one axis
or   take moments about each of two axes.

Note that, because there is a choice of method, readers can, if they wish, use the method not chosen first as a check.

Note also that if only one unknown force has to be found, one only of the above equations may give sufficient information.

Examples 17f

1. The diagram shows a uniform beam of length 3 m and weight 40 N, suspended in equilibrium in a horizontal position by two vertical ropes, one attached at the end A and the other at C, 1 m from the other end B. Find the tension in each rope. (The weight of a uniform beam acts through the midpoint.)

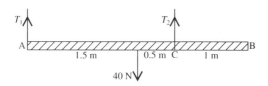

There are no horizontal forces so no information is given by resolving horizontally, but as there are only two unknowns we need only two equations. We will choose to take moments about both A and C.

A⟳	$T_2 \times 2 - 40 \times 1.5 = 0$	[1]
C⟳	$T_1 \times 2 - 40 \times 0.5 = 0$	[2]
From [1]	$T_2 = 30$	
From [2]	$T_1 = 10$	

The tensions in the two ropes are 30 N and 10 N.

Check: Resolving ↑ gives $30 + 10 - 40 = 0$

2. A uniform plank of weight W and length $6a$ rests on two supports at points B and C as shown in the diagram. The plank carries a load $2W$ at the end A and a load $3W$ at the end D. Find, in terms of W, the force exerted by each support.

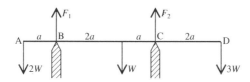

B⟳ gives	$2W \times a + F_2 \times 3a - W \times 2a - 3W \times 5a = 0$	[1]
↑ gives	$F_1 + F_2 - 2W - W - 3W = 0$	[2]
From [1]	$F_2 = 15Wa \div 3a = 5W$	
From [2]	$F_1 = 6W - F_2 = W$	

The supporting forces at B and C are W and $5W$ respectively.

Check: C⟳ gives

$$2W \times 4a - F_1 \times 3a + W \times a - 3W \times 2a = (8 - 3 + 1 - 6)Wa = 0$$

3. **A scaffold board of weight 50 N and length 4 m lies partly on a flat roof and projects 2 m over the edge. A load of weight 30 N is carried on the overhanging end B and the board is prevented from tipping over the edge by a force applied at the other end A. If the weight of the scaffold board acts through a point 1 m from the end A, what is the value of the least force needed?**

The least force will *just* prevent the board from toppling when it is about to lose contact with the roof except at the edge, so the reaction between the board and the roof acts at the edge of the roof.

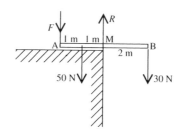

In this problem only the magnitude of the force at A is required and not R. So if we take moments about M, in order to avoid introducing R, only this one equation is needed.

$M\circlearrowright$ gives $30 \times 2 - 50 \times 1 - F \times 2 = 0$

$\Rightarrow$ $\hspace{6.5cm} F = 5$

So the least force needed is 5 N.

Check: $A\circlearrowright$ gives $50 \times 1 - R \times 2 + 30 \times 4 = 0$ $\hspace{1cm} \Rightarrow \hspace{1cm} R = 85$

$\hspace{1.1cm}\uparrow$ gives $\hspace{2cm} R - F - 50 - 30 = 0$ $\hspace{1cm} \Rightarrow \hspace{1cm} F = 5$

EXERCISE 17f

In each question from 1 to 4, a light beam (i.e. the weight is negligible) rests in a horizontal position on two supports, one at A and the other at B, and carries loads as shown. Find the force exerted at each support.

1.

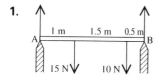

3.

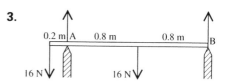

2.

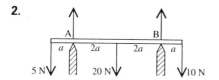

4.

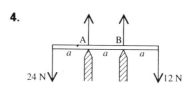

5. A see-saw consists of a light plank of length 4 m, balanced on a fulcrum at the centre. A child of weight 220 N sits on one end.

    (a) How far from the other end should another child, of weight 280 N, sit if the see-saw is to be balanced?

    (b) What force is exerted by the fulcrum?

Questions 6 to 9 concern a horizontal uniform beam supported by vertical ropes. In each case find the tensions in the ropes.

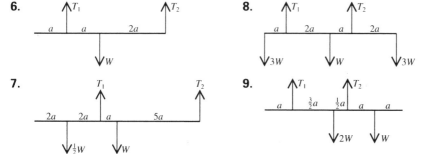

6.

7.

8.

9.

In questions 10 to 12, state any assumptions that are made.

10. A non-uniform plank of wood 3 m long is being carried by two men, one at each end of the plank. Mick is taking a load of 42 N and Tom at the other end is supporting 22 N. Find the distance from Mick's end of the point through which the weight of the plank acts.

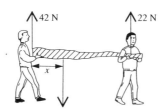

11. A boy builds a simple bridge over a stream by supporting a uniform plank of wood symmetrically on two small brick piers, one on each bank. The piers are 2.6 m apart, the weight of the plank is 300 N and the boy's weight is 420 N.

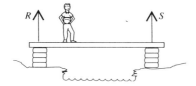

Find the force exerted by each pier when the boy stands

    (a) over one of the piers

    (b) 1 m from one pier

    (c) in the centre of the bridge.

12. A uniform plank AB, of mass 80 kg and length 5 m, overhangs a flat roof by 2 m. A boy can walk to within 0.6 m of the overhanging end B when a load of mass 12 kg is placed on the opposite end A.

(a) What is the mass of the boy?

(b) Find the smallest extra load that must be placed at the end A to enable the boy to walk right to the end B. Do you think it would be sensible for him then to walk to the end B?

13. A uniform beam AB, of weight 50 N and 2.5 m long, has weights of 20 N and 30 N hanging from the ends A and B respectively. Find the distance from A of the point on the beam where a support should be placed so that the beam will rest horizontally.

14. A non-uniform rod of weight W and length $2a$ is suspended by a string attached to the midpoint of the rod. The rod is horizontal when a weight $3W$ hangs from one end of the rod. If this weight is removed, find the supporting force that is needed at the opposite end to maintain the rod in its horizontal position.

15. A rod of length 1.2 m is placed on a table top with part of the rod protruding over the edge. The weight of the rod is 10 N. The rod can just hold a particle of weight 6 N, placed on the overhanging end A, without toppling over the edge.

(a) If the rod is uniform, what length of rod is on the table top?

(b) If, instead, the weight of the rod acts at a point G, and 0.5 m of the rod protrudes over the edge of the table, find the length of AG.

16.

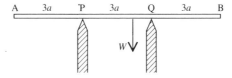

The diagram shows a non-uniform beam AB, of weight W and length $9a$, resting on supports at P and Q. When a load of $\frac{1}{4}W$ is attached at B the beam is just on the point of losing contact with the support at P (i.e. the reaction at P just becomes zero).

(a) Find the distance from Q of the point through which the weight of the beam acts.

(b) The load is removed from B and a load Wx is attached at A. If the beam is now just about to lose contact with the support at Q, find the value of x.

17. A light rod of length 2 m is suspended in a horizontal position by a string attached to a point P on the rod. Particles of masses 0.5 kg, 0.8 kg and 1 kg are attached to the rod at points distant 0.4 m, 1 m and 1.2 m from one end A. If $AP = x$ m, find the value of x.

18. The scales in the diagram consist of a uniform beam ABC of mass 0.5 kg, which is pivoted at B, with AB = 0.5 m and BC = 0.3 m. A scale plan of mass 0.4 kg is attached at C and a sliding counter weight can move on the section AB in such a way that its centre of mass can be placed exactly at A. It is required that masses of up to 10 kg can be weighed when placed centrally in the scale pan.

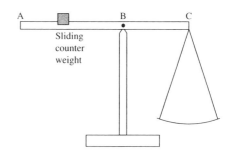

(a) Find the least possible mass for the counter weight.

(b) Find the distance from B of the centre of mass of the counter weight for the scales to balance when the scale pan is empty.

*19. A trolley is made of thin metal bars, with central cross-section ABC. Angle ABC = 90°, AB = 0.3 m and BC = 1.2 m. Its wheels are on an axle passing through B. A packing case, of mass 150 kg has a rectangular cross-section measuring 0.6 m by 0.8 m and it is packed so that the centre of mass is at its centre. The case is placed on the trolley as shown in the diagram and a porter supports the trolley, with BC at an angle θ to the vertical, by exerting a force P newtons vertically upwards at C. Assume that the weight of the trolley is negligible.

(a) If the porter can place the trolley in a position where it just balances, without any force P holding it at C, find the value of θ.

(b) If $\theta = 35°$, find P.

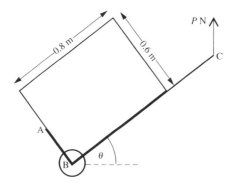

CHAPTER 18

COPLANAR FORCES IN EQUILIBRIUM

GENERAL CONDITIONS FOR EQUILIBRIUM IN A PLANE

We know that if the forces shown
in the diagram act on a body,
that body will be in equilibrium
only if both the resultant force and
the resultant moment are zero.

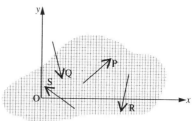

Now the forces acting on the object are not parallel so the resultant force has components in the directions of both Ox and Oy.

Therefore, if the object is to be in equilibrium the sum of the components in each of the directions Ox and Oy must be zero.

So now we can give the general conditions necessary for an object to be in equilibrium under the action of a set of non-parallel coplanar forces, i.e.

the resultant force in the direction Ox is zero

the resultant force in the direction Oy is zero

the resultant moment about any axis is zero

Applying these conditions to a particular problem gives three equations, so three unknown quantities can be found.

In some problems it is more convenient to use an alternative set of three independent equations, i.e.

the resultant in the direction Ox (or Oy but not both) is zero

the resultant moment about a particular axis is zero

the resultant moment about a different axis is also zero

The following Examples and Exercise illustrate the use of both of these methods as applied to a variety of questions.

Note that another way to produce three independent equations is to take moments about each of three axes provided that these axes are not in line.
(If they are collinear, the third resultant moment would simply be a combination of the first two and not an independent fact.)

Examples 18a

1.

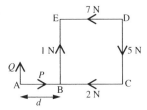

The diagram shows a set of forces in equilibrium.
BCDE is a square of side 0.5 m.
Calculate *P*, *Q* and *d*.

First we will resolve parallel to BC and CD

$\rightarrow$ $\qquad$ $P - 2 - 7 = 0$ $\qquad$ $\Rightarrow$ $\qquad$ $P = 9$

$\uparrow$ $\qquad$ $Q + 1 - 5 = 0$ $\qquad$ $\Rightarrow$ $\qquad$ $Q = 4$

Now taking moments about B gives a simple equation for *d*.

$B\circlearrowleft$ $\qquad$ $Q \times d + 5 \times 0.5 - 7 \times 0.5 = 0$

$\Rightarrow$ $\qquad\qquad\qquad\qquad$ $4d - 1 = 0$ $\qquad$ $\Rightarrow$ $\qquad$ $d = 0.25$

2. A ladder of length 4 m and weight *W* newtons rests in equilibrium with its foot A
on horizontal ground and resting against a vertical wall at the top B. The ladder is
uniform; contact with the wall is smooth but contact with the ground is rough and
the coefficient of friction is $\frac{1}{3}$. Find the angle θ between the ladder and the wall
when the ladder is on the point of slipping.

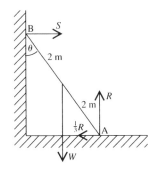

The ladder is about to slip so friction is limiting and $F = \mu R$.

Resolving horizontally and vertically gives

$\rightarrow$ $\qquad$ $S - \frac{1}{3}R = 0$

$\uparrow$ $\qquad$ $R - W = 0$

Hence $R = W$ and $S = \frac{1}{3}W$

The third necessary equation is given by taking moments about *any* axis. The best choice of axis is
through A because both *F* and *R* pass through A and therefore have zero moment.

A↻ $S \times 4 \cos \theta - W \times 2 \sin \theta = 0 \quad \Rightarrow \quad S = \frac{1}{2} W \tan \theta$

∴ $\frac{1}{3} W = \frac{1}{2} W \tan \theta \quad \Rightarrow \quad \tan \theta = \frac{2}{3}$

The angle between the ladder and the wall is $34°$ (nearest degree).

3. **The end A of a uniform rod AB of length $2a$ and weight W is smoothly pivoted to a fixed point on a wall. The end B carries a load of weight $2W$. The rod is held in a horizontal position by a light string joining the midpoint G of the rod to a point C on the wall, vertically above A. The string is inclined at 60° to the wall. Find, in terms of W, the tension in the string and the horizontal and vertical components of the force exerted by the pivot on the rod.**

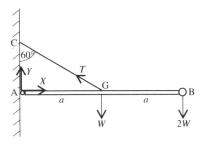

We will represent the components of the force at the pivot by X and Y as shown.

There are three unknown quantities, X, Y and T, so we need three equations and will resolve in two directions and take moments about an axis.

Resolving → $X - T \sin 60° = 0$ [1]

Resolving ↑ $Y + T \cos 60° - W - 2W = 0$ [2]

If we take moments about A, X and Y are not involved.

A↻ $W \times a + 2W \times 2a - T \times a \sin 30° = 0$ [3]

From [3], $Ta\left(\frac{1}{2}\right) = 5Wa \quad \Rightarrow \quad T = 10W$

Using this value of T in [2] gives

$$Y + 10W\left(\tfrac{1}{2}\right) - 3W = 0 \quad \Rightarrow \quad Y = -2W$$

and [1] gives $X - 10W\dfrac{\sqrt{3}}{2} = 0 \quad \Rightarrow \quad X = 5W\sqrt{3}$

Therefore the tension in the string is $10W$.

The vertical component of the pivot force is $2W$ downwards.

The horizontal component of the pivot force is $5W\sqrt{3}$ acting away from the wall.

If required, the magnitude and direction of the resultant force at the pivot can be found from the triangle of forces,

4. A ladder of length $2a$ and weight W rests with its foot, A, on rough horizontal ground where the coefficient of friction is $\frac{3}{4}$. The top of the ladder, B, rests against a vertical wall and a painter, of weight $2W$, is standing at the top of the ladder. By modelling the ladder as a uniform rod, the wall as smooth and the painter as a particle, find the angle θ between the ladder and the wall when the ladder is just about to slip. Are there any respects in which the model is inappropriate?

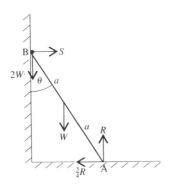

When the ladder is about to slip, the frictional force at A is $\frac{3}{4}R$.

There is no frictional force acting at B.

Resolving horizontally and vertically and taking moments about A give:

$\rightarrow$
$$S - \tfrac{3}{4}R = 0 \qquad\qquad\qquad [1]$$

$\uparrow$
$$R - 3W = 0 \qquad\qquad\qquad [2]$$

A$\circlearrowright$
$$W \times a \sin \theta + 2W \times 2a \sin \theta - S \times 2a \cos \theta = 0 \qquad\qquad\qquad [3]$$

From [1] and [2]
$$S = \tfrac{9}{4} W$$

Then [3] gives
$$5Wa \sin \theta - \tfrac{9}{2} Wa \cos \theta = 0$$

$\Rightarrow$
$$\tan \theta = 0.9$$

To the nearest degree the ladder is inclined at $42°$ to the wall.

Although it is not unreasonable to model the ladder as a uniform rod, many ladders have their centre of mass below the midpoint and so are not uniform.

It is unrealistic to assume the wall to be smooth; the frictional force that almost certainly acts there will help to prevent the ladder from slipping.

Modelling the painter as a particle at the top of the ladder is not appropriate because

> he probably does not stand on the very top rung, but even if he does he is not at the top of the ladder,

> his centre of mass is unlikely to be vertically over his feet.

5. The diagram shows a drawbridge
   over a moat. It is pivoted along
   one end and chains are attached to
   the outside corners of the other
   end. These chains pass over
   pulleys, immediately above the
   gate, to a winch. Normally the
   drawbridge is raised by operating
   the winch, but unfortunately it has
   broken down and soldiers in the
   castle have been detailed to raise
   the bridge by pulling the chains
   vertically downwards.

Choose a model which will allow you to calculate, in terms of the weight W of the
drawbridge, the least total pull needed when the drawbridge has been raised through
30°. State, with comments on their suitability, any assumptions made.

We will model the drawbridge as a uniform plank (this is rough and ready but
not unreasonable), and treat the chains as light strings (this is not very accurate
as the weight of the chains would cause sagging and also increase the pull
needed). We will also assume that equal pulls are exerted on each side so that
the tensions are equal and the resultant pull acts in the middle. Further, it is
assumed that the height of the pulleys is equal to the length of the drawbridge
i.e. that the drawbridge fills the 'hole'.

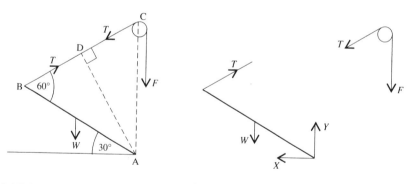

In $\triangle ABC$, $AB = AC = l$ $\Rightarrow$ $A\widehat{B}C = A\widehat{C}B = 60°$ $\Rightarrow$ $B\widehat{A}D = 30°$

The hinge exerts on the drawbridge a force that is unknown both in magnitude
and direction but it can be avoided if we only take moments about A.

A) $\qquad W \times \frac{1}{2} l \cos 30° - T \times l \cos 30° = 0$

$\Rightarrow \qquad\qquad\qquad\qquad T = \frac{1}{2} W$

Now we will assume that the tension is unchanged by passing over the pulley
(not very reliable as there is certain to be some friction at the pulley which would
increase the tension on the other side).

The least total pull, F, on the chains when the drawbridge has been raised through $30°$ is given by

$$F = T = \tfrac{1}{2} W$$

Note that the pull is least when the drawbridge is raised steadily and not at an increasing speed. Note also that the pull of $\tfrac{1}{2} W$ is valid only when the drawbridge has been raised through $30°$.

6. **A wooden plank AB of weight 60 kg and length 4 m rests with A on rough horizontal ground where the coefficient of friction is $\tfrac{1}{2}$. The plank rests in rough contact with the top C of a rail of height 1.5 m, and is just about to slip. Given that AC $= 3$ m and taking g as 10, find**

   (a) **the normal contact forces at A and at C**

   (b) **the coefficient of friction at C.**

   Have any assumptions been made?

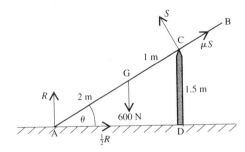

In $\triangle ACD$, $\ \sin \theta = \dfrac{CD}{AC} = \dfrac{1.5}{3} = \dfrac{1}{2}$

$\therefore \quad \theta = 30°$

As the plank is about to slip, friction is limiting at both A and C.

R and S can be found separately by taking moments about A and C; then we resolve only once. In this problem answers will be given to 2 sf, so in intermediate working we will use 3 sf.

(a)   A ↻ $S \times 3 - 600 \times 2 \cos 30° = 0$

    $\therefore$ $S = 400 \cos 30° \quad \Rightarrow \quad S = 346$

    C ↻ $600 \times 1 \cos 30° - R \times 3 \cos 30° + \tfrac{1}{2} R \times 3 \sin 30° = 0$

    $\therefore$ $R(\cos 30° - \tfrac{1}{2} \sin 30°) = 200 \cos 30° \quad \Rightarrow \quad R = 281$

    To 2 sf the normal reaction at A is 350 N and that at C is 280 N

(b)   Resolving ← $S \sin 30° - \mu S \cos 30° - \tfrac{1}{2} R = 0$

    $\Rightarrow$ $\mu S \cos 30° = S \sin 30° - \tfrac{1}{2} R$

    $\therefore$ $\mu = \dfrac{173 - 140}{346 \times 0.866} = 0.110 \quad (3 \text{ sf})$

The coefficient of friction at C is 0.11   (2 sf).

It has been assumed that: the plank is uniform and straight; the top of the rail is small enough to be treated as a point.

EXERCISE 18a

In questions 1 to 6 a uniform ladder of weight 200 N and length 2 m rests with one end A on level ground and the other end B resting against a vertical wall. When the ladder is in limiting equilibrium (i.e. is just about to slip) the angle between the ladder and the wall is θ. Where appropriate give answers corrected to 3 significant figures and angles to the nearest degree.

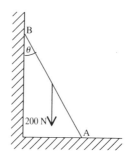

1. Contact with the wall is smooth; contact with the ground is rough and the coefficient of friction is $\frac{1}{3}$. Find θ.

2. Contact with the wall is smooth; contact with the ground is rough and the coefficient of friction is μ. If $\theta = 45°$ find

(a) the normal reaction with the ground

(b) the frictional force

(c) the value of μ.

3. Contact is rough both with the wall and with the ground; the coefficient of friction in both cases is $\frac{1}{4}$. Find θ. (Remember that the ladder will not slip until friction is limiting at both points of contact.)

4. Contact with the wall is smooth and contact with the ground is rough. When $\theta = 60°$ a workman of weight 80 kg can climb one quarter of the way up the ladder before limiting equilibrium is reached. Find the reaction at the wall and the coefficient of friction. Take g as 10 and treat the workman as a point-load.

5. If, in question 4, θ is reduced to $30°$, find how far up the ladder the man can now climb.

6. If, in question 4, contact is rough at both ends of the ladder and $\mu = \frac{1}{3}$ in each case, find θ if the workman can just climb to the top of the ladder. Take g as 10.

In questions 7 to 10 a uniform rod PQ of length 2 m and weight 24 N is hinged at the end P to a fixed point and is in equilibrium. R is a point vertically above P.

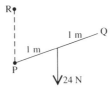

7. PQ is kept horizontal by a support at Q. Find the magnitude of the supporting force and the magnitude and direction of the force exerted on the rod by the hinge.

8. PQ is held at an angle of 60° to the upward vertical through P, by a light string joining Q to R. Given that QR = 2 m, find

(a) the tension in the string

(b) the magnitude of the reaction at the hinge.

9. PQ is held at an angle of 60° to the downward vertical through P by a horizontal force F newtons applied at the end Q. Find the value of F.

10. PQ is horizontal. A light string of length 2.5 m connects Q to R and a load of 20 N is applied to the rod at the end Q. Find

(a) the tension in the string

(b) the magnitude and direction of the force acting on the rod at the hinge.

11. A uniform rod AB of length $4a$ rests with the end A in rough contact with level ground where the coefficient of friction is $\frac{1}{2}$. A point C on the rod, distant $3a$ from A, rests against a smooth peg. The rod is in limiting equilibrium when it is at 30° to the ground. Find, in terms of W

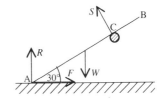

(a) the reaction at the peg

(b) the frictional force.

12. The diagram shows a central cross-section AB of a uniform window which has a mass of 4 kg. The window is held at an angle of 30° to the vertical by a light rod BC, which is attached to the window frame at C. AB = 0.3 m and angle ABC = 90°. Find

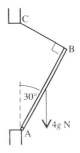

(a) the tension in BC

(b) the magnitude and direction of the reaction at the hinge.

13. A uniform rod XY whose mid point is M, is in equilibrium in a vertical plane as shown in the diagram.

The rod rests on a rough peg at Z and a force F acts at X as shown.
If $YZ = ZM$ and $\tan \alpha = \frac{4}{3}$, find the coefficient of friction at Z and the force F.

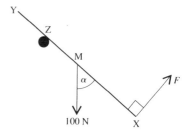

14. In the diagram AB is a uniform table top, of weight 49 N, hinged to a vertical wall at A. The table top is supported by a light rod CD, which is hinged to the wall at D.
$AC = CB = 0.3\,m$ and $AD = 0.4\,m$.
A boy leans on the table at B exerting a force of 8 N vertically downwards. Find

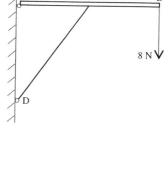

(a) the thrust exerted on the table top by the rod CD

(b) the magnitude and direction of the reaction at the hinge A.

15. A uniform rod XY, whose weight is W, is in equilibrium in a vertical plane. The midpoint of the rod is at M. The rod is supported on a plane inclined at $30°$ to the horizontal, by a string attached to the end X and held vertically. The rod, whose weight is W, is inclined to the plane at $30°$ as shown in the diagram. Find

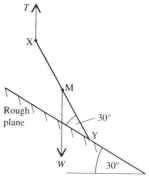

(a) the tension in the string in terms of W

(b) the coefficient of friction between the rod and the plane.

16. A light ladder, of length 4 m stands on rough horizontal ground, with coefficient of friction 0.25, and its upper end rests against a smooth vertical wall. The ladder is inclined at $65°$ to the horizontal. The end rungs are each 0.3 m from an end of the ladder. A man, of mass 80 kg, stands on the top rung and another man, of mass m kg, stands on the bottom rung. Find the least value of m which will prevent the ladder from slipping. State any assumptions made and comment on how reasonable they are.

*17. The diagram shows a uniform rectangular shelf ABCD, of mass 5 kg. It is hinged to a wall along the edge AD and supported by light rods BK and CL, which are attached to the wall at K and L. AB = 0.4 m and AK = DL = 0.3 m. Find

(a)   the tension in each of the rods BK and CL

(b)   the magnitude and direction of the reaction at the hinge.

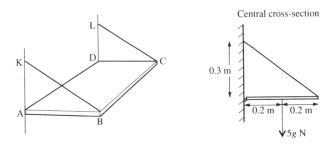

USING 'THREE-FORCE EQUILIBRIUM'

The equilibrium problems in the previous exercise were solved by the general method which uses resolving and taking moments.

There are, however, alternative methods for solving problems involving the equilibrium of a rigid body under the action of three – or sometimes more than three – coplanar forces.

We know that if three forces are in equilibrium they must be concurrent. So if a rigid body is in equilibrium under the action of three forces, and the point of intersection of two of the forces is known, the third force *must* pass through the same point. In this way a force whose direction is not normally known (e.g. the force at a hinge) can be located on a diagram.

This principle can be extended to cover some problems where more than three forces are involved initially but can be reduced in number.

At a point of rough contact, for example, when friction is limiting, the resultant reaction can be used instead of normal reaction plus frictional force, so reducing the number of forces acting.

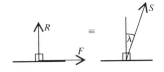

Once it is established that three forces are keeping a body in equilibrium, the triangle of forces or Lami's theorem can be applied, providing some elegant solutions as well as variety of method.

Examples 18b

The first example shows how the resultant reaction at a point of rough contact can be used to reduce a four-force problem to a three-force case allowing the angle of friction to be found.

1. **A uniform ladder AB, resting with B in rough contact with the ground and A in smooth contact with a wall, is just about to slip when inclined at 30° to the wall. Find the coefficient of friction at the ground.**

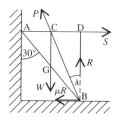

If we use the resultant reaction at B the three forces then acting on the ladder are concurrent at C. CG and DB are parallel and G bisects AB (the rod is uniform), so C bisects AD.

As P is the resultant of R and μR, angle CBD is the angle of friction.

$$\tan \lambda = \frac{CD}{DB} = \frac{\frac{1}{2}AD}{DB} = \frac{1}{2}\tan 30°$$

$$\therefore \quad \tan \lambda = \frac{1}{2} \times \frac{1}{\sqrt{3}} = \frac{\sqrt{3}}{6} = \mu$$

The coefficient of friction is $\frac{\sqrt{3}}{6}$.

The second example shows how the direction of a force that is not generally known can be located in a three-force problem and mensuration used to find this direction.

2. **A uniform rod is hinged at one end A to a wall. The other end B is pulled aside by a horizontal force until the rod is in equilibrium at 60° to the wall. Find the direction of the hinge force.**

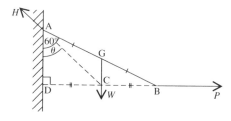

The weight of a *uniform* rod acts through the midpoint of the rod. The horizontal force and the weight of the rod meet at the point C. Three forces in equilibrium must be concurrent, so the hinge force also acts through C, i.e. the direction of H is along CA.

The angle θ can be found from the mensuration of the general diagram. GC is parallel to AD and G bisects AB so C bisects DB (Intercept Theorem).

In $\triangle$ADC $\qquad \tan \theta = \dfrac{DC}{AD} = \dfrac{\frac{1}{2}DB}{AD}$

$$= \tfrac{1}{2} \tan 60° = \tfrac{1}{2}\sqrt{3}$$

$\therefore \qquad\qquad\qquad \theta = 40.9° \;(\text{3 sf})$

The hinge force is inclined at 40.9° to the wall.

The next example again uses the concurrency of three forces in equilibrium and applies Lami's Theorem since all angles are known.

3. **One end A of a uniform rod AB of weight 36 N is hinged to a fixed point. The rod is held in a horizontal position by a string connecting B to a point C vertically above the hinge. If the string is inclined at 45° to the rod, find**

   (a) the direction of the force exerted on the rod by the hinge

   (b) the tension in the string.

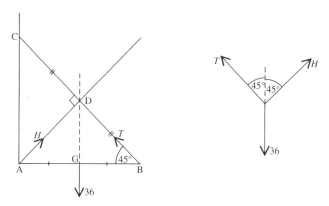

(a) The weight and the tension meet at D. G bisects AB so D bisects BC.

$\therefore \qquad\qquad A\widehat{D}B = 90° \qquad \Rightarrow \qquad D\widehat{A}B = 45°$

The direction of the hinge force is at 45° to the rod.

(b) Using Lami's Theorem,

$$\frac{T}{\sin 135°} = \frac{36}{\sin 90°} \quad \left(= \frac{H}{\sin 135°}, \text{ not wanted} \right)$$

$\Rightarrow \qquad T = 36 \times 0.7071 = 25.5$

The tension is 25.5 N (3 sf)

The questions in the next exercise can each be solved in a variety of ways. The following points may help you to choose the method you wish to use.

If there are only three forces acting, always make sure first that they are concurrent on the diagram. This should be done regardless of the method.

● If what is required is an angle, it may be that it can be found from the diagram without any further work.

● If there is an obvious triangle of forces and its sides are known or easy to find, it is suitable to use the equal ratios of force to length of side.

● If only three forces act and the angles between them are known, Lami's Theorem immediately gives the magnitude of each force separately.

● If there are more than three forces, or you do not care for any of the special methods, the general approach of resolving and taking moments can be applied to any problem. It may sometimes take a little longer but it always works!

EXERCISE 18b

In questions 1 and 2 a uniform rod AB of weight W, is hinged to a fixed point at A and held in the position shown by a force F.

(a) Draw a diagram showing the direction of the force exerted on the rod by the hinge.

(b) Find, in terms of W, the magnitude of the hinge force.

1.
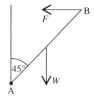

2.
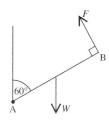

3. The end A of a uniform rod AB is in rough contact with a horizontal surface. The end B rests on the rim of a smooth disc fixed in the same vertical plane as the rod, so that AB is a tangent to the disc and is inclined at 30° to the horizontal. The angle of friction at A is λ and the rod is just about to slip.

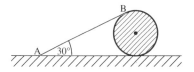

(a) Draw a general diagram and on it mark *three* forces that keep the rod in limiting equilibrium and the angle of friction.

(b) *Write down* Lami's Theorem for the three forces.

4. A uniform rod AB has one end B in rough contact with the ground. The rod rests against a smooth rail C and is about to slip. If $\tan\alpha=\frac{1}{2}$ and $\mu=\frac{1}{2}$, show that the resultant reaction at B and the reaction at C are equal in magnitude.

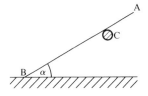

For questions 5 to 8 use this diagram, in which AB is a uniform ladder of weight W and is just about to slip.

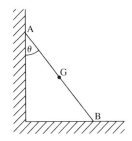

5. Contact at A is smooth, contact at B is rough and $\mu=\frac{3}{4}$. Find θ and the reaction at A.

6. Both contacts are rough and $\theta=45°$. Find μ. (The ladder will not slip until limiting friction is reached at both ends.)

7. Both contacts are rough with $\mu=\frac{1}{2}$. Find θ.

8. Explain why it is impossible for a ladder to rest in equilibrium at any angle to a wall if the wall and the ground are both smooth.

9. The diagram shows a uniform rod AB of weight W, held at right-angles to a wall by a string BC which is inclined at 30° to the wall. Contact with the wall is rough and friction is limiting.

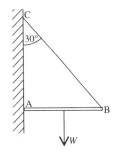

(a) Mark on a diagram the forces acting on the rod, using the resultant reaction at A.

(b) Find the angle of friction λ and hence the value of μ, the coefficient of friction between the rod and the wall. (Remember that $\mu=\tan\lambda$)

(c) Find, in terms of W, the tension in the string.

10. The diagram shows a uniform sphere of weight 20 N, with radius 3 cm, and centre O. The sphere rests in contact with a point A on a smooth wall and is held in position by a string of length 2 cm attached to a point B on the sphere and a point C on the wall.

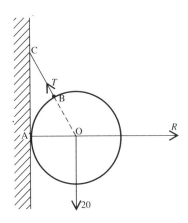

(a) Explain why the direction of the string CB passes through O.

(b) Write down the lengths of AO and CO and find the length of CA.

(c) Identify a suitable triangle of forces and use it to find the tension in the string and the reaction between the sphere and the wall.

FRAMEWORKS

Looking around we can see many structures consisting of metal girders or wooden members jointed to each other at their ends to form a rigid construction, e.g.

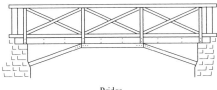

Bridge

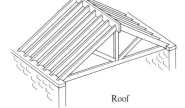

Roof

Such structures are known mathematically as *frameworks of light rods*. It is obvious that in many cases the members are far from light, but in comparison to the loads they bear their own mass is very small. So we model each member as a light rod.

If a framework has external forces acting on it, each rod can carry out one of two operations. It can

> stop the framework from collapsing inwards by exerting an outward force, or thrust, at each end, i.e. it is a strut,

or   prevent the joints from flying apart by exerting an inward pull or tension at each end and is called a *tie*.

Consider a simple triangular framework ABC, supported externally at A and C and carrying a load W at the vertex B.

It *looks as though* AC is a tie and that AB and BC are struts (with equal thrusts because of symmetry) so we will mark the forces in the rods accordingly.

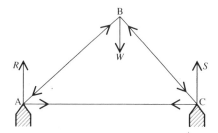

If we consider the framework *as a whole*, the forces in the rods occur in equal and opposite pairs and so do not affect the overall equilibrium. However the weight of the load and the two supporting forces are *external* forces and they must keep the framework in equilibrium. The magnitudes of R and S can be found by taking moments about A and C for the external forces only.

Now at each joint a number of forces act (three in this example) and unless these forces are in equilibrium the joint will break.

At a joint the force at only one end of each rod is involved, so now we must deal with the equilibrium of the concurrent forces that act at each joint individually until all the internal forces have been evaluated.

(It is not always obvious which rods are ties and which are struts; if in doubt mark all forces as thrusts, then any value that turns out to be negative indicates a tension.)

Example 18c

A framework consists of three light rods, each of length 2a, smoothly jointed to form a triangle ABC. The framework is smoothly hinged at B to a smooth vertical wall and carries a weight W at A. It rests in equilibrium with C resting on the wall at a point vertically below A. Find the reaction at C and the force in each rod.

The wall is smooth so the reaction at C is horizontal. The reaction at B is unknown in both magnitude and direction but it is not wanted and can be avoided by taking moments about B. We will mark the forces in each rod as thrusts.

Considering external forces:

$B\circlearrowleft$ gives $W \times 2a \sin 60° - R \times 2a = 0$

$\Rightarrow$ $R = \frac{1}{2}W\sqrt{3}$

Now resolving the forces at C

$\rightarrow$ $R - T_2 \sin 60° = 0$ $\Rightarrow$ $T_2 = W$

$\downarrow$ $T_3 + T_2 \cos 60° = 0$ $\Rightarrow$ $T_3 = -\frac{1}{2}W$

(negative denotes tension)

Now resolving the forces at the joint at A.

$\rightarrow \qquad (T_1 + T_2) \sin 60° = 0$

$\Rightarrow \qquad T_1 = -T_2 = -W$

The reaction at C is $\frac{1}{2} W \sqrt{3}$ away from the wall.

The force in AC is a thrust of magnitude W, the force in AB is a tension of magnitude W and that in BC is a tension of magnitude $\frac{1}{2} W$.

Note that we did not use joint B; we would have done had the reaction at the hinge been wanted.

EXERCISE 18c

State whether each member is a tie or a strut and make use of symmetry whenever this is possible.

1. A lamp of weight 60 N is attached to the outside wall of a pub by two light rods, AB of length 0.8 m and BC of length 0.4 m, smoothly hinged to the wall. The rods are smoothly jointed at B where the lamp is suspended. Find the forces in the rods.

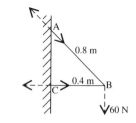

2. The diagram shows a roof truss made up of a triangular framework of smoothly jointed beams, supported at A and C. The weight, 2400 N, of a chimney acts through B. Find the reactions at A and C and the force in each beam.

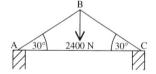

3. One end of a footbridge is supported on the symmetrical structure of smoothly jointed members shown in the diagram. Find, in terms of W, the force in each member.

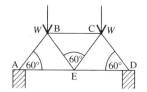

4. The symmetrical roof support shown in the diagram is constructed from jointed members and is supported on a wall at each end. Find the force in each member. (Hint. At B resolve parallel to AB and BF.)

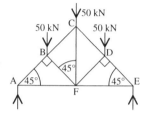

CHAPTER 19

CENTRE OF GRAVITY AND CENTRE OF MASS

CENTRE OF GRAVITY

When we consider a rod of weight W, it is obvious that the rod is made up of a large number of very short lengths of material, each with its own weight.

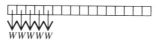

However, when dealing with an equilibrium problem involving a rod we usually mark a singe weight acting at a particular point on the rod which we have frequently referred to as 'the point through which the weight of the rod acts'.

Now if we want to replace all the components of weight by a single weight we must ensure that it has exactly the same effect on the rod as the separate components have, i.e.

the total weight is the sum of all the component weights

and

the single weight acts through a point such that the moment of the single weight about any axis is equal to the resultant moment of the components.

The point through which the resultant weight of a body acts is called the centre of gravity of the body and is very often denoted by G and its coordinates by $(\bar{x}, \bar{y})$.

The position of the centre of gravity of any object can be found by equating the resultant moment of all the parts, to the moment about the same axis of the total weight acting through the centre of gravity.

This principle applies to any object of any size, shape or dimension but we will look at a simple example of a set of separate particles.

Consider three particles, A, B and C, of weights 4 N, 2 N and 3 N respectively, attached to points on a light rod PQ as shown in the diagram below.

We know that, about any axis, the total moment of the weights of the particles is equal to the moment of the total weight.

Choosing to take moments about P we have

$$4 \times 2 + 2 \times 5 + 3 \times 6 = (4 + 2 + 3) \times \bar{x}$$

Hence $\bar{x} = 4$

This approach can be extended to any number of weights, $W_1, W_2, W_3, \ldots,$ at distances $x_1, x_2, x_3, \ldots$ from the chosen axis.

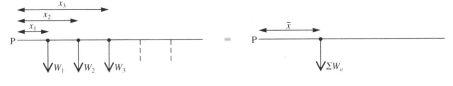

P) $W_1 \times x_1 + W_2 \times x_2 + W_3 \times x_3 + \ldots = \bar{x} \times (W_1 + W_2 + W_3 + \ldots)$

i.e. $\Sigma W_n x_n = \bar{x} \Sigma W_n$

(Σ means 'the sum of terms of this form when n takes values 1, 2, 3 …')

CENTRE OF MASS

Using $W = mg$ in the equation $\Sigma W_n x_n = \bar{x} \Sigma W_n$ gives

$$\Sigma m_n g x_n = \bar{x} \Sigma m_n g$$

If we take the value of g as constant and cancel it from each term in the moment equation above, we get $\Sigma m_n x_n = \bar{x} \Sigma m_n$ in which each term is of the form mass × its distance from a particular axis.

The solution of this equation is the location of the point G which we have so far called the centre of gravity, i.e. the point about which the *weight* of an object is evenly distributed.

However, as we now see, the *mass* also is evenly distributed about G which therefore can also be called the *centre of mass*.

This argument depends on the assumption that the value of g is the same for all the masses (and so can be cancelled). However, provided that we are not dealing with bodies (e.g. a mountain) that are so high that the value of g changes between the highest and lowest parts, we can make this assumption, i.e.

for an object of normal size, the centre of mass coincides with the centre of gravity

Hence, to find the centre of mass of a particular object we form an equation in which, on one side we have the sum of terms like $m_1 \times x_1$, and on the other side the sum of all the masses multiplied by $\bar{x}$. This equation can be written

$$\Sigma m_n x_n = \bar{x} \Sigma m_n$$

Examples 19a

1. The diagram shows a set of three particles of masses 5 kg, 2 kg and 4 kg attached to a light rod at the given positions. Find the distance from O of the centre of mass, G, of the particles.

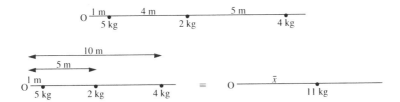

Using $\Sigma m_n x_n = \bar{x} \Sigma m_n$ gives

$$5 \times 1 + 2 \times 5 + 4 \times 10 = \bar{x} \times (5 + 2 + 4)$$

$$\Rightarrow \qquad \bar{x} = 5$$

The distance of G from O is 5 m.

2. Particles of masses $3m$, $2m$, $6m$, and am are attached to a light rod AB at distances $3d$, $4d$, $8d$, and $12d$ from A respectively. If G, the centre of mass of the particles, is distant $7d$ from A find the value of a.

A $\underset{3m}{\overset{3d}{\rule{0pt}{0pt}}} \underset{2m}{\overset{d}{\rule{0pt}{0pt}}} \underset{6m}{\overset{4d}{\rule{0pt}{0pt}}} \underset{am}{\overset{4d}{\rule{0pt}{0pt}}}$ B $\quad \equiv \quad$ A $\overset{7d}{\rule{0pt}{0pt}} \overset{G}{\underset{(11+a)m}{\rule{0pt}{0pt}}}$ B

Using $\Sigma m_n x_n = \bar{x} \Sigma m_n$ gives

$$3m \times 3d + 2m \times 4d + 6m \times 8d + am \times 12d = (11+a)m \times 7d$$

$$(65 + 12a)md = (11+a)m \times 7d$$

$$\Rightarrow \qquad (12-7)a = 77 - 65 = 12$$

$$\Rightarrow \qquad a = 2.4$$

EXERCISE 19a

1. Find the distance from P of the centre of mass, G, of two particles placed at P and Q, where PQ is of length 24 cm, if the masses of the particles are respectively

(a) 1 kg and 2 kg   (c)   3 kg and 8 kg   (e) m and $3m$

(b) 3 kg and 1 kg   (d)   6 kg and 4 kg   (f) $8m$ and $6m$.

In each question from 2 to 10 find the distance from A of the centre of mass of the given set of particles.

2. A——$3a$——o——a——o——$3a$——o——a——B
                  $2m$ $5m$ $2m$

5. A——$2a$——o——$3a$——o——a——B
                $4m$ m

3. A—a—o—a—o—a—o—$2a$—oB
           $2m$ m $2m$ m

6. Ao——a——o——$2a$——oB
           m m $2m$

4. Bo———$3a$———o—a—o—a—A
         m $2m$ $2m$

7. Bo———$4a$———o—a—oA
           $6m$ $4m$ m

8.
```
   o 3m
4a |
   o 6m
2a |
   o 7m
3a |
   o 4m
   A
```

9.
```
   A
 a |
   o 2m
 a |
   o 3m
 a |
   o 4m
 a |
   o 5m
 a |
   o 6m
```

10.
```
   o 2m
5a |
   o 8m
4a |
 a | A
```

11. Three particles whose masses are 2 kg, 5 kg and x kg are placed at points with coordinates $(1, 0)$, $(2, 0)$ and $(6, 0)$ respectively.

(a) If G, the centre of mass of the particles, is at the point $(3, 0)$, find x.

(b) A fourth particle of mass 2 kg is placed at the point $(5, 0)$. Find the centre of mass of the set of four particles.

12. A light rod AB is 80 cm long. Three particles, each with mass 2 kg, are attached at points distant 20 cm, 36 cm and 48 cm from the end A.

(a) Find the distance from A to G, the centre of mass of the particles.

(b) A fourth particle is attached at B. Find its mass if the centre of mass of the four particles is at the midpoint of the rod.

(c) If the particle distant 48 cm from A is then removed, find the distance of the centre of mass of the remaining three particles from A.

THE CENTRE OF MASS OF PARTICLES IN A PLANE

In all the problems above, the particles concerned lay in a straight line. Now we consider particles situated anywhere in a plane. For convenience we will locate the positions of the particles by coordinates in an xy plane.

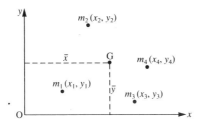

In order to locate G, both its x and y coordinates are required.

The x-coordinate, $\bar{x}$, is given by $\Sigma m_n x_n = \bar{x} \Sigma m_n$

The y-coordinate, $\bar{y}$, is given by $\Sigma m_n y_n = \bar{y} \Sigma m_n$

Examples 19b

1. Particles A, B and C of masses 4 kg, 7 kg and 5 kg are placed respectively at points with coordinates $(4, 2)$, $(0, 6)$ and $(1, 5)$. Find the centre of mass of the particles.

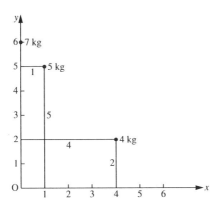

Using $\Sigma m_n x_n = \bar{x} \Sigma m_n$ gives

$$4 \times 4 + 7 \times 0 + 5 \times 1 = \bar{x}(4 + 7 + 5) \qquad \Rightarrow \qquad \bar{x} = 21 \div 16$$

Using $\Sigma m_n y_n = \bar{y} \Sigma m_n$ gives

$$4 \times 2 + 7 \times 6 + 5 \times 5 = 16\bar{y} \qquad \Rightarrow \qquad \bar{y} = 75 \div 16$$

The centre of mass is at the point $\left(\frac{21}{16}, \frac{75}{16} \right)$

2. Particles of masses **4 kg, 2 kg, 5 kg** and **6 kg** are placed respectively at the
vertices **A, B, C** and **D** of a light square lamina of side **2 m.** Find the distances of
the centre of mass of the particles from **AB** and **AD.**

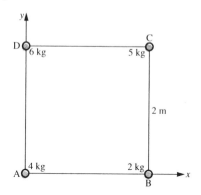

Using $\Sigma m_n x_n = \bar{x} \Sigma m_n$

$$4 \times 0 + 2 \times 2 + 5 \times 2 + 6 \times 0 = \bar{x} \times (4 + 2 + 5 + 6)$$

$\Rightarrow \qquad \bar{x} = 14 \div 17$

Using $\Sigma m_n y_n = \bar{y} \Sigma m_n$

$$4 \times 0 + 2 \times 0 + 5 \times 2 + 6 \times 2 = \bar{y} \times (4 + 2 + 5 + 6)$$

$\Rightarrow \qquad \bar{y} = 22 \div 17$

The distances of the centre of mass are $\frac{14}{17}$ m from AD and $\frac{22}{17}$ m from AB.

Note that from AB (a *horizontal* line) the distance measured is *vertical*, i.e. $\bar{y}$.
(In giving the answer to the question above it is dangerously easy to reverse the
lines from which the distances are given.)

3. Five particles whose masses are **2m, 3m, 2m, 4m** and **m,** are placed at points
with position vectors **3i + j, i − 4j, 5i + 6j, −i + 2j,** and **3i** respectively. Find
the position vector of their centre of mass.

Let the position vector of the centre of mass be $a\mathbf{i} + b\mathbf{j}$.

Using $\Sigma m_n x_n = \bar{x} \Sigma m_n$

$$2m \times 3 + 3m \times 1 + 2m \times 5 + 4m \times (-1) + m \times 3 = 12m \times a$$

$\Rightarrow \qquad a = 18 \div 12 = \frac{3}{2}$

Using $\Sigma m_n y_n = \bar{y} \Sigma m_n$

$$2m \times 1 + 3m \times (-4) + 2m \times 6 + 4m \times 2 + m \times 0 = 12m \times b$$

$\Rightarrow \qquad b = 10 \div 12 = \frac{5}{6}$

The position vector of the centre of mass is $\frac{3}{2}\mathbf{i} + \frac{5}{6}\mathbf{j}$.

Note that, when the particles are located by their position vectors, the separate x and y equations can be combined in the form

$$\Sigma m_n \mathbf{r}_n = \bar{\mathbf{r}} \Sigma m_n$$

where $\quad \bar{\mathbf{r}} = \bar{x}\mathbf{i} + \bar{y}\mathbf{j}$

The solution can then be presented more concisely as

$$2m(3\mathbf{i}+\mathbf{j}) + 3m(\mathbf{i}-4\mathbf{j}) + 2m(5\mathbf{i}+6\mathbf{j}) + 4m(-\mathbf{i}+\mathbf{j}) + m(3\mathbf{i}) = 12m\bar{\mathbf{r}}$$

$\Rightarrow \qquad (18\mathbf{i} + 10\mathbf{j})m = 12m\bar{\mathbf{r}}$

$\Rightarrow \qquad\qquad \bar{\mathbf{r}} = \tfrac{3}{2}\mathbf{i} + \tfrac{5}{6}\mathbf{j}$

EXERCISE 19b

In each question find the coordinates of the centre of mass of the given set of particles.

1.

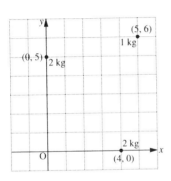

3.

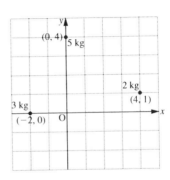

2.

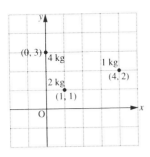

4.

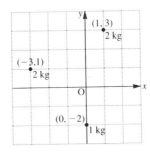

5. The coordinates of the vertices of a light square framework are $(4, 1)$, $(9, 5)$, $(3, 8)$ and $(0, 4)$. Particles of masses 2 g, 4 g, 2 g and 2 g respectively are placed at the vertices.

6. A light L-shaped wire ABC is in the xy plane with particles of equal mass m attached to A, B and C. The coordinates of A, B and C are $(-3, 2)$, $(2, 2)$ and $(2, -1)$ respectively.

THE CENTRE OF MASS OF A UNIFORM LAMINA

The number of particles that make up a lamina is infinitely large, so clearly it is not practical to locate its centre of mass in the same way as for a small number of distinct particles. However there are alternative methods for dealing with some uniform laminas (the mass per unit area of a uniform lamina is constant throughout). One of these methods uses the properties of symmetry.

If a uniform lamina has a symmetrical shape, the mass is equally distributed about the line of symmetry. Therefore the centre of mass must lie somewhere on the line of symmetry. If there is more than one axis of symmetry, it follows that G is located at the point of intersection of these axes.

The Centre of Mass of a Uniform Rod

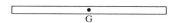

The midpoint of a uniform rod is clearly its centre of mass, as the masses of the two halves are equal and equally distributed.

The Centre of Mass of a Uniform Square Lamina

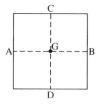

From symmetry, the centre of mass of the square lies somewhere on the line AB that bisects the square because the distribution of the mass is the same on both sides of this line. Similarly the centre of mass lies on CD and therefore is at the midpoint of each of these lines, which is the geometric centre of the square.

The Centre of Mass of a Uniform Rectangular Lamina

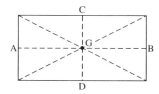

Using symmetry as we did above, the centre of mass is at the point of intersection of AB and CD, i.e. the point where the bisectors of the sides meet.
(Note that this is also the point of intersection of the diagonals of the rectangle.)

The Centre of Mass of a Uniform Triangular Lamina

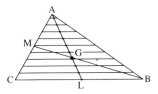

As there is no line of symmetry, we will divide the triangle into strips parallel to one side, BC say, and regard each strip as a 'rod'.

The centre of mass of each strip is at its midpoint, so the centre of mass of the triangle, G, lies on the line passing through all these midpoints, i.e. on the median AL.

Now using strips parallel to AC shows that G also lies on the median BM.

So G is at the point of intersection of the medians of the triangle.

It is a geometric property of a triangle that its medians intersect at a point which is $\frac{1}{3}$ of the way from base to vertex on any median. This point is called the *centroid* of a triangle so we can say that the centre of mass of a uniform triangular lamina is at the centroid of the triangle.

Hence for a right-angled triangle, G is $\frac{1}{3}$ of the way from the right-angle along each of the perpendicular sides, e.g.

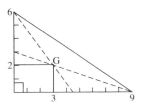

If the lamina is not uniform its centre of mass is unlikely to be at the centroid. In other words, the centroid of a triangle is *always* $\frac{1}{3}$ of the way up any median, but the centre of mass of a triangular lamina is *not necessarily* at the centroid.

Readers may occasionally find centroid used as though it *meant* centre of mass. This is not so. The centroid is a geometric point and *does not change* if the density of the shape varies, whereas the centre of mass is a property of balance and *does* depend on the distribution of the mass.

THE CENTRE OF MASS OF A COMPOUND LAMINA

Now that the position of the centres of mass of a number of common laminas are known, they can be used to find the centre of mass of a uniform lamina that is made up of these shapes.

For each part of the lamina, the mass can be expressed as the product of area and mass per unit area, i.e. mass = area × density.
It follows that the mass of each part, and hence the mass of the whole, is a multiple of ρ, where ρ (pronounced ro) is the symbol for density.

Examples 19c

1. **Find the centre of mass of the uniform lamina OABCDEF.**

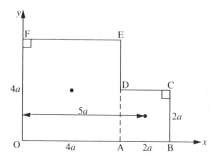

Let ρ be the mass per unit area of the lamina and let $G(\bar{x}, \bar{y})$ be the centre of mass of the whole lamina.

The mass of OAEF is $16a^2\rho$ and its centre of mass is at $(2a, 2a)$

The mass of ABCD is $4a^2\rho$ and its centre of mass is at $(5a, a)$

The mass of OABCDEF is $16a^2\rho + 4a^2\rho = 20a^2\rho$

Using $\Sigma m_n x_n = \bar{x} \Sigma m_n$ gives

$$16a^2\rho \times 2a + 4a^2\rho \times 5a = 20a^2\rho \times \bar{x}$$

Now $4a^2\rho$ cancels giving $\quad\quad \bar{x} = \frac{13a}{5}$

Similarly using $\Sigma m_n y_n = \bar{y} \Sigma m_n$ we have

$$16a^2\rho \times 2a + 4a^2\rho \times a = 20a^2\rho \times \bar{y} \quad \Rightarrow \quad \bar{y} = \frac{9a}{5}$$

The centre of mass of the whole lamina is the point $(\frac{13a}{5}, \frac{9a}{5})$.

The solution of this example can be set down more concisely by using a table as shown below.

Portion	Mass	Coords of G		mx	my
		x	y		
+ OAEF	$16a^2\rho$	$2a$	$2a$	$16a^2\rho \times 2a$	$16a^2\rho \times 2a$
+ ABCD	$4a^2\rho$	$5a$	a	$4a^2\rho \times 5a$	$4a^2\rho \times a$
OABCDEF	$20a^2\rho$	$\bar{x}$	$\bar{y}$	$20a^2\rho \times \bar{x}$	$20a^2\rho \times \bar{y}$

The plus signs are a reminder that the two parts are *added* to give the whole.

Working down the 5th column $16a^2\rho \times 2a + 4a^2\rho \times 5a = 20a^2\rho \times \bar{x}$

$\Rightarrow$ $\bar{x} = \frac{13a}{5}$

Working down the 6th column $16a^2\rho \times 2a + 4a^2\rho \times a = 20a^2\rho \times \bar{y}$

$\Rightarrow$ $\bar{y} = \frac{9a}{5}$

The tabular layout is recommended for all problems of this type.

2. **Find the centre of mass of the uniform lamina ABCDE shown in the diagram.**

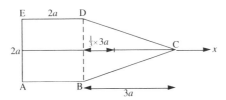

The shape is symmetrical about the line through C that bisects AE so G lies on this line. We will make this line the *x*-axis. Only the *x*-coordinate of the centre of mass is unknown.

Portion	Mass	x-coord. of G	$m_n x_n$
+ ABDE	$4a^2\rho$	a	$4a^2\rho \times a$
+ BCD	$3a^2\rho$	$2a + a$	$3a^2\rho \times 3a$
ABCDE	$7a^2\rho$	$\bar{x}$	$7a^2\rho \times \bar{x}$

$\Sigma m_n x_n = \bar{x} \Sigma m_n$,   i.e. working down the last column gives

$4a^2\rho \times a + 3a^2\rho \times 3a = 7a^2\rho \times \bar{x}$

$\Rightarrow$ $\bar{x} = \frac{13a}{7}$

The centre of mass of the lamina is on the line that bisects AE and DB and distant $\frac{13a}{7}$ from AE.

3.

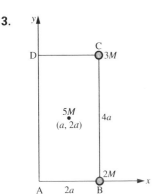

The diagram shows a uniform rectangular lamina ABCD of mass 5*M*. A particle of mass 2*M* is attached to B and a particle of mass 3*M* to C. Find the distances from AB and AD of the centre of mass of the lamina complete with its loads.

		Coords of G			
Portion	Mass	x	y	mx	my
+ ABCD	$5M$	a	$2a$	$5Ma$	$10Ma$
+ Particle B	$2M$	$2a$	0	$4Ma$	0
+ Particle C	$3M$	$2a$	$4a$	$6Ma$	$12Ma$
Loaded lamina	$10M$	$\bar{x}$	$\bar{y}$	$10M\bar{x}$	$10M\bar{y}$

Using $\Sigma m_n x_n = \bar{x}\,\Sigma m_n$ gives

$$5Ma + 4Ma + 6Ma = 10M\bar{x} \qquad \Rightarrow \qquad \bar{x} = 1.5a$$

Using $\Sigma m_n y_n = \bar{y}\,\Sigma m_n$ gives

$$10Ma + 0 + 12Ma = 10M\bar{y} \qquad \Rightarrow \qquad \bar{y} = 2.2a$$

The centre of mass is distant $1.5a$ from AD and $2.2a$ from AB.

4.

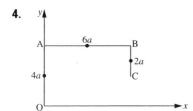

A uniform wire is bent into the shape of three sides of a trapezium, as shown in the diagram. Find the coordinates of the centre of mass of the shape.

The centre of mass of each section of the wire is at the midpoint and the mass of the wire is proportional to its length. We will use k as a constant of proportion.

		Coords of G			
Portion	Mass	x	y	mx	my
+ OA	$4a \times k$	0	$2a$	$4ka \times 0$	$4ka \times 2a$
+ AB	$6a \times k$	$3a$	$4a$	$6ka \times 3a$	$6ka \times 4a$
+ BC	$2a \times k$	$6a$	$3a$	$2ka \times 6a$	$2ka \times 3a$
OABC	$12ak$	$\bar{x}$	$\bar{y}$	$12ak \times \bar{x}$	$12ak \times \bar{y}$

Using $\Sigma mx = \bar{x}\,\Sigma m$, where $\Sigma m = (4 + 6 + 2)ak$, gives

$$4ka \times 0 + 6ka \times 3a + 2ka \times 6a = 12ka\bar{x}$$

$$\Rightarrow \qquad\qquad\qquad\qquad\qquad \bar{x} = \tfrac{5a}{2}$$

Using $\Sigma my = \bar{y}\,\Sigma m$ gives

$$4ka \times 2a + 6ka \times 4a + 2ka \times 3a = 12ka\bar{y}$$

$$\Rightarrow \qquad\qquad\qquad\qquad\qquad \bar{y} = \tfrac{19a}{6}$$

The coordinates of the centre of mass are $(\tfrac{5a}{2}, \tfrac{19a}{6})$.

5.

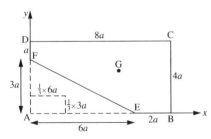

A triangle AEF is cut off from the rectangular lamina ABCD as shown in the diagram. Find the position of G, the centre of mass of the remaining lamina.

The centre of mass for the triangular part is $\frac{1}{3}$ of $6a$ and $\frac{1}{3}$ of $3a$ from A along AB and AD.

Portion	Mass	Coords of G		mx	my
		x	y		
+ ABCD	$32a^2\rho$	$4a$	$2a$	$32a^2\rho \times 4a$	$32a^2\rho \times 2a$
− AEF	$9a^2\rho$	$2a$	a	$9a^2\rho \times 2a$	$9a^2\rho \times a$
EBCDF	$23a^2\rho$	$\bar{x}$	$\bar{y}$	$23a^2\rho \times \bar{x}$	$23a^2\rho \times \bar{y}$

This time the triangle is *taken away* from the rectangle so we subtract the mass of the triangle from that of the rectangle. This is indicated by the negative sign in the second row.

$\Sigma m_n x_n = \bar{x} \Sigma m_n$ gives

$$32a^2\rho \times 4a - 9a^2\rho \times 2a = 23a^2\rho \times \bar{x} \quad \Rightarrow \quad \bar{x} = \frac{110a}{23}$$

$\Sigma m_n y_n = \bar{y} \Sigma m_n$ gives

$$32a^2\rho \times 2a - 9a^2\rho \times a = 23a^2\rho \times \bar{y} \quad \Rightarrow \quad \bar{y} = \frac{55a}{23}$$

The distances of G from AD and AB are respectively $\frac{110a}{23}$ and $\frac{55a}{23}$.

Note that whenever the position of G has been found it should be marked on the diagram, to see whether it looks about right.

EXERCISE 19c

In this exercise all laminas and rods are uniform.

1. State the coordinates of the centre of mass of each lamina.

(a)

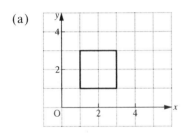

(b)

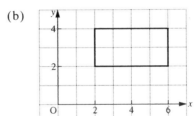

(c) 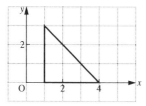   (d)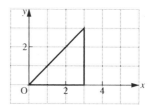

2. Write down the coordinates of the centre of gravity of each section of the given shape.

(a) 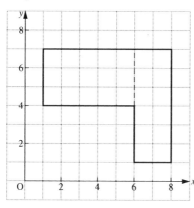   (c)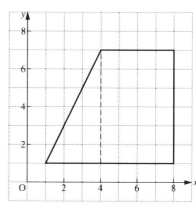

(b) 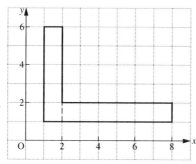   (d)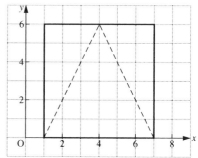

Keep your solutions to questions 3 to 5 as you will need them in Exercise 21a.

In questions 3 to 6, choose axes based on symmetry and find the coordinates of the centre of gravity of each lamina. Take the side of one square as 1 unit.

3. 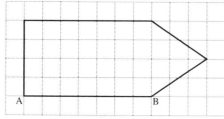   **4.**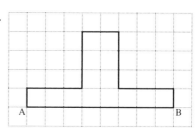

5.

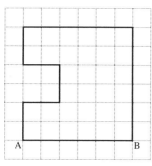

6.

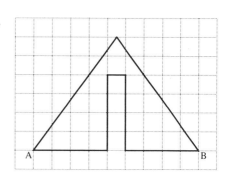

Find the centre of gravity of each framework of rods. All rods are of equal density.

7.

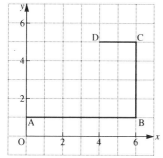

8.

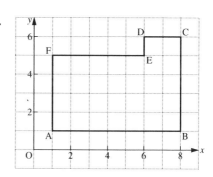

Find the centre of gravity of each lamina.

9.

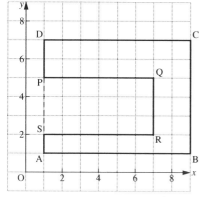

10.

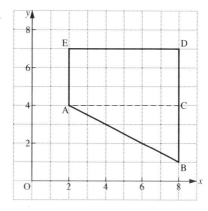

11.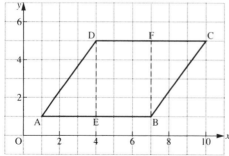

12.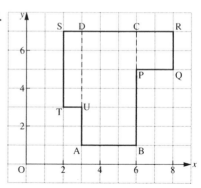

13. The diagram shows a uniform right-angled triangular lamina of mass $3M$, loaded at B with a particle of mass $2M$.
Find the coordinates of the centre of mass of the triangle complete with its load.

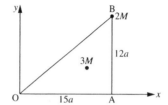

14. Particles of masses M, $3M$ and $4M$ are attached to the vertices A, C and D respectively of a uniform square of side $2a$ and mass $6M$. Find the distances of the centre of mass of the loaded lamina from AB and AD.

15. A 50 cm length of uniform wire is bent into a square shape as shown in the diagram.
A double thickness of wire forms the side AB.
Find the distances of the centre of mass of the wire shape from AB and BC.

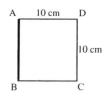

*16. ABCD is a uniform thin rectangular metal plate. $AB = 30$ cm and $BC = 15$ cm. E is the mid point of CD. The plate is folded along the line AE so that corner D then lies at the mid point of AB. Find the distances of the centre of mass of the folded plate from AB and BC.

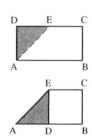

*17. The mass of the earth is 5.976×10^{24} kg and that of the moon is 7.350×10^{22} kg. The distance between the centres at perigee (when they are closest) is 356 400 km. Find the distance of their centre of mass from the centre of the earth at this time.

FURTHER PROBLEMS

In the exercise above all the compound objects involved relatively simple sections.

Now we look at some problems that involve finding the centre of mass of more complex compound bodies.

Examples 19d

1. **The diagram shows a uniform lamina, whose density is 1.2 kg per square metre, in the shape of a square ABCD, of side 2 m. A circular hole of radius 0.5 m is cut from it. The centre of the hole is distant 0.8 m from both AB and AD. A thin uniform strengthening strip is attached to the edge BC. Given that the mass of the strip is 1.5 kg, find the distances of the centre of mass of the lamina from AB and from AD.**

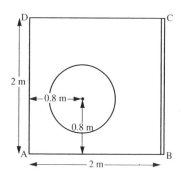

The mass of ABCD is $2^2 \times 1.2$ kg, i.e. 4.8 kg, and the mass removed for the hole is $\pi(0.5)^2 \times 1.2$ kg, i.e. 0.9425 kg (using 4 sf during the working).

		Distance of C of M			
Portion	Mass	x from AD	y from AB	Mass × x	Mass × y
+ABCD	4.8	1	1	4.8	4.8
−Hole	0.9425	0.8	0.8	0.754	0.754
+Strip	1.5	2	1	3	1.5
Whole	5.358	$\bar{x}$	$\bar{y}$	$5.358\bar{x}$	$5.358\bar{y}$

Using $\Sigma mx = \bar{x}\Sigma m$ gives $(4.8 - 0.754 + 3) = 5.358\bar{x}$

$\Rightarrow$ $\bar{x} = 1.315$

Using $\Sigma my = \bar{y}\Sigma m$ gives $(4.8 - 0.754 + 1.5) = 5.358\bar{y}$

$\Rightarrow$ $\bar{y} = 1.035$

The centre of mass is 1.32 m from AD and 1.04 m from AB   (3 sf).

2. A 'modern art' wall plaque, shown in
   the diagram, is made up from an
   isosceles triangular uniform sheet of
   copper, ABC, joined along the edge BC
   to a uniform right-angled triangle of
   aluminium, BCD, where
   BC = BD = 0.6 m. A lead motif
   is attached at C.
   The mass of the copper triangle is 1 kg
   and that of the aluminium triangle is
   0.4 kg.

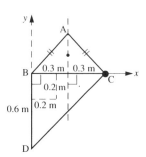

The artist intends the plaque to hang from A with BC horizontal and to achieve this
its centre of gravity must lie on the perpendicular bisector of BC. What must the
mass of the motif be for this to be achieved.

Only the x-coordinate of the centre of mass is needed and this is known to be 0.3.

Portion	Mass (m)	x at C of M	mx
ABC	1	0.3	0.3
BCD	0.4	0.2	0.08
Motif	M	0.6	$0.6M$
Plaque	$(1.4 + M)$	0.3	$(1.4 + M)(0.3)$

Using $\Sigma mx = \bar{x}\Sigma m$ gives $0.3 + 0.08 + 0.6M = (1.4 + M)(0.3)$

$\therefore$ $0.3M = 0.04$ $\Rightarrow$ $M = 0.133\ldots$

The mass of the motif should be 0.133 kg   (3 sf).

EXERCISE 19d

1. A uniform lamina has the shape of a square
   ABCD, of side 30 cm, joined along the side CE to
   an equilateral triangle CDE. Find the distance of
   the centre of mass from AB.

2. The density of the lamina described in question 1 is 0.25 g/cm^2. The centre of
   mass is to be relocated on the line CE by attaching a particle of mass m g at the
   point D. Find m.

3. The centre of mass of the lamina in question 1 is brought on to the line CE by
   making a different adjustment. This time the density of the triangular section is
   increased while the density of the square section remains at 0.25 g/cm^2. Find the
   density of the triangular section.

4. A circular lamina of radius 10 cm, with diameter AB, has a circular hole, of radius r cm, cut in it. The centre of the hole is at a point C on AB, where AC $= x$ cm. It is required that after making this hole the centre of mass of the remainder should lie at a distance of 8 cm from A. By finding the necessary value for x in each case, decide whether this can be done

(a) if $r = 5$ cm,

(b) if $r = 6$ cm.

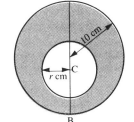

CHAPTER 20

CENTRE OF MASS OF A RIGID BODY

So far we have been finding the centres of mass of laminas, i.e. two dimensional objects. Therefore only two coordinates were needed to locate the centre of mass.

Now we are going to consider rigid three-dimensional objects and the location of their centres of mass clearly requires three coordinates. In all the cases we deal with at this level, however, the body will have either an axis or a plane of symmetry so that at least one of the coordinates is obvious.

For example, a solid right cone has an axis of symmetry from the vertex to the centre of the base so its centre of mass, G, is somewhere on this line; it remains only to find one coordinate, i.e. the distance of G from the vertex (or the base).

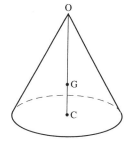

Whenever the centre of mass of a symmetrical solid object is to be found, the line of symmetry should be chosen as one of the axes of coordinates.

FINDING A CENTRE OF MASS BY INTEGRATION

When an object can be divided into a small number of parts, the mass and centre of mass of each part being known, the centre of mass of the whole object can be found by using $\Sigma m_n x_n = \bar{x} \Sigma m_n$ with similar expressions for $\bar{y}$ and $\bar{z}$ when the object is three-dimensional.

Some bodies cannot be divided up in this way, but can be divided into a very large number of very small parts whose masses and centres of mass are known. In cases like this, it may be possible to evaluate $\Sigma m_n x_n$ and Σm_n by using integration.

Suppose that we are asked to find the position of the centre of mass of a uniform solid right circular cone with height h and base radius a.

We will take the axis of symmetry as the x-axis. Therefore the centre of mass G lies on the x-axis and only the x-coordinate of G has to be found. For a reason that will become clear during the calculation we place the origin at the vertex.

The cone can be divided into thin slices, each parallel to the base and each being approximately a thin circular disc with its centre of mass on the x-axis.

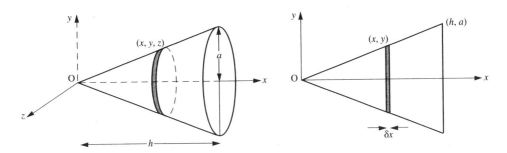

Considering one typical slice, called an *element*, we see that it is approximately a disc for which:

the radius is approximately y,
the thickness is a small increase in x, i.e. δx,

hence the mass, m_n, is approximately $(\pi y^2 \delta x)\rho$, where ρ is the mass per unit volume,

the distance of the centre of mass from O is approximately x_n, hence $m_n x_n$ is approximately $(\pi\rho y^2 \delta x)x$

Therefore $\Sigma m_n x_n = \Sigma \pi\rho y^2 x\delta x$

Now $\lim\limits_{\delta x \to 0} \Sigma \pi\rho y^2 x\delta x = \displaystyle\int \pi\rho y^2 x \; dx$

Therefore as $\delta x \to 0$ $\Sigma m_n x_n \to \displaystyle\int \pi\rho y^2 x \; dx$

The mass of the whole cone is $V\rho$ where V is the volume of the cone,

$\therefore$ $\bar{x} \, \Sigma m_n = V\rho \, \bar{x}$

Then using $\Sigma m_n x_n = \bar{x} \, \Sigma m_n$ gives

$$\int \pi\rho y^2 x \; dx = V\rho \, \bar{x} \tag{1}$$

It is important to appreciate that equation [1] applies to *any* solid that is symmetrical about the x-axis and can be divided into disc-like elements.

For the cone in particular, $V = \frac{1}{3}\pi a^2 h$, and from the similar triangles in the diagram we see that

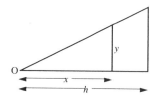

$$\frac{y}{x} = \frac{a}{h} \quad \Rightarrow \quad y = \frac{ax}{h}$$

This relationship is simple because the vertex is the origin of coordinates.

Now when the equation $\int \pi \rho y^2 x \, dx = V\rho \bar{x}$ is applied to the cone,

the LHS becomes $\int \pi \rho \left(\frac{ax}{h}\right)^2 x \, dx$

and the RHS becomes $\frac{1}{3}\pi a^2 h \rho \bar{x}$.

Now the value of x goes from 0 to h, hence [1] becomes

$$\frac{\pi \rho a^2}{h^2} \int_0^h x^3 \, dx = \frac{1}{3}\pi a^2 h \rho \bar{x} \quad \Rightarrow \quad 3\int_0^h x^3 \, dx = h^3 \bar{x}$$

$$\therefore \qquad \frac{3}{4}\left[x^4\right]_0^h = h^3 \bar{x} \quad \Rightarrow \quad \bar{x} = \frac{3}{4}h$$

i.e. **the centre of mass of a uniform solid cone is on the axis of symmetry and three-quarters of the way from the vertex to the base.**

Any uniform solid that is symmetrical about the x-axis, and for which all cross-sections are circular, is a *solid of revolution* and its centre of mass can be found by the process given above. It is only the total volume, and the boundaries of the integration, that differ from one such solid of revolution to another. So $\int \pi \rho y^2 x \, dx = V\rho \bar{x}$ for *any* solid of revolution symmetrical about the x-axis,

i.e.

$$\int \pi y^2 x \, dx = V\bar{x}$$

If a solid of revolution is symmetrical about the y-axis, it can be divided into slices with approximate radius x and thickness δy, leading to a similar result, i.e. $\int \pi x^2 y \, dy = V\bar{y}$

Hence the location of the centre of mass on the y-axis can be found.

These results are quotable but it is advisable to look at a suitable element in each problem as a check on the expression to be integrated.

Examples 20a

1. **The diagram shows a quadrant of a circle of radius a and equation $x^2 + y^2 = a^2$. When this quadrant is rotated through a complete revolution about the x-axis a uniform hemisphere is generated.**

   Find the coordinates of the centre of mass of the hemisphere.

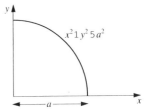

The hemisphere is symmetrical about the x-axis so the centre of mass, G, is on this axis. Therefore the y and z coordinates of G are both zero.

We will divide the hemisphere into slices parallel to the flat face.

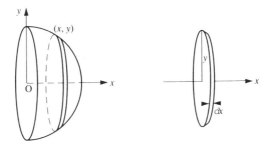

Considering one element as being almost a circular lamina then approximately: the radius is y, the thickness is δx and the distance of the element from O is x.

Hence the mass, m_n, of the element $\approx \pi y^2 \delta x \rho$.

The volume of the whole hemisphere is $\frac{2}{3} \pi a^3$.

Using the quotable result $\int \pi y^2 x \ dx = V \bar{x}$ gives

$$\int_0^a y^2 x \ dx = \frac{2}{3} a^3 \bar{x}$$

The integration cannot be carried out until y is expressed as a function of x.

From the equation of the circle, $y^2 = a^2 - x^2$

$$\therefore \quad \int_0^a y^2 x \ dx = \int_0^a x(a^2 - x^2) \ dx$$

$$= \left[\frac{1}{2} a^2 x^2 - \frac{1}{4} x^4 \right]_0^a = \frac{1}{4} a^4$$

$$\therefore \quad \frac{2}{3} a^3 \bar{x} = \frac{1}{4} a^4 \quad \Rightarrow \quad \bar{x} = \frac{3}{8} a$$

Therefore the centre of mass of the hemisphere is distant $\frac{3}{8} a$ from O on the radius of symmetry.

The coordinates of the centre of mass are $\left(\frac{3}{8} a, 0, 0 \right)$.

2. **The area bounded by the y-axis, the line $y = 2$ and part of the curve with equation $y = 2x^2$ is rotated about the y-axis to give a uniform solid. Find the distance of the centre of mass of the solid from the origin O.**

The solid is symmetrical about the y-axis so the x and z coordinates of its centre of mass are both zero. The elemental slices are perpendicular to the y-axis.

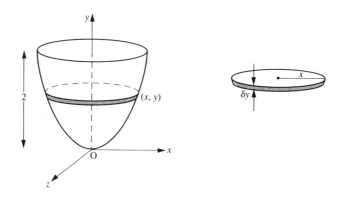

The approximate radius of the element is x, its thickness δy, therefore $m_n \approx \pi x^2 \delta y \rho$.

Quoting $\displaystyle\int \pi \rho x^2 y \ \mathrm{d}y = V \rho \bar{y}$ gives $\displaystyle\pi \int_0^2 x^2 y \ \mathrm{d}y = V \bar{y}$

Using $x^2 = \tfrac{1}{2} y$ the LHS becomes

$$\pi \int_0^2 \tfrac{1}{2} y^2 \ \mathrm{d}y = \pi \left[\tfrac{1}{6} y^3 \right]_0^2 = \tfrac{4}{3} \pi$$

Now there is no formula for the volume of this solid of revolution, so the volume, V, must in this case be found by integration.

The volume of an elemental slice $\approx \pi x^2 \delta y$

$$\therefore \qquad V = \int_0^2 \pi x^2 \ \mathrm{d}y \qquad \left(\text{where } x^2 = \tfrac{1}{2} y \right)$$

$$= \pi \int_0^2 \tfrac{1}{2} y \ \mathrm{d}y = \pi \left[\tfrac{1}{4} y^2 \right]_0^2$$

$$\Rightarrow \qquad V = \pi$$

Returning to $\displaystyle\pi \int_0^2 x^2 y \ \mathrm{d}y = V \bar{y}$ we have

$$\tfrac{4}{3} \pi = \pi \bar{y}$$

Therefore $\bar{y} = \tfrac{4}{3}$

THE CENTRE OF MASS OF A SEMICIRCLE

The integration method can also be used to find the centre of mass of a uniform semicircular lamina.

Consider such a lamina, bounded by the y-axis and part of the curve with equation $x^2 + y^2 = a^2$, divided into vertical strips as shown. The x-axis is a line of symmetry so the centre of mass lies on it.

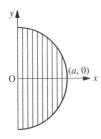
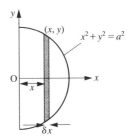

For one elemental strip of width δx,

the length is $2y$ so the area is approximately $2y\delta x$,
the mass is approximately $2y\rho\delta x$ where ρ is the mass per unit *area*,
the distance of the centre of mass from the y-axis is approximately x

Therefore $mx \approx (2y\rho\delta x)x$, i.e. $2xy\rho\delta x$

Now we can sum all the elements using $\Sigma mx = \bar{x}\Sigma m$ where

$$\sum_0^a mx = \sum_0^a 2xy\rho\delta x$$

As the width of the elemental strips approaches zero,

the limit as $\delta x \to 0$ of $\sum_0^a 2xy\rho\delta x = \int_0^a 2xy\rho\,\mathrm{d}x$

and the limit as $\delta x \to 0$ of Σm is the mass of the semicircle, i.e. $\frac{1}{2}\pi a^2\rho$

Now $x^2 + y^2 = a^2 \quad\Rightarrow\quad y = \sqrt{(a^2 - x^2)}$

Therefore $\quad 2\rho\displaystyle\int_0^a x\sqrt{(a^2 - x^2)}\,\mathrm{d}x = \frac{1}{2}\pi\rho a^2\bar{x}$

$\Rightarrow \quad 2\left[-\frac{1}{3}(a^2 - x^2)^{3/2}\right]_0^a = \frac{1}{2}\pi a^2\bar{x}$

$\Rightarrow \quad -\frac{2}{3}[0 - a^3] = \frac{1}{2}\pi a^2\bar{x}$

Hence $\quad \bar{x} = \dfrac{4a}{3\pi}$

i.e.

**the centre of mass of a uniform semicircular lamina
is on the radius of symmetry
and distant $\dfrac{4a}{3\pi}$ from the centre of the plane edge**

This result is quotable.

The centres of mass of other uniform laminas bounded by curves with known equations, can be found in a similar way but it is unlikely that a formula exists for the area of such a lamina. This means that Σm has to be found by integration.

Suppose, for example, that we want to find the centre of mass of a uniform lamina in the shape of the area in the first quadrant bounded by the x and y axes, and the curve $y = 9 - x^2$.

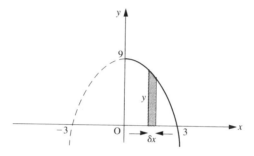

Using vertical elements gives $\bar{x}\Sigma m = \Sigma xy\rho\delta x$ as before.

As before, the r.h.s. becomes $\displaystyle\int_0^3 xy\rho \; dx = \int_0^3 x(9 - x^2)\rho \; dx$

but the l.h.s. cannot be evaluated by formula.

Instead we use $\Sigma m = \Sigma y\rho\delta x$

$\Rightarrow$ the limit as $\delta x \to 0$ of $\displaystyle\sum_{x=0}^{x=3} m = \int_0^3 y\rho \; dx = \int_0^3 (9 - x^2)\rho \; dx$

Hence $\bar{x}\displaystyle\int_0^3 (9 - x^2)\rho \; dx = \int_0^3 x(9 - x^2)\rho \; dx$

i.e. $\bar{x}\displaystyle\int_0^3 (9 - x^2) \; dx = \int_0^3 x(9 - x^2) \; dx$

So we see that two separate integrations have to be carried out before $\bar{x}$ can be found.

EXERCISE 20a

1. The shape of a uniform lamina is that of the area bounded by the x-axis, the curve $y = x^2$ and the line $x = 2$.

(a) Find the area of the lamina

(b) Find the x-coordinate of the centre of mass of the lamina.

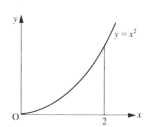

2. For a uniform lamina in the shape of the area in the first quadrant between the curve $y = x^2$ and the line $y = 4$, find

(a) the area

(b) the y-coordinate of the centre of mass (use elemental strips parallel to the x-axis).

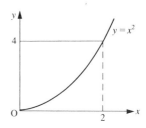

In the questions that follow, assume that each solid of revolution is uniform.

In questions 3 to 10 a solid of revolution is formed when the area bounded by the given lines and part of the line or curve with given equation, is rotated about the x-axis. Find the distance from O of the centre of mass.

3. The area bounded by the line $y = 2x$,   the x-axis and the lines $x = 2$ and $x = 4$.

4. The area between the line $y = ax$ and the x-axis, from $x = h$ to $x = 2h$.

5. The area in the first quadrant bounded by $y = x^2$, the x-axis and the line $x = 2$.

6. The area bounded by $y = x^2 + 2$, the x-axis, the y-axis and the line $x = 1$.

7. The area bounded by $y = \sqrt{x}$, $x = 1$, $x = 4$ and the x-axis.

8. The area bounded by $y = \sqrt{x + 4}$, the x-axis and the lines $x = 0$ and $x = 3$.

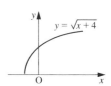

9. The area bounded by $y = e^x$, $x = 1$ and the x and y axes.

10. The area bounded by $y = \dfrac{1}{\sqrt{x + 1}}$, the x-axis, the y-axis and the line $x = 5$.

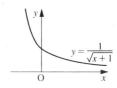

In questions 11 to 14 the area bounded by the given curves and lines is rotated about the y-axis to form a solid of revolution. Find the distance of the centre of mass from the origin.

11. The area in the first quadrant bounded by

$y = \dfrac{1}{x^2}, \quad y = 1, \quad y = 5$ and the y-axis.

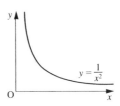

12. The area bounded by $x^2 + y^2 = 64$ and the positive x and y axes.

13. The area bounded by $y = 4 - x^2$, the x-axis and the y-axis.

14. The area in the first quadrant bounded by $x^2 + y^2 = 25$, $y = 3$, $y = 4$ and the y-axis.

15. A solid sphere of radius a is cut into two sections by making a plane cut at a distance $\frac{1}{2}a$ from the centre. Find, by integration, the distance of the centre of mass of the smaller of the two sections from the centre of the original sphere.

*16. The perpendicular height of a solid, right pyramid is h and the base is a square of side $2a$. Find the volume of the pyramid.

The vertex of the pyramid is taken as the origin O, the x-axis is the axis of symmetry and the position of the pyramid is shown in the diagram. The pyramid is divided into slices perpendicular to the x-axis.

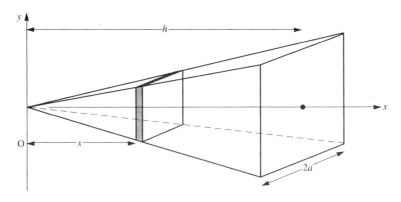

Considering an elemental slice distant x from O,
(a) by using similar figures, express the area of the slice in terms of x, a and h
(b) find the approximate mass, m, of the element
(c) find Σmx
(d) find the position of the centre of mass of the pyramid.

COMPOUND SOLIDS

We are now going to consider finding the centre of mass of a solid body that is made up of two (or more) parts, each of whose volume and centre of mass is known or can easily be found.

The centres of mass you can quote, in addition to the obvious cases of a cuboid and a sphere, are given in the table below.

Body	Position of G
Cylinder	Halfway up the centre line.
Hemisphere	On the radius of symmetry, $\frac{3}{8}$ of the way from the centre of the plane face.
Cone	On the axis of symmetry, $\frac{1}{4}$ of the perpendicular height above the base.
Pyramid	As for a cone.

The method we use when dealing with compound solids is the same as that used for compound laminas, i.e. the mass and location of centre of mass (C of M) of each part, and of the whole body, are tabulated, providing the data necessary to apply $\Sigma m_n x_n = \bar{x} \Sigma m_n$.

Examples 20b

1.

The cross-section of a uniform solid prism of length l, is the trapezium shown in the diagram. Find the centre of mass of the prism.

The plane that divides the prism into equal halves is a plane of symmetry, so the centre of mass G lies in this plane which we will take as the xy plane.

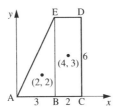

Taking ρ as the mass per unit volume we have:

Portion with cross-section	Mass m	Coords of C of M		mx	my
		x	y		
+ABE	$\frac{1}{2}(3)(6)l\rho$	2	2	$18\,l\rho$	$18\,l\rho$
+BCDE	$(2)(6)l\rho$	4	3	$48\,l\rho$	$36\,l\rho$
+ABCDE	$(9+12)l\rho$	$\bar{x}$	$\bar{y}$	$21\,l\rho\bar{x}$	$21\,l\rho\bar{y}$

Using $\Sigma m_n x_n = \bar{x} \Sigma m_n$ gives

$$18\,l\rho + 48\,l\rho = 21\,l\rho\bar{x} \qquad \Rightarrow \qquad \bar{x} = \tfrac{22}{7}$$

Using $\Sigma m_n y_n = \bar{y} \Sigma m_n$ gives

$$18\,l\rho + 36\,l\rho = 21\,l\rho\bar{y} \qquad \Rightarrow \qquad \bar{y} = \tfrac{18}{7}$$

Therefore G lies on the plane of symmetry at the point distant $\tfrac{22}{7}$ cm and $\tfrac{18}{7}$ cm from A in the directions of AC and CD respectively.

Note that when a prism has a *plane* of symmetry, finding the centre of mass amounts to finding the centre of mass of a lamina in the shape of the cross section.

2. **A child's toy is made up of a uniform solid hemisphere of radius a with its plane face fixed to the base of a solid right circular cone of base radius a and height $4a$. Find the distance of G, the centre of mass of the toy, from the centre of the common face.**

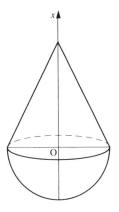

The hemisphere and cone have a common axis of symmetry so the centre of mass lies on it.

Taking O as the centre of the common face and the positive x-axis through the vertex of the cone, the following table can be compiled.

Section	Mass, m	x at C of M	mx
Cone	$\tfrac{1}{3}\pi a^2 (4a)\rho$	$+a$	$\tfrac{4}{3}\pi a^4 \rho$
Hemisphere	$\tfrac{2}{3}\pi a^3 \rho$	$-\tfrac{3}{8}a$	$-\tfrac{1}{4}\pi a^4 \rho$
Whole body	$2\pi a^3 \rho$	$\bar{x}$	$2\pi a^3 \rho\bar{x}$

Using $\Sigma m_n x_n = \bar{x} \Sigma m_n$ gives

$$\tfrac{4}{3}\pi a^4 \rho + \left(-\tfrac{1}{4}\pi a^4 \rho\right) = 2\pi a^3 \rho\bar{x}$$

$$\Rightarrow \qquad \tfrac{13}{12}a = 2\bar{x}$$

The positive sign for $\bar{x}$ shows that G is in the cone and not the sphere.

Therefore G is on the axis of symmetry of the cone, distant $\tfrac{13}{24}a$ from the common face.

3. The diagram shows a uniform solid body formed from a cube of edge 4*a* with a cube of edge 2*a* removed from one corner. Find the position of the centre of mass of the solid.

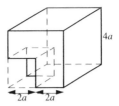

The most convenient choice of origin is the 'corner' that has been removed because this point is at the corner of both of the cubes we consider.

There is an axis of symmetry, passing through O, A and B. The centre of mass, G, is therefore on OAB and is equidistant from the three faces that meet at O. Therefore we need find only one of these distances.

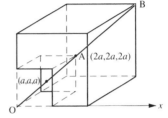

Section	Mass, m	x at C of M	mx
+ Large cube (complete)	$(4a)^3 \rho$	$2a$	$128a^4 \rho$
− Small cube	$(2a)^3 \rho$	a	$8a^4 \rho$
Remaining body	$(4^3 - 2^3)a^3 \rho$	$\bar{x}$	$56a^3 \rho \bar{x}$

Using $\quad \Sigma m_n x_n = \bar{x} \, \Sigma m_n \quad$ gives $\quad 128a^4\rho - 8a^4\rho = 56a^3\rho\bar{x}$

$\Rightarrow \qquad 120a = 56\bar{x} \quad \Rightarrow \quad \bar{x} = \frac{15}{7}a$

The centre of mass of the solid is distant $\frac{15}{7}a$ from the removed corner, along each of the edges.

Note that by using *the ratio of volumes of similar solids*, the table can be made even simpler. The ratio of the volumes, and therefore the masses, of the two cubes is $(4a)^3 : (2a)^3$, i.e. $8 : 1$.

Therefore, taking M as the mass of the removed cube, the table becomes:

Section	Mass m	x at C of M	mx
+ Large cube	$8M$	$2a$	$16aM$
− Small cube	M	a	aM
Remaining body	$7M$	$\bar{x}$	$7M\bar{x}$

giving $\qquad 16aM - aM = 7M\bar{x} \qquad \Rightarrow \qquad \bar{x} = \frac{15}{7}a$

For those readers who are comfortable with using similarity, this method, where appropriate, is obviously neater.

4. A uniform solid right cone is of height **12h**.
The upper part of the cone is removed by a
cut parallel to the base at a distance of **4h**
from the vertex, forming a frustum of a
cone. Find the height above the centre of
the base of the centre of mass of the
frustum.

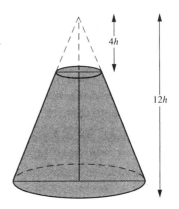

Although we are asked to locate G from the base, we will place O at the original vertex.

The removed cone and the original cone are similar,
with linear a ratio of $4h : 12h$, i.e. $1 : 3$. So the
masses are in the ratio $1 : 27$. If M is the mass of
the cone removed, the mass of the original cone
is $27M$.

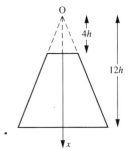

Section	Mass m	x at C of M	mx
+ Large cone	$27M$	$\frac{3}{4}(12h)$	$243Mh$
− Small cone	M	$\frac{3}{4}(4h)$	$3Mh$
Frustum	$26M$	$\bar{x}$	$26M\bar{x}$

Using $\Sigma m_n x_n = \bar{x}\Sigma m_n$ gives

$$243Mh - 3Mh = 26M\bar{x} \quad\Rightarrow\quad \bar{x} = \tfrac{240}{26}h = \tfrac{120}{13}h$$

Therefore the height of G above the base is $12h - \tfrac{120}{13}h = \tfrac{36}{13}h$

There are occasions when a compound object is made of sections which do not all
have the same mass per unit volume. In these cases the separate values of the
densities may be given or, instead, the ratio of the densities of the parts may
be given.

5. A concrete bollard comprises a solid cylinder, of radius 8 cm and height 46 cm, surmounted by a cone of equal radius and height 24 cm. The weight per unit mass of the concrete from which the cylindrical section is made is twice that of the mix in the conical section. Find the height of the centre of mass of the bollard.

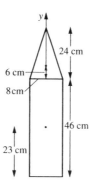

Let the density of the material in the cone be ρ so that the material in the cylinder is of density 2ρ.

Portion	Mass m	y at C of M	my
Cylinder	$\pi(8)^2(46)(2\rho)$	23	$92 \times 23(8)^2 \pi\rho$
Cone	$\frac{1}{3}\pi(8)^2(24)\rho$	$46+6$	$8 \times 52(8)^2 \pi\rho$
Bollard	$(92+8)(8)^2 \pi\rho$	$\bar{y}$	$100(8)^2 \pi\rho\,\bar{y}$

$$\Sigma my^2 = \bar{y}\,\Sigma m$$

$\Rightarrow$ $\quad 92 \times 23(8)^2 \pi\rho + 8 \times 52(8)^2\pi\rho = 100(8)^2\pi\rho\bar{y}$

$\Rightarrow$ $\qquad\qquad\qquad\qquad \bar{y} = 25.32\ldots$

The centre of mass of the bollard is 25.3 cm from the base (3 sf).

Useful Points to Consider When Finding the Centre of Mass of a Solid Body

● If there is a plane of symmetry use it if possible as the xy plane (i.e. the plane containing the x and y axes.)

● If there is an axis of symmetry, use it as the x-axis (or the y-axis)

● Place O at a point from which the centre of mass of each section is easy to locate.

● If O is *inside* the solid, remember that some value(s) of x will be negative.

● If the body is made up of similar sections, the ratio of masses can be found from the ratio of volumes.

EXERCISE 20b

1. A stool has a top that is 30 cm square and 2 cm thick. At each corner there is a leg, 20 cm long and with a cross-section that is a 3 cm square. The top and legs are made from the same material. Find the distance from the top of the stool to the centre of mass of the stool.

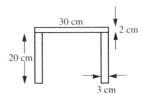

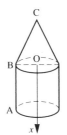

Questions 2 to 4 are about a uniform solid made by joining a right circular cone to a cylinder. Take the centre of the common face as origin and the x-axis along the line of symmetry as shown.

2. Find the distance of the centre of mass from the common face in each case.

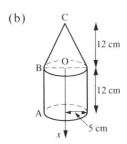

(a)   (b)

3. The centre of mass lies in the common face of the cone and the cylinder. If h is the height of the cone and H is the height of the cylinder, show that $h^2 = 6H^2$.

4. Given the dimensions in the diagram, find the distance of the centre of mass from the common face. Give your answer in terms of h.

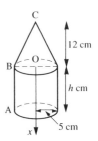

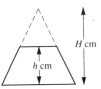

Questions 5 and 6 are about a frustum of a
uniform solid cone. (It is a good idea to take
the vertex of the original cone as origin).

5. Given that $H = 24$ and $h = 12$, find the distance of the centre of mass
from the larger plane face of the frustum.

6. If $H = 3a$ and $h = a$, find the distance of the centre of mass from the
larger plane face of the frustrum.

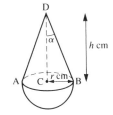

Questions 7 and 8 are about the uniform solid,
shown in the diagram, made by joining the
plane face of a hemisphere to the plane face of
a cone.

7. Find the distance of the centre of mass from the common plane face of the cone
and the hemisphere

(a) if $r = 5$ and $h = 20$ (b) if $r = 15$ and $h = 10$.

8. The centre of mass lies in the common plane face of the cone and the hemisphere.
Find α.

9. A uniform solid is made by joining the plane face of a hemisphere of radius r to
one plane face of a cylinder of radius r and height h.

(a) Find the distance of the centre of mass from the common plane face.

(b) Find r in terms of h given that the centre of mass lies in the common plane
face.

10. A solid cylinder, of height $2a$ and radius $2a$, has a
hemisphere removed from its upper plane end.
The hemisphere has radius a and its centre is on
the axis of the cylinder. Find the depth below the
top of the centre of mass of the remaining solid.

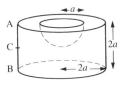

*11. A solid consists of a cone, of height $80\,$cm, which
has had a cone of the same radius and of height h
centimetres removed from its base. The centre of
mass of the remaining solid is $25\,$cm above the
base. Find the value of h.

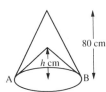

*12. An A level Maths student is planning a cycle tour. As a modelling exercise he decides to investigate which of two methods of carrying his luggage will have the lower centre of mass. He has a saddlebag, which he models as a cylinder, and two pairs of pannier bags, one pair larger than the other, which he models as prisms with a trapezium cross-section. He makes the following simplifying assumptions.

- The masses of the means of attachment are negligible.
- All the bags are fully packed to a uniform density.

The two methods he considers are as follows.

Method A Two large panniers suspended from a carrier above the rear wheel. AD is at a height of 70 cm above the ground.

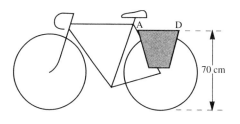
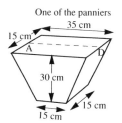

One of the panniers

(a) Find the distance below AD of the centre of mass of a pannier and hence the height of this centre of mass above ground level.

(b) Find the volume of luggage in the panniers.

Method B Two small panniers attached to carriers on the front forks plus a saddlebag. AD is at a height of 50 cm above the ground. The saddlebag is cylindrical, radius 10 cm, length 30 cm and its axis of symmetry is at a height of 80 cm above the ground.

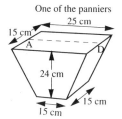
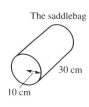

One of the panniers   The saddlebag

(c) Find
    (i)   the distance below AD of the centre of mass of a pannier
    (ii) the volume of luggage in the panniers
    (iii) the volume of luggage in the saddlebag
    (iv) the height above ground level of the centre of mass of all the luggage.

(d) Which of the two methods
    (i)   carries the most luggage
    (ii) has the lower centre of mass?

CHAPTER 21

SUSPENDED BODIES
SLIDING AND TOPPLING

SUSPENDED LAMINAS

When an object is suspended by a string attached
to one point of the object, two forces act on the
object; the tension in the string vertically upwards
and the weight of the object vertically downwards.
If the object hangs at rest it is in equilibrium and
the two forces must therefore be in the same line,
i.e. the centre of gravity, and therefore the centre
of mass, is vertically below the point of
suspension.

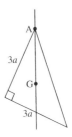

Now if the object is one whose centre of mass is
known, a right-angled triangular lamina for
example, the position in which the lamina will
hang can be found by joining the point of
suspension, A say, to G.
In equilibrium, AG is vertical.

Suppose that we are asked to find the angle between the vertical and one of the
sides of the lamina.

It is not always easy to see how to calculate this angle from the diagram of the
lamina in its suspended position. It is often more straightforward, if it is possible,
to draw the lamina so that two of its sides are horizontal and vertical, and on this
diagram mark the line which would be vertical.

As an example consider the right-angled triangle shown above.

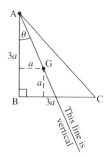

We know that AG is vertical and that G is distant a from both BA and BC.

Therefore to find the angle θ between AB and the vertical we use

$$\tan \theta = \frac{a}{2a} = \frac{1}{2}$$

$$\Rightarrow \qquad \theta = 27° \quad (\text{nearest degree})$$

In this example, because the centre of mass of a triangular lamina is quotable, the position of G is known at the outset.

Now consider the compound lamina given in the first worked problem in Examples 19c.

Clearly the suspended position cannot be dealt with until G is located, i.e. all the work done in the example has to be carried out before we can begin to consider how it hangs in equilibrium when suspended freely from a specified point, O say.

Now we have already drawn a diagram to locate G and, *on the same diagram,* we can draw OG, i.e.

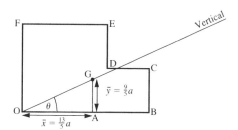

It is immediately clear that $\tan \theta = \dfrac{\overline{y}}{\overline{x}} = \dfrac{9a}{5} \div \dfrac{13a}{5} = \dfrac{9}{13}$

Although there may be some readers who prefer to work with the 'suspended position' diagram, many will find it easier to use the approach given above.

Examples 21a

1. **A uniform lamina ABCD is in the shape of a trapezium in which** $AB = 4a$, $AD = 3a$, $DC = a$ **and angle DAB is** $90°$.

    (a) **Find the distance of G, the centre of mass of the lamina, from AB and AD.**

    (b) **The lamina is suspended from A and hangs freely. What is the angle between AB and the vertical?**

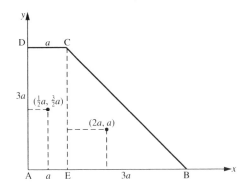

(a)

| Portion | Mass | Coords of Centre of Mass | | mx | my |
		x	y		
+ AECD	$3a^2\rho$	$\frac{1}{2}a$	$\frac{3}{2}a$	$3a^2\rho \times \frac{1}{2}a$	$3a^2\rho \times \frac{3}{2}a$
+ EBC	$\frac{9}{2}a^2\rho$	$2a$	a	$\frac{9}{2}a^2\rho \times 2a$	$\frac{9}{2}a^2\rho \times a$
ABCD	$\frac{15}{2}a^2\rho$	$\bar{x}$	$\bar{y}$	$\frac{15}{2}a^2\rho \times \bar{x}$	$\frac{15}{2}a^2\rho \times \bar{y}$

Using $\Sigma m_n x_n = \bar{x} \Sigma m_n$

$$3a^2\rho \times \tfrac{1}{2}a + \tfrac{9}{2}a^2\rho \times 2a = \tfrac{15}{2}a^2\rho \times \bar{x}$$

$\Rightarrow$ $$\bar{x} = \frac{7a}{5} = \text{distance of G from AD}$$

Using $\Sigma m_n y_n = \bar{y} \Sigma m_n$

$$3a^2\rho \times \tfrac{3}{2}a + \tfrac{9}{2}a^2\rho \times a = \tfrac{15}{2}a^2\rho \times \bar{y}$$

$\Rightarrow$ $$\bar{y} = \frac{6a}{5} = \text{distance of G from AB}$$

(b) When ABCD is hanging freely from A, AG is vertical, so the angle we want is GÂB.

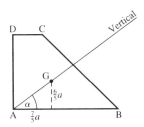

$$\tan \alpha = \frac{\bar{y}}{\bar{x}} = \frac{6a}{5} \div \frac{7a}{5} = \frac{6}{7}$$

$\Rightarrow$ $\alpha = 41°$ (nearest degree)

SUSPENDED SOLID BODIES

Just as in the case of a suspended lamina, a body suspended from a point P hangs in equilibrium with its centre of mass, G, vertically below P.

If the body has a plane of symmetry in which P is located and the position of G is known, the equilibrium position of the body can be found in the same way as if that section were a lamina.

2. **A solid right circular cone, of height 4***a* **and base radius** *a*, **is suspended freely from a point P on the circumference of the base. Find the angle** α **between PO and the vertical, where O is the centre of the base.**

The centre of mass of a solid cone is known to be on the line of symmetry, $\frac{3}{4}$ of the way from the vertex to the base.

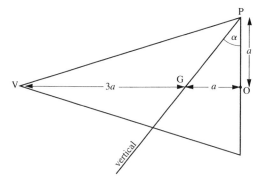

When the cone hangs in equilibrium, PG is vertical.

G is distant 3*a* from the vertex and *a* from O.

Therefore $\qquad \tan \alpha = \dfrac{GO}{OP} = \dfrac{a}{a} = 1$

The angle between OP and the vertical is 45°.

3. **A sculpture is in the form of a uniform solid cylinder of radius 2 cm and mass 4***M*, **with a small lead bead of mass** *M* **let in at a point A on the rim of the base. If the sculpture is suspended from the point B, directly above A on the upper rim, AB is inclined to the vertical at an angle** α **whose tangent is** $\frac{1}{3}$. **Find the height, AB, of the sculpture.**

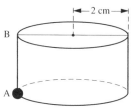

Let the height of the cylinder be 2*h* centimetres.

First we need to find the centre of mass of the sculpture.

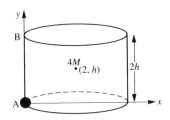

Portion	Mass	Coords of C of M		Σmx	Σmy
		x	y		
Cylinder	$4M$	2	h	$8M$	$4Mh$
Lead bead	M	0	0	0	0
Sculpture	$5M$	$\bar{x}$	$\bar{y}$	$5M\bar{x}$	$5M\bar{y}$

Using $\Sigma mx = \bar{x}\,\Sigma m$ gives $8M = 5M\bar{x}$ $\Rightarrow$ $\bar{x} = \frac{8}{5}$

Using $\Sigma my = \bar{y}\,\Sigma m$ gives $4Mh = 5M$ $\Rightarrow$ $\bar{y} = \dfrac{4h}{5}$

Now consider the plane of symmetry through A.

$B\widehat{G}$ is vertical therefore $A\widehat{B}G$ is α.

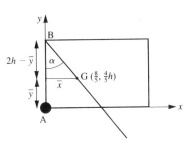

$\therefore$ $\quad \tan \alpha = \dfrac{\bar{x}}{2h - \bar{y}} = \dfrac{\frac{8}{5}}{\frac{6h}{5}} = \dfrac{4}{3h}$

But $\quad \tan \alpha = \frac{1}{3}$ $\Rightarrow$ $h = 4$

The height of the sculpture is 8 cm.

EXERCISE 21a

1. When it is freely suspended from A, find the angle between AB and the vertical, for the lamina given in Exercise 19c

   (a) question 3   (b) question 4   (c) question 5.

2. When the lamina shown in the diagram is freely suspended from A, AB hangs at an angle α to the vertical where $\tan \alpha = \frac{10}{9}$.

   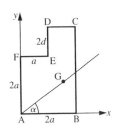

   (a) Find, in terms of a and d the coordinates of the centre of the mass of the lamina.

   (b) Express d in terms of a.

Most of the remaining questions in this exercise refer to objects whose centres of mass were found in Exercise 20b and the position found for the centre of mass of each body is given. This is to save you having to begin by calculating the position of G in each case. The centre of mass of each solid is on the axis of symmetry. (For interest, the question reference is given.)

3. The centre of mass of this solid (question 2a) is $\frac{5}{3}$ cm from O. If the solid is suspended from A, find the angle between AB and the vertical.

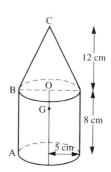

4. The centre of mass of the frustum of a cone shown in the diagram is 4.7 cm above the base. Find the angle between the axis of symmetry and the vertical when the frustum is suspended freely from the point P as indicated.

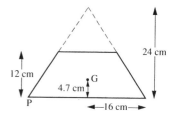

5. The solid shown in the diagram is freely suspended from C, the midpoint of AB. Find the angle between AB and the vertical given that the height above the base of the centre of mass is $\frac{83}{88}a$ (question 10).

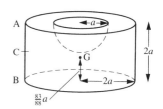

6. This solid is suspended from B and hangs freely. The distance below O of the centre of mass is given by $\dfrac{h^2 - 24}{2(h+4)}$ (question 4). Find h given that

   (a) AB is horizontal

   (b) BC is horizontal.

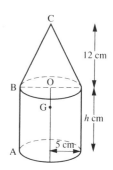

BODIES RESTING IN EQUILIBRIUM ON A HORIZONTAL PLANE

Consider a lamina or a solid body resting with one face on a horizontal plane. Depending on the shape of the cross-section, the body may have a tendency to topple over. The solid whose cross-section is shown in the following diagrams, for example, could topple over about B to rest on the edge through C, but whether or not it does fall over depends on the relative positions of B and G.

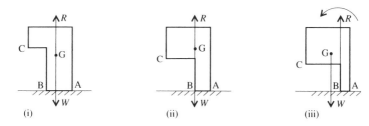

(i)   (ii)   (iii)

In diagram (i) the weight W of the body acts through a point within the face through AB and the normal contact force R also acts through the same point, so the body can remain in equilibrium.

In the extreme position, diagram (ii), W and R act through B itself, keeping the body *just* in equilibrium.

If we consider moments about an axis through B, both W and R have zero moment. In this case we can see that the weight of the shaded part exerts an overturning moment, while the weight of the unshaded portion exerts a restoring moment about B and these must be equal.

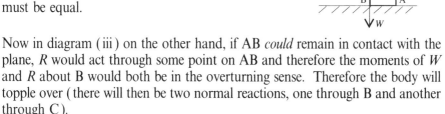

Now in diagram (iii) on the other hand, if AB *could* remain in contact with the plane, R would act through some point on AB and therefore the moments of W and R about B would both be in the overturning sense. Therefore the body will topple over (there will then be two normal reactions, one through B and another through C).

So the question of whether the body will rest in the given position can be answered either by checking the resultant moment about B or by checking that the overall weight passes through a point within the base of contact (this may involve having to find the centre of gravity of the object).

Examples 21b

1. **A uniform prism whose cross-section is the trapezium shown in the diagram, is placed with the side PQ on a horizontal plane. Find the range of values of k for which the prism will rest in equilibrium in this position.**

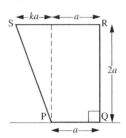

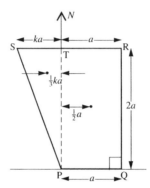

Consider the extreme position when the normal reaction passes through P.

The weight of the portion PQRT exerts a clockwise moment about an axis through P, i.e. in the sense to preserve equilibrium. The moment of the weight of the portion PTS is anticlockwise and tends to cause overturning.

The weight of the portion PTS is $ka^2\rho$ and that of PQRT is $2a^3\rho$.

The resultant clockwise moment, M, about an axis through P is given by

$$M = 2a^2\rho\left(\tfrac{1}{2}a\right) - ka^2\rho\left(\tfrac{1}{3}ka\right) = \tfrac{1}{3}(3 - k^2)a^3\rho$$

Equilibrium will be maintained as long as M does not become negative,

i.e. for equilibrium $M \geqslant 0$

$\Rightarrow$ $3 - k^2 \geqslant 0$

In this problem k cannot be negative so, for equilibrium,

$$0 \leqslant k \leqslant \sqrt{3}$$

2. **A man has invented a game that is played with a wood in the shape of a frustum of a cone. The wood is bowled with its curved surface in contact with the ground so that it describes a circle. The man is cutting his wood from the solid cone with base radius 12 cm and height 16 cm, shown in the diagram.**

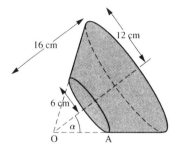

He plans to make a cut 8 cm from the vertex so as to remove the top half of the cone and wants to check that the wood will not fall over on to its smaller face when it is placed on horizontal ground.

(a) Find the length of OA.

(b) Write down, as a fraction, the value of cos α.

(c) Find the length of OG where G is the centre of mass of the wood.

(d) Given that the weight of the wood passes through the point H on the plane, find the length of OH.

(e) Determine whether the wood will topple on to its smaller face.

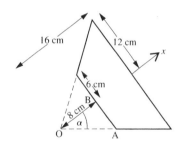

(a) OB $= \frac{1}{2}$ (height of whole cone)

$\Rightarrow$ AB $= \frac{1}{2}$ (radius of whole cone) $= 6$ cm

Using Pythagoras' theorem in $\triangle$OAB gives
OA $= 10$ cm

(b) cos α $=$ OB/OA $= \frac{8}{10} = \frac{4}{5}$

(c) The linear ratio of the similar cones is $8:16$, i.e. $1:2$, so the ratio of their volumes, and therefore their masses, is $1:8$. Let their masses be M and $8M$.

Portion	Mass	x coord of centre of mass	Σmx
+ Large cone	$8M$	$\frac{3}{4} \times 16$	$96M$
− Small cone	M	$\frac{3}{4} \times 8$	$6M$
Wood	$7M$	$\bar{x}$	$7M\bar{x}$

Using $\Sigma mx = \bar{x}\,\Sigma m$ gives $96M - 6M = 7M\bar{x}$

$\Rightarrow$ $\bar{x} = \frac{90}{7} = 12.9\ldots$

i.e. OG $= 13$ cm (2 sf)

(d) The weight of the wood acts vertically
     downwards through G and passes
     through the point H,

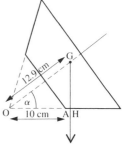

$\therefore$ $OH = OG \cos \alpha$

i.e. $OH = \frac{90}{7} \times \frac{4}{5} = 10.3 \, cm$

(e) $OH = 10.3 \, cm$ and $OA = 10 \, cm$,   i.e. $OA < OH$

     so H is a point inside the line of contact between the wood and the plane.

     Therefore the wood will not topple over when placed on the plane.

The examples we have looked at so far have had either a plane or a line in
contact with the horizontal plane. If an object with a spherical surface rests in
contact with a plane, there is only a point of contact. The normal contact
force therefore must pass through this point so the situation is similar to that
of suspension, i.e. the point of contact must be vertically below the centre of
gravity. Additionally we know that, because the horizontal plane is tangential
to the circular cross-section, the normal reaction must pass through the centre
of the circle.

Examples 21b (continued)

3. **The diagram shows the central cross-section of a
    casting in the shape of a hemisphere of radius
    8 cm, with a hemispherical depression, of radius
    4 cm. When the casting rests in equilibrium with
    its curved surface touching a horizontal surface,
    find the inclination to the horizontal of the line
    of symmetry (OAB) of the plane face.**

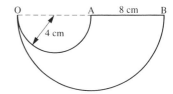

First we will find the centre of mass of the casting, knowing that it lies in the section of symmetry shown
in the diagram.

The two hemispheres are similar with a linear ratio $2:1$, so their masses are in
the ratio $8:1$.

Portion	Mass	Coords of G		mx	my
		x	y		
+ Large hemisphere	$8M$	8	$\frac{3}{8} \times 8$	$64M$	$24M$
− Small hemisphere	M	4	$\frac{3}{8} \times 4$	$4M$	$\frac{3}{2}M$
Casting	$7M$	$\bar{x}$	$\bar{y}$	$7M\bar{x}$	$7M\bar{y}$

Using $\Sigma mx = \bar{x}\,\Sigma m$ gives $64M - 4M = 7M\bar{x}$

Using $\Sigma my = \bar{y}\,\Sigma m$ gives $24M - \frac{3}{2}M = 7M\bar{y}$

$\Rightarrow$ $\bar{x} = \frac{60}{7}$ and $\bar{y} = \frac{45}{14}$

When the casting rests in equilibrium on a horizontal plane, touching at P, GP is vertical. Also as the plane is tangential to the large hemisphere, GP passes through its centre A.

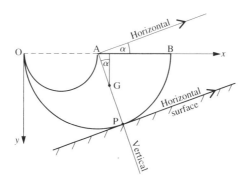

Therefore $\tan \alpha = \dfrac{\bar{x} - 8}{\bar{y}} = \dfrac{4}{7} \div \dfrac{45}{14} = \dfrac{8}{45}$

$\therefore$ $\alpha = 10°$ to the nearest degree.

EXERCISE 21b

1. The diagram shows the cross-section through the centre of mass of a uniform prism. The prism has been placed with the face containing AB in contact with a horizontal plane.

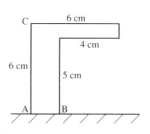

(a) Find the distance of the centre of mass of the prism from AC.

(b) Can the prism rest in equilibrium without toppling over?

2. ABCDE is a cross-section through the centre of mass of a uniform prism, which is placed with the face containing AB in contact with a horizontal plane.

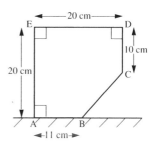

(a) Find the distance of the centre of mass of the prism from AE.

(b) Determine whether the prism will rest in equilibrium or will topple.

3. The prism described in question 2 is altered by reducing AB to 8 cm without changing the other dimensions. Find out whether it can now rest in equilibrium with AB on the horizontal plane.

4. The diagram shows a piece of a wooden puzzle.

   (a) Find, in terms of d, the distance from AF of its centre of mass.

The piece of puzzle is now placed in a vertical plane with AB in contact with a horizontal plane.

   (b) Will the piece topple when $d = 2$?

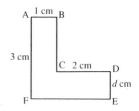

5. Calculate the range of values of d for which the wooden puzzle piece in question 4 can rest in equilibrium with AB on a horizontal plane.

6. ABCDEF is a cross-section through the centre of mass of a prism. The prism is placed with this cross-section in a vertical plane and the face containing AB in contact with a horizontal plane.

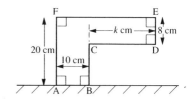

   (a) Find, in terms of k, the distance of the centre of mass of the prism from AF.

   (b) Determine whether the prism can stay in equilibrium in this position without toppling if (i) $k = 5$ (ii) $k = 20$

   (c) Find the range of values of k for which equilibrium is possible.

7. The solid in the diagram consists of a cylinder and a cone attached to each other at their common plane faces. The centre of mass is at the point G, where $OG = \frac{15}{4}$ cm. Determine whether the solid will remain in equilibrium when it is placed with BC in contact with a horizontal plane, or will topple over on to AB.

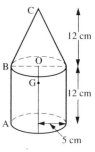

Questions 8 and 9 refer to this diagram of a uniform solid in the shape of a frustum of a cone.

8. (a) Find the distance from V of its centre of mass.

   (b) The frustum is placed with its curved surface in contact with a horizontal plane. Will it remain in equilibrium in this position if $\alpha = 35°$?

*9. Use the result of 8a to show that the frustum is on the point of toppling when $\cos^2 \alpha = \frac{13}{30}$.

EQUILIBRIUM ON AN INCLINED PLANE

One of the properties necessary for the equilibrium of a body resting on a horizontal plane, applies also to a body in equilibrium on an inclined plane, namely,

if there is a plane or a line of contact, the weight, which acts vertically downwards through G, must pass through a point within that region of contact; the normal reaction acts through that point.

On an inclined plane however, the normal reaction, which is perpendicular to the inclined plane, is not vertical and is therefore not collinear with the weight. The normal reaction has a horizontal component but the weight of the body does not. Therefore equilibrium is not possible unless another force with a horizontal component acts on the body. So, unless an extra supporting force is applied to the body, there must be friction between the body and the plane.

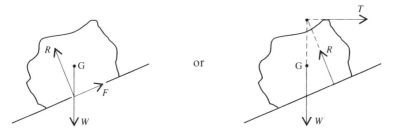

or

Note that, in either case, the three forces are concurrent at a point on the line of action of the weight.

Note also that if a body has a spherical surface, it has contact with the plane at only one point so the weight must pass through that point.

When friction maintains equilibrium, the *resultant* contact force S (i.e. the resultant of the friction and the normal reaction) balances the weight; it is therefore vertical and must pass through G.

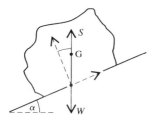

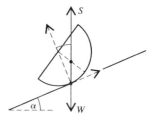

In general, then, if a body is placed on an inclined plane without any supporting force, its equilibrium depends not only upon the position of G relative to the contact region, but also on whether slipping can occur.

When slipping is about to occur, it can be seen from the diagram that the angle of friction (between R and S) is equal to α.

Examples 21c

1. **A uniform solid cylinder, with radius a and height $3a$, is resting in equilibrium with one end on a rough plane inclined at an angle α to the horizontal. The inclination of the plane is gradually increased until the cylinder is just on the point of toppling over.**

   (a) **Find the greatest possible value of α.**

   (b) **Find the least value of the coefficient of friction between the plane and the cylinder.**

(a)

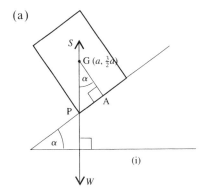

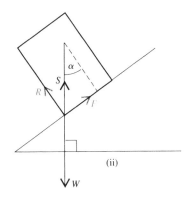

(i)   (ii)

The cylinder is in equilibrium under the action of two forces, its weight and the resultant contact force, therefore these two forces are collinear. The cylinder is on the point of toppling over so the contact forces act through the lowest point P of its plane face.

From diagram (i) we see that $A\widehat{G}P = \alpha$

In $\triangle AGP$, $\tan \alpha = a \div \frac{3}{2}a = \frac{2}{3}$ $\Rightarrow$ $\alpha = 33.6\ldots^\circ$

Therefore, to the nearest degree, the largest angle at which the plane can be elevated without making the cylinder topple over is $33°$ ($34°$ would cause toppling).

(b) In diagram (ii) we see that $F = R \tan \alpha$

But $F \leqslant \mu R$ $\Rightarrow$ $\mu \geqslant \tan \alpha$

Therefore just to prevent slipping at the maximum elevation, $\mu = \frac{2}{3}$

2. **At a warehouse, packages of goods are delivered from the first floor store to the ground floor for despatch, by placing them on a moving ramp. All the goods in one section are packaged in boxes that are cuboids, all with a base 1.6 m square but of varying heights. The ramp, which is rough enough for there to be no possibility of a box slipping, slopes down at 40° to the horizontal. By modelling the boxes as uniform cuboids, find the maximum permitted height of a box for safe delivery.**

State, with reasons, whether you think that the model is appropriate and suggest ways, if any, in which it might be improved.

The boxes cannot slip, so the only restriction necessary is to ensure that no box topples over.

The weight acts through the centre of mass G and, for equilibrium, the resultant contact force must act in the same line.

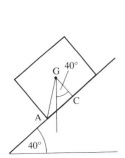 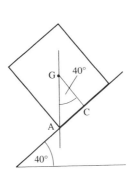 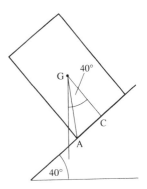

For equilibrium, the angle AGC must be greater than or equal to $40°$

i.e. $\tan AGC \geqslant \tan 40°$

Taking the height of a box as $2h$ gives $\tan AGC = \dfrac{0.8}{h}$ $\Rightarrow$ $\dfrac{0.8}{h} \geqslant \tan 40°$

$\Rightarrow$ $h \leqslant 0.95$ (to 2 sf, rounded down for safety)

Therefore the maximum permitted height should be 1.9 m.

The suitability of the model depends on whether the contents of a box are always such that the centre of mass is halfway up. Also the conveyor belt may not run smoothly and jerks could make toppling more likely even at 'safe' heights.

Taking a centre of mass at, say, 60% of the height of the box might give more reliable results.

3. The diagram shows the cross-section of a uniform solid brass ornament in the shape of a hemisphere with a cone attached to its plane face. The radius of both hemisphere and cone is 2*a*.

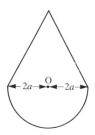

(a) Sketch the ornament if the height of the cone is (i) *a* (ii) 6*a* and mark on each sketch an estimate of the position of the centre of mass, G, of the ornament.

(b) It is intended to display the ornament with its hemispherical surface resting on a velvet covered inclined shelf as shown.

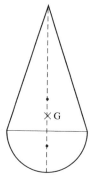

Use your estimated positions for G to explain, by marking on a diagram the forces that act on the ornament in each case, that equilibrium is possible in only one of these cases.

(a) (i)

(ii)

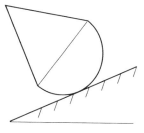

Mass of cone < mass of hemisphere
G lies within the hemisphere less
than $\frac{3}{4}a$ from the centre.

Mass of cone > mass of hemisphere
G lies within the cone,
less than $\frac{3}{2}a$ from its base.

(b) Assume that contact with the velvet is rough enough to prevent slipping.

The normal reaction passes through the centre of the common base and friction acts up the plane. So the resultant contact force, *S*, is inclined 'uphill' from P.

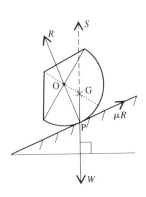

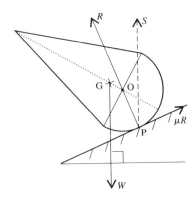

In case (i) G is right of, and below, the centre O so the resultant contact force *can* pass through G producing equilibrium.

In case (ii) G is left of, and above, the centre O so the resultant contact force *cannot* pass through G and equilibrium is impossible.

EXERCISE 21c

1. A cube with sides of length 20 cm is placed with its plane face in contact with an inclined plane, which is rough enough to prevent sliding. Determine whether the cube will rest in equilibrium, or topple.

(a)

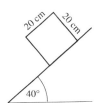

(b)

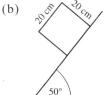

2. A uniform rectangular lamina ABCD is such that $AB = 0.5$ m and $BC = 0.3$ m. The lamina is placed with one of its sides on a rough inclined plane. The plane of the lamina is vertical.

Find the maximum angle of inclination of the plane to the horizontal, and the least coefficient of friction between the lamina and the plane for which the lamina can rest in equilibrium without toppling or sliding, if the side in contact with the plane is

(a) AB   (b) BC.

3. The uniform lamina ABC is placed, as shown in the diagram, on an inclined plane rough enough to prevent the lamina from sliding down. When the angle of inclination, α, of the plane to the horizontal is such that $\tan \alpha = \frac{3}{4}$, the lamina is on the point of toppling about A. Find the value of h.

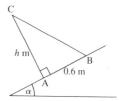

In questions 4 to 6 a uniform solid cone, with dimensions shown in the diagram, rests with its plane face in contact with a plane that is rough enough to prevent sliding. Determine whether the cone will topple.

4.

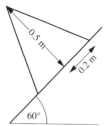

5.

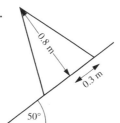

6.

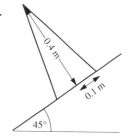

7. A uniform solid cylinder of base radius 10 cm and height h cm is just on the point of toppling when it is placed with one end on a rough plane inclined at an angle of 30° to the horizontal. Find the value of h.

8. ABC is a uniform lamina in the shape of an isosceles triangle in which AB = AC = 0.6 m and the length of BC is $2a$ metres. The lamina is placed, in a vertical plane, with BC in contact with a rough inclined plane for which the coefficient of friction is $\frac{1}{2}$. If the lamina is on the point of sliding and toppling simultaneously find

(a) the angle at which the plane is inclined to the horizontal

(b) the value of a.

9. The larger circular face of the frustum of a cone shown in the diagram, is placed on a rough plane inclined at θ to the horizontal. The inclination of the plane is steadily increased until, when $\theta = 60°$, the frustum is on the point of toppling. Find the radius of each of the plane faces.

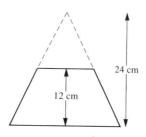

10. The frustum given in question 9 is now placed with its smaller circular face on the inclined plane and θ is again increased gradually until the frustum is about to topple. Find the value of θ at this instant.

*11. The diagram shows a solid formed by joining the plane faces of a cone, of radius a and height a, to a hemisphere of radius a.

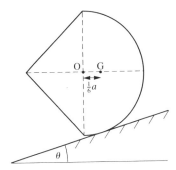

The centre of mass of the compound body is distant $\frac{1}{6}a$ from O. The body rests in equilibrium with its axis of symmetry horizontal on a rough inclined plane as shown.

Find the value of θ.

FURTHER PROBLEMS

In this section we look at some problems involving rigid bodies which, while using the same mechanical principles as have been applied so far, involve more analysis and are a little harder. Anyone who enjoys dipping a little deeper into a subject will find them interesting and rewarding.

We give a few examples to illustrate some of the possibilities, and the exercise that follows includes more variations.

Examples 21d

1. A uniform solid consists of a hemisphere of radius r and a right circular cone of base radius r fixed together so that their plane faces coincide. If the solid can rest in equilibrium with *any* point of the curved surface of the hemisphere in contact with a horizontal plane, find the height of the cone in terms of r.

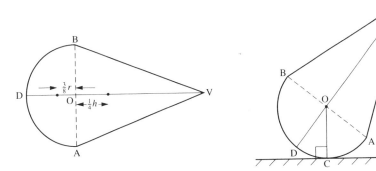

The plane is tangential to the hemisphere so the normal reaction acting at the point of contact, C, passes through O no matter where C is on the surface of the hemisphere.

The only two forces acting on the body are its weight and the normal reaction.

As two forces that are in equilibrium must be collinear, the weight of the solid must also pass through O for all positions of C,

i.e.   the centre of mass of the solid *must be at O.*

Therefore, if we take moments about an axis through O, the moments of the cone and the hemisphere must be equal and opposite,

i.e. $\left(\frac{1}{3}\pi r^2 h\rho\right)\left(\frac{1}{4}h\right) = \left(\frac{2}{3}\pi r^3\rho\right)\left(\frac{3}{8}r\right)$ $\Rightarrow$ $h^2 = 3r^2$

∴ $h = r\sqrt{3}$

2. **A circular disc with centre O, radius 2*a* and weight *W*, rests in a vertical plane on two rough pegs A and B. OA and OB are inclined to the vertical at 60° and 30° respectively. Given that the coefficient of friction is $\frac{1}{2}$ at each peg, find the greatest force that can be applied tangentially at the highest point of the disc without causing rotation. Give the answer, in surd form, in terms of *W*.**

When rotation is about to take place the disc is about to slip at both pegs, i.e. friction is limiting at both pegs.

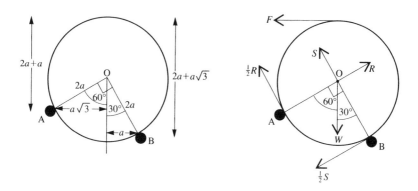

Resolving in any direction is not particularly simple. On the other hand it *is* simple to take moments about A, B and O. As these are not collinear we will use this method to form the three independent equations we need.

A↺ $F \times 3a + S \times 2a - W \times a\sqrt{3} - \frac{1}{2}S \times 2a = 0$

$\Rightarrow$ $S = \sqrt{3}W - 3F$ [1]

B↺ $F \times (2 + \sqrt{3})a + W \times a - R \times 2a - \frac{1}{2}R \times 2a = 0$

$\Rightarrow$ · $3R = W + F(2 + \sqrt{3})$ [2]

O↺ $F \times 2a - \frac{1}{2}S \times 2a - \frac{1}{2}R \times 2a = 0$

$\Rightarrow$ $2F = S + R$ [3]

Using [1] and [2] in [3] × 3 gives

$$6F = 3\sqrt{3}W - 9F + W + 2F + \sqrt{3}F$$

$$\Rightarrow \qquad F(13 - \sqrt{3}) = W(3\sqrt{3} + 1)$$

The greatest force is $\dfrac{(3\sqrt{3} + 1)\,W}{(13 - \sqrt{3})}$

Note that taking moments about three different axes gives three independent facts *provided that the axes are not collinear.* In this problem A, O and B are not collinear.

3. **The cross-section of a uniform prism is a trapezium ABCE. This trapezium is formed from a square lamina, ABCD, with the portion CDE removed. The side of the square is 2 m and the length of ED is 1.5 m.**

   (a) **If the prism is placed with ABCE in a vertical plane and AE on a rough horizontal plane, show that it will topple about the edge through E.**

   (b) **Find the least force that must be applied at C to prevent toppling.**

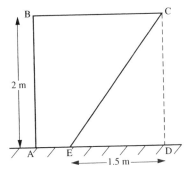

The prism will topple if the line of action of the weight does not pass through a point within AE; to check this we need only the distance from AB of the centre of mass of the cross-section.

Portion	Mass (m)	Distance from AB of C of M (x)	mx
ABCD	4ρ	1	4ρ
$-$DCE	1.5ρ	1.5	2.25ρ
ABCE	2.5ρ	$\bar{x}$	$2.5\rho\bar{x}$

Using $\Sigma mx = \bar{x}\,\Sigma m$ gives

$$4\rho - 2.25\rho = 2.5\rho\bar{x} \qquad \Rightarrow \qquad \bar{x} = \frac{1.75}{2.5} = 0.7$$

(a) AE $= 0.5$ which is less than $\bar{x}$, therefore the line of action of the weight of the prism does not intersect AE and the prism will topple about the edge through E.

(b) Toppling about E is caused by the moment of the weight about E. In order just to prevent toppling, the moment about E of the force F newtons applied at C must counterbalance the moment of the weight. The moment of the force is given by $F \times$ perpendicular distance from E. Therefore F will be least when this distance is greatest and this is when the line of action of F is perpendicular to EC.

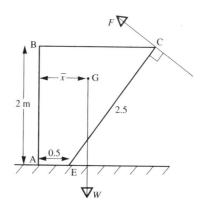

Taking W newtons as the weight of the prism,

E⤴ $W \times (\bar{x} - 0.5) - F \times EC = 0$

$\Rightarrow$ $2.5F = W(0.7 - 0.5) = 0.2W$

$\Rightarrow$ $F = \frac{2}{25} W$

The least force required is $0.08W$ newtons.

4. The diagram shows the central cross-section of a uniform cube which is placed on a rough plane inclined at α to the horizontal where $\tan \alpha = \frac{1}{2}$. A horizontal force P of gradually increasing magnitude is applied at D as shown. If $\mu = \frac{2}{3}$, show that the cube will begin to turn about the edge through B before it begins to slide up the plane.

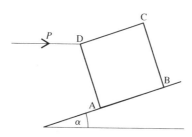

Consider separately the values of P for which sliding or overturning would begin.

Suppose that sliding up is about to begin when $P = P_1$.

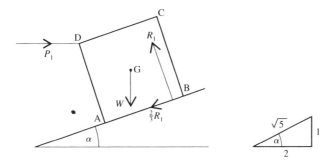

$\uparrow \qquad R_1 \cos \alpha - \dfrac{2}{3} R_1 \sin \alpha - W = 0 \quad \Rightarrow \quad R_1 \left(\dfrac{2}{\sqrt{5}} - \dfrac{2}{3} \times \dfrac{1}{\sqrt{5}} \right) = W$ [1]

$\rightarrow \qquad P_1 - R_1 \sin \alpha - \dfrac{2}{3} R_1 \cos \alpha = 0 \quad \Rightarrow \quad R_1 \left(\dfrac{1}{\sqrt{5}} + \dfrac{2}{3} \times \dfrac{2}{\sqrt{5}} \right) = P_1$ [2]

From [1] $\qquad R_1 = \dfrac{3\sqrt{5}W}{4}$

From [2] $\qquad P_1 = \dfrac{7R_1}{3\sqrt{5}} \qquad \Rightarrow \qquad P_1 = \dfrac{7}{4}W$

Now suppose instead that $P = P_2$ and the cube is just about to overturn about **B**.

To make taking moments easier, both W and P_2 are replaced by their components parallel and perpendicular to the plane as shown.

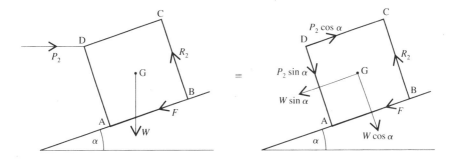

Friction at B is not necessarily limiting but we can avoid the unknown frictional force by taking moments about B.

$\text{B}) \qquad P_2 \cos \alpha \times 2a - P_2 \sin \alpha \times 2a - W \cos \alpha \times a - W \sin \alpha \times a = 0$ [3]

$\therefore \qquad P_2 \dfrac{(4-2)}{\sqrt{5}} = W \dfrac{(2+1)}{\sqrt{5}} \qquad \Rightarrow \qquad P_2 = \dfrac{3W}{2}$

The values of P_1 and P_2 are different so the cube cannot begin to slide and topple simultaneously.

As soon as the lower of these values is reached, equilibrium will be broken.

$P_2 < P_1$ so the cube begins to turn about B before it can slide up the plane.

EXERCISE 21d

1. A packing case, of mass 40 kg, is in the shape of a cuboid measuring 2 m by 1 m by 1.5 m. It can be assumed that it is packed to a uniform density. A man places it on a moving ramp which is inclined at an angle θ to the horizontal and is rough enough to prevent slipping. The 2 m by 1 m face is in contact with the ramp so that, in cross-section, the case appears as in the diagram.

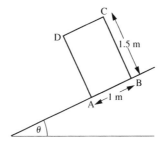

(a) Show that it cannot rest in equilibrium in this position without toppling if $\tan \theta > \frac{2}{3}$.

(b) If $\theta = 35°$ and the man applies a force P newtons at D in the direction DC, find P if

    (i) P is just great enough to prevent the case from toppling down the ramp

    (ii) P is so great that the case is just on the point of toppling up the ramp.

2. When the uniform solid shown in the diagram is placed with a point of its hemispherical surface in contact with a horizontal plane as shown in the diagram, one of the following things will happen:

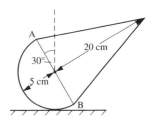

    (i) it will remain in equilibrium in this position

    (ii) it will rotate until AB is horizontal

    (iii) it will rotate until it topples about B

Find which of these three occurs.

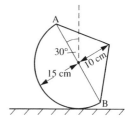

3. Carry out the same investigation as that required in question 2, for another solid of the same type but with different dimensions as shown.

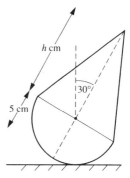

4. The dimensions of a third solid of the same type are shown in the diagram.

  (a) Find the value of h so that the solid will remain in equilibrium in the position shown.

  (b) If h has the value found in part (a), describe what will happen if the solid is placed on the plane with its axis of symmetry at $80°$ to the vertical and then released.

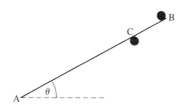

5. A uniform rod AB, of weight W and length $4a$, rests in limiting equilibrium at an angle θ to the horizontal in rough contact with two pegs, as shown in the diagram. One peg is at B and the other at point C on the rod where $AC = 3a$. Find, in terms of θ, the coefficient of friction, which is the same at both pegs.

6. A uniform solid cone, of weight W, base radius a and height $4a$, is placed with its plane face in contact with a rough horizontal plane. The coefficient of friction between the cone and the plane is $\frac{3}{4}$. A horizontal force P is applied to the cone half-way up its height.

  (a) Assuming that it does not topple first, find P when the cone is just on the point of sliding.

  (b) Assuming that it does not slide, find P when the cone is just on the point of toppling.

  (c) If initially $P = 0$ and then the value of P is gradually increased, in what way will equilibrium be broken?

7. A uniform solid cone, of weight W, base radius a and height $2a$, is placed with its plane surface in contact with a rough plane which can be inclined at an angle θ to the horizontal. The coefficient of friction between the cone and the plane is $\frac{1}{4}$.

If initially $\theta = 0$ and then the plane is gradually tilted so that the value of θ increases, in what way will equilibrium be broken?

8. The diagram models a tower crane which consists of a gantry ABCD, of length $16a$, which rests on top of a vertical tower. The gantry is of mass $10M$. It has a counter weight centred on end A and a trolley of mass M can move along section CD. Loads are carried suspended from the trolley on a cable.

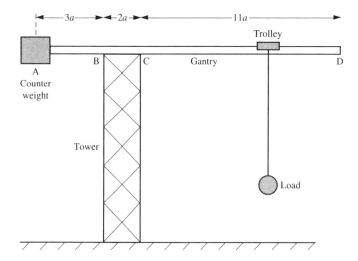

(a) The counter weight is such that, if the trolley were not fitted, it would be on the point of pulling end A of the gantry downwards, turning about point B. Find the mass of the counter weight.

(b) The theoretical maximum load that the crane could lift is determined by considering the load suspended below D which would bring end D of the gantry to the point of turning downwards about point C. Find this theoretical load.

  (In practice the trolley cannot reach this position and other safety margins would be built in.)

CONSOLIDATION E

SUMMARY

Equilibrium of Rigid Bodies

Three conditions are necessary for the equilibrium of coplanar forces.

Either

the resultant force in a specified direction, Ox say, is zero,

the resultant force in the perpendicular direction, Oy, is zero,

the resultant moment about any one axis perpendicular to the plane is zero.

Or

the resultant force in a direction parallel to either Ox or Oy is zero,

the resultant moment about an axis through A is zero,

the resultant moment about another axis through B is zero.

Or

the resultant moment about each of three axes through non-collinear points is zero.

Centre of Mass of a Rigid Body

The C of M of a uniform solid of revolution lies on the axis of rotation, which is an axis of symmetry. Its position on that axis can be found by integration provided that the equation of the curve used to generate the solid is known.

If the x-axis is the axis of symmetry, then $\bar{x}$, the x-coordinate of the C of M, is given by

$$\int \pi y^2 x \ dx = V\bar{x} \qquad \text{where } V \text{ is the volume of the solid}$$

Similarly, if the body is symmetrical about the y-axis,

$$\int \pi x^2 y \ dy = V\bar{y}$$

Quotable centres of mass are:

cylinder	midpoint of centre line
cone of height h	on axis of symmetry and $\frac{1}{4}h$ above the base
pyramid of height h	on axis of symmetry and $\frac{1}{4}h$ above the base
hemisphere of radius a	on radius of symmetry and $\frac{3}{8}a$ from the centre

Suspended Solids

When a body whose centre of mass is G is suspended from a point P, PG is vertical.

Bodies Resting on a Horizontal Plane

For equilibrium the vertical through G must pass through a point within the base of contact. If the surface of the body in contact with the plane is spherical, the weight must pass through the point of contact.

Bodies Resting on an Inclined Plane

As for a horizontal plane, the weight must pass through a point within the base of contact. In the case of a body with circular cross-section there is contact at only one point so the weight must pass through that point.

If no extra supporting force is applied to the body, there must be friction between the plane and the body in order to maintain equilibrium.

This means that equilibrium depends not only upon the position of G relative to the contact region but also on whether slipping can occur.

MISCELLANEOUS EXERCISE E

1. A uniform ladder of length 5 m and weight 80 newton stands on rough level ground and rests in equilibrium against a smooth horizontal rail which is fixed 4 m vertically above the ground.

   If the inclination of the ladder to the *vertical* is θ, where $\tan \theta \leqslant \frac{3}{4}$, find expressions in terms of θ for the vertical reaction R of the ground, the friction F at the ground and the normal reaction N at the rail.

   Given that the ladder does not slip, show that F is a maximum when $\tan \theta = \dfrac{1}{\sqrt{2}}$, and give this maximum value.

   The coefficient of friction between the ladder and the ground is $\frac{1}{5}$. How much extra weight should be added at the bottom of the ladder so that the ladder will not slip when $\tan \theta = \frac{3}{4}$. (OCSEB)

2. A smooth horizontal rail is fixed at a height of 3 m above a horizontal playground whose surface is rough. A straight uniform pole AB, of mass 20 kg and length 6 m, is placed to rest at a point C on the rail with the end A on the playground. The vertical plane containing the pole is at right angles to the rail. The distance AC is 5 m and the pole rests in limiting equilibrium.

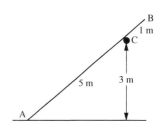

Calculate

(a) the magnitude of the force exerted by the rail on the pole, giving your answer to the nearest N,

(b) the coefficient of friction between the pole and the playground, giving your answer to 2 decimal places,

(c) the magnitude of the force exerted by the playground on the pole, giving your answer to the nearest N. (ULEAC)

3. The diagram shows a ladder AB, of mass 15 kg and length 8 m, whose lower end A rests on rough horizontal ground. The ladder is inclined at 60° to the horizontal supported by a smooth rail at C, where AC = 6 m. By modelling the ladder as a uniform rod, find

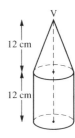

(a) the force exerted on the ladder by the rail at C,

(b) the vertical and the horizontal components of the force exerted on the ladder by the ground at A.

Given further that the ladder is in limiting equilibrium,

(c) find the coefficient of friction between the ladder and the ground. (ULEAC)ₛ

In questions 4 to 7 a problem is set and is followed by a number of suggested responses. Choose the correct response.

4. The diagram shows a solid cylinder with one of its plane faces joined to the plane face of a solid cone. The distance from V of the centre of mass of the combined uniform solid is

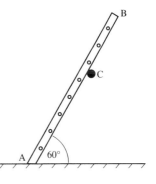

A 9 cm   **B** 12 cm   **C** 18 cm   **D** $15\frac{3}{4}$ cm

5. A uniform solid cone, of base radius a and height $4a$, rests with its flat face on an inclined plane that is rough enough to prevent slipping. The cone will be about to topple when

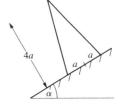

  A $\alpha = 45°$ **B** $\tan\alpha = \frac{1}{4}$ **C** $\tan\alpha = \frac{3}{4}$

  D $\alpha = 90°$ **E** $\tan\alpha = \frac{1}{2}$

6. A container consists of a hollow cylinder joined to a solid hemisphere as shown. When it is placed on a horizontal plane and tilted, it always returns to the position where AC is vertical. The centre of mass of the container is

  A   between B and C   **B**   at A   **C**   at B

  D   between B and A

7. In this question a situation is described and is followed by several statements. Decide whether each of the statements is true (T) or false (F).

A uniform solid body consists of a hemisphere and a cone joined together as shown. The centre of mass of the body is at O, the centre of the common plane face. When placed on an inclined plane, sufficiently rough to prevent slipping, the solid can rest in equilibrium on the plane in each of the following positions.

(i)   (ii)   (iii)   (iv)

8. A uniform solid paperweight is in the shape of a frustum of a cone, as shown in the diagram. It is formed by removing a right circular cone of height h from a right circular cone of height $2h$ and base radius $2r$.

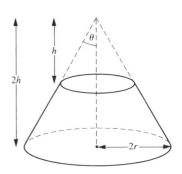

  (a)   Show that the centre of mass of the paperweight lies at a height of $\frac{11}{28}h$ from its base.

When placed with its curved surface on a horizontal plane, the paperweight is on the point of toppling.

  (b)   Find θ, the semi-vertical angle of the cone, to the nearest degree.

                                                           (ULEAC)

9. The diagram shows a man suspended by means of a
rope which is attached at one end to a peg at a
fixed point A on a vertical wall and at the other to
a belt round his waist. The man has weight $80g$ N,
the tension in the rope is T and the reaction of the
wall on the man is R. The rope is inclined at 35° to
the vertical and R is inclined at $\alpha°$ to the vertical as
shown. The man is in equilibrium.

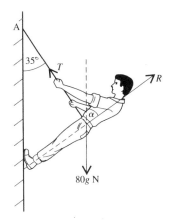

(i) Explain why $R > 0$.

(ii) By considering his horizontal and vertical
equilibrium separately, obtain two
equations connecting T, R and α.

(iii) Given that $\alpha = 45°$, show that T is
about 563 N and find R.

(iv) What is the magnitude and the direction of the force on the peg at A?

The peg at A is replaced by a smooth pulley. The rope is passed over the pulley
and tied to a hook at B directly below A. Calculate

(v) the new value of the tension in the rope section BA,

(vi) the magnitude of the force on the pulley at A.   (MEI)

10.

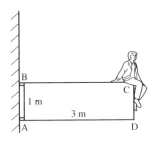

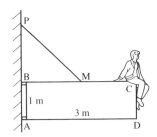

A rectangular gate ABCD, where $AB = 1$ m and $AD = 3$ m, is supported
by smooth pins at A and B, where B is vertically above A. The pins are located
in such a way that the force at B is always horizontal. The gate has mass 120 kg
and it can be modelled by a uniform rectangular lamina. A boy, of mass 45 kg,
sits on the gate with his centre of mass vertically above C. Find the magnitudes
of the forces on the gate at B and at A.

In order to support the gate, the owner fits a cable attaching the mid-point M of
BC to a point P, vertically above B and such that $BP = 1.5$ m. The boy once
again sits on the gate at C, and it is given that there is now no force acting at B.
Find the tension in the cable and the magnitude of the force now acting at A.

                                                                                      (UCLES)$_s$

11. A uniform right cylinder has height 40 cm and base radius r cm. It is placed with its axis vertical on a rough horizontal plane. The plane is slowly tilted, and the cylinder topples when the angle of inclination θ (see diagram) is 20°. Find r.

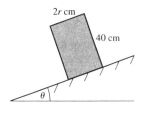

What can be said about the coefficient of friction between the cylinder and the plane?

(UCLES)$_s$

12. A uniform wooden 'mushroom', used in a game, is made by joining a solid cylinder to a solid hemisphere. They are joined symmetrically, such that the centre O of the plane face of the hemisphere coincides with the centre of one of the ends of the cylinder. The diagram shows the cross-section through a plane of symmetry of the mushroom, as it stands on a horizontal table.

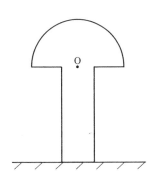

The radius of the cylinder is r, the radius of the hemisphere is $3r$, and the centre of mass of the mushroom is at the point O.

(a) Show that the height of the cylinder is $r\sqrt{\left(\frac{81}{2}\right)}$.

The table top, which is rough enough to prevent the mushroom from sliding, is slowly tilted until the mushroom is about to topple.

(b) Find, to the nearest degree, the angle with the horizontal through which the table top has been tilted. (ULEAC)

13. (i) An object is in equilibrium under the action of exactly three forces. What must be true about the forces?

(ii) The figure shows a cross-section ABCD of a uniform rectangular box of weight W. The centre of mass of the box lies in the plane ABCD and AB = a, BC = $2a$. The box rests with CD on a rough horizontal floor and is pulled with force P by a rope attached at A. The rope is at an angle α to the horizontal, as shown in the figure, and the box is on the point of rotating about C. Draw a diagram showing all the forces acting on the box.

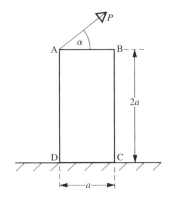

By taking moments about C, prove that $P = W/(2 \sin \alpha + 4 \cos \alpha)$.

Hence, or otherwise, show that P is a minimum when $\tan \alpha = \frac{1}{2}$. State this minimum value of P.

(iii) When $\tan \alpha = \frac{1}{2}$ and the box is on the point of rotating about C without first sliding along the floor, show that μ, the coefficient of friction between the box and the floor, must be at least $\frac{2}{9}$.

What is the magnitude of the reaction at C?

(iv) If $\mu = \frac{1}{5}$, what is the minimum value of P necessary to cause the box to rotate about C without sliding, and what is the corresponding value of α?

(OCSEB)

14.

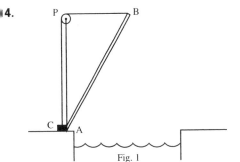

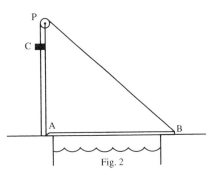

Fig. 1   Fig. 2

The diagrams show a simple mechanism by which a bridge over a canal is raised and lowered. The bridge AB is hinged at A, and a rope attached to B passes over a pulley P located vertically above A, at the top of a fixed vertical structure. A counterweight C is attached to the rope.

Figure 1 shows the bridge in the raised position, with C on the ground and B at the same horizontal level as P, and Fig. 2 shows the bridge lowered to its horizontal position. The mass of the bridge AB is 300 kg, and when raised (as in Fig. 1) the angle PAB is 30°.

Making suitable assumptions, which should be stated, find

(i)   the least mass needed for the counterweight C if it is to be capable of holding the bridge in the raised position,

(ii) the extra force that needs to be applied to start raising the bridge from the horizontal position by pulling on the rope, if the mass of C is the minimum value found in (i).

(UCLES)s

15. The foot of a uniform ladder of mass m rests on rough horizontal ground and the top of the ladder rests against a smooth vertical wall. When a man of mass $4m$ stands at the top of the ladder the system is in equilibrium with the ladder inclined at 60° to the horizontal. Show that the coefficient of friction between the ladder and the ground is greater than or equal to $\frac{3}{10}\sqrt{3}$.

16. The diagram shows a sketch of the region R bounded by the curve with equation $y^2 = 4x$ and the line with equation $x = 4$. The unit of length on both the x-axis and the y-axis is the centimetre. The region R is rotated through π radians about the x-axis to form a solid S.

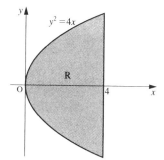

(a) Show that the volume of S is $32\pi\,\text{cm}^3$.

Given that the solid S is uniform,

(b) find the distance of the centre of mass of S from O.　　　　(ULEAC)

17.

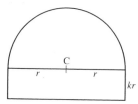

The diagram shows a cross-section containing the axis of symmetry of a uniform body consisting of a solid right circular cylinder of base radius r and height kr surmounted by a solid hemisphere of radius r. Given that the centre of mass of the body is at the centre C of the common face of the cylinder and the hemisphere, find the value of k, giving your answer to 2 significant figures.

Explain briefly why the body remains at rest when it is placed with any point of its hemispherical surface in contact with a horizontal plane.　　　　(ULEAC)

18. A mould for a right circular cone, base radius r and height h, is produced by making a conical hole in a uniform cylindrical block, base radius $2r$ and height $3r$. The axis of symmetry of the conical hole coincides with that of the cylinder, and AB is a diameter of the top of the cylinder, as shown in the diagram.

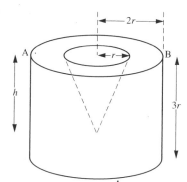

(a) Show that the distance from AB of the centre of mass of the mould is

$$\frac{216r^2 - h^2}{4(36r - h)}.$$

The mould is suspended from the point A, and hangs freely in equilibrium.

(b) In the case $h = 2r$, calculate, to the nearest degree, the angle between AB and the downward vertical..　　　　(ULEAC)

19. Three particles of masses 0.1 kg, 0.2 kg and 0.3 kg are placed at the points with position vectors $(2\mathbf{i} - \mathbf{j})\,\text{m}$, $(2\mathbf{i} + 5\mathbf{j})\,\text{m}$ and $(4\mathbf{i} + 2\mathbf{j})\,\text{m}$ respectively. Find the position vector of the centre of mass of these particles.　　　　(ULEAC)s

0.

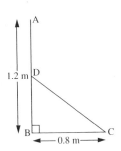

A uniform bracket consists of three straight pieces AB, BC and CD, where D is the mid-point of AB, as shown in the diagram. The lengths of AB and BC are 1.2 m and 0.8 m respectively. Calculate the distance of the centre of mass of the bracket from

(a) BC,   (b) AB.

The bracket is freely suspended from the point A and hangs in equilibrium.

(c) Calculate the acute angle between BC and the horizontal, giving your
      answer to the nearest degree.   (ULEAC)

21. A uniform rectangular sheet of metal ABCD, of mass 10 kg, is suspended from A. In equilibrium AB, which has length 0.3 m, is inclined at 20° to the vertical. Find the length of AD.   (UCLES)$_s$

22. The figure shows a triangular framework ABC
    which consists of three equal light rods smoothly
    jointed together. The joint A is attached to a
    smooth pivot and a load of weight 30 N hangs
    from C. The framework is held in a vertical
    plane with AC horizontal and B uppermost
    by a horizontal force of magnitude P applied
    at B.

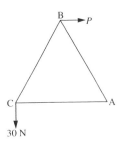

Neglecting the weight of the rods,

(a) show that the value of P is $20\sqrt{3}$ N;

(b) find the magnitudes of the forces in the rods;

(c) find the magnitude of the reaction at A.   (AEB)$_s$

CHAPTER 22

MOTION IN A PLANE

MOTION WITH VARIABLE ACCELERATION

We saw in Chapter 2 that, when an object is moving in a straight line with constant acceleration, there are a number of standard relationships linking velocity, time, displacement and acceleration. It is very important to appreciate that these formulae can be used *only for motion with constant acceleration*. There are many different types of motion in which the acceleration is not constant and we must now investigate ways in which such motion can be analysed.

We know that velocity is the rate at which displacement varies and that acceleration is the rate at which velocity varies, so we can write

$$v = \frac{ds}{dt} \quad \text{and} \quad a = \frac{dv}{dt}$$

Using the dot notation introduced in Chapter 11, we have

$$v = \dot{s} \quad \text{and} \quad a = \dot{v}$$

Further,

$$a = \frac{dv}{dt} = \frac{d}{dt}\left(\frac{ds}{dt}\right) = \frac{d^2s}{dt^2} \quad \text{so} \quad a = \ddot{s}$$

These relationships provide the means to solve problems in which the motion varies with time and are equally valid when a, v and s are given in vector form.

Conversely, if we start with the acceleration, then

$$a = \frac{dv}{dt} \quad \Rightarrow \quad v = \int a \, dt$$

and

$$v = \frac{ds}{dt} \quad \Rightarrow \quad s = \int v \, dt$$

i.e. if the acceleration of a moving body is a function of time,

and

velocity can be found by integrating a with respect to t

displacement can be found by integrating v with respect to t.

476

Examples 22a

1. A body moves along a straight line so that its displacement, s metres, from a
   fixed point O on the line after t seconds, is given by $s = t^3 - 3t^2 - 9t$

   (a) Find the velocity after t seconds.

   (b) Find the time(s) when the velocity is zero.

   (c) Sketch the velocity–time graph.

   (a) $$s = t^3 - 3t^2 - 9t$$

   $\therefore$ $$v = \dot{s} = 3t^2 - 6t - 9$$

   (b) When $v = 0$, $\quad 3t^2 - 6t - 9 = 0$

   $\therefore$ $$3(t^2 - 2t - 3) = 0 \qquad \Rightarrow \qquad 3(t - 3)(t + 1) = 0$$

   $$\Rightarrow \qquad t = 3 \quad \text{or} \quad -1$$

   Therefore the velocity is zero *after* 3 seconds; it was also zero 1 second
   before the body reached O.

   (c) The expression for the velocity is a quadratic function for which the graph is a parabola crossing
   the t-axis where $t = 3$ and $t = -1$.

   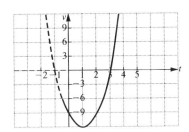

2. A particle P moving in a straight line has an initial velocity of 2 m s^{-1} at a
   point O on the line. The particle moves so that its acceleration t seconds later is
   given by $(2t - 6)\text{ m s}^{-2}$.

   Find expressions for (a) the velocity and (b) the displacement of P from O
   when $t = 5$ and comment on your answers.

   (a) $$v = \int a \, dt = \int (2t - 6) \, dt = t^2 - 6t + K_1$$

   $$v = 2 \quad \text{when} \quad t = 0 \quad \text{therefore} \quad K_1 = 2$$

   $\therefore$ $$v = t^2 - 6t + 2$$

   When $t = 5$, $\quad v = 25 - 30 + 2 = -3$

   The velocity when $t = 5$ is -3 m s^{-1}

   i.e.   P is moving with speed 3 m s^{-1} *towards* O

(b) $s = \int v \, dt = \int (t^2 - 6t + 2) \, dt = \frac{1}{3}t^3 - 3t^2 + 2t + K_2$

$s = 0$ when $t = 0$ therefore $K_2 = 0$

$\therefore$ $s = \frac{1}{3}t^3 - 3t^2 + 2t$

When $t = 5$, $s = \frac{125}{3} - 75 + 10 = -23\frac{1}{3}$

When $t = 5$ the displacement of P from O is $-23\frac{1}{3}$ m, i.e. P has changed direction, moving back towards O, and then passes through O to the opposite side.

3. **The velocity, v metres per second, of a particle moving in a straight line, is given by $v = (3t^2 - 12t + 9)$ where t is the number of seconds after the particle passes through O, a point on the line. Find the times(s) when the direction of motion of the particle is reversed.**

The *direction of motion* is determined by the sign of the velocity (*not* by the sign of the displacement). Hence, whenever the direction of motion is reversed, the velocity is momentarily zero.

When $v = 0$, $3t^2 - 12t + 9 = 0$ $\Rightarrow$ $3(t^2 - 4t + 3) = 0$

$\Rightarrow$ $3(t - 1)(t - 3) = 0$

$\therefore$ $v = 0$ when $t = 1$ or   3.

Now we must check that the sign of the velocity *changes* at these values of t and can do this quickly by finding the sign of v on either side of the values of t where $v = 0$.

t	0	1	2	3	4
v	9	0	-3	0	33
sign of v	+	0	$-$	0	+

At $t = 1$ and $t = 3$ the velocity becomes zero and changes sign.

$\therefore$ the particle's direction is reversed after 1 second and again after another 2 seconds.

Note that when we locate a value of t where v, i.e. $\dfrac{ds}{dt}$,   is zero, we have found a stationary point on the curve $s = f(t)$. Now a stationary point may be a point of inflexion where the sign of $\dfrac{ds}{dt}$ might not change, so when we are looking for a change in direction of motion, it is essential to check that the sign of $\dfrac{ds}{dt}$ *does* change, i.e. that we have found a turning point.

Any method for identifying the nature of a stationary point can be used but the numerical check used in the example above is usually quick and easy.

4. **A particle P is moving on a straight line through a fixed point O. The displacement, s metres, of P from O at time t is given by $s = 5 + 9t^2 - 2t^3$. Find the *distance* covered in the first 4 seconds.**

Distance and displacement are equal only if the direction of motion does not change within the time interval concerned, so first check whether there are values of t when the direction of motion changes, i.e. when $v = 0$ momentarily and also changes sign.

$$s = 5 + 9t^2 - 2t^3 \quad \Rightarrow \quad v = 18t - 6t^2 = 6t(3 - t)$$

$$\therefore \qquad v = 0 \quad \text{when} \quad t = 0 \quad \text{and when} \quad t = 3$$

When $t = 0$ the motion starts, so we need only check that v changes sign when $t = 3$.

When $t = 2$, $v > 0$ and when $t = 4$, $v < 0$, so v does change sign when $t = 3$.

The direction of motion changes when $t = 3$ so the distance travelled in the first 4 seconds is not equal to the corresponding increase in displacement.

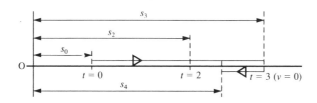

The distance travelled from $t = 0$ to $t = 3$ is $(s_3 - s_0)$ metres.

$$s_3 = 5 + 81 - 54 = 32 \quad \text{and} \quad s_0 = 5 \quad \Rightarrow \quad s_3 - s_0 = 27$$

$$\therefore \qquad \text{the distance travelled in the first 3 seconds is 27 m.}$$

The distance from when $t = 3$ to when $t = 4$ is travelled in the opposite direction so is $(s_3 - s_4)$ m, where $s_3 - s_4 = 32 - (5 + 144 - 128) = 11$

$$\therefore \qquad \text{the distance travelled in the fourth second is 11 m.}$$

The distance travelled in the first 4 seconds is $(27 + 11)$ m, i.e. 38 m.

Variable motion is not always defined by a given function for acceleration, velocity or displacement. Instead we may be given information from which a formula for a relationship can be found.

5. **The motion in a straight line of a particle P is such that the acceleration, a, is proportional to $(t + 1)$ at any time t seconds. Initially P has a velocity of 2 m s^{-1} and when $t = 4$ the velocity is 26 m s^{-1}.**
   Express the velocity and the acceleration as functions of t.

$$a \propto (t + 1) \quad \Rightarrow \quad a = k(t + 1) \quad \text{where k is a constant of proportion}$$

$$v = \int a \, dt \quad \Rightarrow \quad v = \int k(t + 1) \, dt = k(\tfrac{1}{2}t^2 + t) + K$$

When $t = 0$, $v = 2$ $\Rightarrow$ $K = 2$

When $t = 4$, $v = 26$ $\Rightarrow$ $26 = k(8 + 4) + 2$ $\Rightarrow$ $k = 2$

Hence $a = 2(t + 1)$ and $v = t^2 + 2t + 2$

EXERCISE 22a

In each question from 1 to 10, a particle P is moving on a straight line and O is a fixed point on that line. After a time of t seconds the displacement of the particle from O is s metres, the velocity is v metres per second and the acceleration is a metres per second^2.

1. Given that $s = 4t^3 - 5t^2 + 7t + 6$, find v when $t = 3$.

2. Given that $v = 9t^2 + 14t + 6$, find a when $t = 2$.

3. If $s = t^3 - 2t^2 + 9t$, find a when $t = 5$.

4. Given that P is at O when $t = 0$, and that $v = 2t^2 + 3t + 4$, find s when $t = 4$.

5. P starts from O with velocity 3 m s^{-1}. If $a = 12t - 5$, find v and s when $t = 2$.

6. If $s = t^3 - 9t^2 + 24t - 11$, find the time(s) when $v = 0$.

7. Find the times when the direction of motion of P changes given that $s = 6t^3 - 9t^2 + 4t$.

8. P starts from rest at O and moves with an acceleration given by $2t \text{ m s}^{-2}$. Find v and s in terms of t.

9. When $t = 0$, P passes with a velocity of 4 m s^{-1} through a point with a displacement of 2 m from O. Given that $a = t^2 + 1$, find the velocity and displacement when $t = 4$.

10. When $t = 0$, P passes through O with velocity -4 m s^{-1}. If $a = 8 - 6t$, find

    (a) the times when P is instantaneously at rest

    (b) the displacement of P from O at these times.

11. A particle P moves in a straight line and O is a fixed point on the line. The displacement, s metres, of P from O at any time t seconds is given by $s = t^3 + t^2 + 12t - 23$. Show that the motion is always in the same direction.

12. A particle moves in a straight line with an acceleration given at any time by $(3t - 1) \text{ m s}^{-2}$. If the particle has a velocity of 3 m s^{-1} and is 7 m from a fixed point O on the line when $t = 2$, find

    (a) its velocity when $t = 5$

    (b) its displacement from O when $t = 4$.

13. The acceleration of a particle P after t seconds is proportional to $(3t^2 + 1)$. When $t = 3$, the acceleration is 14 m s^{-2} and the speed is 25 m s^{-1}. Find

(a) the acceleration as a function of t (b) the initial velocity.

14. Given that $a = -\dfrac{1}{t^3}$ and that $v = 3$ when $t = 1$,

(a) find the velocity when $t = 4$

(b) show that, as the value of t becomes large, the velocity approaches a particular value (called the terminal velocity) and state this value.

15. A particle starts from rest at a point A and moves along a straight line AB with an acceleration after t seconds given by $a = (8 - 2t^2)$. Find

(a) the greatest speed of the particle in the direction $\overrightarrow{AB}$

(b) the time when this greatest speed occurs

(c) the distance travelled in this time.

16. At any time t, the acceleration of a particle P, travelling in a straight line, is inversely proportional to $(t + 1)^3$. Initially, when $t = 0$, P is at rest at a point O and 3 seconds later it has a speed of 2 m s^{-1}. Find, in terms of t, the displacement of P from O at any time.

17. A particle travelling in a straight line passes initially through a fixed point O on the line with a velocity u. The acceleration of the particle has a constant value a. By using integration find expressions for the velocity v and the displacement s from O, after t seconds.
Compare these results with the standard formulae for motion with constant acceleration.

18. A particle starts from rest at a fixed point A and moves in a straight line with an acceleration which, t seconds after leaving A, is given by $a = 4t$. After 2 seconds the particle reaches a point B and the acceleration then ceases. Find

(a) the velocity when the particle reaches B   (b) the distance AB.

Immediately the particle moves on with acceleration given by $-3t$ until it comes to rest at a point C. Find

(c) the value of t when the particle reaches C   (d) the distance AC.

19. The displacement, s metres, of a body from a point O after t seconds is given by $s = t^2 + \dfrac{1}{t}$.

(a) Find in terms of t an expression for the acceleration of the body.

(b) Given that the mass of the body is 3 kg, use Newton's Second Law to find the force acting on the body after 5 seconds.

*20. An object is moving in a straight line under the action of a force whose value at any time t seconds is given by $F = (12t + 20)$ newtons. When $t = 2$ the object, whose mass is 4 kg, passes through a point A on the line with a velocity of 22 m s^{-1}. Find, as a function of t, the displacement of the object from A at any time.

*21. A wagon whose mass is 200 kg is pulled by a cable along a straight level track. Contact between the wagon and the track is smooth and the tension in the cable is directly proportional to the time. The wagon starts from rest and, 10 seconds later, its speed is 20 m s^{-1}. How far has the wagon been pulled?

VARIABLE MOTION IN THE *X-Y* PLANE

When a particle is moving in a plane it can be convenient to consider separately its motion in two perpendicular directions. If displacement, velocity and acceleration are functions of time, then the calculus methods used so far can be applied to the components in each direction.

Consider, for example, the motion of the particle shown in the diagram.

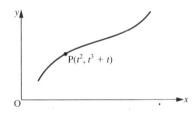

In the direction Ox, at any time t, $x = t^2$

i.e.   the displacement from O is t^2
∴   the velocity is $2t$
and   the acceleration is 2

In the direction Oy, $y = t^3 + t$

i.e.   the displacement from O is $t^3 + t$
∴   the velocity is $3t^2 + 1$
and   the acceleration is $6t$

Now we can express these components in terms of unit vectors **i** and **j** in the chosen directions and, by adding them, form a resultant vector.

For example, the position vector of P is denoted by $\mathbf{r}$, and is given by

$$\mathbf{r} = t^2\mathbf{i} + (t^3 + t)\mathbf{j}$$

Similarly the velocity vector, $\mathbf{v}$, is $2t\mathbf{i} + (3t^2 + 1)\mathbf{j}$ and the acceleration vector, $\mathbf{a}$, is $\quad 2\mathbf{i} + 6t\mathbf{j}$

As each component of $\mathbf{v}$ is obtained by differentiating the corresponding component of $\mathbf{s}$ with respect to t, we can say

$$\mathbf{v} = \frac{d\mathbf{s}}{dt} \quad \text{and} \quad \mathbf{s} = \int \mathbf{v} \, dt$$

Similarly

$$\mathbf{a} = \frac{d\mathbf{v}}{dt} \quad \text{and} \quad \mathbf{v} = \int \mathbf{a} \, dt$$

Examples 22b

In these examples, $\mathbf{i}$ and $\mathbf{j}$ are perpendicular unit vectors and t is the elapsed time. Acceleration, velocity and displacement are all to be expressed as vectors in $\mathbf{ij}$ form. All quantities are measured in units based on metres and seconds.

1. **A particle is moving in a plane in such a way that its velocity at any time t is given by $2t\mathbf{i} + 3t^2\mathbf{j}$. Initially the position vector of the particle, relative to a fixed point O in the plane, is $5\mathbf{i} - 8\mathbf{j}$. Find, when $t = 3$,**

   (a) **the acceleration of P**   (b) **the position vector of P.**

   (a) $$\mathbf{v} = 2t\mathbf{i} + 3t^2\mathbf{j}$$

   $$\mathbf{a} = \frac{d\mathbf{v}}{dt} = 2\mathbf{i} + 6t\mathbf{j}$$

   When $t = 3$, $\mathbf{a} = 2\mathbf{i} + 18\mathbf{j}$

   (b) $$\mathbf{r} = \int \mathbf{v} \, dt = \int (2t\mathbf{i} + 3t^2\mathbf{j}) \, dt$$

   When a function is integrated, a constant of integration must be added. In this problem we are integrating a *vector function* so the constant of integration must also be a vector quantity. We will denote it by $\mathbf{A}$.

   $\therefore \qquad \mathbf{r} = t^2\mathbf{i} + t^3\mathbf{j} + \mathbf{A}$

   Initially (i.e. when $t = 0$) $5\mathbf{i} - 8\mathbf{j} = 0\mathbf{i} + 0\mathbf{j} + \mathbf{A}$

   $\Rightarrow \qquad \mathbf{A} = 5\mathbf{i} - 8\mathbf{j}$

   $\therefore \qquad \mathbf{r} = t^2\mathbf{i} + t^3\mathbf{j} + 5\mathbf{i} - 8\mathbf{j} = (t^2 + 5)\mathbf{i} + (t^3 - 8)\mathbf{j}$

   When $t = 3$, $\mathbf{r} = (9 + 5)\mathbf{i} + (27 - 8)\mathbf{j} = 14\mathbf{i} + 19\mathbf{j}$

2. **At any time _t_, the position vector of a particle moving in a plane, relative to a fixed point O in the plane, is $10t\mathbf{i} + (t^4 - 4t)\mathbf{j}$.**

   (a) **Show that the particle has no acceleration in the direction of i.**

   (b) **Find the time when the velocity is perpendicular to the acceleration.**

   (c) **Find the _distance_ from O of the particle when $t = 2$.**

   (d) **Find the angle between the vector i and the direction of motion when $t = 2$.**

   (a) $\mathbf{r} = 10t\mathbf{i} + (t^4 - 4t)\mathbf{j}$

         $\mathbf{v} = \dfrac{d\mathbf{r}}{dt} = 10\mathbf{i} + (4t^3 - 4)\mathbf{j}$

         $\mathbf{a} = \dfrac{d\mathbf{v}}{dt} = 12t^2\mathbf{j}$

   a has no term in **i**, therefore the acceleration has no component in the direction of **i**.

   (b) The acceleration is always in the direction of **j**. Therefore, in order to be perpendicular to the acceleration, the velocity must be parallel to **i**.

   $$\mathbf{v} = 10\mathbf{i} + (4t^3 - 4)\mathbf{j}$$

   v is perpendicular to **a** when the coefficient of **j** is zero,

   i.e when $4(t^3 - 1) = 0 \quad \Rightarrow \quad t = 1$

   The velocity is perpendicular to the acceleration after 1 second.

   (c) When $t = 2$, $\mathbf{r} = 20\mathbf{i} + (16 - 8)\mathbf{j}$

                 $= 20\mathbf{i} + 8\mathbf{j}$

   The distance between O and P is

             $\sqrt{(20^2 + 8^2)}\,\text{m} = 21.5\,\text{m} \quad (3\,\text{sf})$

   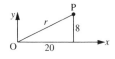

   (d) The direction of motion depends upon the components of the velocity.

   When $t = 2$, $\mathbf{v} = 10\mathbf{i} + (32 - 4)\mathbf{j}$

                 $= 10\mathbf{i} + 28\mathbf{j}$

   ∴ the direction of motion makes an angle α with **i**,

   where $\tan \alpha = \frac{28}{10} \quad \Rightarrow \quad \alpha = 70.3° \quad (3\,\text{sf})$.

3. A force, in newtons, is expressed at any time t seconds by $\quad \mathbf{F} = 2\mathbf{i} + 3(t^2 - 1)\mathbf{j}$.
The force acts on a particle **P**, of mass 2 kg, moving in the xy plane.
When $t = 0$, **P** is at rest at the point with position vector $\mathbf{i} + \mathbf{j}$.

(a) Find **(i)** the acceleration vector **(ii)** the position vector of **P** at time t.

(b) Write down separate equations for the x and y coordinates of **P** at time t.

(c) By eliminating t from these two equations, find the Cartesian equation of the
path of **P** and sketch the path.

(a) **(i)** Using Newton's Law, $\mathbf{F} = m\mathbf{a}$, gives $\quad 2\mathbf{i} + 3(t^2 - 1)\mathbf{j} = 2\mathbf{a}$

$$\Rightarrow \qquad \mathbf{a} = \mathbf{i} + \tfrac{3}{2}(t^2 - 1)\mathbf{j}$$

(ii) First find the velocity vector.

$$\mathbf{v} = \int \mathbf{a}\ dt = \int [\mathbf{i} + \tfrac{3}{2}(t^2 - 1)\mathbf{j}]\ dt = t\mathbf{i} + (\tfrac{1}{2}t^3 - \tfrac{3}{2}t)\mathbf{j} + \mathbf{A}$$

When $\quad t = 0$, $\quad \mathbf{v} = \mathbf{0}\quad$ therefore $\quad \mathbf{A} = \mathbf{0}$

$$\therefore \qquad \mathbf{v} = t\mathbf{i} + (\tfrac{1}{2}t^3 - \tfrac{3}{2}t)\mathbf{j}$$

$$\mathbf{r} = \int \mathbf{v}\ dt = \tfrac{1}{2}t^2\mathbf{i} + (\tfrac{1}{8}t^4 - \tfrac{3}{4}t^2)\mathbf{j} + \mathbf{B}$$

When $\quad t = 0$, $\quad \mathbf{r} = \mathbf{i} + \mathbf{j}\quad$ therefore $\quad \mathbf{i} + \mathbf{j} = \mathbf{B}$

$$\therefore \qquad \mathbf{r} = \tfrac{1}{2}t^2\mathbf{i} + (\tfrac{1}{8}t^4 - \tfrac{3}{4}t^2)\mathbf{j} + \mathbf{i} + \mathbf{j}$$

i.e. $\qquad \mathbf{r} = (\tfrac{1}{2}t^2 + 1)\mathbf{i} + (\tfrac{1}{8}t^4 - \tfrac{3}{4}t^2 + 1)\mathbf{j}$

(b) $\qquad\qquad x = \tfrac{1}{2}t^2 + 1$ [1]

$\qquad\qquad y = \tfrac{1}{8}t^4 - \tfrac{3}{4}t^2 + 1$ [2]

(c) From [1] $\quad t^2 = 2(x - 1)$

Substituting in [2] gives $\quad y = \tfrac{1}{8}[4(x - 1)^2] - \tfrac{3}{4}[2(x - 1)] + 1$

Multiplying throughout by 2 gives $\quad 2y = (x - 1)^2 - 3(x - 1) + 2$

i.e. $\quad 2y = x^2 - 5x + 6 = (x - 2)(x - 3)$

This is the Cartesian equation of the path of **P** which can be recognised as a
parabola.

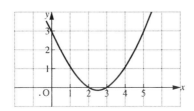

4. At any time t seconds, the position vector of a particle of mass 9 kg is given in metres by $\mathbf{r} = 2t^2\mathbf{i} + 5t\mathbf{j}$. Find the kinetic energy of the particle when
(a) $t = 3$ (b) $t = 0$.

The velocity, $\mathbf{v}$, of the particle is given by

$$\mathbf{v} = \frac{d\mathbf{r}}{dt} = 4t\mathbf{i} + 5\mathbf{j}$$

(a) When $t = 3$, $\mathbf{v} = 12\mathbf{i} + 5\mathbf{j}$

The speed of the particle is given by $|\mathbf{v}|$

i.e.   the speed is 13 m s^{-1}

$\therefore$ K.E. $= (\frac{1}{2} \times 9 \times 13^2)$ J $= 760.5$ J

(b) When $t = 0$, $\mathbf{v} = 5\mathbf{j}$ $\Rightarrow$ the speed is 5 m s^{-1}

$\therefore$ K.E. $= (\frac{1}{2} \times 9 \times 5^2)$ J $= 112.5$ J

EXERCISE 22b

In these questions, a particle P is moving in the xy plane and O is the origin. Unit vectors in the directions Ox and Oy are $\mathbf{i}$ and $\mathbf{j}$ respectively. For P the position vector relative to O at time t is $\mathbf{r}$, the velocity vector is $\mathbf{v}$ and the acceleration vector is $\mathbf{a}$. All units are consistent and based on metres, seconds and newtons.

1. Find, in $\mathbf{ij}$ form, expressions for $\mathbf{v}$ and $\mathbf{a}$
(i)   at any time t (ii) at the specified times,   if

(a) $\mathbf{r} = 2t^3\mathbf{i} + 3t^2\mathbf{j}$; $t = 2$, $t = 3$

(b) $\mathbf{r} = t(t+1)\mathbf{i} + (4 - t^2)\mathbf{j}$; $t = 1$, $t = 4$

(c) $\mathbf{r} = \frac{2}{t}\mathbf{i} + 3\left(1 + \frac{2}{t^2}\right)\mathbf{j}$; $t = 1$, $t = 2$

2. Find the Cartesian equation of the path traced out by P if

(a) $\mathbf{r} = 2t\mathbf{i} + 3t^2\mathbf{j}$ (b) $\mathbf{r} = (t+1)^2\mathbf{i} + 4t\mathbf{j}$ (c) $\mathbf{r} = 2t\mathbf{i} + \frac{3}{t}\mathbf{j}$

3. The angle between $\mathbf{i}$ and the direction of motion of P is α. Find $\tan\alpha$,
(i)   at any time t (ii) at the specified times.

(a) $\mathbf{r} = 6t\mathbf{i} + (12t - 3t^2)\mathbf{j}$; $t = 0$, $t = 1$, $t = 2$

(b) $\mathbf{r} = t^2\mathbf{i} + (1 + t^3)\mathbf{j}$; $t = \frac{1}{3}$, $t = \frac{2}{3}$, $t = 1$.

4. Initially P is at rest at a point with position vector $3\mathbf{i} + \mathbf{j}$. Given that the acceleration of P after t seconds is $\mathbf{i} - 2\mathbf{j}$, find in terms of $\mathbf{i}$ and $\mathbf{j}$, expressions for $\mathbf{v}$ and $\mathbf{r}$ at any time t.

5. At any time t, $\mathbf{v} = 3t^2\mathbf{i} + (t - 1)\mathbf{j}$. Given that P is initially at O find in $\mathbf{ij}$ form,

 (a) the initial velocity

 (b) $\mathbf{a}$ when $t = 3$

 (c) $\mathbf{r}$ when $t = 2$.

6. When $t = 1$, $\mathbf{v} = \mathbf{i} + 3\mathbf{j}$ and $\mathbf{r} = 4\mathbf{i} - \mathbf{j}$. If $\mathbf{a} = t\mathbf{i} + (2 - t)\mathbf{j}$, find $\mathbf{r}$ when $t = 4$. What is the distance of P from O at this time?

7. The acceleration vector of P is constant and given by $\mathbf{a} = p\mathbf{i} + q\mathbf{j}$. When $t = 0$ the velocity vector is zero and when $t = 1$, $\mathbf{v} = 3\mathbf{i} - 2\mathbf{j}$. Find $\mathbf{v}$ at any time t. What is the *speed* of P when $t = 3$?

8. The coordinates of P at any time t are $(t + t^2, 3t^2 - 2)$. Prove that P has a constant acceleration and give its magnitude.

9. A force $\mathbf{F}$ acts on particle of mass 2 kg. Given that $\mathbf{F} = 4t\mathbf{i} + 6\mathbf{j}$, and that P is initially at O with velocity $5\mathbf{j}$, find $\mathbf{v}$ and $\mathbf{r}$ when $t = 3$.

10. $\mathbf{a} = \dfrac{32}{t^3}\mathbf{j}$ and, when $t = 2$, $\mathbf{v} = 9\mathbf{i} - 4\mathbf{j}$ and $\mathbf{r} = 18\mathbf{i} + 8\mathbf{j}$.

 (a) Find $\mathbf{v}$ in terms of t.

 (b) Find $\mathbf{r}$ in terms of t. ·

 (c) Find the value of t when $\mathbf{v}$ is perpendicular to $\mathbf{r}$.

 (d) Show that, as t increases, $\mathbf{v}$ approaches a constant value and state the magnitude and direction of this terminal velocity.

11. Given that $\mathbf{r} = (2t - 1)\mathbf{i} - t^2\mathbf{j}$,

 (a) find, in $\mathbf{ij}$ form, the direction of motion
 (i) initially (ii) at time t

 (b) explain why the direction in which P moves can never be perpendicular to the initial direction of motion

 (c) show that $\mathbf{a}$ is constant and give its magnitude.

12. Initially P is at O with velocity vector $(V \cos \alpha)\mathbf{i} + (V \sin \alpha)\mathbf{j}$. Given that $\mathbf{a} = -g\mathbf{j}$, where g is the acceleration due to gravity,

 (a) find, at any time t, expressions for (i) $\mathbf{v}$ (ii) $\mathbf{r}$

 (b) hence derive the equation of the path of a particle projected from O with speed V at an angle α to the horizontal.

13. A force represented by $12\mathbf{j}$ is the only force acting on P. Initially P, whose mass is 3 kg, is at O with a speed of 6 m s⁻¹ in the direction O*x*.

(a) Express in **ij** form   (i) the initial velocity
    (ii) the acceleration vector   (iii) **v** and **r** after *t* seconds.

(b) Show that P moves on a parabola.

14. Two forces **P** and **Q**, of magnitudes P newtons and Q newtons, act on a particle A of mass 2 kg. $\mathbf{P} = 6\mathbf{i} - \mathbf{j}$ and $\mathbf{Q} = 2\mathbf{i} + 5\mathbf{j}$.

(a) Find the resultant of **P** and **Q** and hence the acceleration of A.

(b) Given that A is initially at rest at a point with position vector $3\mathbf{j}$
   ($|\mathbf{j}| = 1$ m), find the position vector of P after *t* seconds.

15. A force $\mathbf{F}_1$, of magnitude 10 N, acts in the direction of $4\mathbf{i} + 3\mathbf{j}$; the direction of another force $\mathbf{F}_2$, of magnitude $3\sqrt{5}$ N, is $2\mathbf{i} + \mathbf{j}$. Both forces act on a particle P of mass 2 kg, which is initially at rest at the origin.
Express in **ij** form

(a) $\mathbf{F}_1$, $\mathbf{F}_2$ and their resultant

(b) the acceleration of P and the velocity of P after *t* seconds

(c) the position vector of P after 2 seconds.

16. A particle P of mass 2 kg is at rest, when a force $4\mathbf{i} + 6t\mathbf{j}$ acts on it for one second. Find, at the end of this time,

(a) the velocity of P   (b) the kinetic energy of P.

17. At any time *t* seconds, the position vector of a particle P of mass 4 kg is given by $\mathbf{r} = 6t\mathbf{i} - 4t^2\mathbf{j}$. Only one force, **F**, acts on P. Find

(a) the velocity and acceleration vectors of P at time *t*

(b) the force **F**

(c) the kinetic energy of P after 4 seconds

(d) the work done by **F** in the first 4 seconds.

18. A force represented by $4\mathbf{i} - 6\mathbf{j}$ acts on a particle P of mass 2 kg which is initially at rest at the origin O. Find

(a) the acceleration of P

(b) the velocity of P after *t* seconds

(c) the kinetic energy of P after 2 seconds.

During the first 2 seconds, what is

(d) the work done by the force

(e) the average power exerted by the force.

USING DIFFERENTIAL EQUATIONS

The problems in Exercises 22a and 22b were solved by interpreting given information about motion by means of a differential equation in which velocity and acceleration were denoted by derived functions such as $\dfrac{ds}{dt}$ and $\dfrac{dv}{dt}$.

These particular derived functions were suitable because both s and v were functions of t and the resulting equation therefore contained only two variables (either s and t, or v and t).

In practice the acceleration of a moving object is more likely to depend on the distance or speed rather than the time of motion. For instance, the tension in an elastic string depends on the extension, so the force acting on a particle attached to the end, and hence the acceleration of that particle, depends on its displacement from the natural length position. In such cases, where the acceleration is not a function of time, we may have to find an alternative derived function to represent acceleration.

ACCELERATION AS A FUNCTION OF DISPLACEMENT

The general case of motion with an acceleration that is a function of displacement is represented by $a = f(s)$.

The expression so far used for acceleration is $\dfrac{dv}{dt}$, but when acceleration is a function of s this gives $\dfrac{dv}{dt} = f(s)$.

There are three variables in this differential equation, v, s and t, so at this stage the equation cannot be solved and we must look for another way of expressing the acceleration.

Using the chain rule gives $\dfrac{dv}{dt} = \dfrac{dv}{ds} \times \dfrac{ds}{dt}$

Now $\dfrac{ds}{dt} = v$ therefore   acceleration $= \dfrac{dv}{dt} = v\dfrac{dv}{ds}$

Hence, when $a = f(s)$ we have $f(s) = v\dfrac{dv}{ds}$

Separating the variables in this differential gives $\displaystyle\int f(s)\,ds = \int v\,dv$

If $f(s)$ is known (and can be integrated) a solution can be found.

Examples 22c

1. A particle P is moving along a straight line with an acceleration that is proportional to s^2 where s metres is the displacement of P from a fixed point A on the line.

   (a) Find a general relationship between the displacement and the velocity, v metres per second.

   Given that v and s are equal in magnitude when $s = 0$ and when $s = 4$,

   (b) find the speed when the displacement is 1.5 m

   (c) find the displacement when the velocity is $2\,\mathrm{m\,s}^{-1}$

   (d) sketch the graph of velocity against displacement for $0 \leqslant s \leqslant 4$.

   (a) Acceleration $\propto s^2$ i.e. $a = ks^2$ where k is a constant

   Using $a = v \dfrac{dv}{ds}$ gives

$$v \frac{dv}{ds} = ks^2 \quad \Rightarrow \quad \int v \, dv = \int ks^2 \, ds$$

$$\therefore \qquad \tfrac{1}{2}v^2 = \tfrac{1}{3}ks^3 + K$$

   This relationship between s and v is general because it contains two unknown constants.

   $v = 0$ when $s = 0$ so $K = 0$ $\Rightarrow$ $3v^2 = 2ks^3$

   $v = 4$ when $s = 4$ $\Rightarrow$ $48 = 128k$ $\Rightarrow$ $k = \tfrac{3}{8}$

   $\therefore \qquad\qquad\qquad\qquad 4v^2 = s^3$ •

   (b) When $s = 1.5$; $4v^2 = (1.5)^3$

      $\Rightarrow$ $v^2 = 0.843\ldots$ $\Rightarrow$ $v = \pm 0.918\ldots$

      The speed is $0.92\,\mathrm{m\,s}^{-1}$ (2 dp)

   (c) When $v = 2$, $16 = s^3$ $\Rightarrow$ $s = 2.519\ldots$

      The displacement is $2.5\,\mathrm{m}$ (2 sf)

   (d) Because $v = \pm\tfrac{1}{2}\sqrt{s^3}$,   there are two values of v for each value of s.

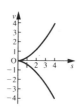

2. A particle P is moving in a straight line and O is a fixed point on the line. The magnitude of the acceleration of P is proportional to the distance of P from O; the direction of the acceleration is always towards O. Initially the particle is at rest at a point A where the *displacement* of A from O is *l*.

(a) Using x and v for the displacement and velocity of P, express v^2 in terms of x, l and a constant of proportion, k.

(b) Show that the speed of P is greatest when P is at O.

(c) Determine a position other than A where the velocity of P is zero.

(d) Describe briefly the motion of P.

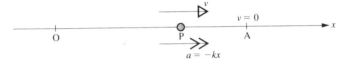

In the diagram the acceleration of P is towards O, i.e. in the negative direction, so $a = -kx$. If P is left of O, the acceleration is in the positive direction but x is negative, so again we have $a = -kx$.

(a) $\qquad a = -kx$

Using $a = v\dfrac{dv}{dx}$, gives $v\dfrac{dv}{dx} = -kx$

$\Rightarrow \qquad\qquad \displaystyle\int v\,dv = -\int kx\,dx$

$\Rightarrow \qquad\qquad \tfrac{1}{2}v^2 = -\tfrac{1}{2}kx^2 + K$

$v = 0$ when $x = l$ $\qquad\Rightarrow\qquad K = \tfrac{1}{2}kl^2$

$\qquad\qquad\qquad\qquad\qquad v^2 = k(l^2 - x^2)$

(b) The value of x^2 can never be negative, so the greatest value of $(l^2 - x^2)$ is $(l^2 - 0)$

The greatest speed occurs when $x = 0$, i.e. when P is at O.

(c) When $v = 0$, $k(l^2 - x^2) = 0$ $\qquad\Rightarrow\qquad l^2 = x^2$

$\qquad\qquad\qquad\qquad\qquad\qquad\qquad\Rightarrow\qquad x = \pm l$

i.e. the displacement of P from O is $\pm l$.

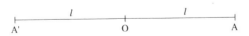

$\therefore$ the speed of P is zero when P is at A and at a point A$'$ where OA$' = l$ and A$'$ is on the opposite side of O from A.

(d) Wherever P is on the line it is accelerating towards O, so P oscillates between A and A$'$.

Note that in Example 2, displacement is represented by the symbol x as an alternative to s. The two symbols are equivalent and either may appear in questions.

EXERCISE 22c

In questions 1 to 10 a particle P is moving along a straight line and O is a fixed point on that line. After a time of t seconds the displacement of P from O is s metres, the velocity is $v \, \text{m} \, \text{s}^{-1}$ and the acceleration is $a \, \text{m} \, \text{s}^{-2}$.

1. $a = 2s + 5$ and $v = 2$ when $s = 0$.

    (a) Find v^2 in terms of s.

    (b) Find the speed when $s = 1$.

    (c) Find s when $v = 4$.

2. $a = 6s + 4$ and $v = 3$ when $s = 0$.

    (a) Find v^2 in terms of s.

    (b) Find the speed when $s = 2$.

    (c) Find s when $v = 5$.

3. $a = s - 4$ and $v = 12$ when $s = 4$.

    (a) Find v^2 in terms of s.

    (b) Find the speed when $s = 0$ and when $s = 9$.

    (c) For what value of s is $\dfrac{dv}{ds}$ zero?

    (d) Sketch the graph of v against s for $0 \leqslant s \leqslant 9$.

4. P is moving in the positive direction with an acceleration given by $a = 8s^3$, and when $s = 0$, $v = 0$.

    (a) Find v in terms of s.

    (b) Find the value of s for which $v = 50$.

5. P is moving in the positive direction with an acceleration proportional to s^3. When $s = 0$, $v = 0$ and when $s = 1$, $v = 9$.

    (a) Find v in terms of s.

    (b) Express a in terms of s.

    (c) Find the value of s for which $v = 900$.

6. When $v = 7$, $s = 5$ and when $v = 18$, $s = 6$. Given that a is proportional to s and $s > 0$ for all values of t,

    (a) find v^2 in terms of s

    (b) find the value of s when $v = 10$

    (c) find the least distance of P from the origin.

7. $a = 36 - 12s^2$ and $v = 0$ when $s = 0$.

    (a) Find v^2 in terms of s.

    (b) Find the speed when $s = 1$.

    (c) Find all the values of s for which $v = 0$.

    (d) By considering the sign of v^2, show that motion can take place on the section of the line for which $0 \leqslant s \leqslant 3$.

    (e) Find the acceleration   (i) when $s = 0$ (ii) when $s = 3$.

    (f) Find the value of s when v is a maximum and find the maximum speed.

    (g) Describe the motion of P.

8. $a = e^s$ and $v = 2$ when $s = 0$.

    (a) Find v^2 in terms of s.

    (b) Find v when $s = 4$.

    (c) Find s when $v = 20$.

9. $a = 40e^{-s}$ and $v = 8$ when $s = 0$.

    (a) Find v^2 in terms of s.

    (b) Find s when $v = 11$.

    (c) Find the terminal speed, i.e. the speed which is approached as s increases indefinitely.

10. $a = \dfrac{10}{s + 1}$ for $s \geqslant 0$, and, when $s = 0$, $v = 4$.

    (a) Find v^2 in terms of s.

    (b) Find v when $s = 4$.

    (c) Find s when $v = 12$.

11. A particle moves in a straight line with an acceleration $12s^2$ m s^{-2} where s metres is the displacement of the particle from O, a fixed point on the line, after t seconds. The particle has zero velocity when its displacement from O is -2 m. Find the velocity of the particle as it passes through O.

12. A spacecraft is moving in a straight line directly away from the centre of the earth. When it is at a distance x kilometres from the centre of the earth its acceleration, which is due to gravity, is $\dfrac{4 \times 10^5}{x^2}$ km s^{-2} towards the centre of the earth. When it is at 8000 km from the centre of the earth its speed is 11 km s^{-1}.

    (a) Find v^2 in terms of x.

    (b) Find v when $x = 10\,000$.

    (c) Find the terminal velocity.

VELOCITY GIVEN AS A FUNCTION OF DISPLACEMENT

In the previous section we saw that when acceleration is expressed in terms of displacement, using $v \dfrac{dv}{ds}$ for the acceleration results in a relationship between velocity and displacement.

So if the motion of a particle is defined by $v = f(s)$, it seems logical that we should be able to reverse the process to find the acceleration from this relationship.

Finding Acceleration as a Function of Displacement

If $v = f(s)$ and we want to find the acceleration, using $a = \dfrac{dv}{dt}$ is of no help as we cannot differentiate $f(s)$ with respect to t.

However, as we saw earlier in the chapter, $\dfrac{dv}{dt} \equiv v \dfrac{dv}{ds}$, and this form for the acceleration is useful again here.

Examples 22d

1. The velocity, in metres per second, of a particle P moving in a straight line is given by $v = x + \dfrac{1}{x}$ where x metres is the displacement of P from a fixed point O on the line. Find the acceleration of P when $x = 2$.

$$v = x + \frac{1}{x} \qquad \Rightarrow \qquad \frac{dv}{dx} = 1 - \frac{1}{x^2}$$

Using $a = v \dfrac{dv}{dx}$ gives $a = \left(x + \dfrac{1}{x} \right)\left(1 - \dfrac{1}{x^2} \right)$

$\Rightarrow$
$$a = x - \frac{1}{x^3}$$

When $x = 2$, $a = 2 - \frac{1}{8} = 1\frac{7}{8}$

When $x = 2$ the acceleration is $1.875 \, \mathrm{m \, s^{-2}}$.

Relating Displacement and Time

For motion defined by $v = \mathrm{f}(s)$, in order to relate displacement and time we can use $v = \dfrac{\mathrm{d}s}{\mathrm{d}t}$, giving $\dfrac{\mathrm{d}s}{\mathrm{d}t} = \mathrm{f}(s)$.

This differential equation contains only two variables so it provides a suitable method for trying to find a relationship between s and t.

Consider a particle, travelling in a straight line, whose velocity v is given by $v = \sqrt{s}$, where s is the displacement of the particle from a fixed point on the line.

Using $v = \dfrac{\mathrm{d}s}{\mathrm{d}t}$ gives $\dfrac{\mathrm{d}s}{\mathrm{d}t} = s^{\frac{1}{2}}$

$\Rightarrow \qquad\qquad \displaystyle\int s^{-\frac{1}{2}}\,\mathrm{d}s = \int \mathrm{d}t \quad \left(\text{i.e.} \int 1\,\mathrm{d}t \right)$

$\Rightarrow \qquad\qquad 2s^{\frac{1}{2}} = t + K$

To find the exact relationship between s and t, further information is needed to give the value of K.

In general, if $v = \mathrm{f}(s)$ we have

$$\frac{\mathrm{d}s}{\mathrm{d}t} = \mathrm{f}(s) \qquad \Rightarrow \qquad \int \frac{1}{\mathrm{f}(s)}\,\mathrm{d}s = \int \mathrm{d}t$$

Whether we can go any further depends on whether it is possible to perform the integration on the left hand side of this equation. In Example 22c/2, for example, from which we find that $v = \sqrt{k(l^2 - x^2)}$, many readers would be unable to carry on because they have not yet seen a method for integrating

$\dfrac{1}{\sqrt{(l^2 - x^2)}}$ with respect to x.

In this book, however, any question requiring solution, in which a type of motion is defined by a differential equation, will lead to an expression that can be integrated at this stage.

Examples 22d (continued)

2. The velocity, $v\,\mathrm{m\,s^{-2}}$, of a particle P at any time t is proportional to the square of the displacement, s metres, of P from a fixed point A. P is moving in a straight line through A. Initially, i.e. when $t = 0$, P is 2 m from A and, 3 seconds later, $AP = 1.25\,\mathrm{m}$.

(a) Find s when $t = 2$

(b) Show that, as the time increases indefinitely, the displacement of P from A approaches a particular value and state the position that P is then approaching.

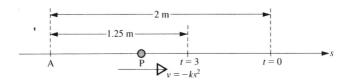

In this example it is clear from the positions of P when $t = 0$ and $t = 3$, that P is travelling towards O, i.e. v is negative.

$$\therefore \qquad v \propto s^2 \qquad \Rightarrow \qquad v = -ks^2$$

$$\text{Using} \quad v = \frac{ds}{dt} \quad \text{gives} \quad \frac{ds}{dt} = -ks^2$$

$$\text{Hence} \qquad -\int \frac{1}{s^2}\,ds = \int k\,dt \qquad \text{i.e.} \quad -\int s^{-2}\,ds = \int k\,dt$$

$$s^{-1} = kt + K$$

When $t = 0$, $s = 2$, therefore $K = 0.5$

$$\Rightarrow \qquad\qquad\qquad s^{-1} = kt + 0.5$$

When $t = 3$, $s = 1.25$, therefore $0.8 = 3k + 0.5 \qquad \Rightarrow \qquad k = 0.1$

$$\therefore \qquad\qquad\qquad s^{-1} = 0.1t + 0.5$$

(a) When $t = 2$, $\qquad s^{-1} = 0.2 + 0.5 = 0.7 \qquad \Rightarrow \qquad s = 1.43 \quad (3\,\mathrm{sf})$

(b) First we will express s as a simplified function of t.

$$s^{-1} = 0.1t + 0.5 \qquad \Rightarrow \qquad 10s^{-1} = t + 5$$

$$\therefore \qquad\qquad\qquad s = \frac{10}{t + 5}$$

As t becomes very large, s becomes very small, i.e. s approaches 0. When t is very large, P approaches A.

Note that it was not necessary to notice that v was negative; the calculation of the constant k sorts out the correct sign.

3. A particle is travelling along the line ABC as shown in the diagram.

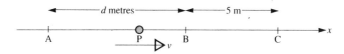

The velocity, $v\,\mathrm{m\,s}^{-1}$, is given by $v = \dfrac{4}{2x+1}$ when the particle is x m from A.

Given that the particle travels from B to C, a distance of 5 m, in 12 seconds, find the distance from A to B.

Using $\quad v = \dfrac{\mathrm{d}x}{\mathrm{d}t} \quad$ gives $\quad \dfrac{\mathrm{d}x}{\mathrm{d}t} = \dfrac{4}{2x+1}$

$\therefore \qquad \displaystyle\int (2x+1)\,\mathrm{d}x = \int 4\,\mathrm{d}t \qquad \Rightarrow \qquad x^2 + x = 4t + K$

As x is measured from A, it is convenient to measure t from the time when the particle is at A, so that $x = 0$ when $t = 0$.

When the particle is at A, $\quad x = 0 \quad$ and $\quad t = 0 \qquad \Rightarrow \qquad K = 0$

$\therefore \qquad 4t = x^2 + x$

Let the distance AB be d m and the time taken to travel from A to B be T seconds.

When the particle is at B, $\quad x = d \quad$ and $\quad t = T$

$\therefore \qquad 4T = d^2 + d$ $\hfill$ [1]

When the particle is at C, $\quad x = (d+5) \quad$ and $\quad t = (T+12)$

$\therefore \qquad 4(T+12) = (d+5)^2 + (d+5)$

$\Rightarrow \qquad 4T + 48 = d^2 + 10d + 25 + d + 5$

$\Rightarrow \qquad 4T = d^2 + 11d - 18$ $\hfill$ [2]

$[2] - [1] \quad$ gives $\quad 0 = 10d - 18$

$\therefore \qquad\qquad d = 1.8$

The distance from A to B is 1.8 m.

EXERCISE 22d

In questions 1 to 6 a particle P is moving along a straight line and O is a fixed point on that line. After a time of t seconds the displacement of P from O is s metres, the velocity is $v\,\mathrm{m\,s}^{-1}$ and the acceleration in $a\,\mathrm{m\,s}^{-2}$.

1. If $\quad v = \dfrac{1}{s^2}, \quad$ find

(a) a in terms of s $\qquad$ (b) a when $s = 0.5$.

Given that v is proportional to s, show that a is also proportional to s.

Given that $v = 6$ when $s = 2$, find a when $s = 2$.

3.

4. Given that $v = \dfrac{5}{1 + 2s}$ find a in terms of s, and the value of a when $s = 2$.

Given that $v = \sqrt{s}$ and $s = 0$ when $t = 0$, find

(a) s in terms of t

(b) the times taken to travel   (i) the first $25\,\mathrm{m}$ (ii) the next $25\,\mathrm{m}$

5. (c) v when $t = 5$.

If $s = 5$ when $t = 0$, and $v = -\frac{1}{4}s$, find

(a) s in terms of t

(b) s when $t = 2$

6. (c) the value approached by s as t increases.

The table shows corresponding values of v and s.

s	0	1	3
v	12	10	6

(a) Show that all these values fit a relationship of the form $v = cs + d$ and state the values of c and d.

(b) Show that $\ln(6 - s) = A - 2t$, where A is a constant.

(c) Given that $s = 0$ when $t = 0$, express s in terms of t.

7. (d) Find the maximum distance of P from O.

A body is moving on a horizontal straight line through a liquid. It passes through a fixed point O on the line and t seconds later its displacement from that point is s metres, its velocity is $v\,\mathrm{m\,s}^{-1}$ and its acceleration is $a\,\mathrm{m\,s}^{-2}$. The motion of the body is modelled by the equation

$$v = pe^{ks}, \quad \text{where } p \text{ and } k \text{ are constants.}$$

(a) Show that this is consistent with the hypothesis that the acceleration is given by $a = kv^2$.

(b) Find t in terms of s, p and k given that, at $t = 0$, $s = 3$.

(c) Find p and k given that $v = 20$ when $s = 0$ and $v = 10$ when $s = 3$.

(d) Find the *distance* travelled when the velocity has fallen to $5\,\mathrm{m\,s}^{-1}$.

8. A particle P is moving along a straight line and O is a fixed point on that line. After t seconds the displacement of P from O is s m and P's velocity is $v\,\text{m s}^{-1}$. Throughout the motion v is inversely proportional to s and $s > 0$. When $t = 0$, $s = 4$ and P takes 7 seconds to move from the point where $s = 6$ to the point where $s = 8$. Find

(a) s in terms of t (b) v in terms of s (c) v when $t = 2$.

9. The motion of a body falling vertically through a liquid is modelled by the equation

$$v^2 = \frac{g}{k}\left(1 - e^{-2ks}\right)$$

where, at time t seconds, $v\,\text{m s}^{-1}$ is its velocity and s metres is its displacement from its initial position and k is a constant.

(a) It is observed that, after falling several metres, its velocity starts to approach a constant value which is estimated to be $7\,\text{m s}^{-1}$.
Taking $g = 9.8$, find the value of k.

(b) Find the acceleration in terms of s. .

(c) Show that the acceleration can be written in the form $a = g - 0.2v^2$.

*10. A cyclist approaches a hill at a speed of $9\,\text{m s}^{-1}$ but slows down gradually as he climbs it. After climbing for t seconds his displacement from a point O at the bottom of the hill is s metres and his speed is $v\,\text{m s}^{-1}$. He thinks that his velocity may possibly be modelled by one of the following equations, in which λ and μ are constants.

Model 1 $v = 9 - \lambda s$

Model 2 $v = 9 - \mu^2 s^2$

(a) Show that model 1 gives $t = \dfrac{1}{\lambda}\ln\left\{\dfrac{9}{9 - \lambda s}\right\}$

(b) Show that model 2 gives $t = \dfrac{1}{6\mu}\ln\left\{\dfrac{3 + \mu s}{3 - \mu s}\right\}$

(c) When he reaches the top of the hill, which is $100\,\text{m}$ long, his speed has dropped to $1\,\text{m s}^{-1}$. Find the values of λ and μ.

(d) As one check on whether either of these models is suitable he measures the time taken over the first $50\,\text{m}$ and finds that it is 7 seconds. For each model find the time predicted for the cyclist to cover the $50\,\text{m}$. Does either model give a result consistent with the measured value?

(e) He reaches the top of the hill in 27 seconds. Does this time strengthen the case for either model?

CHOOSING SUITABLE DERIVED FUNCTIONS

When attempting a solution to a problem on variable motion, it is important to represent velocity and/or acceleration in a way that leads to a differential equation that can be solved, i.e. not more than two variables appear in the equation. A summary of suitable differential equations for most situations is given below.

When a, v or s is a function of time, $f(t)$, use:

$$v = \frac{ds}{dt} \qquad a = \frac{dv}{dt}$$

$$s = \int v \, dt \qquad v = \int a \, dt$$

When a is a function of displacement, $f(s)$, use

$$a = v \frac{dv}{ds} \quad \Rightarrow \quad \int f(s) \, ds = \int v \, dv \quad (\text{giving } v \text{ as a function of } s)$$

When v is a function of displacement, $f(s)$, use

$$a = v \frac{dv}{ds} = f(s) \times \frac{d}{ds} f(s)$$

$$v = \frac{ds}{dt} = f(s) \quad \Rightarrow \quad \int \frac{1}{f(s)} \, ds = \int dt$$

Remember that these relationships apply when a, v and s (or r) are given in cartesian vector form.

An opportunity to practise making the best choice is given in the following exercise of mixed questions.

EXERCISE 22e

In questions 1 to 5 a particle P is moving along a straight line and O is a fixed point on that line. After a time of t seconds the displacement of P from O is s metres, the velocity is $v \, \text{m s}^{-1}$ and the acceleration is $a \, \text{m s}^{-2}$.

1. The motion satisfies the equation $v = \dfrac{k}{s^2}$. Further, $s = 2$ when $t = 1$ and $s = 3$ when $t = 10.5$.

   (a) Find s in terms of t.   (b) Find t when $s = 4$.

2. If $v = 8 - 6e^{-2t}$ and $s = 0$ when $t = 0$, find

    (a) s in terms of t (b) s when $v = 7.5$

3. The motion is described by the equation $v = -12 \sin 3t$

    (a) Find a in terms of t.

    (b) Find a when $t = 0.5$.

    (c) Find s in terms of t given that $s = 4$ when $t = 0$.

    (d) Show that $a = -n^2 s$ where n is a constant and state the value of n.

4. (a) If $a = 10t - 40$, for $0 \leqslant t \leqslant 10$, and $v = 0$ when $t = 0$,

        (i) find v in terms of t

        (ii) find t when the velocity is zero

        (iii) find the greatest speed in the positive direction and the greatest speed in the negative direction and the times at which they occur

        (iv) sketch a graph of v against t for $t \geqslant 0$.

    (b) If, instead, $a = 40 - 10t$ for $0 \leqslant t \leqslant 10$, sketch a graph of v against t for $t \geqslant 0$.

5. Given that $a = 8 - 2t^2$ and, when $t = 0$, $v = 0$ and $s = 0$, find

    (a) the greatest speed in the positive direction

    (b) the distance covered by the particle in the first two seconds of its motion.

In questions 6 to 8, a particle P is moving in the xy plane. O is the origin and, at any time t, the displacement of P from O is $\mathbf{r}$, the velocity vector is $\mathbf{v}$ and the acceleration vector is $\mathbf{a}$. All units are consistent and based on metres and seconds.

6. Given that $\mathbf{v} = 3t^2 \mathbf{i} - 4t \mathbf{j}$

    (a) find $\mathbf{a}$ by differentiating $\mathbf{v}$ with respect to time

    (b) show that $\mathbf{r} = t^3 \mathbf{i} - 2t^2 \mathbf{j} + \mathbf{R}$, where $\mathbf{R}$ is a constant vector.

    If also $\mathbf{r} = 8\mathbf{i}$ when $t = 2$, find $\mathbf{R}$.

7. When $\mathbf{r} = e^{2t} \mathbf{i} + t^2 \mathbf{j}$, find

    (a) $\mathbf{v}$ in terms of t

    (b) $\mathbf{a}$ in terms of t

    (c) the initial velocity and acceleration

    (d) the time at which $\mathbf{a} = 20\mathbf{i} + 2\mathbf{j}$ and the velocity at this time.

8. If $\mathbf{v} = 10 \cos 2t\,\mathbf{i} - 10 \sin 2t\,\mathbf{j}$ and $\mathbf{r} = 5\mathbf{j}$ when $t = 0$

(a) find $\mathbf{r}$ in terms of t

(b) find $\mathbf{a}$ in terms of t

(c) find $\mathbf{r}$ when $t = \frac{\pi}{4}$

(d) show that $|\mathbf{r}|$ is constant and state its value

(e) show that $|\mathbf{v}|$ is constant and state its value

(f) show that $\mathbf{a} = \lambda\mathbf{r}$ and state the value of λ

(g) find the cartesian equation of the path of P.

9. After a car has travelled from rest along a straight level road, its acceleration, $a\,\mathrm{m\,s^{-1}}$, is given by $a = \dfrac{200 - s}{60}$ where s is the distance travelled in metres.

(a) Find the speed of the car after it has travelled 25 metres.

(b) Find the maximum speed it achieves and the distance it has then travelled.

(c) How far from the starting point is it when it comes to rest again?

10. At any time t seconds, the position vector of a particle P, relative to a fixed origin O, is $\mathbf{r}$ metres where

$$\mathbf{r} = 3t^2\mathbf{i} - 4t^{3/2}\mathbf{j}, \qquad t \geqslant 0$$

When $t = 4$ find

(a) the speed of P

(b) the acceleration vector of P.

*11. At a time t seconds the velocity of a particle moving on the positive x-axis is given by $v = 8 - 3e^{-2t}$

(a) By expressing the acceleration in two different ways, show that

$$\int \frac{v}{8 - v}\,\mathrm{d}v = \int 2\,\mathrm{d}s$$

(b) Hence deduce that $s = K - v - 8 \ln(8 - v)$ where K is a constant.

(c) Find the distance travelled by the particle as the speed increases from zero to $7\,\mathrm{m\,s^{-1}}$.

*12. An automatic conveyor is designed for a factory. It moves forwards and backwards along a straight track between two points A and B which are 81 m apart. It comes to instantaneous rest at A and B, and the journey between these points takes 6 seconds. A mathematical model for the twelve-second return journey, from A to B and back to A, is to be considered.

Take an origin at A. Let the displacement t seconds after leaving A be s metres and the velocity be v metres per second.

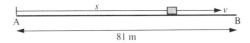

It is thought that the motion can be modelled by an equation of the form

$$v = kt(t-6)(t-12), \quad \text{for} \quad 0 \leqslant t \leqslant 12, \quad \text{where } k \text{ is a constant.}$$

(a) Verify that this is consistent with the given information about times and velocities.

(b) Find s in terms of k and t and use the given information about the distance AB to find the value of k.

(c) The model will not be suitable unless it gives $s = 0$ when $t = 12$. Is it suitable in this respect?

(d) Find the maximum speeds predicted by the model on the outward and return journeys and the times at which these occur.

CHAPTER 23

SIMPLE HARMONIC MOTION

One particular type of variable acceleration is important in its own right because it occurs frequently in everyday life. This is motion in a straight line in which the acceleration is proportional to the distance from a fixed point on the line, and is always directed towards that point; it is called *simple harmonic motion, SHM*. A weight attached to the end of a spring, for example, moves in this way; this, and other real-life cases of SHM, will be covered in the next chapter.

Properties of Simple Harmonic Motion

Consider a particle P moving in a straight line with an acceleration that is directed towards O, a fixed point on the line, the magnitude of the acceleration being proportional to the distance OP.

When the displacement of P from O is x, the acceleration is in the negative direction, i.e. $\ddot{x}$ is negative and we can write $\ddot{x} = -n^2x$. (We use n^2 for the constant of proportion because n^2 cannot be negative.)

Similarly if P is at a point where x is negative, $\ddot{x}$ is in the positive direction so again $\ddot{x} = -n^2x$.

The equation $\ddot{x} = -n^2x$ is the basic equation of SHM.
Any motion that satisfies this equation is known to be simple harmonic.

Further, if the velocity, $\dot{x}$, is zero at a point A, distant a from O, the following diagram shows the basic information about SHM.

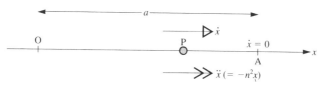

As can be seen from the description of the motion, the symbol a is used for a distance in this topic. This means that we cannot use a to indicate the acceleration so $\ddot{x}$ is usually used instead.

The acceleration is a function of x, so we will use $v \dfrac{dv}{dx}$ and because it is directed in the negative sense we have

$$\ddot{x} = -n^2 x \qquad \Rightarrow \qquad v\frac{dv}{dx} = -n^2 x \tag{1}$$

Hence

$$\int v\ dv = -n^2 \int x\ dx$$

$$\Rightarrow \qquad \tfrac{1}{2}v^2 = -\tfrac{1}{2}n^2 x^2 + K$$

$v = 0$ when $x = a,$ $\qquad \Rightarrow \qquad K = \tfrac{1}{2}n^2 a^2$

$$\therefore \qquad\qquad v^2 = n^2(a^2 - x^2) \tag{2}$$

From [2] we can see that the greatest value of v is given when $x = 0$

i.e. $\qquad\qquad\qquad\qquad v_{max} = na$

Equation [2] also shows that $v = 0$ when $x = \pm a$, so the particle oscillates between two points A and A', on opposite sides of O and each distant a from O.

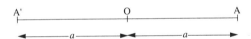

The point O is called the centre, or mean position, of the SHM and the distance OA is called the *amplitude* of the motion.

Examples 23a

1. A particle is describing SHM with amplitude $2\,m$. If its speed is $3\,m\,s^{-1}$ when it is $1\,m$ from the centre of the path find

   (a) the basic equation of the SHM being described,

   (b) the maximum acceleration,

   (c) the speed when the particle is $1.5\,m$ from the centre of the path.

   (a) We know that $a = 2$ and $v = 3$ when $x = 1$.

   Using $\qquad v^2 = n^2(a^2 - x^2)\quad$ gives

   $$9 = n^2(4-1) \qquad \Rightarrow \qquad n^2 = 3$$

   $\therefore\quad$ the basic equation of the SHM is $\ddot{x} = -3x$

(b)   From $\ddot{x} = -3x$ we see that $\ddot{x}$ is greatest when x is greatest.

The greatest value of x is 2

∴   the maximum acceleration is $6\,\mathrm{m\,s}^{-2}$ towards O.

(c)   When $x = 1.5$, $v^2 = 3(2^2 - 1.5^2) = 5.25$

∴   the speed when $x = 1.5$ is $2.29\,\mathrm{m\,s}^{-1}$ (3 sf)

ASSOCIATED CIRCULAR MOTION

There is a very interesting link between circular motion and SHM which we are now going to look at.

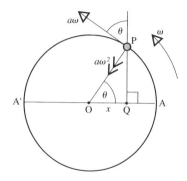

The diagram shows a particle P travelling at a constant angular velocity ω, round a circle with centre O and radius a. Q is the projection of P on the diameter AA′ (i.e. Q is the foot of the perpendicular from P to AA′).

We know from the work on circular motion that the acceleration of P is $a\omega^2$ towards the centre.

Now the component in the direction AA′ of this acceleration gives the acceleration of Q.

Therefore the acceleration of Q is $a\omega^2 \cos\theta$ in the direction $\overrightarrow{QO}$

From triangle OPQ, $\cos \theta = \dfrac{x}{a}$

Therefore the acceleration of Q towards O is $a\omega^2 \left(\dfrac{x}{a}\right)$,   i.e. $\omega^2 x$.

But ω^2 is constant so we see that the acceleration of Q is proportional to the distance of Q from O, and is always towards O.

Therefore Q describes SHM about O as centre and with amplitude a.

As a point P travels round a circle at constant angular speed ω, its projection on a diameter of the circle describes SHM with equation $\ddot{x} = -\omega^2 x$

Comparing this equation of SHM with the standard equation $\ddot{x} = -n^2 x$ we see that ω is equivalent to the constant n, used earlier.

Further properties of SHM can now be discovered by considering the associated circular motion.

● We can find period of the oscillations.

The time, T, taken to describe one complete oscillation (from A to A′ and back to A) is called the *periodic time*, or simply *the period*.

As Q describes a complete oscillation, P performs one complete revolution of the circle.

The angular velocity of P is ω radians per second.

In one revolution P turns through 2π radians.

So the time taken to describe the revolution is $\dfrac{2\pi}{\omega}$

Therefore T, the period or periodic time of the SHM, is given by $T = \dfrac{2\pi}{\omega}$

But as $\omega = n$,   the period is also given by

$$T = \frac{2\pi}{n}$$

Note that the period is independent of the amplitude of the motion.

● We can find an expression for x in terms of time.

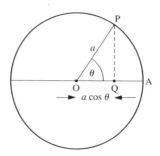

If we measure time from when P is at A, i.e. $t = 0$ when $x = a$, then, t seconds after leaving A, P will have turned through ωt radians, i.e. $\theta = \omega t$.

By this time Q has reached the point where $x = a \cos \theta$.

Therefore, t seconds after leaving the end of the path, P is in a position where $x = a \cos \omega t$. But $\omega = n$ therefore

$$x = a \cos nt$$

Remember that, in this formula, x is measured from O but t is measured from A.

● The expression $x = a \cos nt$ can be used to check the expression derived for the velocity of P on p. 505.

Differentiating x with respect to time gives $\dot{x} = -an \sin nt$

In triangle OPQ, $PQ^2 = a^2 - x^2$

therefore $\sin \omega t = \dfrac{\pm\sqrt{(a^2 - x^2)}}{a}$ $\quad \Rightarrow \quad$ $\dot{x} = v = \pm n\sqrt{(a^2 - x^2)}$

$\therefore$ $\qquad\qquad v^2 = n^2(a^2 - x^2)$

Note that v can be either positive or negative. This is because P passes through any particular point in both the positive and the negative direction.

● A formula can be recognised as SHM when time is not measured from the end of the path.

Suppose that a particle is moving on a straight line such that its distance from a point O on the line is given by $x = a \cos(\omega t + \alpha)$ where α is a constant.

First we note that $x_{max} = a$ and $x = a \cos \alpha$ when $t = 0$.
Hence, on the auxiliary circle, P is at P_0 when $t = 0$, where $\angle P_0OA = \alpha$.

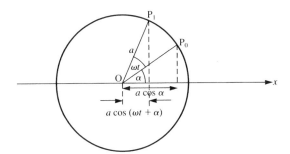

At any subsequent time t, $x = a \cos(\omega t + \alpha)$ therefore $\angle P_0 O P_1 = \omega t$.

Differentiating gives $\dot{x} = -a\omega \sin(\omega t + \alpha)$

and $\ddot{x} = -a\omega^2 \cos(\omega t + \alpha)$

showing that $\ddot{x} = -\omega^2 x$

This confirms that a particle, moving so that $x = a \cos(\omega t + \alpha)$, is describing SHM with amplitude a and period $\frac{2\pi}{\omega}$ but that the motion is not timed from the end of the path.

The basic properties of SHM, but not necessarily their derivations, should be known and can be quoted.

Summary

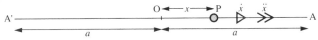

O is the centre, or mean position and a is the amplitude.

An oscillation is the journey from A to A′ and back to A.

The period of an oscillation is T where $T = \frac{2\pi}{n}$

When the particle P is distant x from O

$\ddot{x} = -n^2 x$ where n is a constant

$x = a \cos nt$ if $x = a$ when $t = 0$

$v^2 = n^2(a^2 - x^2)$

$x = a \cos(\omega t + \alpha)$ if $x = a \cos \alpha$ when $t = 0$

Maximum acceleration, at A and A′, is $n^2 a$ and maximum speed, at O, is na.

Dealing with SHM by using the relationship to motion in a circle with constant angular velocity, however, should not be underestimated. This approach often provides a simple solution to a problem.

Examples 23a (continued)

2. A particle P is describing SHM of amplitude 2.5 m. When P is 2 m from the centre of the path, the speed is $3\,\text{m s}^{-1}$. Find

   (a) the periodic time of the oscillations
   (b) the greatest speed
   (c) the magnitude of the greatest acceleration.

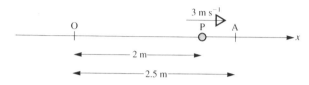

First we will find the value of n as this occurs in every formula for SHM.

Using $\quad v^2 = n^2(a^2 - x^2)$ with $a = 2.5$ and $v = 3$ when $x = 2$

gives $\quad 3^2 = n^2(2.5^2 - 2^2) \quad \Rightarrow \quad n^2 = \dfrac{9}{2.25} \quad \Rightarrow \quad n = 2$

(a) $\quad T = \dfrac{2\pi}{n} \quad \Rightarrow \quad T = \pi$

   The periodic time is π seconds.

(b) The speed, v, is greatest when $x = 0$, i.e. $v_{\text{max}} = na$

   The greatest speed is $5\,\text{m s}^{-1}$.

(c) $\ddot{x} = -n^2 x$ so the acceleration is greatest when $x = a$

   The greatest acceleration is $2^2 \times 2.5$, i.e. $10\,\text{m s}^{-2}$.

3. A piston is performing SHM at the rate of 4 oscillations per minute. The maximum speed of the piston is $0.3\,\text{m s}^{-1}$. By modelling the piston as a particle, find the amplitude of the motion. Find also the speed and the acceleration of the piston when it is 0.5 m from the centre of the path.

   Give answers corrected to 2 significant figures.

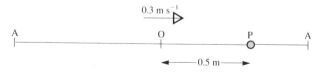

There are 4 oscillations per minute so 1 oscillation takes 15 seconds

$\therefore \quad T = 15 \quad \Rightarrow \quad \dfrac{2\pi}{n} = 15 \quad \Rightarrow \quad n = \dfrac{2\pi}{15} = 0.4188\ldots$

In this type of solution where the value of n is often used in subsequent calculations, you can, as an alternative to storing it in your calculator or retaining 4 significant figures, use $\frac{2\pi}{15}$ whenever n occurs. This has the added benefit of avoiding the introduction of rounding errors into subsequent calculations.

The maximum speed is given by na where a is the amplitude.

Therefore $\quad 0.3 = \left(\dfrac{2\pi}{15}\right)a \quad \Rightarrow \quad a = \dfrac{2.25}{\pi} = 0.7161\ldots$

The amplitude is $0.72\,\text{m}$ (2 sf).

The speed $v\,\text{m s}^{-1}$ is given by $\quad v = n\sqrt{(a^2 - x^2)}$

When $\quad x = 0.5, \quad v = \dfrac{2\pi}{15}\sqrt{\left(\left[\dfrac{2.25}{\pi}\right]^2 - 0.5^2\right)} = 0.214\ldots$

and the acceleration is $\quad n^2 x$, i.e. $\quad \left(\dfrac{2\pi}{15}\right)^2 \times 0.5$

The acceleration is $0.088\,\text{m s}^{-2}$ (2 sf) and the speed is $0.21\,\text{m s}^{-1}$ (2 sf).

4. **A, B and C, in that order, are three points on a straight line and a particle P is moving on that line with SHM. The velocities of P at A, B and C are zero, $2\,\text{m s}^{-1}$ and $-1\,\text{m s}^{-1}$ respectively. If AB $= 1\,\text{m}$ and AC $= 4\,\text{m}$, find the amplitude, a metres, of the motion and the periodic time.**

The velocity at A is zero so A must be at one end of the path (we will take it as the left-hand end). The signs of the velocities at B and C show that P is moving away from A when at B and back towards A when at C. *Speed* at B > *speed* at C, so B is nearer to the centre than C is. The length of the path is greater than 4 cm, so the amplitude is greater than 2 cm. From all these facts we see that B and C are on opposite sides of the centre O.

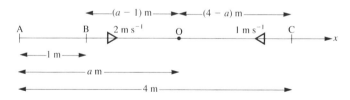

$v = 2 \quad$ when $\quad x = -(a - 1)$

$v = -1 \quad$ when $\quad x = 4 - a$

Using $\quad v^2 = n^2(a^2 - x^2) \quad$ gives

$\qquad 4 = n^2[a^2 - (1-a)^2] \quad \Rightarrow \quad 4 = n^2[2a - 1] \qquad$ [1]

and $\qquad 1 = n^2[a^2 - (4-a)^2] \quad \Rightarrow \quad 1 = n^2[8a - 16] \qquad$ [2]

[1] $- 4 \times$ [2] gives $\quad n^2[2a - 1] - 4n^2[8a - 16] = 0 \quad \Rightarrow \quad 30a = 63$

The amplitude is $2.1\,\text{m}$.

From [1], $4 = n^2(4.2 - 1)$ $\Rightarrow$ $n = 1.118\ldots$

The period, T, is given by $\dfrac{2\pi}{n}$

$\therefore$ $T = \dfrac{2\pi}{1.118\ldots} = 5.619\ldots$

The periodic time is 5.62 seconds   (3 sf).

5. **A particle P is moving with SHM of period 3π seconds, on a path with centre O and amplitude 6 m. The particle starts from A, one end of the path.**

   (a) **Find the time taken for P to travel from A to B, a distance of 3 m.**

   (b) **Find the distance AC if P travels from A to C in $\frac{2}{3}\pi$ seconds.**

   (a) 1st Method – using associated circular motion.

SHM from A to B corresponds to
circular motion from A to B$'$.

When P is at B, $\cos\theta = \frac{3}{6} = \frac{1}{2}$

$\Rightarrow$ $\theta = \frac{1}{3}\pi$

which is $\frac{1}{6}$ of a revolution.

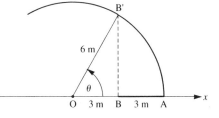

The periodic time is equal to the time it takes B$'$ to complete 1 revolution.

$\therefore$ the time taken to travel arc AB$'$ is $\frac{1}{6}$ of the periodic time,

i.e. $\frac{1}{6} \times 3\pi$ seconds

The time it takes P to travel along AB with SHM is $\frac{1}{2}\pi$ seconds.

2nd Method – using the formula $x = a \cos nt$

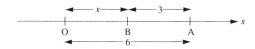

$x = $ OB $= 6 - 3 = 3$

$T = \dfrac{2\pi}{n}$ $\Rightarrow$ $n = \dfrac{2\pi}{3\pi}$ $\Rightarrow$ $n = \dfrac{2}{3}$

$x = a\cos nt$ gives $3 = 6\cos\dfrac{2t}{3}$

$\therefore$ $\cos\dfrac{2t}{3} = \dfrac{1}{2}$ $\Rightarrow$ $\dfrac{2t}{3} = \dfrac{\pi}{3}$ $\Rightarrow$ $t = \dfrac{\pi}{2}$

$\therefore$ time taken to travel AB with SHM is $\frac{1}{2}\pi$ seconds.

(b) 1st Method

Time to travel round the circumference $= 3\pi$ s.

Time taken to travel round arc AC$'$ $= \frac{2}{3}\pi$ s

$= \frac{2}{9}$ of 3π s

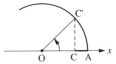

$\therefore$ arc AC$'$ $= \frac{2}{9}$ of the circumference

$\therefore$ A$\hat{O}$C$'$ $= \frac{2}{9}$ of 2π rad $= 4\pi/9$ rad

OC $= 6 \cos$ AOC$'$ $= 1.0418\ldots$ $\Rightarrow$ AC $= 6 - $ OC $= 4.958\ldots$

$\therefore$ the distance AC is 4.96 m (3 sf).

2nd Method

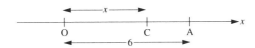

As found in part (a), $n = \frac{2}{3}$

Using $x = a \cos nt$ when $n = \frac{2}{3}$ and $t = \frac{2}{3}\pi$

gives $x = 6 \cos \left(\frac{2}{3}\right)\left(\frac{2}{3}\pi\right) = 6 \cos \frac{4}{9}\pi = 1.0418\ldots$

But x m is the distance OC

$\therefore$ the distance AC is $(6 - x)$ m, i.e. 4.96 m (3 sf).

EXERCISE 23a

In questions 1 to 8 a particle P is performing simple harmonic motion in a straight line as shown.
All units are consistent and based on metres and seconds.

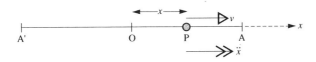

1. When $\ddot{x} = -9x$ and the amplitude is 5 m, find

(a) the period   (b) the maximum speed   (c) the speed when $x = -2$.

2. If the maximum acceleration is 10 m s^{-2} and the maximum speed is 8 m s^{-1}, find

(a) the period   (b) the amplitude   (c) the speed when $x = 4$

3. Given that $v = 4.8$ when $x = 0.7$ and $v = 3$ when $x = 2$,   find

   (a)   the amplitude

   (b)   the period

   (c)   the maximum speed

   (d)   the maximum acceleration.

4. If the period is 5π seconds and the maximum speed is $2\,\mathrm{m\,s^{-1}}$, find

   (a)   the amplitude

   (b)   the speed and the acceleration when $x = -1$.

5. Given that the frequency of oscillation is 0.25 oscillations per second, and that $v = 0.5$ when $x = 0.2$ find

   (a)   the period

   (b)   the amplitude

   (c)   the speed and the acceleration when $x = 0.15$.

6. The motion is defined by $x = 4 \cos 3t$.

   (a)   Find   (i) $\dot{x}$ (ii) $\ddot{x}$.

   (b)   Use your result from (a) to express $\ddot{x}$ in terms of x.

   (c)   What is the period of the motion?

7. Given that $\ddot{x} = -\left(\dfrac{\pi^2}{16}\right) x$ and the amplitude is $0.5\,\mathrm{m}$,

   (a)   write down an expression in the form $x = a \cos nt$ for this motion,

   (b)   by differentiation, find v as a function of t,

   (c)   complete this table and hence sketch the graph of v against x for the range $1 \geqslant x \geqslant -1$.

t	0	1	2	3	4	5	6	7	8
x									
v									

8. When the period of the motion is 8 seconds, use $x = a \cos nt$ to find the time taken to go from the point where $x = a$ to the point where $x = \frac{1}{2}a$.

9. A particle performs two SHM oscillations each second. Its speed when it is 0.02 m from its mean position is half the maximum speed. Find the amplitude of the motion, the maximum acceleration and the speed at a distance 0.01 m from the mean position.

10. A particle P is moving on a straight line; O is a fixed point on the line and x is the displacement of P from O at a time t. When $t = 0$, $x = \sqrt{3}$ m and after $\frac{1}{6}\pi$ seconds, $x = 1$ m.

(a)   The equation of the motion of P can be expressed in the form

$$x = a \cos(t + \alpha)$$

Show that P travels with simple harmonic motion.

(b)   State the amplitude and period of the SHM and illustrate the motion on the auxiliary circle.

The solutions to questions 11 to 13 may be based either on the standard formulae for SHM or on the use of the associated circular motion.

11. A particle is travelling between two points P and Q with simple harmonic motion. If the distance PQ is 6 m and the maximum acceleration of the particle is $16 \, \mathrm{m \, s^{-2}}$, find the time taken to travel

(a)   a distance 1.5 m from P

(b)   from P to the midpoint O of PQ

(c)   from the midpoint of PO to the midpoint of OQ.

12. A particle describes simple harmonic motion between two points A and B. The period of one oscillation is 12 seconds. The particle starts from A and after 2 seconds has reached a point distant 0.5 m from A. Find

(a)   the amplitude of the motion

(b)   the maximum acceleration

(c)   the velocity 4 seconds after leaving A.

13. A particle is performing simple harmonic motion of amplitude 0.8 m about a fixed point O. A and B are two points on the path of the particle such that OA = 0.6 m and OB = 0.4 m. If the particle takes 2 seconds to travel from A to B find, correct to one decimal place, the periodic time of the SHM if

(a)   A and B are on the same side of O

(b)   A and B are on opposite sides of O.

14. The prongs of a tuning fork, which sounds middle C, are vibrating at a rate of 256 oscillations per second. Assuming that the motion of the prongs is simple harmonic and that the amplitude of the end of a prong is 0.1 mm, find

(a)   the maximum velocity and the maximum acceleration of the end of a prong

(b)   the velocity and acceleration of the end of a prong when its displacement from the centre of its path is 0.05 mm.

Mixed Problems

The questions in this exercise involve a variety of methods and ideas.

EXERCISE 23b

1. A cylindrical buoy, of length l metres, is held at
    rest, with its lower plane face touching the surface
    of the water and is then released. After t seconds
    its lower plane face is at a depth x metres below
    the surface, its velocity is $v\,\mathrm{m\,s}^{-1}$ and its
    acceleration is $a\,\mathrm{m\,s}^{-2}$. Because the upward force
    exerted by the water is known to be proportional
    to x, the motion may be modelled by the equation

    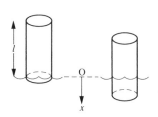

    $$a = g\left(1 - \frac{k}{l}x\right), \quad \text{where, for this particular buoy,} \quad k = 3.$$

    (a) Find v^2 in terms of x.

    (b) Find the value of x when $v = 0$.

    (c) This method of placing the buoy in the water is unsatisfactory as it springs
         back up. If it is to be placed gently at a lower depth where it will remain in
         equilibrium, what is the value of x in this position?

2. A body is fired vertically into space from the surface of the earth.
    After t seconds its displacement from the centre of the earth is s metres and its
    velocity is v metres per second. Its acceleration $a\,\mathrm{m\,s}^{-2}$ is given by

    $$a = -\frac{k}{s^2} \quad \text{where } k \text{ is a constant.}$$

    Initially $s = r$ and $v = u$.

    (a) Show that

    $$\frac{1}{2}(v^2 - u^2) = k\left(\frac{1}{s} - \frac{1}{r}\right)$$

    (b) Given that $k = 4 \times 10^{14}$ and $r = 6.4 \times 10^6$, find the change in kinetic
         energy of a satellite of mass $2000\,\mathrm{kg}$ in going from the surface of the earth
         to a point where $s = 6.8 \times 10^6$.

3. It is thought that the motion of a man running a $100\,\mathrm{m}$ race may be modelled by
    the equation $a = \lambda - \mu t$ where $a\,\mathrm{m\,s}^{-2}$ is his acceleration after t seconds
    from the start and λ and μ are constants.

    (a) Find expressions for his speed and the distance covered at time t.

    (b) Suggest a practical way in which values for all the unknown constants
         (λ, μ and any others) may be found.

    (c) Suggest a way in which the model could then be assessed for validity.

4. A student is considering a model for the motion of a train on a light railway between two stopping points A and B. He has the following data.

1. The journey time from A to B is 120 seconds.
2. The train accelerates from rest for 60 seconds to a speed of $30\,\text{m s}^{-1}$ and then decelerates for 60 seconds, coming to rest at B.
3. The distance between A and B is 1900 metres.

On the basis of this data he constructs this model.

First model   The acceleration is constant and the deceleration is also constant.

Another piece of data is then obtained,

4.   30 seconds after leaving A the train has a speed of $17\,\text{m s}^{-1}$ and has travelled 200 metres.

As a result the student constructs a different model.

Second model   The displacement of the train from A after t seconds is s metres and the motion is modelled by the equation

$$k \frac{ds}{dt} = t^2 (120 - t)^2, \quad \text{where } k \text{ is a constant.}$$

(a)   Verify that the second model gives zero acceleration when $t = 60$.
(b)   By considering the maximum velocity ($30\,\text{m s}^{-1}$), find the value of k.
(c)   Find the values given for the distance AB
        (i)   by the first model
        (ii)   by the second model.
(d)   Find the values given for the velocity after $30\,\text{s}$
        (i)   by the first model
        (ii)   by the second model.
(e)   Find the values given for the distance travelled after $30\,\text{s}$
        (i)   by the first model
        (ii)   by the second model.
(f)   Do you think that either of these models fits the data reasonably well or would you look for a better model?

5. On a particular day in a harbour, high tide at its entrance occurs at noon and the water depth is then 11 m. Low tide occurs $6\frac{1}{4}$ hours later and the water depth is then 5 m. The motion of the water level is modelled as simple harmonic motion.

(a)   Find the amplitude and period of this motion.
(b)   Find the time when the water level will be falling at its maximum rate and find this rate in metres per minute.
(c)   A ship needs a depth of 7 m to enter the harbour. Find the latest time after noon at which it can enter without having to wait for low tide to pass.
(d)   If it arrives after this time, what is the next time when it can enter harbour?

6. An astronomer observes a small body which moves towards the planet Jupiter and eventually passes in front of it. This body appears to be moving in simple harmonic motion along the straight line AB. He thinks that it is a moon orbiting the planet and that he is viewing it along the plane of the orbit. He knows that the radius of the planet is 71 600 km.

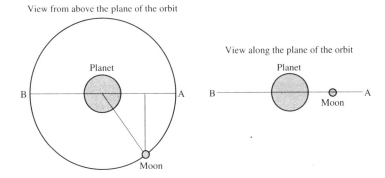

Waiting until it reaches a position in front of point C, which is the centre of the planet, he then records the time it takes to travel from there to two other positions. These are

(i) in front of point D, which is at the edge of the planet,

(ii) at B, just before it appears to return along the line BA.

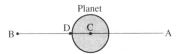

His results are in the table, where t is the time in minutes from when the moon appeared in front of C.

Position	C	D	B
t	0	69	637

(a) Find the period in minutes.

(b) Find the angular velocity of the moon about the planet in radians per minute.

(c) Find the radius of the orbit.
    (This is also the amplitude of the apparent oscillation.)

CHAPTER 24

VARIABLE FORCES

THE RELATIONSHIP BETWEEN FORCE AND ACCELERATION

Newton's Law of Motion applies to *any* motion, however it is caused. It applies whether the force producing the motion is constant or variable, i.e.
for a body of mass m, moving under the action of *any* force F, then $F = ma$.

When F varies in a specified way, it may be possible to represent the motion by a differential equation.

If F is a function of time, i.e. $F = f(t)$, then using $a = \dfrac{dv}{dt}$ gives

$$f(t) = m \frac{dv}{dt}$$

$\Rightarrow$
$$\int f(t) \; dt = \int m \; dv$$

If F is a function of displacement, i.e. $F = f(s)$, then using $a = v\dfrac{dv}{ds}$ gives

$$f(s) = mv \frac{dv}{ds}$$

$\Rightarrow$
$$\int f(s) \; ds = \int mv \; dv$$

The Work Done by a Variable Force

Suppose that a particle moving under the action of a force F, where $F = f(s)$, has a velocity u when $s = 0$ and a velocity v after covering a distance s.

The relationship $F = mv\dfrac{dv}{ds}$ becomes $\displaystyle\int_0^s F \; ds = \int_u^v mv \; dv$

giving
$$\int_0^s F \; ds = \tfrac{1}{2}mv^2 - \tfrac{1}{2}mu^2$$

Now $\frac{1}{2}mv^2 - \frac{1}{2}mu^2$ is the increase in KE of the particle and this is equal to the amount of work done in causing it.

Therefore $\int F\,\mathrm{d}s$ represents the work done by the force F, i.e.

when a variable force F, where $F = f(s)$, moves its point of application through a distance s, the work done by the force is given by $\int f(s)\ \mathrm{d}s$

Note that, when F is constant, this result gives Work done $= Fs$ which was derived in Chapter 7.

The Impulse Exerted by a Variable Force

Again we consider a force F that causes the velocity of a particle to increase from u to v, but this time the increase takes place over an interval of time t.

Using the equation $F = m\dfrac{\mathrm{d}v}{\mathrm{d}t}$ gives $\displaystyle\int_0^t F\ \mathrm{d}t = \int_u^v m\ \mathrm{d}v$

giving $$\int_0^t F\ \mathrm{d}t = mv - mu$$

But $mv - mu$ is the increase in momentum of the particle over the time t and we know that this is equal to the impulse of the force producing it.
Therefore $\int F\,\mathrm{d}t$ represents the impulse of the force F, i.e.

when a variable force F, where $F = f(t)$, acts on an object for a time t the impulse exerted by the force is given by $\int f(t)\ \mathrm{d}t$

When F is constant, this gives Impulse $= Ft$, as used in Chapter 13.

Examples 24a

1. **A particle P is moving in a straight line under the action of a variable force F. The particle passes through a point O on the line and t seconds later its displacement s from O is given by $s = r\sin 4t$. Show that the force is proportional to the displacement of P from O and describe the behaviour of the force.**

$$s = r \sin 4t$$

$\therefore \qquad\qquad v = \dfrac{ds}{dt} = 4r \cos 4t \quad \text{and} \quad a = \dfrac{dv}{dt} = -16r \sin 4t$

Newton's Law gives $\quad F = m(-16r \sin 4t) \qquad \Rightarrow \qquad F = (-16m)\,s$

As $16m$ is constant, $\quad F$ is proportional to s

The relationship between F and s can be expressed as $\quad F = 16m(-s)$. In this form, as $16m$ is positive, we see that F and s are of opposite sign.

Therefore as P moves away from O in either direction, the force F acts towards O and is proportional to the distance of P from O.

2. **A particle P, of mass 1 kg, is moving horizontally along the x-axis and passes through the origin O with speed $2\,\mathrm{m\,s}^{-1}$. A force F acts on P in the positive direction and F is equal to $2x$. When P has moved through 3 m, find**

    (a)   the work done by F (b)   the speed of P.

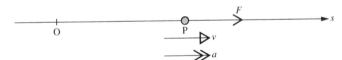

(a)   The work done by F is given by $\int F\ dx$, where $F = 2x$.

$\qquad \therefore \quad$ when $x = 3,\qquad$ work done $= \displaystyle\int_{0}^{3} 2x\ dx$

$$= \left[x^2 \right]_{0}^{3} = 9$$

The work done by F is 9 J

(b)   P is moving horizontally so there is no change in its PE

Therefore using   Work done $=$ Increase in ME   gives

$\qquad 9 = \tfrac{1}{2}mv^2 - \tfrac{1}{2}mu^2$

$\qquad\quad = \tfrac{1}{2}v^2 - \tfrac{1}{2}(4)$

$\Rightarrow \qquad v^2 = 22$

The speed of P. is $4.69\,\mathrm{m\,s}^{-1}$ (3 sf).

3. A body of mass 1.4 kg falls from rest in a medium which exerts a resistance of $(2t + k)$ N, where k is a constant. The speed of the body after falling for 4 seconds is 18 m s^{-1}. Find

(a) the value of k (b) the speed after a further 3 seconds.

The resultant force F acting downwards on the body is $mg - R$

i.e. $F = 1.4 \times 9.8 - (2t + k) = 13.72 - 2t - k$

(a) When $t = 4$, $v = 18$
     so using impulse = increase in momentum   gives

$$\int_0^4 (13.72 - 2t - k)\, dt = 1.4(18 - 0)$$

$\therefore$ $\left[13.72t - t^2 - kt \right]_0^4 = 25.2$

$\Rightarrow$ $4k = 13.68$

$\therefore$ $k = 3.42$

(b) After a further 3 seconds, $t = 7$ and the velocity is v m s^{-1}.

Using impulse = increase in momentum   gives

$\therefore$ $\left[13.72t - t^2 - 3.42t \right]_0^7 = 1.4v$ $\Rightarrow$ $v = 16.5$

The speed after 7 seconds is 16.5 m s^{-1}

EXERCISE 24a

In questions 1 to 5 a particle P, of mass m, is moving along a straight line and O is a fixed point on that line. At a time t the force acting on P is F, the acceleration of P is a, the velocity is v and the displacement from O is s. All units are consistent.

1. If $v = 15t - e^{3t}$ and $m = 2$, find

(a) F in terms of t

(b) the value of F when $t = 0$.

2. Given that $s = \dfrac{1}{t+1}$ and $m = 4$, find

(a) v in terms of t (b) F in terms of t.

3. The force F is given by $F = 24 - 6e^{-2t}$, and $v = 5$ when $t = 0$.
If $m = 3$ find

(a) a in terms of t (b) v in terms of t.

4. Given that $m = 3$ and that $v = s + \dfrac{1}{s}$ for $s > 0$,

(a) find F in terms of s
(b) find a positive constant b such that $F = 0$ when $s = b$
(c) show that F opposes the motion when $0 < s < b$, and is in the direction of the motion when $s > b$.

5. The mass is $3\,\text{kg}$ and when $t = 2$, $v = 14$ and $s = 20$. The force F newtons produces power of $120\,\text{W}$. (Power is the rate of doing work, i.e. $F\dfrac{ds}{dt}$ or Fv.)

(a) Express F in terms of v.   (b) Find v in terms of   (i) t (ii) s.

6. A particle P, of mass $5\,\text{kg}$ is moving along a straight line and O is a fixed point on that line. Initially P is at rest at a point A on the line where $OA = 2\,\text{m}$. After t seconds the displacement of P from O is x metres and its velocity is $v\,\text{m}\,\text{s}^{-1}$. A force F acts on P, where $F = -90 \cos 3t$.

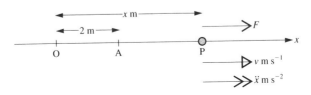

(a) Find v in terms of t.
(b) Find x in terms of t.
(c) Express the acceleration, $\ddot{x}$, in terms of x and hence identify the motion of P.

7. Find the work done by a force $F\,\text{N}$ which moves a particle from the origin O along the x-axis to the point $5\,\text{m}$ from O if

(a) F is of constant magnitude 10,   (b) $F = x^2$.

8. A wagon whose mass is $200\,\text{kg}$ is pulled by a cable along a straight level track. Contact between the wagon and the track is smooth and the tension in the cable is directly proportional to the time. The wagon starts from rest and, 10 seconds later, its speed is $20\,\text{m}\,\text{s}^{-1}$. How far has the wagon been pulled?

9. A body of mass m falls from rest against a resistive force equal to $\frac{1}{10}s$ where s is the distance fallen. Find the work done by the resistance when the body has fallen a distance d. What is the kinetic energy of the body in this position?

10. A force acts on a particle of mass m causing its speed to increase from u to $2u$ in T seconds. The force acts in a straight line in the direction of motion and is of magnitude kt, where $t = 0$ when the speed is u. Find T in terms of m, u and k.

11. A particle P of mass m passes through the origin O with speed V and moves along the positive x-axis. It is subjected to a retarding force R which causes an acceleration of magnitude kx towards O. When P reaches the point where $x = 4$ find

(a) the work done by R

(b) the speed of P, giving your answers in terms of k and V.

12. A parachutist, of total mass 75 kg, descends vertically in free fall for a period before opening his parachute. Take his initial velocity as zero and model the air resistance by the expression $0.5v^2$.

(a) Find the greatest speed he can reach in free fall.

(b) Find s in terms of v.

(c) Find the distance he has fallen when his speed is $30\,\mathrm{m\,s^{-1}}$.

FORMING, TESTING AND IMPROVING MODELS

A model is formed by making assumptions in order to make use of known mathematical formulae to solve a practical problem. The formulae can then be used to predict results for varying values of the parameters involved in the problem.

Testing a Model

The results predicted by a model are of little value unless the model is known to be reasonably valid over the range to which it is being applied. The reliability of a model can only be tested by experiment and measurement, allowing comparison between the measured results and those predicted by the model. If they are reasonably close we would be encouraged to think that the model is fairly satisfactory; further practical measurements may confirm the reliability of the model; on the other hand we might find that the correlation between predicted and observed values is not good enough.

Improving a Model

If the predicted and observed results are not close enough for the accuracy required, the next step is to identify the feature(s) in the model that may be the cause of the discrepancy and to make modifications where these seem necessary.

Whether or not the model has been improved by the adjustment can only be determined by testing the new model and comparing its predictions against observed values.

The process of testing and refining a model can be shown by a flowchart.

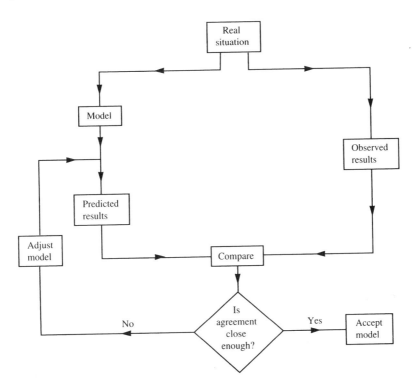

Sometimes when a mathematical model of a situation is constructed, the results it gives appear to correlate well with observed results over the range of circumstances within which it can be tested. If, at a later stage, that range is widened, the model may not be so reliable in the extended conditions and an adjustment may then be necessary.

Consider the following example.

For a project a group of students investigated the vertical motion of a ball of mass 1 kg. They borrowed a ball-projecting machine from the local cricket club, which can give the ball known initial speeds. For each speed the students measured the time taken for the ball to reach its highest point and the following table gives the values recorded, corrected to 2 significant figures.

Initial speed $V\,\mathrm{m\,s^{-1}}$	8	12	15	21	26	35
Time, Ts, to highest point	0.8	1.1	1.2	1.7	2.1	2.7

They took $10\,\mathrm{m\,s^{-2}}$ as the acceleration and their first model assumed zero air resistance.

Using $v = u + at$ they calculated the value of T given by this model for each value of V and entered these results in the table for comparison with the observed results, inserting these values of T in a third row.

Initial speed $V\,\mathrm{m\,s^{-1}}$	8	12	15	21	26	35
Time, Ts, to highest point	0.8	1.1	1.2	1.6	2.0	2.7
Time given by first model	0.8	1.2	1.5	2.1	2.6	3.5

The students were not satisfied with this correlation because, although there was good agreement between the calculated and observed values for the lower values of V, the correlation decreased as V increased, so they tried another model.

The second model assumed that air resistance is given by kv where k is a constant.

Using $F = ma$ and $a = \dfrac{dv}{dt}$ gives

$$-(g + kv) = \frac{1\,dv}{dt} \quad \Rightarrow \quad -\int \frac{1}{g + kv}\, dv = \int dt$$

$$-\left[\frac{1}{k} \ln (g + kv) \right]_V^0 = \left[t \right]_0^T$$

$$\therefore \quad T = \frac{-1}{k} [\ln g - \ln (g + kV)] = \frac{1}{k} \ln \left(\frac{g + kV}{g} \right)$$

Using the observed value of T when $V = 21$, the students found, by trial and improvement, that 0.21 is a reasonable value for k. The resulting values of T were compared with the observed results as before.

Initial speed $v\,\mathrm{m\,s^{-1}}$	8	12	15	21	26	35
Time, Ts, to highest point	0.8	1.1	1.2	1.7	2.1	2.7
Time given by second model	0.7	1.1	1.3	1.7	2.1	2.6

The students were satisfied with the agreement between the observed figures and those predicted by the second model, for the range of values of V tested.

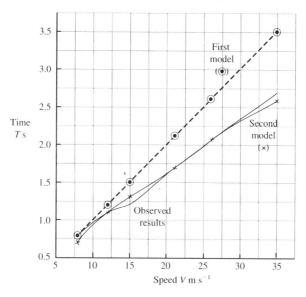

EXERCISE 24b

1. The velocity–time graph illustrates the motion of a car, of mass $800 \, \text{kg}$, whose velocity is $v \, \text{m s}^{-1}$ at time t seconds. The car starts from rest, accelerates to a speed of $15 \, \text{m s}^{-1}$ in $20 \, \text{s}$ and then continues at this speed. Its acceleration decreases as the speed of $15 \, \text{m s}^{-1}$ is approached. A mathematical model is sought that will give the value of the resultant force, F newtons, at various times. A suitable model is thought to be $F = p - qt$ and the values of p and q that fit the given data are required.

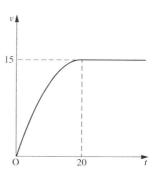

(a) Using the fact that the acceleration approaches zero as $t \to 20$, find a relationship between p and q.

(b) By integrating $F = p - qt$, find v in terms of p, q and t.

(c) Using the result of part (b), together with the data given on velocity and time, find another relationship between p and q.

(d) Find the values of p and q using the results of parts (a) and (c).

(e) Find the value predicted by this model for the velocity $10 \, \text{s}$ after the start and comment on how well this fits the given data.

(f) Do you consider that this model would be appropriate for predicting the value of F when the car accelerates from rest to $30 \, \text{m s}^{-1}$ in 40 seconds?

2. A cyclist is travelling along a horizontal road exerting a constant forward force of 80 N. The total mass including the bicycle is 80 kg. His velocity at t seconds after he started is $v \, \text{m s}^{-1}$. He starts from rest when $t = 0$. The total resistance to his motion is to be modelled as proportional to v.

(a) Given that his speed cannot exceed $8 \, \text{m s}^{-1}$, find the resistance in terms of v.

(b) Find v in terms of t.

(c) Find his velocity after 5 s.

(d) How long does it take him to reach a speed of
   (i) $7 \, \text{m s}^{-1}$ (ii) $7.9 \, \text{m s}^{-1}$ (iii) $8 \, \text{m s}^{-1}$

(e) Using the result of part (b), find s in terms of t and hence find the distance the cyclist travels in the first 5 seconds.

(f) State, with reasons, whether you think that the model is reliable.

3. Aircraft windscreens were being tested to determine the maximum impulse they could withstand when in collision, in flight, with small objects such as birds. A machine fired objects, of known mass and velocity, at the windscreen and the impulse was measured each time by sensors attached to the screen.

Suspecting that the sensors might not be accurate, a 'real-life' test was carried out using oven-ready chickens, of standard mass 3 kg, as the missiles. The results given by the sensors were then compared with the values of J calculated by using the model $J = mv$.

Speed v (m s^{-1})	20	40	80	100	150	200
J (Ns) from sensor	67	135	273	342	509	681
J (Ns) from model	60	120	240	300	450	600

The results did not correlate well so the test was carried out with new sensors. The results from the second set of sensors, however, were very close to those given by the first sensors, so the suitability of the model came under scrutiny instead.

Careful observation of what actually happened at impact revealed that the 'birds' bounced off the screen. The model was then adapted to take this into account, by estimating the coefficient of restitution as 0.1 for birds of the same mass.

(a) Form a new model for J, allowing for the bounce.

(b) Use this model to calculate J for the speeds given in the table.

(c) Comment on the correlation between the values predicted by the new model and those given by the sensors.

(d) Can you suggest a further refinement to the model that might give closer correlation?

4. A material has the property that, when a bullet is fired into it, the resistance to the motion of the bullet increases as the bullet penetrates further. It is thought that this resistance, R newtons, can be modelled by either $R = kx$ or $R = kx^2$, where x is the depth of penetration and k is a constant.
Tests are carried out with bullets of mass 0.02 kg which are fired horizontally into a fixed block of the material. It is found that a bullet entering at 400 m s^{-1} penetrates to a depth of 0.1 m and one entering at 800 m s^{-1} penetrates 0.16 m. These test data allow two values for k to be calculated for each model. By doing this, or otherwise, decide which model fits the data more closely. Give an estimate of the value of k for that model.

*5. The acceleration from rest of a car on a level road is being tested. The car is of mass 800 kg and its engine is working at a constant rate of 40 kW. It is found that after 5 seconds the speed is 20 m s^{-1}, and the speed reached 30 m s^{-1} after 12 seconds.

In order to predict the time, t seconds, taken to reach a speed of v m s^{-1}, a model is formed assuming a constant resistance, of 800 N, to the motion of the car.

(a) (i) Find the maximum speed the car can reach.

(ii) Show that $\int \dfrac{v}{50 - v} \, dv = \int dt$

(iii) Express t in terms of v. $\left(\text{Write } \dfrac{v}{50 - v} \text{ as } \dfrac{(v - 50) + 50}{50 - v} \right)$

(iv) Find the times predicted by this model for the car to reach 20 m s^{-1} and 30 m s^{-1}.

A second model is considered, taking the resistance as kv where k is constant.

(b) (i) Using the maximum speed found in part (i) above, find the value of k.
(ii) Express t in terms of v.
(iii) Find the times predicted by this model for the car to reach 20 m s^{-1} and 30 m s^{-1}.

(c) By comparing the results predicted by each model with the measured results, comment on the suitability of each model.

A further experimental measurement gives the time taken to reach 35 m s^{-1} as 19 seconds.

(d) (i) Calculate the time predicted by each model to reach 35 m s^{-1}.
(ii) How does this result affect your answer to part (c).

THE LAW OF UNIVERSAL GRAVITATION

Yet another law formulated by Newton concerns the forces of attraction that arise between any two bodies in the universe. It states that if the centres of the bodies are distant r apart and their masses are m_1 and m_2, the force F that attracts each to the other is given by

$$F = \frac{Gm_1 m_2}{r^2}$$

The value of the constant G is approximately 6.7×10^{-11} and rearranging the formula as $G = \dfrac{Fr^2}{m_1 m_2}$ shows that G is measured in $m^3\,kg^{-1}\,s^{-2}$ units.

G is called the *universal gravitational constant.*

Consider, for example, the gravitational force exerted by the earth on an object of mass m on the earth's surface. Taking the mass of the earth as approximately $6 \times 10^{24}\,kg$ and its radius as $6.4 \times 10^6\,m$ we have

$$F \approx \frac{Gm\,(6 \times 10^{24})}{(6.4 \times 10^6)^2} = \frac{(6.7 \times 6 \times 10^{13})\,m}{6.4^2 \times 10^{12}} = 9.8\,m$$

This confirms that the weight of an object of mass m at the surface of the earth is mg.

Note that the size of the object was not taken into account. This is because, unless the object were enormous, its 'radius' would make negligible difference to its distance from the centre of the earth.

The effective radius of the earth, however, does vary a little between sea-level and the tops of high mountains. It follows from the formula for F that the weight of an object should decrease slightly as its height above sea-level increases and this is consistent with practical observation.

When assessing the motion of an object relative to the earth, or some other massive body of mass m_1, the formula $F = \dfrac{Gm_1 m_2}{r^2}$ is often simplified by combining Gm_1 into a single constant k say, so that $F = \dfrac{km}{r^2}$. In this case the unit for k is $m^2\,s^{-2}$.

Newton's Law of Universal Gravitation is, like all physical 'laws', no more than a model based on the evidence available at the time. With the advance of astronomical observation and the study of the motion of atomic particles, it became apparent that some errors arose when using Newton's law in situations where velocities approached the speed of light. Einstein's theory of relativity (early C20) resolved these discrepancies but, for ordinary circumstances, applying Newton's laws to the motion of moving objects is quite accurate enough.

Examples 24c

1. A satellite of mass M is orbiting the earth at a height of 2×10^5 m above the earth's surface. The force of attraction, F newtons, exerted by the earth on the satellite is given by $F = \dfrac{4 \times 10^{14}\, M}{r^2}$, where r metres is the distance of the satellite from the centre of the earth. By modelling the earth as a sphere of radius 6.4×10^6 m and the satellite as a particle,

(a) find the acceleration of the satellite towards the earth.

Assuming that the satellite performs a circular orbit, find

(b) its angular velocity   (c) the time taken to complete one orbit.

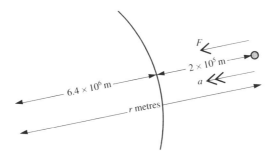

$$r = 6.4 \times 10^6 + 2 \times 10^5 = 6.6 \times 10^6$$

$$F = \frac{4 \times 10^{14}\, M}{r^2} = \frac{4 \times 10^{14}\, M}{(6.6 \times 10^6)^2} = 9.182\ldots M$$

$$F = 9.18\, M \quad (3\text{ sf})$$

(a) Using Newton's law $\;F = ma\;$ gives

$$(9.18\ldots)\,M = Ma \quad \Rightarrow \quad a = 9.182\ldots$$

The acceleration towards the earth is $9.18 \,\mathrm{m\,s^{-2}}$ (3 sf)

(b) For a particle travelling in a circle the acceleration towards the centre is $r\omega^2$ where ω is the angular velocity.

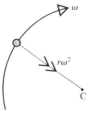

$$a = r\omega^2 \quad \Rightarrow \quad \omega^2 = \frac{9.182\ldots}{6.6 \times 10^6}$$

The angular velocity of the satellite is $1.18 \times 10^{-3} \,\mathrm{rad\,s^{-1}}$ (3 sf)

(c) The time of one revolution is $2\pi/\omega$ seconds,

i.e. $2\pi \div (1.18\ldots \times 10^{-3})\,\mathrm{s} \quad \Rightarrow \quad 5330\,\mathrm{s}$ (3 sf)

One orbit takes approximately 1.5 hours.

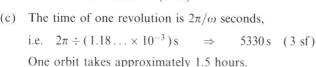

2. A rocket of mass M is fired vertically from the surface of the earth with a speed V and moves under the action of gravity only. The speed V is not great enough for the rocket to 'escape' from the earth's gravitational field.

Use the law of gravitation in the form $F = \dfrac{kM}{x^2}$ where x is the distance at any time between the rocket and the centre of the earth.

(a) Express k in terms of g and R, the radius of the earth at the launch site.

(b) Find the greatest distance from the centre of the earth reached by the rocket, giving your answer in terms of g and R.

(a) At the surface of the earth, $x = R$ and the gravitational force is Mg

$$\therefore \quad Mg = \frac{kM}{R^2} \quad \Rightarrow \quad k = gR^2$$

(b)

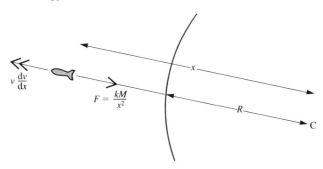

The force acting on the rocket at any time is $\dfrac{kM}{x^2} = \dfrac{gR^2M}{x^2}$

Using $v\dfrac{dv}{dx}$ for the acceleration, we have

$$-\frac{gR^2M}{x^2} = Mv\frac{dv}{dx}$$

$$\Rightarrow \quad -\int \frac{gR^2}{x^2}\,dx = \int v\,dv$$

$$\therefore \quad \frac{gR^2}{x} = \frac{1}{2}v^2 + A$$

When $x = R$, $v = V$ therefore $A = gR - \tfrac{1}{2}V^2$

$$\Rightarrow \quad \frac{1}{2}v^2 = \frac{gR^2}{x} - gR + \frac{1}{2}V^2$$

When the greatest distance D is reached, the velocity becomes zero,

$$\therefore \quad 0 = \frac{gR^2}{D} - gR + \frac{1}{2}V^2 \quad \Rightarrow \quad gR^2 = D\left(gR - \tfrac{1}{2}V^2\right)$$

$$\therefore \quad D = \frac{2gR^2}{2gR - V^2}$$

EXERCISE 24c

In this exercise take the value of G as $6.7 \times 10^{-11} \, \text{m}^3 \, \text{kg}^{-1} \, \text{s}^{-2}$ unless another instruction is given.

1. Using the method of the Cavendish experiment of 1798, the gravitational force between two lead spheres is found. The masses of the spheres are 0.008 kg and 12 kg and the distance between their centres is 0.01 m. The force of attraction between them is $6.36 \times 10^{-8} \, \text{N}$. Use the data to find a value for G.

2. Use the Law of Universal Gravitation to estimate the mass of the earth by considering the gravitational force on a particle of mass m kg at the earth's surface. Treat the earth as a sphere of radius 6.4×10^6; take the acceleration due to gravity at the surface of the earth as $9.8 \, \text{m s}^{-2}$.

3. The moon takes 27.3 days to orbit the earth. Using the mass of the earth calculated in question 2, and assuming that the moon's orbit is a circle, find

   (a)   the angular velocity of the moon about the earth

   (b)   the radius of the moon's orbit and hence the distance of the moon from the surface of the earth.

4. A spacecraft is travelling from the earth to the moon on a straight line joining their centres. Find the distance of the spacecraft from the centre of the earth when it reaches the point at which the gravitational attraction of the earth is equal to that of the moon. The mass of the earth is 81 times the mass of the moon and the distance between their centres is $3.8 \times 10^8 \, \text{m}$.

5. A spacecraft is put into a circular orbit about the moon. It takes 109 minutes to complete each orbit at a height of $5 \times 10^4 \, \text{m}$ above the surface. The radius of the moon is $1.7 \times 10^6 \, \text{m}$. Find

   (a)   the radius of the orbit

   (b)   the angular velocity of the spacecraft

   (c)   the mass of the moon.

6. A man of mass 80 kg is standing on the surface of the moon. He accidently releases a piece of moon rock which he is studying and it drops from a height of 1.4 m and lands on his foot. Take the mass of the moon to be $7.4 \times 10^{22} \, \text{kg}$ and its radius to be $1.7 \times 10^6 \, \text{m}$. Find

   (a)   the weight (in newtons) of the man on the moon

   (b)   the acceleration due to the moon's gravity

   (c)   the speed with which the rock hits his foot.

7. A rocket is fired vertically from the surface of the earth and attains a velocity $u\,\mathrm{m\,s^{-1}}$ near the surface. It then moves acted on by gravity only, assuming that air resistance is negligible.
After t seconds its distance from the centre of the earth is x metres and its velocity is $v\,\mathrm{m\,s^{-1}}$.
The radius of the earth is R metres.

Use the law of gravitation in the form $F = \dfrac{km}{x^2}$.

(a) Write down the equation of motion of the rocket and by integration show that

$$v^2 = u^2 + 2k\left(\frac{1}{x} - \frac{1}{R}\right)$$

(b) Find the value which v^2 approaches as x increases.

(c) Using the result of part (b) find, in terms of k and R, the value which u must exceed if the rocket is never to return to earth (i.e. find the escape velocity).

(d) Evaluate the escape velocity using $k = 4 \times 10^{14}\,\mathrm{N\,m^2\,kg^{-1}}$ and $R = 6.4 \times 10^6$.

*8. A space station is set up on Mars with the ability to launch spacecraft from the surface of that planet. A rocket is fired vertically from the surface with initial velocity $u\,\mathrm{m\,s^{-1}}$ and it reaches a greatest height h metres.
For a body of mass m kilograms the gravitational attraction to Mars is F newtons when the body is at a distance x kilometres from the centre of Mars.
The mass of Mars is M kilograms where $M = 6.4 \times 10^{23}$,
its radius is R metres where $R = 3.4 \times 10^6$
and the acceleration due to gravity at its surface is $g_1\,\mathrm{m\,s^{-2}}$

(a) find the value of g_1.

If the law of gravitation is to be used in the form $F = \dfrac{km}{x^2}$

(b) find the value of k.

(c) express k in terms of g_1 and R.

(d) find h in terms of g_1, R and u.

(e) by considering the values of u for which h is large, deduce the 'escape' velocity from Mars

    (i) in terms of g_1 and R,

    (ii) evaluated corrected to 2 significant figures.

THE SIMPLE PENDULUM

When a heavy particle, attached to one end of a light string whose other end is fixed, oscillates through a *small* angle, the system is called a *simple pendulum*.

In the following diagram we will take m as the mass of the particle P, l as the length of the string and θ as the angle between the string and the vertical at any time t. So that quantities can be marked clearly on the diagram, the angle is shown larger than it should be.

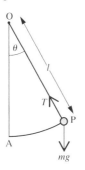 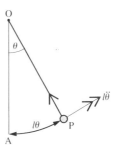

The length of the arc AP is $l\theta$, so the acceleration in the direction of the tangent is $\dfrac{d}{dt}(l\dot\theta)$, or $l\ddot\theta$, away from A.

The force along the tangent is $mg \sin \theta$ towards A.

Therefore using $F = ma$ gives $mg \sin \theta = -ml\ddot\theta$

Now θ is at all times a small angle, so $\sin \theta \approx \theta$.

Therefore $\qquad\qquad\qquad mg\theta \approx -ml\ddot\theta$

which can be written $\qquad l\ddot\theta \approx \dfrac{-g}{l}(l\theta)$ [1]

From the diagram we see that $\qquad x = l \sin \theta \approx l\theta$

$\therefore \qquad\qquad\qquad\qquad\qquad \ddot{x} \approx l\ddot\theta$

Equation [1] then becomes $\qquad \ddot{x} \approx \dfrac{-g}{l}x$

showing that the acceleration $\ddot{x}$, is proportional to the displacement and in the opposite direction.

Hence, as motion along the arc is approximately equivalent to motion along A'P,

to a good approximation the particle describes simple harmonic motion.

Comparing $\ddot{x} \approx -\dfrac{g}{l}x$ with the standard equation of SHM, i.e. $\ddot{x} = -n^2x$,

we see that the constant n^2 is equal to $\dfrac{g}{l}$.

Therefore the period T of a complete oscillation is $\dfrac{2\pi}{n}$ where $n^2 = \dfrac{g}{l}$

i.e. $$T = 2\pi\sqrt{l/g}$$

This formula can be quoted unless its derivation is asked for.

Note that T depends on the length of the string but not on the mass of the particle (often called the *pendulum bob*).

The Seconds Pendulum

A simple pendulum which swings from one end of its path to the other end in exactly one second is called a *seconds pendulum* and is said to *beat seconds*.

Since each half oscillation takes 1 second, the period of oscillation is 2 seconds, i.e. $T = 2$.

The length of string, l, required for a seconds pendulum can then be calculated using $T = 2\pi\sqrt{l/g}$

giving $2 = 2\pi\sqrt{l/g}$ $\quad\Rightarrow\quad$ $l = \dfrac{g}{\pi^2}$ $\;(\approx 0.99\,\text{m})$

Examples 24d

1. **The period of a simple pendulum is $3T$. If the period is to be reduced to $2T$, state whether the length of the pendulum should be increased or decreased and find the percentage change required.**

Taking the original length of the pendulum as l, we have

$$3T = 2\pi\sqrt{\dfrac{l}{g}} \tag{1}$$

The period increases as l increases so, to reduce the period, the length should be reduced.

Let kl be the required reduction in length.

Then $\quad 2T = 2\pi\sqrt{\dfrac{(l - kl)}{g}} \quad\Rightarrow\quad \left(\dfrac{T}{\pi}\right)^2 = \dfrac{l(1 - k)}{g}$

From [1] $\left(\dfrac{T}{\pi}\right)^2 = \dfrac{4l}{9g}$

∴ $\dfrac{4l}{9g} = \dfrac{l(1-k)}{g}$ ⇒ $9(1-k) = 4$

∴ $k = \tfrac{5}{9}$ and the reduction in length is $\tfrac{5}{9}l.$

The percentage reduction in length is $\dfrac{kl}{l} \times 100,$ i.e. 56% (2 sf).

2. **At a location where $g = 9.81\,\text{m}\,\text{s}^{-2}$ a seconds pendulum beats exact seconds. If it is taken to a place where $g = 9.80\,\text{m}\,\text{s}^{-2}$ by how many seconds per day will it be wrong?**

If l is the length of the pendulum then

$$\pi\sqrt{\dfrac{l}{9.81}} = 1 \quad \text{giving} \quad \sqrt{l} = \dfrac{\sqrt{9.81}}{\pi}$$

When $g = 9.80$ the time, t, of one beat is given by $\pi\sqrt{\dfrac{l}{9.8}}$

and is therefore $\sqrt{\dfrac{9.81}{9.8}}$ seconds.

The number of beats in 24 hours is now $(24 \times 60 \times 60) \div \sqrt{\dfrac{9.81}{9.8}}$

The number of beats lost in 24 hours is therefore

$$24 \times 60 \times 60\left(1 - \sqrt{\dfrac{9.8}{9.81}}\right) = 24 \times 60 \times 60\,(0.0005)$$

$$= 44$$

Therefore, where $g = 9.80$ the pendulum will lose 44 seconds per day.

EXERCISE 24d

1. A simple pendulum is 2 m in length and the time it takes to perform 50 complete oscillations is measured.

   (a) The pendulum is on the earth and the time taken is 142 s. Find g.

   (b) The pendulum is on the moon and the time taken is 341 s. Find the acceleration due to gravity on the moon.

2. Two simple pendulums have periods 1.6 s and 2.4 s. They are set oscillating, initially in step.

   (a) Find the length of each pendulum.

   (b) Find the interval after which they are next in step.

3. The length of a simple pendulum is shortened by 54 cm and this has the effect of reducing the period by 20%. Find the length of the original pendulum.

4. By what length should a simple pendulum be shortened if it is meant to beat seconds but loses 40 seconds in 12 hours?

5. A pendulum which beats seconds where $g = 9.81\,\text{m s}^{-2}$ is taken up a mountain to a place where it loses 30 seconds per day. What is the value of g at the new location?

6. If a simple pendulum which beats exact seconds has its length changed by 1%, find the number of seconds per day by which it will be inaccurate if

(a) the length is increased,

(b) the pendulum is shortened.

FORCES THAT PRODUCE SIMPLE HARMONIC MOTION

When a particle performs SHM, its acceleration is always directed towards, and is proportional to, its distance from a fixed point. It follows that SHM is produced by the action of *a force directed towards a fixed point and proportional to the distance from that point.*

The tension in an elastic string is proportional to the extension and always acts to restore the string to its natural length. These are the conditions in which SHM is likely, and we are now going to consider the motion of a particle attached to the end of an elastic string.

Examples 24e

1. **A particle P of mass 2 kg is lying at a point A on a smooth horizontal table. P is attached to one end of a light elastic string of length 1 m and modulus of elasticity 8 N, whose other end is fixed at a point O on the table. Initially the string is just taut. P is pulled away from O until it is 2 m from O and is then released.**

(a) **Show that, at first, P performs SHM and state the position of P when SHM ceases.**

(b) **Find the time taken to reach the point found in part (a) and the speed of P at this point.**

(c) **Describe briefly the motion of P after it passes through the position found in part (a) and give an assumption that is made.**

(a)

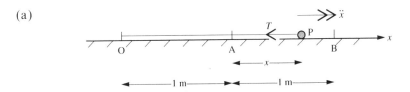

Consider the particle in a general position, distant x from A, where the string is taut.

As long as the string is taut, Hooke's Law $\left(\text{tension} = \dfrac{\lambda \times \text{extension}}{\text{natural length}} \right)$

gives $T = 8x$

Then using $F = ma$ gives $T = -2\ddot{x}$

$\therefore$ $8x = -2\ddot{x}$ $\Rightarrow$ $\ddot{x} = -4x$

This is the equation of SHM in which $n^2 = 4$

$\therefore$ P performs SHM as long as the string is taut.

SHM ceases when the string becomes slack, i.e. when P reaches A.

(b)   At A, $x = 0$ therefore A is the centre of the SHM that has been performed so far and at this stage P has covered one quarter of an oscillation from the end B to the centre A.

$\therefore$ the time taken to reach A is given by $\dfrac{1}{4} \left(\dfrac{2\pi}{n} \right)$

Therefore, as $n = 2$,   the time taken is $\frac{1}{4}\pi$ seconds

The amplitude, a, is AB, i.e. 1 m.

The speed of P at the centre of the SHM is given by an, and is $2\,\text{m s}^{-1}$.

(c)   After P passes through A the string is slack so there is no horizontal force acting on it. Therefore P travels on with constant velocity $2\,\text{m s}^{-1}$ until the string becomes taut again. This will happen when P is at a point A' on the opposite side of O such that $OA = OA'$. Then P will again move with SHM.

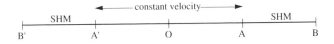

This assumes that the slack string does not interfere with the motion.

2. **A light elastic string of natural length $2a$ and modulus of elasticity λ is stretched between two points A and B, distant $4a$ apart on a smooth horizontal table. A particle of mass m is attached to the midpoint of the string and is pulled towards A through a distance a and then released. Show that the particle describes SHM; state the centre and amplitude, and find the period of the motion.**

When the particle is attached to the centre of the string, two independent strings are created, each with length a and modulus of elasticity λ.

Consider the motion of the particle as it passes through a general point P which is distant x ($x < a$) from M, the midpoint of AB.

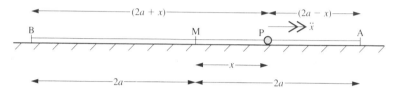

For the string BP, natural length $= a$ and stretched length $= 2a + x$

$\therefore$ extension $= a + x$

The tension T_B is given by Hooke's Law, i.e. $T_B = \dfrac{\lambda}{a}(a + x)$

Similarly for the string AP, for which the extension is $(2a - x) - a = a - x$, Hooke's Law gives $T_A = \dfrac{\lambda}{a}(a - x)$

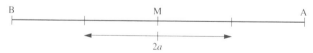

The resultant force acting on the particle is $T_B - T_A$ towards M.

Then using $F = ma$ gives $\dfrac{\lambda}{a}(a + x) - \dfrac{\lambda}{a}(a - x) = -m\ddot{x}$

$\therefore \qquad \dfrac{2\lambda x}{a} = -m\ddot{x} \quad \Rightarrow \quad \ddot{x} = -\dfrac{2\lambda}{ma}x$

This is the equation of SHM in which $n^2 = \dfrac{2\lambda}{ma}$

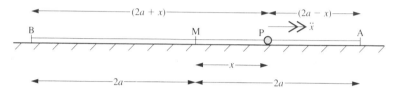

$\ddot{x} = 0$ when $x = 0$ and this is at M, so M is the centre of oscillation.

The length of the path is $2a$, so the amplitude is a.

The period of the motion is $\dfrac{2\pi}{n}$, i.e. $2\pi\sqrt{\dfrac{ma}{2\lambda}}$

Note that neither section of the string goes slack during the motion.

3. One end of a light elastic string AB, of length $4a$ and modulus of elasticity $4mg$, is fixed at a point B. A particle of mass m attached to the other end of the string is hanging in equilibrium at a point E.

(a) Find the length of the stretched string.

The particle is then pulled vertically downwards from E to a point C, which is distant a below E, and then released

(b) Show that the particle performs SHM, giving the centre and amplitude and period of the motion.

If, instead, the particle were pulled down to a point distant $2a$ below E and released, explain why the subsequent motion of the particle is not entirely SHM.

(a) The particle is in equilibrium at E

$\therefore$ $T = mg$

Using Hooke's Law

$$T = \frac{4mg\,x}{4a} \quad \Rightarrow \quad x = a$$

$\therefore$ the stretched length is $5a$

(b) Consider the particle at a general point P distant y from E.

The extension in the string is $a + y$

Therefore Hooke's Law gives

$$T = \frac{4mg(a+y)}{4a}$$

Using Newton's Law gives

$$mg - T = m\ddot{y}$$

i.e. $mg - mg\,\dfrac{(a+y)}{a} = m\ddot{y}$

$\Rightarrow$ $\ddot{y} = -\dfrac{g}{a}\,y$

Comparing $\ddot{y} = -\dfrac{g}{a}y$ with $\ddot{x} = -n^2 x$ shows that this represents SHM

in which $n^2 = \dfrac{g}{a}$

$\ddot{y} = 0$ when $y = 0$, so E is the centre of the SHM.

The greatest value of y is EC, i.e. a, so the amplitude is a.

∴ the distance that the particle rises above E is a, showing that the string does not go slack.

The period of oscillation is $\dfrac{2\pi}{n}$, i.e. $2\pi\sqrt{\dfrac{a}{g}}$

(c) If the particle were released from a point distant $2a$ below E, the amplitude of the SHM would be $2a$ so the particle *would* rise above A, the natural end of the string. But above A the string is slack and no tension acts.

Therefore the only force acting on the particle is its weight, producing motion with constant acceleration due to gravity, not SHM.

The speed of the particle when it reaches A on its SHM path is given by $n\sqrt{(2a)^2 - a^2}$, i.e. $na\sqrt{3}$, and this is the initial upward speed for the motion under gravity.

(SHM begins again as the particle passes through A with speed $na\sqrt{3}$ on its downward path.)

4. **The diagram shows a fairground 'Test your Strength' machine consisting of a long spring fixed at one end A, and with a platform of mass 2 kg attached to the other end.**

Competitors strike the platform vertically downwards with a mallet, as hard as they can. The depth to which the platform descends is recorded on a scale and the person who attains the greatest depth wins. The natural length of the spring is 2 m and its modulus of elasticity is $40g$ N. The winner causes the platform to descend by 0.5 m.

By modelling the platform as a particle,

(a) **find the length of the spring when the platform hangs freely at E,**

(b) **find the initial speed of the platform when struck (use $g = 10$),**

(c) **find the impulse of the blow from the mallet,**

(d) **show that the particle describes simple harmonic motion.**

State any assumptions that have been made and suggest any way in which you think the model might be improved.

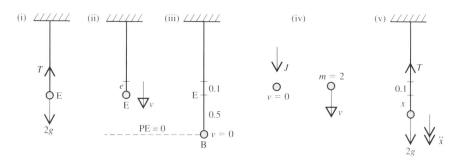

(a) When the platform hangs at rest, (diagram i) $\quad T = 2g$

Using Hooke's Law (diagram ii) gives $\qquad T = 40g \dfrac{e}{2}$

Hence $\quad e = 0.1$

The length of the spring is 2.1 m

(b) Conservation of mechanical energy can be used from E (ii) to B (iii).

At E, $\quad$ PE $= 2g(0.5) \quad$ KE $= \dfrac{1}{2}(2)v^2 \quad$ EPE $= \dfrac{40g}{4}(0.1)^2$

At B, $\quad$ PE $= 0 \quad$ KE $= 0 \quad$ EPE $= \dfrac{40g}{4}(0.6)^2$

Conservation of ME gives $\quad g + v^2 + 0.1g = 3.6g$

$\Rightarrow \qquad v^2 = 2.5g = 25 \qquad \Rightarrow \qquad v = 5$

The speed at E is $5\,\mathrm{m\,s}^{-1}$

(c) Taking $J\,\mathrm{N\,s}$ as the impulse and using $\quad$ impulse $=$ increase in momentum
(diagram iv) we have $\qquad J = 2v \quad \Rightarrow \quad J = 10$
The impulse is $10\,\mathrm{N\,s}$

(d) When the platform is in a general position P, where the depth below E is x (diagram v), using Newton's Law gives

$$2g - T = 2\ddot{x}$$

Hooke's Law gives $\qquad T = \dfrac{40g}{2}(0.1 + x) = 2g(1 + 10x)$

$\therefore \qquad 2g - 2g(1 + 10x) = 2\ddot{x} \quad \Rightarrow \quad \ddot{x} = -100x$

This is the equation of SHM for which $n = 10$ and about E as centre.

The *spring* can push as well as pull so the equation $\ddot{x} = -100x$ applies in all positions so in theory the platform performs SHM throughout the whole of its path.

Assumptions are:

> The spring is light and cannot distort. The machine has no device for 'damping down' the motion as the platform descends. There is no air resistance to the motion of the platform.

Possible improvements:

> The surface area of the platform does offer some air resistance so a term reflecting this could be incorporated into the equation of motion.

5. **A firm making small fine porcelain ornaments, wishes to ensure that the packaging used for exporting the goods is completely satisfactory. To test for the effect of vibration, a typical package is placed on a belt that is kept taut by passing round two pulleys as shown.**

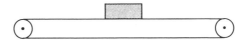

The belt can be made to oscillate along its length with simple harmonic motion performing 4 oscillations per second. If the mass of the package is 0.6 kg and the coefficient of friction with the belt is 0.9, find the greatest amplitude permissible if the package must not slip on the belt.

Four oscillations are performed per second, therefore the period of the motion is 0.25 seconds

The period is given by $\dfrac{2\pi}{n}$ therefore $\dfrac{2\pi}{n} = 0.25$ $\Rightarrow$ $n = 8\pi$

The maximum acceleration of the belt occurs at each end of the oscillation and its magnitude is $n^2 a$, where a metres is the amplitude.

$\therefore$ the magnitude of the greatest acceleration is $(8\pi)^2 a$

If the package is not to slip, it must move with the same acceleration as the belt, so the maximum acceleration of the package is also $(8\pi)^2 a$.

The force that must produce this acceleration is the frictional force exerted by the belt on the package.

Using $F = m\ddot{x}$ gives $F = 0.6 \times (8\pi)^2 a$

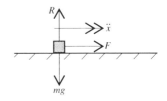

The greatest value of F is μR, i.e. μmg.

When F has this value $\mu mg = 0.6 \times (8\pi)^2 a$

i.e. $\qquad 0.9 \times 0.6 \times 9.8 = 0.6 \times 64\pi^2 a$

$\Rightarrow \qquad a = 0.013\,96\ldots$

Therefore the amplitude must not exceed 1.40 cm (3 sf)

EXERCISE 24e

In questions 1 to 4 a particle P is moving with SHM on a smooth horizontal surface under the action of a horizontal force. The path lies between A and B, and O is the midpoint of AB.

1. If the mass of P is 2 kg, the amplitude of the motion is 5 m and the period is $\dfrac{2\pi}{3}$ s, find the magnitude of the horizontal force acting on P

(a) when P is 4 m from O

(b) when the horizontal force has its greatest magnitude.

2. When P is 0.32 m from O its speed is $3.6\,\mathrm{m\,s^{-1}}$ and the horizontal force acting on it is 18 N. Given that the mass of P is 0.25 kg find

(a) the period   (b) the amplitude of the motion.

3. The mass of P is 0.2 kg and it is projected from O towards A with speed $3\,\mathrm{m\,s^{-1}}$. The motion takes place under the action of a horizontal force of $20x$ newtons, directed towards O, where x metres is the distance of P from O. Find

(a) the period

(b) the amplitude

(c) the magnitude of the maximum horizontal force on P.

4. When P is at A, the horizontal force acting on it is 90 N towards O. After P has moved a distance of 2 m towards B, this force is 54 N.
If the mass of P is 0.5 kg find

(a) the period   (b) the amplitude.

5. The motion of the piston in a car engine is approximately SHM with an amplitude of 5 cm. The mass of the piston is 0.5 kg.
When the piston is making 60 oscillations per second find

(a) the maximum force required to give the piston this motion

(b) the maximum speed of the piston.

6. A particle P of mass 2 kg is placed on a rough horizontal surface. The coefficient of friction between the particle and the surface is 0.4. The surface moves horizontally with SHM of period T seconds and amplitude a metres. Find whether the particle will slip on the surface

(a) if $T = 4$ and $a = 1.2$ (b) if $T = 2$ and $a = 0.7$.

7. A fairground ride consists of a rough horizontal surface which is made to oscillate horizontally with approximate SHM of amplitude 0.5 m. A girl of mass 50 kg stands on the ride. The coefficient of friction between the girl and the surface is 0.3. Find the smallest possible value of the period of oscillation if the girl is not to slip.

8. A particle P of mass m is placed on a horizontal surface which oscillates vertically up and down in SHM with amplitude a and period $\dfrac{2\pi}{\omega}$. Take an origin O at the centre of the motion and an x-axis with positive direction vertically upwards. When the displacement of P from O is x, the normal reaction is R.

(a) Write down the equation of motion.

(b) Find R in terms of m, ω and x.

(c) Find (i) the least value of R (ii) the greatest value of R.

(d) Find the greatest amplitude the motion can have if P is to remain in contact with the platform.

9. During a storm a boat is riding up and down on the waves in approximate SHM of amplitude 4.9 m. At one extreme position of the motion the normal reaction exerted on a passenger of mass 60 kg, by his seat, has half the value it would have if the boat were stationary. Find

(a) the period of an oscillation

(b) the normal reaction from the seat when the boat is at the opposite extreme position.

10. A particle P, of mass 0.5 kg, is lying on a smooth horizontal table and is attached to one end of a light elastic spring of natural length 0.8 m and modulus of elasticity 40 N. The other end of the spring is fixed to a point A on the table. P is held 0.6 m from A, thus compressing the spring, and is then released.

(a) Show that P performs SHM and find the period.

(b) Find the distance AP when P first comes to instantaneous rest.

(c) Will P perform complete oscillations in SHM?

11. A particle P, of mass 2 kg is lying on a smooth horizontal surface. P is attached to a point A on that surface by a light elastic string of modulus 18 N and natural length 1 m. Initially P lies at rest at point B, which is 1 m from A. P is then projected in the direction AB with velocity 1.8 m s^{-1}.

(a) Show that, at first, P performs SHM.

(b) Find the distance AP when P first comes to instantaneous rest.

(c) Find the time taken for P to return to B.

(d) Find the velocity of P when (i) AP = 1.4 m (ii) AP = 0.6 m.

12. A particle P of mass 0.4 kg is lying on a smooth horizontal surface. P is attached to a point A on that surface by a light elastic spring of modulus 5 N and natural length 0.5 m. Initially P lies at rest at point B, which is 0.5 m from A. P is then projected in the direction AB with velocity u metres per second.

    (a) Use conservation of energy to find the length AP when P is first in a position of instantaneous rest   (i) if $u = 1$ (ii) if $u = 3$. State in each case whether complete SHM oscillations are possible and also any assumptions which you need to make.

    (b) In the case in part (a) where complete oscillations are possible, find the period and amplitude.

13. Two points A and B are distant $5l$ apart on a smooth horizontal table. A particle P, of mass m, is attached to A by an elastic string of natural length l and modulus of elasticity mg and to B by an elastic string of natural length l and modulus of elasticity $2mg$. Initially P lies in equilibrium at point O on the line AB.

    (a) Find the lengths of AO and OB.

P is then moved along the line AB until $AP = 2l$ and is released from rest at that point.

    (b) Taking O as the origin and an x-axis in the direction OA, consider the particle when its displacement from O is x metres. (Put P on the positive part of the x-axis.)
        (i) Find in terms of x the extensions in AP and BP.
        (ii) Write down the equation of motion and, by simplifying it, deduce that P performs SHM with centre at O.
        (iii) Find the period and amplitude.

14. A particle of mass 0.5 kg is attached to one end of a light elastic string of natural length 2 m and modulus of elasticity 7 N. The other end of the string is fixed to a point A and the particle hangs in equilibrium at a point E.

    (a) Find the distance AE.

The particle is then pulled down a further distance a metres and released from rest.

    (b) Taking E as the origin and an x-axis vertically downwards, consider the particle when it is at a distance x below E.
        (i) Write down the equation of motion and, by simplifying it, deduce that P performs complete oscillations of SHM with centre at E, provided that $a \leqslant 1.4$.
        (ii) Find the period.
        (iii) Describe the motion if $a = 1.8$.

15. A particle of mass $2m$ is attached to one end of a light elastic string of natural length l. The other end of the string is fixed to a point A and the particle hangs in equilibrium at a point E, where $AE = 2l$. It is then projected vertically downwards from E with an initial velocity $\sqrt{gl}$.

(a) Find the modulus of elasticity.

(b) Use conservation of energy to find the depth below A at which the particle next comes to instantaneous rest.

(c) Show that the particle performs SHM and find the period and amplitude.

(d) Describe the motion if the initial velocity is changed to a value
      (i) less than $\sqrt{gl}$ (ii) greater than $\sqrt{gl}$

*16. 'Springmakers' Ltd have been asked to provide a spring, of length $30\,\text{cm}$, for an application where it is required to make a mass of $0.5\,\text{kg}$ oscillate vertically at a rate of 2 oscillations per second.

(a) Find the modulus of elasticity which the spring should have.

When the spring is fitted the rate of oscillation is found to be 2.2 oscillations per second. It is decided to correct the rate to the required 2 oscillations per second by attaching an extra, separate mass to the end of the spring.

(b) Find the extra mass needed to make this adjustment.

CONSOLIDATION F

SUMMARY

Variable Motion

Take s, v, a and t to represent displacement, velocity, acceleration and time, when a, v or s is a function of time, $f(t)$, use:

$$\dot{s} = v = \frac{ds}{dt} \qquad \dot{v} = a = \frac{dv}{dt}$$

$$v = \int a \ dt \qquad s = \int v \ dt$$

When a is a function of displacement, $f(s)$, use:

$$a = v\frac{dv}{ds} \quad \Rightarrow \quad \int f(s) \ ds = \int v \ dv \quad (\text{giving } v \text{ as a function of } s)$$

When v is a function of displacement, $f(s)$, use:

$$a = v\frac{dv}{ds} = f(s) \times \frac{df(s)}{ds}$$

$$v = \frac{ds}{dt} = f(s) \quad \Rightarrow \quad \int \frac{1}{f(s)} \ ds = \int 1 \ dt$$

Simple Harmonic Motion

SHM is motion in a straight line in which the acceleration is proportional to the distance from a fixed point on the line, O say, and is always directed towards that point.

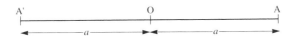

AA′ is the path

O is the centre, or mean position

a is the amplitude (the distance OA)

An oscillation is the journey from A to A′ and back to A.

T is the period of an oscillation

For a general position of the particle P where $OP = x$,

$$\ddot{x} = -n^2 x \quad \text{where } n \text{ is a constant}$$

$$\dot{x} = -n\sqrt{(a^2 - x^2)}$$

$$x = a \cos nt \quad \text{where} \quad t = 0 \quad \text{when} \quad x = a$$

or $\quad x = a \cos(\omega t + \alpha) \quad \text{where} \quad t = 0 \quad \text{when} \quad x = a \cos \alpha$

$$T = \frac{2\pi}{n}$$

The maximum acceleration occurs at A and A' and its magnitude is $n^2 a$.

The maximum speed is na, occurring at O.

Associated Circular Motion

As a point P travels round a circle at constant angular speed ω, its projection on a diameter of the circle describes SHM with equation

$$\ddot{x} = -\omega^2 x$$

The Simple Pendulum

When a heavy particle, attached to one end of a light string whose other end is fixed, oscillates through a *small* angle, the system is called a simple pendulum. To a good approximation the particle describes simple harmonic motion.

The period T of a complete oscillation, i.e. a swing forward and back, is given by $\quad T = 2\pi\sqrt{l/g}$.

A 'seconds pendulum' is designed to take exactly 1 second to swing through half an oscillation.

Forces Producing Simple Harmonic Motion

SHM is produced by the action of a force that is directed towards a fixed point and is proportional to the distance from that point. The commonest example of such a force is the tension in a stretched elastic string acting on a particle attached to the end of the string.

Properties of a Variable Force

When the force acting on an object of mass m is a function of time, i.e. $F = f(t)$, Newton's Law of Motion can be applied using $a = dv/dt$ giving

$$f(t) = m\frac{dv}{dt} \quad \Rightarrow \quad \int f(t)\ dt = \int m\ dv$$

When the force is a function of displacement, i.e. $F = f(s)$, Newton's Law of Motion can be applied using $a = v\ dv/ds$ giving

$$f(s) = mv\frac{dv}{ds} \quad \Rightarrow \quad \int f(s)\ ds = \int mv\ dv$$

When the point of application of a force F, where $F = f(s)$, moves through a displacement s, the **work done** is given by $\int f(s)\ ds$.

If the PE does not change then $\int f(s)\ ds = \frac{1}{2}mv^2 - \frac{1}{2}mu^2$.

The **impulse** exerted over an interval of time t, by a force F where $F = f(t)$, is given by $\int f(t)\ dt$.

It follows that $\int f(t)\ dt = mv - mu$.

Law of Universal Gravitation

This law states that if the centres of any two bodies in the universe are at a distance r apart and their masses are m_1 and m_2, the force F that attracts each to the other is given by $F = \dfrac{Gm_1 m_2}{r^2}$

The value of the constant G is approximately 6.7×10^{-11} and G is measured in $m^3\,kg^{-1}\,s^{-2}$ units.

MISCELLANEOUS EXERCISE F

In questions 1 to 4 a problem is set and is followed by a number of suggested responses. Choose the correct response.

For questions 1 and 2, use this diagram of a particle P moving along a straight line.

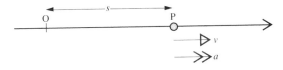

1. Given that $v = s^2$, the acceleration is given by

A $2s$ B 2 C $2s^3$ D $\frac{1}{3}s^3$

2. If $v = e^{2t}$ and $s = 0$ when $t = 0$, the value of s at any time t is given by

A $\frac{1}{t}e^{2t}$ B $\frac{1}{2}(e^{2t} - 1)$ C $2(e^{2t} - 1)$ D $\frac{1}{2}e^{2t}$

3. A particle moves along a straight line with an acceleration of $\frac{2}{v}$ where v is the velocity at any instant. Initially the particle is at rest. The velocity of the particle at time t is:

A $2t$ B $4t$ C $\frac{2t}{v}$ D $2\sqrt{t}$

4. A particle moves in a straight line with an acceleration $2s$ where s is its displacement from a fixed point on the line, and $v = 0$ when $s = 0$. Its speed when its displacement is s is:

A s B s^2 C $s\sqrt{2}$ D $\sqrt{2s}$

5. A particle of mass $3\,\text{kg}$ moves along a straight line Ox under the action of a force F such that at time t, $x = t^2 + 3t$. The magnitude of F at time t is given by

A 0 B $5\,\text{N}$ C $3(2t + 3)$ D $6\,\text{N}$

6. A particle P moves in a straight line Ox so that at time t seconds, its acceleration in the direction of increasing x is $8te^{2t}\,\text{m s}^{-2}$. Given that P starts from rest at O when $t = 0$, find

(a) the speed of P when $t = \frac{1}{2}$,

(b) the distance of P from O when $t = \frac{1}{2}$.   (ULEAC)$_s$

7. A dot moves on the screen of an oscilloscope so that its position relative to a fixed origin is given by

$$\mathbf{r} = 2t\mathbf{i} + \sin\left(\frac{\pi t}{2}\right)\mathbf{j}$$

(a) Sketch the path of the dot for $0 \leqslant t \leqslant 4$.

(b) Find the velocity and acceleration of the dot when $t = 3$. Draw vectors on your diagram to show these two quantities.   (AEB)

8. A particle P, of mass 2 kg, is moving under the influence of a variable force **F**. At time t seconds, the velocity $\mathbf{v}\,\mathrm{m\,s}^{-1}$ of P is given by

$$\mathbf{v} = 2t\mathbf{i} + e^{-t}\mathbf{j}.$$

   (a) Find the acceleration, $\mathbf{a}\,\mathrm{m\,s}^{-2}$, of P at time t seconds.

   (b) Calculate, in N to 2 decimal places, the magnitude of **F** when $t = 0.2$.
   (ULEAC)

9. A particle P moves along the x-axis passing through the origin O at time $t = 0$. At any subsequent time t seconds, P is moving with a velocity of magnitude $v\,\mathrm{m\,s}^{-1}$ in the direction of x increasing where

$$v = 2t^3 + 2t + 3, \qquad t \geqslant 0.$$

   (a) Find the acceleration of P when $t = 3$.

   (b) Find the distance covered by P between $t = 0$ and $t = 4$.

   A second particle Q leaves O when $t = 1$ with constant velocity of magnitude $10\,\mathrm{m\,s}^{-1}$ in the direction of the vector $3\mathbf{i} - 4\mathbf{j}$, where **i** and **j** are unit vectors parallel to Ox and Oy respectively.

   Find, as a vector in terms of **i** and **j**,

   (c) the velocity of Q

   (d) the velocity of P relative to Q at the instant when $t = 1$.

   Hence

   (e) find the magnitude of the velocity of P relative to Q when $t = 1$

   (f) find the angle between the relative velocity and the vector **i** at this instant.
   (ULEAC)

10. A vector **v** is given by $\mathbf{v} = (t^2 - t)\mathbf{i} + (2t - t^2)\mathbf{j}$, where **i** and **j** are constant perpendicular unit vectors and t is a variable scalar.
    Find expressions, in terms of t, for

    (a) $\dfrac{d\mathbf{v}}{dt}$, (b) $\dfrac{d^2\mathbf{v}}{dt^2}$, (c) $\left|\dfrac{d\mathbf{v}}{dt}\right|^2$.
    (ULEAC)

11. A particle A has mass 0.5 kg and is acted on by two forces $\mathbf{F}_1$ N and $\mathbf{F}_2$ N. At time $t = 0$ the particle is at rest at the point whose position vector relative to a fixed origin O is $(0.5\mathbf{i})$ m. Given that $\mathbf{F}_1 = 25\mathbf{i} + 20\mathbf{j}$, $\mathbf{F}_2 = 15\mathbf{i} - 20\mathbf{j}$ find the position vector of A at time t seconds.

    At time t seconds the position vector of a second particle B relative to O is $(0.75\mathbf{i} + 30t^2\mathbf{j})$ m. Find the position vector of A relative to B at time t seconds and hence, or otherwise, find the time T seconds at which the particles are closest together.

    Determine the work done on A by each of the forces $\mathbf{F}_1$ and $\mathbf{F}_2$ during the interval $0 \leqslant t \leqslant T$.
    (AEB)

12. A particle of mass $3\,\text{kg}$ moves under the action of a force $\mathbf{F}\,\text{N}$. At time t seconds the velocity $\mathbf{v}\,\text{m s}^{-1}$ of the particle is given by $\mathbf{v} = 3\mathbf{i} + 2t\mathbf{j}$.

   (a) Find $\mathbf{F}$.

   (b) Find the kinetic energy of the particle at time t seconds.

   (c) Given that the position vector of the particle when $t = 0$ is $\mathbf{i} + \mathbf{j}$, find its position vector when $t = 2$. (NEAB)

13. The position vector $\mathbf{r}$ metres of a particle P at time t seconds is given by

$$\mathbf{r} = (\cos 2t)\mathbf{i} - (\sin 2t)\mathbf{j}.$$

   (a) Find the velocity of P at time t seconds.

   (b) Show that the speed of P is constant and find its value. (ULEAC)

14. A particle P, of mass $0.2\,\text{kg}$, moves in a straight line through a fixed point O. At time t seconds after passing through O, the distance of P from O is x metres, the velocity of P is $v\,\text{m s}^{-1}$ and the acceleration of P is $(x^2 + 4)\,\text{m s}^{-2}$.

   (a) Use the information given to form a differential equation in the variables v and x only for the motion of P.

   Given that $v = 2$ when $x = 0$,

   (b) show that $3v^2 = 2x^3 + 24x + 12$.

   (c) Find, in J, the work done on P by the force producing its acceleration as P moves from $x = 0$ to $x = 9$. (ULEAC)

15. A steel ball of mass $0.1\,\text{kg}$ falls vertically through thick oil and, in addition to a constant gravitational force, it is subject to a resistance, the magnitude of which is $49v\,\text{N}$, where $v\,\text{m s}^{-1}$ is the speed of the ball.

   (a) At time $t = 0$, the ball is released from rest.

   (i) Show that the speed of the ball at time t is given by

$$v = 0.02(1 - e^{-490t}).$$

   (ii) Draw a speed–time graph and discuss the motion for large values of t.

   (b) Describe the motion when the ball is projected downwards with speed $0.02\,\text{m s}^{-1}$. (AEB)$_s$

16. An aeroplane of mass $M\,\text{kg}$ moves along a horizontal runway, starting from rest. The aeroplane's engines exert a constant thrust of T newtons and, when the speed of the aeroplane is $v\,\text{m s}^{-1}$, the magnitude of the resistance to motion is kv^2 newtons, where k is a positive constant.

   Show that, to reach a speed of $V\,\text{m s}^{-1}$ on the runway, the aeroplane travels a

   distance $\dfrac{M}{2k} \ln \left(\dfrac{T}{T - kV^2} \right)$ metres. (ULEAC)

17. A child drops an air-filled balloon of mass 30 grams from a bridge 6 metres above a river. The balloon is slowed by air resistance of magnitude R where $R = 0.6v$ newtons and $v\,\mathrm{m\,s^{-1}}$ is the speed of the balloon at time, t. Using $g = 10$ and assuming that the upthrust of the air can be neglected,

   (a) write down an equation of motion for the balloon's descent;

   (b) hence show that the terminal speed of the balloon is $0.5\,\mathrm{m\,s^{-1}}$;

   (c) show that an expression for the velocity in terms of time is given by

$$v = \frac{1 - e^{-20t}}{2};$$

   (d) find how long it takes for the balloon to reach half its terminal speed. (SMP)$_s$

18. The engine of a train of total mass M kilograms generates P watts of power. Resistance to motion is Mkv^2 newtons where v is the speed of the train in metres per second and k is a positive constant. The train starts from rest and continues to accelerate under full power.

Find

   (i) its terminal speed,

   (ii) an expression for the train's acceleration in terms of M, P, k and v,

   (iii) an expression for its speed v in terms of the distance x travelled from its initial position. (MEI)

19. A particle of mass m moves along the positive x-axis under a single force, directed towards the origin O and of magnitude $\dfrac{km}{x}$, where k is a constant.

The points A and B lie on the positive x-axis at distances a and $3a$ from O, respectively. Show that the work done by the force as the particle moves from A to B is

$$-km \ln 3.$$

Given that the particle has speed $2u$ at A and speed u at B, express u^2 in terms of k. (NEAB)

20. A space ship S, of mass M, near the moon, experiences a gravitational force, of magnitude F, which is directed towards O, the centre of the moon. It is known that

$$F = \frac{Mk}{r^2},$$

where $\mathrm{OS} = r$ and k is a constant. By modelling the moon as a sphere of radius $2 \times 10^6\,\mathrm{m}$ and by taking the acceleration due to gravity at the moon's surface to be of magnitude $1.6\,\mathrm{m\,s^{-2}}$,

   (a) find the value, in $\mathrm{m^3\,s^{-2}}$, of k.

The space-ship S moves around the moon in a circular orbit, centre O and radius $3 \times 10^6\,\mathrm{m}$.

   (b) Estimate the speed of S, giving your answer in $\mathrm{m\,s^{-1}}$. (ULEAC)$_s$

21. The magnitude of the gravitational force between two uniform spherical bodies of mass M and m with centres A and B respectively is

$$\frac{GMm}{r^2}$$

where $r = AB$ and $G = 6.67 \times 10^{-11}\,\text{m}^3\,\text{kg}^{-1}\,\text{s}^{-2}$ is the universal gravitational constant.

(a) Given that the moon is a uniform sphere of mass $7.36 \times 10^{22}\,\text{kg}$ and radius $1.74 \times 10^6\,\text{m}$ find, to 2 decimal places, the magnitude of the acceleration due to gravity on the surface of the moon.

(b) Deduce that an astronaut, weighing 750 N on the surface of Earth, will weigh approximately 124 N on the surface of the moon. (AEB)$_s$

22. Assume that the gravitational attraction of the earth on an object of mass m at a distance r from the centre of the earth is $\dfrac{km}{r^2}$, where k is a positive constant. A rocket is launched from the earth's surface, and it travels vertically upwards. When the fuel is exhausted, the distance of the rocket from the centre of the earth is a and the speed of the rocket is u. Some time later, the distance of the rocket from the centre of the earth is x and the speed of the rocket is v. Neglecting any forces other than the gravitational attraction of the earth, find an expression for v.

Deduce that, if $u^2 \geqslant \dfrac{2k}{a}$, the rocket will never fall back to the earth.

(UCLES)$_s$

In questions 23 to 28 a problem is set and is followed by a number of suggested responses. Choose the correct response.

23. A particle P describes SHM of amplitude 1 m. In performing one complete oscillation, P travels a distance:

A 2 m   **B** 0   **C** 4 m   **D** −2 m

24.

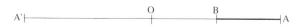

A particle travels between A and A′ with SHM of period 24 seconds. O is the centre and B is the midpoint of AO. The time taken to travel from A to B is

A 3 s   **B** 8 s   **C** 6 s   **D** 4 s

In questions 25 and 26 a particle is moving in a straight line AA′ with SHM. The equation of motion is $\ddot{x} = -4x$ and the amplitude of the motion is 3 m.

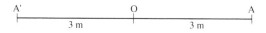

25. The greatest speed, in $m\,s^{-1}$, is

    A   36   **B**   6   **C**   12   **D**   18

26. The time, in seconds, taken to travel from O to A is

    A $\frac{1}{8}\pi$ **B** π **C** $\frac{1}{2}\pi$ **D** $\frac{1}{4}\pi$

27. The period of a simple pendulum at a place where $g = 9.8\,m\,s^{-2}$ is 3 s. This pendulum is taken to another planet and, there, its period is 6 s. The value of g, in $m\,s^{-2}$, on that planet is

    A $\frac{1}{2}(9.8)$ **B** $(9.8)^2$ **C** $\frac{1}{4}(9.8)$ **D** $2(9.8)$

28. If a man has weight W on the surface of a spherical planet of radius R, then his weight at a height R above the surface of that planet is

    A $4W$ **B** $\frac{1}{2}W$ **C** $\frac{1}{4}W$ **D** $2W$

In each question from 29 to 34 a statement is made. Decide, giving reasons for your decision where you can, whether the statement is true (T) or false (F).

29. A particle whose acceleration is proportional to its displacement from a fixed point is moving with SHM.

30. A particle hanging at the end of an elastic string is pulled down and then released. The motion of the particle must be entirely SHM.

31. A particle describing linear SHM on a path AB with midpoint O has its greatest acceleration at either A or B.

32. The work done by any force F in moving an object a distance d is Fd.

33. A particle which is oscillating is not necessarily performing SHM.

34. A particle is moving along a straight line with variable acceleration. If, at some instant, the particle has a maximum velocity, the acceleration at that instant is zero.

35. A particle P, of mass 0.3 kg, moves in a horizontal straight line with simple harmonic motion of period 2 s and maximum speed $4\,m\,s^{-1}$. The centre of the path is O and the point A, on the path of P, is $\dfrac{2}{\pi}$ m from O.

    Find

    (a)   the speed of P as it passes through A,

    (b)   the magnitude of the force acting on P as it passes through A.   (ULEAC)$_s$

36. A particle P describes SHM centre O, period $2\pi/3$ seconds and its maximum speed is $24\,\text{m}\,\text{s}^{-1}$. Find

   (i) the amplitude of the motion,

 (ii) the time taken for P to travel from O directly to a point $4\,\text{m}$ from O.

Given that the particle is of mass $0.25\,\text{kg}$, find

(iii) the rate at which the force acting on P is working when $t = \pi/9$.

                                                                  (WJEC)ₛ

37. The motion of the top of a piston is modelled as Simple Harmonic with a period of $0.1\,\text{s}$ and an amplitude $0.2\,\text{m}$ about a mean position A.

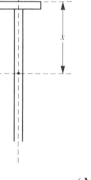

Piston top

   (i) Show that x, the displacement of the top of the piston from A after t seconds, is given by the equation $x = 0.2\sin(20\pi t)$, given that $x = 0$ when $t = 0$.

Where appropriate your answers to the following questions should be expressed in terms of π.

 (ii) What is the greatest piston speed?

(iii) What is the greatest magnitude of the acceleration of the piston?

(iv) For what fraction of the period is $x > 0.1$?   (MEI)

38. A simple pendulum has a period of 1 second. It is tested in two towns, A and B. It has 3601 oscillations in an hour in town A and 3599 oscillations in an hour at town B.

 (a) Compare the values of g for the two towns.

 (b) Give a possible explanation for the difference in g between the two towns.

                                                                 (NEAB)

39. A particle P, of mass $0.01\,\text{kg}$, moves along a straight line with simple harmonic motion. The centre of the motion is the point O. At the points L and M, which are on opposite sides of O, the particle P has speeds of $0.09\,\text{m}\,\text{s}^{-1}$ and $0.06\,\text{m}\,\text{s}^{-1}$ respectively and $2\text{OL} = \text{OM} = 0.02\,\text{m}$.

 (a) Show that the period of this motion is $2\pi\sqrt{\left(\frac{1}{15}\right)}\,\text{s}$.

Find

 (b) the greatest value of the magnitude of the force acting on P, giving your answer to 2 significant figures,

 (c) the time for P to move directly from L through O to M, giving your answer to 2 significant figures.   (ULEAC)

40. Some students are modelling the last 2 seconds of the flight of a bird as it lands at its nest. At time $t = 0$ the bird is moving with speed $3\,\text{m}\,\text{s}^{-1}$ and when $t = 2$ it is at rest. One student proposes a model of the form

$$v = a + bt, \quad 0 \leqslant t \leqslant 2,$$

where v is the speed of the bird at time t seconds and a and b are constants.

(a) Find the values of a and b.

The teacher points out that a further feature of the problem that the student should try to encompass in the model is that as the bird approaches the nest the magnitude of its deceleration decreases. A refined model of the form

$$v = k(2 - t)^2, \quad 0 \leqslant t \leqslant 2,$$

where k is a constant, is proposed.

(b) Find the value of k.

(c) Find the acceleration of the bird as it approaches the nest and show that this extra feature is included.

A camera is set up to film the flight of the bird during these last two seconds and a trigger is to be set up so that when, at $t = 0$, the bird passes the trigger the camera is automatically switched on.

(d) Find how far from the nest the trigger should be placed. (ULEAC)

41. A cyclist moves against a total resisting force of magnitude $4v\,\text{N}$, where $v\,\text{m}\,\text{s}^{-1}$ is the speed of the cyclist. The total mass of the cyclist and cycle is $100\,\text{kg}$. Given that the cyclist is working at a rate of $56\,\text{W}$ at all times, find the maximum speed which the cyclist reaches when travelling down a slope of inclination θ, where $\sin\theta = \frac{1}{49}$. (AEB)$_s$

42. A space-ship S, of mass M, is orbiting the moon $10^6\,\text{m}$ above its surface with constant speed $v\,\text{m}\,\text{s}^{-1}$. In a preliminary model of this situation the moon is modelled as a sphere of radius $2 \times 10^6\,\text{m}$, the space-ship as a particle and the acceleration due to gravity of the moon is modelled by the constant value $1.6\,\text{m}\,\text{s}^{-2}$. The space-ship travels round the moon in a circular orbit.

(a) Estimate, to 3 significant figures, the value of v.

A more refined model of the gravitational force F, experienced by S, is

$$F = \frac{Mk}{r^2},$$

where F is directed towards O, the centre of the moon, and k is a constant. Given that the acceleration due to gravity at the moon's surface is of magnitude $1.6\,\text{m}\,\text{s}^{-2}$,

(b) show that $k \doteqdot 6.4 \times 10^{12}\,\text{m}^3\,\text{s}^{-2}$.

(c) Find a revised estimate for v. (ULEAC)

43. A light elastic string, of natural length L and modulus $2mg$, has one end attached to a fixed point A. A particle of P, of mass m, is attached to the other end of the string and hangs freely in equilibrium at the point O, vertically below A.

(a) Find, in terms of L, the length OA.

The particle P is pulled down a vertical distance h below O, and released from rest at time $t = 0$. At time t, the displacement of P from O is x.

(b) Show that, while the string is taut,

$$\frac{d^2x}{dt^2} = -\frac{2gx}{L}.$$

(c) State the set of values of h for which P performs complete simple harmonic oscillations.

Given that $h = \frac{1}{3}L$,

(d) find the time at which P first comes instantaneously to rest,

(e) find the greatest speed of P.   (ULEAC)

44. A spaceship of mass 10^4 kg is in a circular orbit 10^6 m above the surface of a planet whose diameter is 4×10^6 m.
The mutual force of attraction between the spaceship and the planet can be written as $\dfrac{k}{r^2}$, where k is a constant with units N m^2 and r is the distance in metres between the centre of the planet and the spaceship.

(a) The acceleration due to gravity on the planet's surface is $1.5 \, \text{m s}^{-2}$. Show that $k = 6 \times 10^{16}$.

(b) Find

    (i) the period of the orbit

    (ii) the speed $v \, \text{m s}^{-1}$ of the spaceship.

(c) The spaceship uses its engines to escape from its orbit about the planet. The engines apply an impulse of magnitude I N s at angle α to the spaceship's present direction of motion, as illustrated.

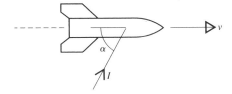

    The result is that the speed of the spaceship is doubled and its new direction of motion is at an angle θ to its original direction of motion.

    (i) Show that $I = 10^4 v\sqrt{(5 - 4\cos\theta)}$.

    (ii) Hence find the minimum value of I as θ varies.

    (iii) State the direction of the resulting velocity, in the case when $I = 10^4 v\sqrt{3}$ (UODLE)

45. A particle of mass m falls from rest, under gravity, in a medium in which the resistance to its motion is mkv, where k is a constant and v is the speed of the particle. Write down the equation of motion for the particle. If the motion were to continue indefinitely, v would approach a constant value V.

Show that $k = \dfrac{g}{V}$. Hence show that $\dfrac{dv}{dt} = \dfrac{g}{V}(V - v)$.

Show that the particle is moving with velocity $\dfrac{V}{2}$ after a time $\left(\dfrac{V}{g}\right) \ln 2$.

Show also that, during this time interval, the particle has fallen a distance s given by

$$\frac{V}{g} \int_0^{\frac{V}{2}} \frac{v}{V - v}\, dv$$

Hence show that

$$s = \frac{V^2}{g} \left(\ln 2 - \frac{1}{2} \right)$$

Find the average speed, in the form λV, during this time interval, expressing λ correct to 2 decimal places. (NEAB)

46. A cyclist is travelling along a road.

(a) What can be deduced about the resultant force on the cycle if they are travelling at top speed
   (i) along a straight road,
   (ii) along a winding road?

A cyclist whose maximum rate of working is 600 W can reach a top speed of $10\,\mathrm{m\,s^{-1}}$ on a level road. The combined mass of the cycle and cyclist is 90 kg.

(b) By assuming that the resistance forces of the cycle and cyclist are proportional to their speed, find a simple model for the total resistance force.

(c) By assuming that the forward force on the cyclist is constant, show that

$$\frac{dv}{dt} = \frac{10 - v}{15}$$

where $v\,\mathrm{m\,s^{-1}}$ is the speed of the cyclist at time t seconds.

(d) Find an expression for the speed of the cyclist in terms of time, if the cyclist starts at rest.

(e) Criticise your model for the resistance on the cyclist. (AEB)

47. The diagram shows a framework that is smoothly hinged to a fixed point at D. The points A, B, C and D form a square that lies in a vertical plane. A vertical force of 100 N is applied at A as shown in the diagram.

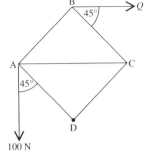

(a) What two key assumptions need to be made if you are to find the force in each member of the framework?

(b) Find the magnitude, Q, of the horizontal force that acts at B if the framework is to remain at rest.

(c) Show that the magnitude of the force in BC is $25\sqrt{2}\,N$ and find the magnitudes of the forces in the other members of the framework.

(d) Which of the members could be replaced by ropes? (AEB)

ANSWERS

CHAPTER 1

Exercise 1a – p.5

1. (a) 6 m (b) 2 m/s (c) 1.5 m/s
(d) 3 s (e) 1.2 m/s

2.

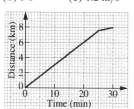

4.45 m/s

3. (a)

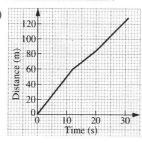

(b) 5 m/s; 3 m/s; 4 m/s
(c) (i) 4.2 m/s (ii) 4.13 m/s (3 sf)

4. (a) 5.2 s
(b) (i) 1.4 cm/s (ii) 3.4 cm/s
(c) (i) 1.6 cm/s (ii) 2.6 cm/s

5.

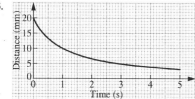

(a) 0.8 m/s (b) (i) 1.8 m/s (ii) 2.4 m/s
(c) 5.5 s

6.

(a) 1.6 mm/s (b) 1.5 s
(c) 3.3 mm/s

Exercise 1b – p.12

1. (a) 8 mph/minute (b) 70 mph
(c) 3 minutes
(d) 4.3 mph/minute

2.

(a) 0.8 m/s² (b) 2.5 m/s²
(c) 240 m

3. (a) (i) 20 s (ii) 15 s (iii) 65 s
(b) (i) 0.5 m/s² (ii) 0.67 m/s²
(c) 825 m

4. 15 s; 495 m

5. (a) (i) 17.6 m/s (ii) 1 m/s
(b) decelerating (c) 4 m/s²
(d) 84 m

6.

(a) (i) ≈ 4 m/s² (ii) −3.5 m/s²
(b) 4 s
(c) 75 m; an underestimate because the
area of each trapezium is less than the
area under the corresponding curve

7. (a) 26 km/min² (b) 21 km (c) 17 km

8. **B**;
A false as gradient of tangent varies
C false as gradient not horizontal when
$t = 0$

9. **C**;
A speed decreases then increases
B gradient is constant
D graph is above the time axis

10. C;
   A line drawn at 40 does not bisect the area
   B gradient varies
   D area under graph not a triangle
11. B   **12.** B
13. D;
   A $v = 15$
   B v still positive
   C tangent inclined at about 45° but scales
   different.

Exercise 1c – p. 18
1. (a), (d) and (f) are vectors
2. (a) (i) 2.3 m   (ii) 0   (iii) −1.2 m
   (b) (i) 3.4 m   (ii) 14 m
   (c) 0.08 m/s

3.
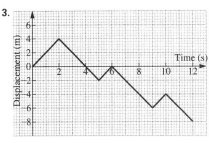

Exercise 1d – p. 20
1. (a) Correct
   (b) Incorrect; the direction is changing all
   the time
   (c) Incorrect; we do not know whether
   the speed is constant
2. (a) 1 m/s   (b) −2 m/s
   (c) 2 m/s   (d) −0.5 m/s
3. −4 m/s²
4. (a) −2 m/s²   (b) −3 m/s²
5. (a) 23 m/s   (b) −7 m/s
   (c) 5 m/s

Exercise 1e – p. 25
1. (a) (i)   15 m/s   (ii) −8 m/s   (iii) 0
   (b) (i) 0
      (ii) Ball hits wall and its direction is
         reversed
   (c) After 3 s

2. (a) (i) 2 m/s   (ii) −2 m/s
      (iii) 0   (iv) −4 m/s
   (b) (i) 2 m/s   (ii) 2 m/s
      (iii) 2 m/s   (iv) 4 m/s
   (c) ≈ 4 m/s   (d) $t = 2$

2.
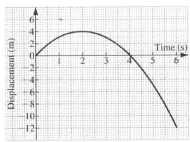

3. (a) s   0   4   6   6   4   0   −6
   (b)
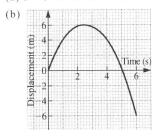

   (c) 2.5 s   (d) −5 m/s
   (e) (i) −1 m/s   (ii) 3 m/s
4. (a) 1 s, 1.75 s, 2.5 s, 3 s, 3.5 s, 3.8 s
   (b) (i) 1 m/s   (ii) 0.8 m/s
      (iii) 0.7 m/s
   (c) (i) −1 m/s   (ii) 0   (iii) 0
   (d) −0.22 m/s
   (e) (i) −0.85 m/s   (ii) 0.5 m/s

Exercise 1f – p. 29
1. (a) 3 m/s²   (b) 1 m/s²
   (c) (i) 152 m   (ii) 400 m   (iii) 200 m;
      the car travels in the same direction all
      the time.

2.
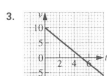
   360 m

3.
   (a) $t = 5$
   (b) 16 m
   (c) 34 m

4. (a) (i) Accelerates at a reducing rate to zero acceleration
   (ii) Constant velocity
   (iii) Accelerates at varying rate
   (b) 185 m; less, because trapeziums have smaller area than that under the curve
5. (a) (i) after $8\frac{1}{2}$ s (ii) after 20 s
   (b) ≈ 1 m/s^2; ≈ -0.5 m/s^2
   (c) 6 s, 15 s after starting
   (d) The girl stops accelerating and begins to slow down.
6. C 7. B 8. C 9. A

CHAPTER 2

Exercise 2a – p. 38
1. 5 2. -1.2 3. 6 4. 60
5. 2 6. -2 7. 6 8. 1.5
9. -54 10. -8
11. No. He needs 85 m to come to rest and is travelling at 17 m/s at collision
12. 0.32 m/s^2; 4.38 m/s
13. (a) 16 m (b) 7 m
14. (a) $t = 2$ and 5 (b) 8.22 s
15. About 7 yards (0.0042 miles)
16. $-\frac{1}{15}$ m/s^2
17. (a) 6.5 m/s (b) -1.5 m/s^2
   (c) 8.7 s
18. 48 m 19. 612 m
20. 12.4 s; 617 m
22. Yes, after 0.78 s

Exercise 2b – p. 44
1. (a) 120 m (b) 11 m/s
2. (a) 31 m (b) 5 s
3. (a) 4.9 m (b) 19.6 m (c) 24.5 m
4. (a) 11.0 m (b) 14.7 m/s
5. (a) 3.2 s (b) 31 m/s
6. (a) (i) 62.7 m/s (ii) 62.5 m/s
   (b) The value of g depends upon the distance from the centre of the earth and increases as the distance decreases. Because the earth is not a perfect sphere, this distance is smaller at the poles than at the equator.
.7. 50 m 8. 28 m/s
9. (a) 1.1 s
   (b) 2.6 s; 18 m/s in each case
10. 18 m
11. (a) 11 m (b) 2.6 s
12. 0.4 m 13. 440 m
14. 13 m/s; 2 s 15. 37 cubits/s

CHAPTER 3

Exercise 3a – p. 52
1. (a) $-\mathbf{a} + \mathbf{b}$ (b) $-\mathbf{b} + \mathbf{a}$
2. (a)   (b)
   (c)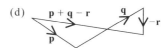
   (d)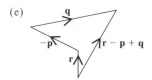
   (e)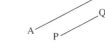

3. (a) $-\mathbf{p} + \mathbf{q}$ (b) $-\mathbf{q} + \mathbf{r}$
   (c) $-\mathbf{p} + \mathbf{r}$
4. (a) $-\mathbf{a} + \mathbf{b}$ (b) $\frac{1}{2}(\mathbf{b} - \mathbf{a})$
   (c) $\frac{1}{2}(\mathbf{a} - \mathbf{b})$
5. (a) $\mathbf{a} + \mathbf{b}$ (b) $\mathbf{b} + \mathbf{c}$
   (c) $-(\mathbf{b} + \mathbf{c})$ (d) $-(\mathbf{a} + \mathbf{b} + \mathbf{c})$
6. (a)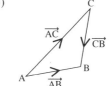
   (b)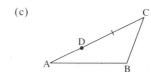
   (c)

8. (a) $\mathbf{b}$ (b) $-\mathbf{b}$ (c) $2(\mathbf{a} - \mathbf{b})$
10. (a) (i) $\overrightarrow{AD}$ (ii) $\overrightarrow{AC}$ (iii) $\overrightarrow{EB}$
   (b) $\overrightarrow{AE} + \overrightarrow{ED}$, $\overrightarrow{AB} + \overrightarrow{BC} + \overrightarrow{CD}$, $\overrightarrow{AB} + \overrightarrow{BD}$, etc
11. 2

12. (a) $\overrightarrow{BE} = \mathbf{b} - 2\mathbf{a}$, $\overrightarrow{CF} = \mathbf{a} - 2\mathbf{b}$
     (b) $h = \frac{1}{3} = k$
     (c) $2:1$

Exercise 3b – p. 59

1. 13 km, 157°
2. 13 km, 157°, No
3. 25 m/s, 016°
4. 20 N, 217°
5. 20.3 m, 065°
6. 7.81 m/s, 146°
7. 96.4 N, 39° to 40 N force
8. 5.2 m/s, 30° to 6 m/s
9. 14.9 m, 314°
10. 4.95 km, 046°
11. 444 km/h, 005°
12. 1100 m, 103°; 283°
13. Downriver at 53° to bank; 5 km/h
14.

                         2.65 km/h

15. 166 km/h, 102°
16. (a) at 27° to PQ out of harbour
     (b) at 30° to PQ into harbour

Exercise 3c – p. 63

1. (a) $6 \sin 25°$ (b) $10 \sin 20°$
    (c) $52 \cos 20°$ (d) $20 \cos 60°$
    (e) $2 \cos 50°$ (f) $8 \sin 40°$

2.   **3.**

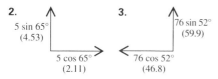

4.   **5.**

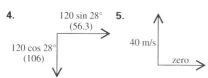

6.   **7.**

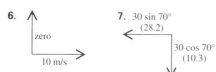

8.   **9.**

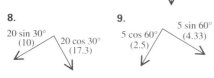

10.

11.

     $7 \sin 68°$
     (6.49) $7 \cos 68°$
                 (2.62)

12.

     $8 \sin 45°$ $8 \cos 45°$
     (5.66)   (5.66)

13.

     $100 \cos 40°$ $100 \sin 40°$
     (76.6)   (64.3)

14.

     $20 \cos 20°$ $20 \sin 20°$
     (18.8)   (6.84)

15.

     $45 \sin 10°$ $45 \cos 10°$
     (7.81)   (44.3)

16.

     32.9 m/s
                12.0 m/s

17. 50.8 mph parallel
     114 mph perpendicular
18. 68.8 m/s parallel
     162 m/s perpendicular
19. 80.6 N parallel
     59.1 N perpendicular

Exercise 3d – p. 68

1. (a) $4\mathbf{i} + 3\mathbf{j}$ (b) $4\mathbf{j}$
    (c) $-2\mathbf{i} + 4\mathbf{j}$ (d) $-4\mathbf{j}$
    (e) $-5\mathbf{i}$ (f) $-2\mathbf{i} - 3\mathbf{j}$
2. (a) 5   (b) 4   (c) $\sqrt{20}$
    (d) 4   (e) 5   (f) $\sqrt{13}$
3. (a) $12.1\mathbf{i} + 7\mathbf{j}$ (b) $-20\mathbf{j}$
    (c) $-53\mathbf{i} - 53\mathbf{j}$ · (d) $240\mathbf{i} + 320\mathbf{j}$
4. (a) $9\mathbf{i} - 4\mathbf{j}$

(b) $3\mathbf{i} + 6\mathbf{j}$

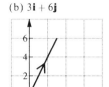

5. (a) $\mathbf{i} - \mathbf{j}$ (b) $\mathbf{i} + \mathbf{j}$

6. (a) $-\mathbf{i} - 2\mathbf{j}$ (b) $-3\mathbf{i}$

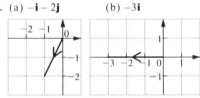

7. (a) 5   (b) 13   (c) $\sqrt{2}$
8. $24\mathbf{i} - 18\mathbf{j}$
9. (a) $-11\mathbf{i} + 4\mathbf{j}$ (b) $\sqrt{137}$
10. $8\mathbf{i} - 6\mathbf{j}$ or $-8\mathbf{i} + 6\mathbf{j}$
11. (a) $-35\mathbf{i}$ (b) $8\mathbf{i} - 4\mathbf{j}$

CHAPTER 4

Exercise 4a – p. 75

1.

2.

3.

4.

5.

6.

7.
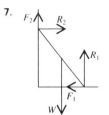

8. (a)   (b)   (c)

9. (a)   (b)   **10.**

11.

12.
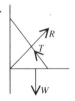

13.
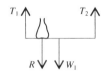

14.

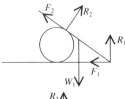
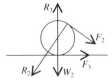

15. (a)   (b)

Exercise 4b – p. 82
1. 7.17 N, 33°
2. 1.43 N, 170°
3. 22.9 N, 234°
4. 12.1 N, 38°
5. 8.28 N, 51°
6. 17.2 N, 63°
7. (a)

4A3

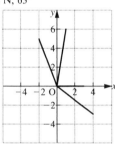

(b) 6**i** + 8**j**   (c) 10 N, 53°

8. (a)

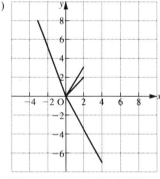

(b) 12**i** + 5**j**   (c) 13 m/s, 23°

9. (a)

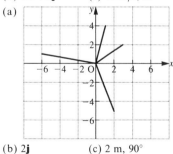

(b) 2**j**   (c) 2 m, 90°

10. (a)

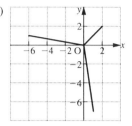

(b) −3**i** − 4**j**
(c) 5 N, 127°

11. 8.22 N, 49°
12. 261 N, bearing 305°
13. 292 N, 6°
14. 9**i** + 5**j**
15. $(3 − 2\sqrt{2})\mathbf{i} + (5 − 2\sqrt{2})\mathbf{j}$; 2.18 m/s
16. 7.08 km/h, 133°
17. (a)

(b) 0.364**i** + 7.71**j**   (c) 003°
(d) 7.71 m/s
18. 1.01 N, 97°

CONSOLIDATION A

Miscellaneous Exercise A – p. 87
1. C   2. A
3. C   4. B
5. $|−2\mathbf{i} + 3\mathbf{j}| = 3.61$ N; 124°
6. $\overrightarrow{OC} = 2\mathbf{i} + 14\mathbf{j}$, $\overrightarrow{OD} = 6\mathbf{i} + 12\mathbf{j}$

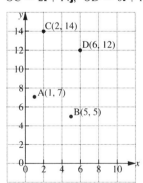

7. (a) 9**i** + 17**j**   (b) 19.2 N
(c) 0.468 ($9/\sqrt{370}$)
8. The midpoint of XY

9.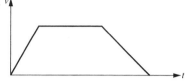
(a) 40 km/h
(b) 12 minutes   (c) 16 km
10. $s = 19.6(t - 2) - 4.9(t - 2)^2$,
  $s = 19.6t - 4.9t^2$; 3, 9.8 m/s, 9.8 m/s
11. 2
12. (a) 5 min 20 s
  (b) increase of 2 min 20 s
13. (a) 1.875 m/s²   (b) $\frac{1}{2}$
14.

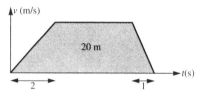

15. (a) 50 s   (b) 24.2 m/s
16. (a)

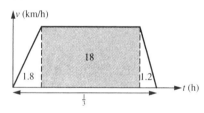

(b) 72 km/h   (c) 2160 km/h²
17. (a) 5.55 km   (b) 2.12 minutes
18. $(k - 4)\mathbf{i} + 4\mathbf{j}$; $\sqrt{(k^2 - 8k + 32)}$; 1, 7; $\frac{4}{3}$
19. $4v$, v, $v^2 - 60v + 800 = 0$, $v = 20$
  (for $v = 40$, it takes 160 s to accelerate,
  which is greater than the total time, so
  $v \neq 40$)
20. $\frac{2}{9}\mathbf{a}$
21. F   **22.** F   **23.** F   **24.** F
25. (b) $\overrightarrow{PN} = \mathbf{q} - 2\mathbf{p}$, $\overrightarrow{QM} = \mathbf{p} - 2\mathbf{q}$
  (d) $2\mu\mathbf{q} + (1 - \mu)\mathbf{p}$
26.

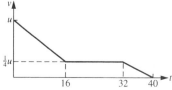

(a) $\frac{3}{64}u$ m/s²   (b) $\frac{1}{32}u$ m/s²
(c) 3 m/s
27.

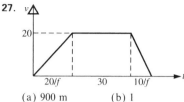

(a) 900 m   (b) 1
28. (a)

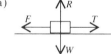

(b) $V = 2.3T$

CHAPTER 5

Exercise 5a – p. 94
1. $P = 10$
2. $P = 20$, $Q = 90$
3. $P = 12$, $Q = 4$
4. $P = 36$, $Q = 28$
5. $P = 18$
6. $P = 12$, $Q = 26$
7. Yes, horizontal
8. Yes, vertical
9. No
10. Yes, neither vertical nor horizontal
11. No
12. Yes, neither horizontal nor vertical
13. (a)

(b) (i) $T > F$ (ii) $T = F$
14. Yes ∥ to $\mathbf{j}$
15. (a) $R \cos 30° + N - W$
  (b) $R \sin 30° - F$
  (c) $R\frac{\sqrt{3}}{2} + N - W = 0$, $\frac{1}{2}R - F = 0$
16.

$S - F = 0$
$R - 2W = 0$

17.

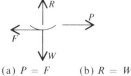

(a) $P = F$ (b) $R = W$

Exercise 5b – p. 99

1. 1.5 m/s^2
2. 28 N
3. 40 kg
4. $\dfrac{5\mathbf{i} - 12\mathbf{j}}{26}$, $\frac{1}{2} \text{ m/s}^2$
5. $5(7\mathbf{i} + 2\mathbf{j}) = 35\mathbf{i} + 10\mathbf{j}$ (in newtons)
6. (a) 38 N
   (b) $P = 10 \text{ N}$, $Q = 8 \text{ N}$
   (c) $P = 34 \text{ N}$, $Q = 30 \text{ N}$
7. (a) $6 \text{ m/s}^2 \rightarrow$ (b) $16 \text{ m/s}^2 \rightarrow$
   (c) $\frac{20}{3} \text{ m/s}^2 \rightarrow$
8. (a) $m = 4$, $P = 23 \text{ N}$
   (b) $m = 5$, $P = 30$
   (c) $m = 10$, $P = 40$
9. $P = 8$, $Q = 31$
10. 8 N
11. $156\frac{1}{4}$ m
12. 30 N
13. (a) $-0.4\mathbf{i} + \mathbf{j}$ (b) $-0.8\mathbf{i} + 2\mathbf{j}$
14. 16 kg
15. 100 m
16. 14 m/s

Exercise 5c – p. 102

1. (a) 49 N (b) 15 kg (c) 0.59 N
2. (a) 6 N (b) 122.5 kg (c) 0.072 N
3. 31.2 N
4. (a) 28.8 N (b) 101 N (c) 58.8 N
5. (a) 5.47 kg (b) 87.5 kg (c) 7.14 kg
   (d) 7.14 kg
6. (a) (i) 18 000 N (ii) 6790 N
   (b) 750 kg
7. 10 920 N
8. (a) 59.2 N (b) 19.2 N

Exercise 5d – p. 106

1. (a) $(g + 1) \text{ N}$ (b) $(g + 5) \text{ N}$
2. (a) 583 kg (b) 5070 N

3. (a) (b) (c)

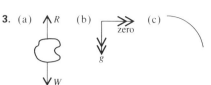

zero

4.

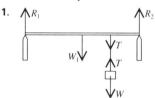

18   66 N   5g

5. 41 m

Exercise 5e – p. 109

1.

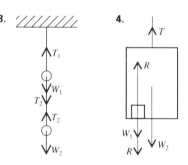

2.

3.

4.

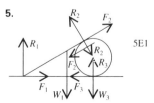

5.

5E1

6.

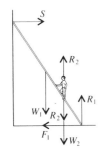

Exercise 5f – p. 115

1. (a) $a = g/2$, $T = 3g$
   (b) $a = 5E^2 g/5$, $T = 12g$
   (c) $a = \left(\dfrac{M-m}{M+m}\right)g$, $T = \dfrac{2Mmg}{M+m}$
2. (a) $a = g/3$, $T = 20g/3$
   (b) $a = g/5$, $T = 48g/5$
   (c) $a = g/3$, $T = 4Mg/3$
3. (a) $g/3$ (b) $25g/6$
4. (a) $(\sqrt{3}-1)g/3$ (b) $2(\sqrt{3}+2)g/3$
5. (a) $3g/8$ (b) $15g/8$
6. $g(\sqrt{2}+\sqrt{6})$ N, bisecting angle between faces
7. g metres; g m/s
8. (a) $g/6$
   (b) $T_1 = 10mg/3$, $T_2 = 7mg/3$
9. (a) $g/9$ (b) $\sqrt{2g}/3$
   (c) $\frac{1}{3}\sqrt{\dfrac{2}{g}}$
10. (a) $g/3$ (b) $g/18$
11. (a) $\sqrt{g/2}$ (b) g N
12. (a) g (b) 0
13. (a) 3.3 m/s^2 ($\frac{1}{3}g$) (b) 2.8 m/s, ($\sqrt{0.8g}$)
14. (a) freely under gravity
   (b) 1.25 m/s

Exercise 5g – p. 121

Model each object as a particle, ignore any resistance unless specifically mentioned. Assume cables, tow-bars etc. to be straight and inextensible.

1. (a) $\frac{1885}{3}g$ N, $580g$ N, $\frac{260}{3}g$ N
   (b) $522g$ N, $58g$ N, $72g$ N
2. (a) 1480 N   (b) 3330 N
3. (a) 3.08 m/s^2 (b) −0.25 m/s^2
   (c) 80 s
4. $a = 2$ m/s^2, $T = 600$ N
5. (a) (i) $35(2g+1)$ N; a particle or a
            small block   (ii) $435(2g+1)$ N
   (b) (i) $35(2g-1)$ N   (ii) $435(2g-1)$ N

CHAPTER 6

Exercise 6a – p. 125

1. (a) $P = 30\sqrt{3}$, $Q = 30$
   (b) $P = 12$, $Q = 12\sqrt{3}$
   (c) $P = 100 \sin 70° = 34.2$,
       $Q = 100 \cos 20° = 94.0$
2. (a) $P = 24$, $Q = 22\sqrt{3}$
   (b) $P = 24$, $Q = 15\sqrt{3}$
   (c) $P = 8$, $\theta = 60°$
3. (a) $P = 13\sqrt{3}$, $Q = 0$
   (b) $\theta = 60°$, $P = 4\sqrt{3}$
   (c) $P = 10$, $Q = 8$
4. $5g$ N
5. $90°$; $T_1 = 12$ N, $T_2 = 16$ N
6. $w/\sqrt{3}$ $(w\sqrt{3}/3)$
7. $w/2$
8. (a) $30°$ (b) $18°$
9. (anticlockwise from P) $P\mathbf{i}$, $Q\mathbf{j}$,
   $\frac{5\sqrt{3}}{2}\mathbf{i} - \frac{5}{2}\mathbf{j}$, $\frac{-3\sqrt{3}}{2}\mathbf{i} - \frac{3}{2}\mathbf{j}$, $2\sqrt{3}\mathbf{i} - 6\mathbf{j}$;
   $P = -3\sqrt{3}$, $Q = 10$
10. (a) $q = -16$, $p = 0$
   (b) $p = 4$, $q = -7$
11. (a) $a = 7$, $b = -11$ (b) $-11\mathbf{i} + 2\mathbf{j}$

Exercise 6b – p. 132

1. $\sqrt{3}/3$ (0.577)   **2.** $\frac{1}{3}$
3. $6(2+\sqrt{3})$ N (22.4 N)
4. $\frac{1}{3-\sqrt{3}}$ (0.789)
5. $12(1+\sqrt{3}/5)$ N (16.2)
6. $\frac{1}{2}$
7. (a) 17.4 N   (b) 9.18 N
8. (b) $11w/23$
   (c) $\left(\dfrac{2+5\sqrt{3}}{5-2\sqrt{3}}\right)w$ ($\approx 6.94w$)
9. (a) 10 N, no
   (b) 40 N, no (just on the point of
       slipping)
   (c) 40 N, yes
10. 0.309
Assume no other resistance to motion; model porter and trolley each as a particle
11. $3g/10$; $7Mg/10$
12. $g\sqrt{2}/15$; $2mg\sqrt{2}/3$

CHAPTER 7

Wherever it is appropriate in Exercises 7a, 7b and 7c all large objects are treated as particles and air resistance, unless specified, is ignored.

Exercise 7a – p. 139

1. (anticlockwise from 20 N)
   $20\sqrt{3}$ J, 0, -24 J, 0
2. (anticlockwise from 8 N)
   -24 J, 0, 48 J
3. (anticlockwise from 7 N)
   28 J, 0, -8 J, -16 J
4. 63 J
5. 6100 J; assume all bales raised to exactly loft floor level
6. 3500 J   **7.** 24 kJ   **8.** 110 kJ
9. (a) 180 J   (b) 290 J
   Assume constant speed
10. 2800 kJ
11. (a) 1100 J   (b) 8300 J
12. -4400 J; assume rope doesn't stretch
13. 1400 J; assume steady speed and rope doesn't stretch
14. 25 N; assume steady speed and rope doesn't stretch
15. 100 J; assume steady speed and rope doesn't stretch
16. $\frac{1}{4}$; assume steady speed and rope doesn't stretch
17. 11 kJ; 14 kg; assume constant speed
18. (a) 270 N   (b) 170 N
    (c) 250 J   (d) 930 J
    Assume steady speed and rope doesn't stretch
19. 7500 J   (a) (i) 500 N↓   (ii) 1300 N↓
    (b) 12 kJ
20. (a) $\frac{1}{4}g$ N   (b) $\frac{1}{2}g$ N   (c) $\frac{1}{2}$ kg
    (d) $\frac{1}{2}g$ J   (e) $\frac{1}{2}g$ J

Exercise 7b – p. 146

1. 270 W   **2.** 2 kW
3. 36 W   **4.** 4.8 W
5. 8.6 W
6. (a) 24 kW   (b) 1800 N
   (c) 20 m/s
7. 1000 kW   **8.** 1400 N
9. 360 W; 4.6 m/s
10. (a) 17 m/s
    (b) 40 m/s; e.g. resistance unlikely to be constant in different conditions
11. (a) 31 m/s   (b) 23 m/s
    (c) 26 m/s

12. (a) 29 m/s   (b) 40%
13. (a) 96 kW   (b) 6 kN
    (c) 16 m/s
14. (a) 500/16   (b) 33 m/s   (c) 2°

Exercise 7c – p. 150

1. (a) 38 kW   (b) 0.61 m/s²
2. 0.65 m/s²   **3.** 2.2 m/s²   **4.** 0.13 m/s²
5. (a) 30 kW   (b) 1.2 m/s²
6. (a) 19   (b) 0.48 m/s²
7. (a) 14 m/s   (b) 0.55 m/s²
8. (a) 310 N   (b) 0.5 m/s²
9. (a) $R = 1000$, $H = 21$
   (b) 0.25 m/s²
10. 1.8 m/s²

CHAPTER 8

Exercise 8a – p. 156

1. (a) 431   (b) 0.816
2. (a) 320 J   (b) 60 000 J
   (c) 2000 J   (d) 65 J
3. 14 J, 4.24 m/s, 22 kg
4. Treat woman as a particle
   (a) (i) 0   (ii) 4200 J
   (b) 4800 J   (c) 1900 J
   (d) 1400 J
5. (a) 144 J   (b) 598 J
6. (a) 79 kJ   (b) 25 m/s (90 k/h)
7. (a) 10 m/s   (b) 25 m
8. Treat 45 g of the water as a particle
   (a) 5300 J   (b) 3200 J
9. Treat volume of water discharged per second as a particle
   (a) 0.48 m³   (b) 480 kg
   (c) 39 kJ

Exercise 8b – p. 160

1. 22.9 N; assume string light and inextensible
2. 576 J
3. 1.64 N; 6.41 m/s; assume resistance constant
4. 11 J; 6.2 N (assume zero resistance)
5. 12.4 m/s
6. (ignore air resistance)
   (a) 6.20 m/s   (b) 4.77 m/s
7. 21.6 kN
8. (a) 137 J   (b) 148 J
9. (a) 16.1 m   (b) 2.99 m/s
10. 26.1 m/s; ignore air resistance
11. 3.4 N; assume resistance constant

12. (a) 0.5 m^3 (b) 5000 N ($g = 10$)
(c) 64 kJ
13. (a) 2 (b) 10 kW
14. 3.8 m/s
15. (a) 16 J (b) 36 J
16. (a) 203 J (b) 34 J (c) 0.077
17. (a) $\mu mgd \cos \theta$ (b) $mgd \sin \theta$
(c) $mgd (\mu \cos \theta + \sin \theta) + \frac{1}{2} m (v^2 - u^2)$

Exercise 8c – p. 165

1. 1.84 m
2. 9.86 m/s
3. (a) 4.3 m (b) 6.1 m/s
(c) 9.1 m/s
4. 1.63 m **5.** 4.85 m/s
6. 3.93 m/s **7.** 3.13 m/s
8. 0.3 m **9.** 5.6 m/s
10. $\sqrt{gl}$ **11.** $\sqrt{2gl/3}$
12. 4.8 m/s; treat Sue as a particle, rope as light and inextensible, no air resistance
13. (a) 6.3 m/s (b) 3.1 m/s
Assumptions as in Question 12.
14. 1.9 m. Model girl plus seat as a particle – a very rough approximation. Other assumptions as in Question 12
15. 18 m/s

CONSOLIDATION B

Miscellaneous Exercise B – p. 171

1. $P \cos \theta = 4$, $P \sin \theta = \sqrt{3}$,
$\tan \theta = \frac{1}{4}\sqrt{3}$, $P = \sqrt{19}$
2. (a) 57.3 N (b) 53.9 N
3. 256 s, 7440 m **4.** $\dfrac{u^2}{g}$, $\dfrac{2u^2}{3g}$
5. (a) 15 m/s (b) 112 N
6. (a) 4.4 N in direction of Ox
(i) 4.4 N in direction of xO
(ii) 44 m/s^2
(b) 22.8 N at $-21.6°$ to PA
7. (a) 2.7 N (b) 10.6 N
8. 1.4 m/s^2, 0.672 N **9.** $\sqrt{2gl/3}$
10. $a = \frac{1}{3}g$, $T = \frac{4}{3}mg$
11. (b) 0.632 s, 3.79 m/s (c) 3.12 m
12. (a) 0.61 m/s^2 (b) 27.57 N
13. C **14.** A **15.** B **16.** C
17. (a) 20 m/s (b) 11.44 m/s
18. 700 N, 38°; 1400 J; 140 kg
19. 9 770 000 J
20. (a) 11 000 J (b) 122 N (c) 32.1 m

21. (a) 11.2 m (b) 9.5 J
22. 20 N
23. T; any value up to μR
24. F **25.** T
26. T; there is a resultant force upwards so $T - mg > 0$
27. T; no acceleration so resultant force, $T - R$, must be zero.
28. T; if a vector is constant its direction is constant.
29. F; total energy is constant but KE and PE can interchange.
30. T; a force perpendicular to the direction of motion does no work.
31. 0.164 m/s^2
32. 450 N; $\frac{9}{16}$ m/s^2
33. 350 kW (a) 400 s (b) 8 km
34. (b) 17 000 J
35. (b) 18 m/s
36. (a) 0.25 m/s^2
(b) 23 m/s; 12 s; 188 m
37. 4000 kg, 0.25
38. (b) 20 s (c) 885 W (d) 0.5 m/s^2
39. (a) 2000 J; 2000 J; $2\sqrt{10}$ m/s (6.32 m/s)
(b) 40.1 m/s; 64 360 J

CHAPTER 9

Exercise 9a – p. 183

1. 10 N **2.** 5.14 N **3.** 0.6 m
4. (a) 2.88 m (b) 2.24 m (c) 1.92 m
5. (a) 0.45 m (b) 0.27 m (c) 0.72 m
6. (a) 7.2 (b) 3.6 (c) 1.8
7. (a) 1.38 m (b) 1.1 m (c) 0.688 m
8. (a) 10 N (b) 15 m/s^2
(c) (i) 4 N; 10 m/s^2
(ii) 2 N; 5 m/s^2 (iii) 0; 0
9. 150 N
10. Ben: 780 N; Tony: 810 N;
Tony is the stronger.

Exercise 9b – p. 187

1. 4.70 **2.** 0.823 m
3. 4.90 **4.** 58.8
5. 0.316 m **6.** 1.43
7. 1.5 kg **8.** $\frac{16}{7}a$
9. 0.5 m **10.** $\frac{5}{3}a$
11. $7\frac{9}{13}$ N (7.69 2 sf)
12. Natural length 2 m
Modulus of elasticity 40 N
13. 21.1 N **14.** (a) 30° (b) $\frac{1}{2}mg$
15. (a) $1 - x$ (b) $\frac{2}{3}$ (c) 2.95 kg

Exercise 9c – p. 191

1. (a) 6 N   (b) $50 \, \text{m/s}^2$
2. 200 N
3. (a) 3.3 cm   (b) 6.5 N
4. (a) 1440 N   (b) 0.24 m
5. (a) 39.5 cm   (b) 21.3 cm   (c) 11.2 cm
6. (a) $\frac{77}{8}g$ N
   (b) 0.27 m (2 sf)
   Assumptions: light seat, light ropes that obey Hooke's Law and do not reach their elastic limit.
7. (a) 588 N   (b) 1176 N;
   assumptions as in Q. 6
8. 1.7 N (2 sf)

CHAPTER 10

Exercise 10a – p. 197

1. (a) 0.533 J   (b) 1.2 J   (c) 2 J
2. (a) $\frac{1}{5}mga$ (b) mga (c) $\frac{5}{4}mga$
3. (a) 2.4 J   (b) 3 J
4. (a) 96 N   (b) 24 J
5. 0.01π J
6. (a) 4.83 m   (b) 33.6 J
7. (a) 56.3 J   (b) 2.81 W
8. (a) 11.8 N   (b) 0.443 J
9. $4\lambda a$

Exercise 10b – p. 205

1. (a) 4.5 J   (b) 0.125 J
   (c) 4.375 J   (d) $2.09 \, \text{m s}^{-1}$
2. (a) 0   (b) 100 J
   (c) 3.92 J   (d) $9.8 \, \text{m s}^{-1}$
3. (a) 0   (b) 600 J   (c) 150 J
   (d) 450 J   (e) 0.5 m
4. (a) 100 J   (b) 25 J   (c) 24 J
   (d) 51 J   (e) 0.58
5. (a) (i) 6.4 J   (ii) 7.84 J   (iii) 0
   (b) 0   (c) $3.77 \, \text{m s}^{-1}$
6. (a) 0, (i) 0,   (ii) $10(2 + x)$
   (b) 0; $30 x^2$ (c) 3 m
7. (a) 14 N   (b) $3.17 \, \text{m s}^{-1}$
8. (a) $6a$
9. (a) $(2ga\sqrt{2})^{\frac{1}{2}}$ (b) $2\sqrt{ga}$
   (c) $2[ga(\cos \alpha - \cos 2\alpha)]^{\frac{1}{2}}$
10. (a) 0.3 J   (b) 0.3 J   (c) $2.24 \, \text{m s}^{-1}$
   (d) 0.225 J   (e) 23.7 cm
11. (a) 0.49 m   (b) $3.49 \, \text{m s}^{-1}$
   (c) 2.59 m
12. (b) mg
13. (a) 1.3 m
   (b) (i) $16 \, \text{m s}^{-1}$ (ii) 2.1 m
14. 445 kN

Exercise 10c – p. 210

1. Assume climber as a particle; no air resistance or wind
   (a) (i) 41 kN   (ii) $31 \, \text{m s}^{-1}$
   (b) Rope likely to have a significant weight; air resistance or wind quite likely.
2. (a) $d = v\sqrt{Ma/\lambda}$
   (d) When $R = 0$, $d = v\sqrt{Ma/\lambda}$
3. Hooke's Law valid throughout; no wind or other resistance.
   (a) $\lambda = 68\,600\,\dfrac{a}{x^2}$ (b) $e = 686\,\dfrac{a}{\lambda}$
   (c) $\sqrt{\dfrac{\lambda e}{70a}(2a + e)}$
   (d)

a	10	20	30	40	45
x	40	30	20	10	5
λ	429	1520	5150	27 400	123 000
e	16.0	9.03	4.00	1.00	0.028
$2a + e$	36.0	49.0	64.0	81.0	90.0
V	18.8	21.9	25.0	28.2	29.7

   (e) Hooke's Law unlikely to apply throughout for shorter ropes.
   (f) Allow for weight of rope.

CHAPTER 11

Exercise 11a – p. 218

1. (a) 21 m/s ↗   (b) 23 m/s ↘
2. (a) 12 m   (b) 15 m
   (c) 7 m
3. (a) 25 m/s; (24, 13)
   (b) 24 m/s; (48, 16)
4. 1 s; 3.7 m
5. (a) $20\sqrt{2}$ m/s (28 m/s, 2 sf)
   (b) $56\sqrt{2}$ m (79 m, 2 sf)
6. $20\mathbf{i} + 10\mathbf{j}$
7. $x = 4.33$, $y = 1.25$; 4.5 m
8. $9.6\mathbf{i} - 11.5\mathbf{j}$
9. (a) $\mathbf{v} = \mathbf{i} + (2 - 10t)\mathbf{j}$;
   $\mathbf{r} = t\mathbf{i} + (2t - 5t^2)\mathbf{j}$
   (b) $\mathbf{v} = \mathbf{i} - 13\mathbf{j}$; $\mathbf{r} = 1.5\mathbf{i} - 8.3\mathbf{j}$
10. (a) $\mathbf{v} = 5\mathbf{i} + (3 - gt)\mathbf{j}$;
   $\mathbf{r} = 5t\mathbf{i} + (3t - \frac{1}{2}gt^2)\mathbf{j}$
   (b) $\mathbf{v} = 4\mathbf{i} - (1 + gt)\mathbf{j}$;
   $\mathbf{r} = 4t\mathbf{i} - (t + \frac{1}{2}gt^2)\mathbf{j}$
   (c) $\mathbf{v} = 10\mathbf{i} + (10\sqrt{3} - gt)\mathbf{j}$;
   $\mathbf{r} = 10t\mathbf{i} + (10\sqrt{3}t - \frac{1}{2}gt^2)\mathbf{j}$
   (d) $\mathbf{v} = 10\mathbf{i} + (20 - gt)\mathbf{j}$;
   $\mathbf{r} = 10t\mathbf{i} + (20t - \frac{1}{2}gt^2)\mathbf{j}$

11. (a) $t = 0.8$, $t = 4$ (b) 32 m
12. (a) 15 m/s, 9.5° below $\rightarrow$
    (b) 10 m/s, 14° below $\rightarrow$
13. $20/g$ s; $200/g$ m
14. $\frac{8}{3}\mathbf{i} + \frac{49}{6}\mathbf{j}$
15. (a) 2.6
    (b) No, the ball lands 24 m from the wall
16. $a = 40$, $b = 110$
17. 2.6 s
18. (a) $6\mathbf{i} + 12\mathbf{j}$ (b) $6\mathbf{i} - 8\mathbf{j}$
19. 39 m
20. Yes by almost 1 m
21. (a) 35 m (b) 215 m
22. 63°
23. (a) $4\mathbf{i} + 4\mathbf{j}$ (b) $4t\,\mathbf{i} + (4t - 5t^2)\mathbf{j}$
    (c) $t = 1$ (d) 4 m
24. 54 m
25. (a) $u\mathbf{i} + u\mathbf{j}$
    (b) (i) $u\mathbf{i} + (u - 20t)\mathbf{j}$
        (ii) $u\mathbf{i} + (u - 20)\mathbf{j}$
    (c) $2u\mathbf{i} + (2u - 20)\mathbf{j}$
26. (a) 3.5 s (b) 1.5 s
27. (a) $V \cos \theta$ (b) $V \sin \theta + gt$
    (c) $Vt \cos \theta$ (d) $Vt \sin \theta + \frac{1}{2}gt^2$
28. (a) 35 m/s (b) 53 m
29. Between 4.7 m and 15 m
    (Assume no air resistance)

Exercise 11b – p. 230

1. 7.2 m; 50 m
2. 22° or 68°
3. 20 m/s; 40 m
4. 26 m/s; 26 m
5. 90 m
6. 27°; 130 m
7. (a) 48 m (b) 52 m
8. $y = 0.839x - 0.003\,41x^2$; 15 m
9. 12°, 5.9 m
10. (a) $y = \frac{4}{3}x - \frac{5}{9}x^2$
    (b) $y = -x - \frac{1}{90}x^2$
11. 1.9 s; 25 m
12. 31°
13. (a) $h\sqrt{3}$ (b) $2h$ (c) $h\sqrt{3}$
14. 15° 15. 45° to 53°
16. (a) $\frac{11}{8}h$ (b) $\frac{3}{2}h$ (c) $\frac{3}{8}h$ (d) $-2h$
17. 74° (not 18° as this is not a "skyer")
18. 45°; 155 m/s
19. $y = 1.4x - 0.2x^2$
20. No; hits net 0.14 m below cord
21. First cleared by 3.2 m; second one is hit 11 m above the ground
22. (a) 3 s (b) 15 m (c) 45°

Exercise 11c – p. 236

1. $36\mathbf{i} + 8\mathbf{j}$ 2. 2.9 s
3. (a) 37° (b) 100 m
4. (b) $V^2\sqrt{3}/g$
5. (b) 1 s after Q is projected
   (c) P at 14° below $\rightarrow$; Q at 19° above $\rightarrow$
6. (a) 15 m (b) $\sqrt{3}$ s (1.7 s)
7. (a) 45 m (b) 3 s
8. 50 m
9. 42° to the fairway; 230 m

CHAPTER 12

Exercise 12a – p. 241

1. (a) -10 mph (b) 90 mph
2. (a) -22 mph (b) 42 mph
3. (a) $13\mathbf{i} + 19\mathbf{j}$ (b) $-13\mathbf{i} - 19\mathbf{j}$
4. (a) $18\mathbf{i} - \mathbf{j}$ (b) $5\sqrt{13}$ units
5. $9\mathbf{i} + 13\mathbf{j}$
6. (a) $8\mathbf{j}$ $\frac{5}{\sqrt{2}}(\mathbf{i} - \mathbf{j})$ (b) $\frac{5}{\sqrt{2}}\mathbf{i} + (8 - \frac{5}{\sqrt{2}})\mathbf{j}$
7. (a) $2\mathbf{i} + 6\mathbf{j}$ (b) $5\mathbf{i} + \mathbf{j}$
   (c) $\mathbf{i} - 4\mathbf{j}$
8. 87 km/h (2 sf), 279°
9. 38 km/h (2 sf), 168°
10. 267° or 353°

Exercise 12b – p. 243

1. (a) $3\mathbf{i} - 4\mathbf{j}$ (b) $-3\mathbf{i} + 4\mathbf{j}$
2. (a) $-10\mathbf{i} + 4\mathbf{j}$ (b) $10\mathbf{i} - 4\mathbf{j}$
3. (a) $-12\mathbf{i} + 10\mathbf{j}$ (b) $12\mathbf{i} - 10\mathbf{j}$
4. (a) $12\mathbf{i} + 8\mathbf{j}$ (b) $-12\mathbf{i} - 8\mathbf{j}$
5. (a) $P: t(6\mathbf{i} - \mathbf{j})$, $Q: t(3\mathbf{i} + 7\mathbf{j})$
   (b) $-9\mathbf{i} + 24\mathbf{j}$
6. (a) $\mathbf{v} = \frac{1}{4}t^2\mathbf{i}$, $\mathbf{r} = \frac{1}{12}t^3\mathbf{i}$
   (b) $\mathbf{v} = 3\mathbf{i} + 4\mathbf{j}$, $\mathbf{r} = 3\mathbf{i} + t(3\mathbf{i} + 4\mathbf{j})$
   (c) $-\mathbf{i} + 4\mathbf{j}$ (d) $\frac{29}{3}\mathbf{i} + 16\mathbf{j}$
7. (a) (i) $t(10\mathbf{i} + 4\mathbf{j})$
       (ii) $36\mathbf{i} + 2\mathbf{j} + t(-8\mathbf{i} + 3\mathbf{j})$
       (iii) $36\mathbf{i} + 2\mathbf{j} + t(-18\mathbf{i} - \mathbf{j})$
   (b) they collide
8. (a) $10\mathbf{i} + t(12\mathbf{i} + 5\mathbf{j})$
   (b) $20\mathbf{i} - 4\mathbf{j} + t(-3\mathbf{i} + 10\mathbf{j})$
   (c) $-10\mathbf{i} + 4\mathbf{j} + t(15\mathbf{i} - 5\mathbf{j})$
   (d) 2.40 pm
9. (a) $\mathbf{i} - 2\mathbf{j} + 4t\mathbf{i} + 3t^2\mathbf{j}$ (b) $\sqrt{181}$
   (c) $(2t^2 + t - 1)\mathbf{i} + (t^3 - 2t)\mathbf{j}$
   (d) 29 km

Exercise 12c – p. 249

1. 12.27 2. 2.3 km (2 sf)
3. 038° 4. 214°
5. 6.9 km/h or 9.23 km/h
6. 440 km/h 7. 34 km; 047°

CONSOLIDATION C

Miscellaneous Exercise C – p. 253

1. D
2. A
3. (i) T   (ii) F   (iii) F   (iv) T   (v) F
4. At 40.5° to OE, i.e. on bearing 130.5°; 33 s (2 sf)
5. (a) 10 N   (c) $1.33\,\mathrm{m\,s^{-1}}$ (3 sf)
6. 112.0 m
7. (a) $3Mg$ (b) 48°
8. (b) $6.64\,\mathrm{m\,s^{-1}}$ (c) 4 m
9. (a) $\frac{7}{g}$ s (0.714), $\frac{35}{g}$ s (3.57)
   (b) 10 m and 50 m →, 12.5 m ↑ for both
10. 3.05 s, 18.5 m
11. A
12. (a) 10 m   (b) 2 s   (c) $24.9\,\mathrm{m\,s^{-1}}$
13. (a) 48 m   (b) 116 m   (c) $49.5\,\mathrm{m\,s^{-1}}$
14. 020° (nearest degree)
15. (a) 11.8 kN   (b) 27.4 m   (c) $20\,\mathrm{m\,s^{-1}}$ no air resistance; Hooke's Law valid throughout
16. (a) $5Mg$ (b) $20Mg$
17. (a) 3 s   (b) 16 m   (c) 0.75 rad
18. $H = \dfrac{V^2}{2g}$; $T = \dfrac{V}{g}$; $V = \frac{3}{8}U$
19. (b) 78°, 10°   (c) 2 s
20. (a) $265\,\mathrm{km\,h^{-1}}$
    (b) 344°
    (c) 43 : 53 (1 : 1.23)
21. 125 s, 100 s, 300 m
22. (a) 16.6 m
    (b) Yes, by 0.11 m
23. (a) $\mathbf{i} - \mathbf{j}$ (b) $3\mathbf{i} - 4\mathbf{j}$ (c) 25 J
24. 29.5 s
25. (b) 2.86 s   (c) $17.5\,\mathrm{m\,s^{-1}}$
26. An overestimate; work done by air resistance assists in stopping the descent.
27. $T = \lambda$;   EPE $= \lambda a$; $v = \sqrt{\dfrac{2\lambda a}{m}}$
28. $mg - \dfrac{\lambda}{\sqrt{3}}(2 - \sqrt{3})$
29. No air resistance
    (i) $5.5\,\mathrm{m\,s^{-1}}$ (ii) 1.9 m
30. (a) EPE – work done by friction > 0
    (b) $\frac{1}{3}\sqrt{ga}$
31. 19 kN (2 sf)
32. (i) The point B is in equilibrium.
    (ii) AB = 0.8; $T = 6$ N
    (iii) 2.7 J   (iv) $10\,\mathrm{m\,s^{-1}}$
33. (a) 92 kg   (b) $28\,\mathrm{m\,s^{-1}}$

CHAPTER 13

Exercise 13a – p. 266

1. (a) 120 N s   (b) 24 000 N s
   (c) 11 040 N s   (d) 1177×10^4 N s
   (e) 4 N s
2. (a) 84 N s   (b) 72×10^4 N s
   (c) 88 N s   (d) 6000 N s ($g = 10$)
3. 2   4. −2   5. 25
6. −2   7. $\frac{8}{3}$ 8. 3 s
9. (a) 26$\mathbf{i}$ (b) $-6\mathbf{i}$ 10. 7 N
11. 32 N
12. (a) $\frac{76}{5}\mathbf{i}$ (b) 12$\mathbf{i}$; 17.5 s, 22.5 s
13. (a) 160 N   (b) 20 N
14. (a) 31 200 N s   (b) 31 200 N s
    (c) 15 600 N

Exercise 13b – p. 268

1. 12.1 N s   2. 10.2 N s   3. 12.5 N s
4. 0.72 N s   5. 108 N s

Exercise 13c – p. 272

1. $3\,\mathrm{m\,s^{-1}}$ 2. 0   3. $-1.5\,\mathrm{m\,s^{-1}}$
4. 0   5. 6 kg   6. 1.5 kg
7. $3\,\mathrm{m\,s^{-1}}$ 8. $6\,\mathrm{m\,s^{-1}}$
9. $u = 2$, $v = 7$
10. (a) $4\,\mathrm{m\,s^{-1}}$ (b) 2.4 N s   (c) 240 N
11. (a) $0.57\,\mathrm{m\,s^{-1}}$ (b) $0.73\,\mathrm{m\,s^{-1}}$
    (c) 36 N s. Stones treated as particles; ignore friction and air resistance
12. $6\frac{2}{3}\,\mathrm{m\,s^{-1}}$ 13. 1 kg
14. (a) $\sqrt{2gl}$ (b) $\frac{1}{2}\sqrt{2gl}$

Exercise 13d – p. 276

1. (a) (i) 12 J   (ii) 6 N s
   (b) (i) 58.5 J   (ii) 18 N s
   (c) (i) 240 J   (ii) 30 N s
   (d) (i) 96 J   (ii) 24 N s
   (e) (i) 45 J   (ii) 30 N s
   (f) (i) 3.6 J   (ii) 1.8 N s
2. $13\frac{1}{3}$ N s; 4 N s
3. (a) 600 kg   (b) 480 N s   (c) 480 J
4. (a) (i) u (ii) $3mu$ (iii) $6mu^2$
   (b) (i) $3u$ (ii) $3mu$ (iii) $6mu^2$
5. (a) $2.6\,\mathrm{m\,s^{-1}}$ downstream
   (b) $0.9\,\mathrm{m\,s^{-1}}$ upstream
6. (a) $2\sqrt{10}\,\mathrm{m\,s^{-1}}$ ($6.3\,\mathrm{m\,s^{-1}}$)
   (b) $6\sqrt{10}$ N s (19 N s)
   (c) $2\sqrt{7}\,\mathrm{m\,s^{-1}}$ ($5.3\,\mathrm{m\,s^{-1}}$)
   (d) $6(\sqrt{10} + \sqrt{7})$ N s (35 N s)
7. $1.12\,\mathrm{m\,s^{-1}}$, 3.36 N s (2 sf)

CHAPTER 14

Exercise 14a – p. 281
1. (a) $4\,\mathrm{m\,s^{-1}}$ (b) $32\,\mathrm{N\,s}$
2. (a) $8\,\mathrm{m\,s^{-1}}$ (b) $120\,\mathrm{J}$
3. (a) 0 (b) $14\,\mathrm{N\,s}$ (c) $98\,\mathrm{J}$
4. (a) $6\,\mathrm{m\,s^{-1}}$ (b) $48\,\mathrm{N\,s}$ (c) 0
5. (a) $2.8\,\mathrm{m\,s^{-1}}$ (b) $1.4\,\mathrm{m\,s^{-1}}$
   (c) $\frac{1}{2}$ (d) $0.1\,\mathrm{m}$
6. $2.5\,\mathrm{m}$; $42\,\mathrm{N\,s}$
7. (a) $-6\mathbf{i}$

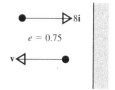

(b) 0.4

8. (a) $-20\mathbf{i}$

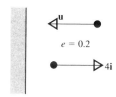

(b) $3\mathbf{i} - 5\mathbf{j}$

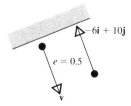

9. (a) 0

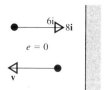

(b) 0.5

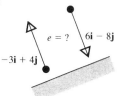

10. (a) $5\,\mathrm{m}$, $1.25\,\mathrm{m}$, $0.3125\,\mathrm{m}$
   (b) $r = 0.25$
   (c) $33.125\,\mathrm{m}$
11. 0.6; ignore friction and air resistance
12. (a) $\sqrt{ga}$
   (b) $\frac{3}{4}\sqrt{ga}$
   (c) $\frac{9}{32}a$
13. (a) 0.78
   (b) $2.04 \leqslant v \leqslant 2.36$

Exercise 14b – p. 287
1. (a) $u = 7\frac{1}{3}$, $v = 9\frac{1}{3}$; $J = 5\frac{1}{3}\,\mathrm{N\,s}$
2. (a) 11
   (b) $\frac{11}{13}$; $16\,\mathrm{J}$
3. (a) 2
   (b) 3
   (c) $25\,\mathrm{J}$
4. (a) 3 (b) $\frac{2}{3}$
5. $0.2\,\mathrm{kg}$; $2.5\,\mathrm{m\,s^{-1}}$; $0.5\,\mathrm{N\,s}$
7. 0.8
8. 0.217%
9. (c) $\frac{1}{2}mu^2(9 - k^2)$
10. (a) $3\,\mathrm{m\,s^{-1}}$ (b) 5.1
   Treat ball and stone as particles; assume that velocity of stone at impact can be horizontal; ignore friction.
11. Spacecraft $16\,050\,\mathrm{m\,s^{-1}}$;
   rocket $15\,970\,\mathrm{m\,s^{-1}}$
   (4 sf used in order to show the difference)
12. (a) $u = \frac{1}{7}(60 - 10e)$, $v = \frac{1}{7}(4e + 60)$
   (b) $\frac{60}{7} \leqslant v \leqslant \frac{64}{7}$
13. (a) $\dfrac{(k - 3)u}{3(k + 1)}$

   (c) $\dfrac{4u}{3(k + 1)}$

   (d) $\dfrac{4kmu}{3(k + 1)}$
14. $\frac{1}{8}u(9 - k)$
15. (a) $k = 3$
   (b) $2\sqrt{gl}$, $6\sqrt{gl}$
   (c) 0
   (d) $6m\sqrt{gl}$

Exercise 14c – p. 293

1. $u_1 = -\frac{4}{3}$, $v_1 = \frac{2}{3}$, $v_2 = \frac{1}{3}$; A travelling away from the wall faster than B so B will not catch A to collide again.

2. (a) $u_1 = 3$, $v_1 = 12$
  (b) A: $-3\,\mathrm{m\,s^{-1}}$, B: $6\,\mathrm{m\,s^{-1}}$
  (c) B will collide with the wall again.

3.

After A strikes B

After B strikes wall

After B strikes A

After B strikes wall again

There will be no more collisions.

4. After 1st collision

After 2nd collision

5. 1st:

  2nd:

  3rd:

6. (a)

  (b)

  (c) (i) Velocity of B is negative so B hits A.
      (ii) Velocity of B is positive and < velocity of C; no further collisions

7. (a) $\dfrac{m_1}{m_2}u$ (b) $\dfrac{m_1}{m_2}$; (c) $\dfrac{(m_2)^2}{m_1}$

8. $6.12\,\mathrm{m\,s^{-1}}$ $\left(\frac{300}{49}\right)$

9. $11.4\,\mathrm{m\,s^{-1}}$ $\left(\frac{343}{30}\right)$

10. (a) P $= 0.25\,(2e - 1)\,\mathrm{m\,s^{-1}}$, Q: $0.25\,(e + 1)\,\mathrm{m\,s^{-1}}$
  (b) $e = \frac{4}{5}$; P: $0.15\,\mathrm{m\,s^{-1}}$, Q: $0.45\,\mathrm{m\,s^{-1}}$
  (c) P: $0.57\,\mathrm{m\,s^{-1}}$, Q: $0.09\,\mathrm{m\,s^{-1}}$

11. (a) e^2h, e^4h, e^6h (b) GP in which $r = e^2$ (c) $\dfrac{h}{1 - e^2}$

Exercise 14d – p. 300

1. (a) $\frac{3}{4}\sqrt{13}$ m s^{-1} at $16°$ to the wall
   (b) 3 m s^{-1} at $30°$ to the wall
   (c) $\frac{3}{2}\sqrt{3}$ m s^{-1} along the wall.
2. $\frac{1}{3}$
3. (a) $-2\mathbf{i} + 2\mathbf{j}$, $\mathbf{j}$
   (b) $-3\mathbf{i} + 2\mathbf{j}$, $\mathbf{i} + \mathbf{j}$
   (c) $-\mathbf{i} + 2\mathbf{j}$, $-\mathbf{i} + \mathbf{j}$
4. $\frac{1}{2}$, $\mathbf{i} + 3\mathbf{j}$
6. (a) $36.9°$ (3 sf)
   (b) $\frac{1}{5}u\sqrt{58}$, $\frac{1}{5}u\sqrt{61}$
7. (a) 0.2 m s^{-1} (b) 0.4 s (c) 1.6 m
8. 0.2
9. $75\mathbf{i} - 130\mathbf{j}$

CHAPTER 15

Exercise 15a – p. 304

1. (a) 1.91 (b) 10.5
2. 0.00175
3. (a) 6.94×10^{-4} (b) 7.27×10^{-5}
4. (a) 1.2 (b) 12.5 (c) 4
5. 8.38 m s^{-1} 6. 1680 km h^{-1}
7. 0.393 m s^{-1}
8. (a) 1.6 rad s^{-1} (b) 1.6 m s^{-1}
9. (a) 0.628 m s^{-1}
   (b) 15.9 cm; rope is of negligible thickness.

Exercise 15b – p. 308

1. 51.2 2. 432
3. (i) $\overrightarrow{PO}$ (ii) $\overrightarrow{QO}$ 4. 30
5. 13.9
6. (a) 0.0338 m s^{-2} towards the centre of earth
   (b) 0.0169 m s^{-2} towards, and at right angles to, the earth's axis

7. $\frac{48\,200}{g}$ m 8. 4.2 rad s^{-1} 9. 9.49 m s^{-1}
10. 8 m s^{-1} (below 8.94 m s^{-1} for safety)

Exercise 15c – p. 312

1. 10 N 2. 16.7 m
3. 2.19 m s^{-1} 4. 112.5 N
5. 8 N
6. (a) 42.7 N, 3.92 N (b) 14.4 rad s^{-1}
7. Model passenger as a particle and assume plane's angular velocity constant.
   (a) horizontal: 1130 N; vertical: 588 N
   (b) 1280 N at $27°$ to the horizontal

8. Model satellite as particle, assume earth spherical, assume satellite orbiting steadily.
   (a) $9.5m$ N (b) 7860 m s^{-1}
   (c) 1.21×10^{-3} rad s^{-1} (d) 1.44 hours

Exercise 15d – p. 317

1. (a) 288 N (b) $84°$
2. (a) 21.6 N (b) 3.54 rad s^{-1}
3. 0.392 m 4. (a) 3 N (b) 1.73 rad s^{-1}
5. (a) 9.8 N (b) 4.9 m (c) 1.15 m
6. (a) $T = 2ml\omega^2$, $R = mg$
   (b) $T = 2ml\omega^2$, $R = m(g - l\omega^2)$
7. (a) (i) $\frac{2}{3}$ (ii) $1.5a\cos\theta$
   (b) towards A: $\frac{15}{4}mg$; towards B: $\frac{9}{4}mg$
8. $\frac{2}{3}\sqrt{30ga}$
9. (a) 2.29 N (b) 1.71 m s^{-1}
   (c) 32.6 m (d) 26.0 N
10. (a) $60°$ (b) 3.95 N (c) 5.87 N
11. (a) $\sqrt{gr\tan\theta}$ (b) 94 km h^{-1}
    (c) Pendulum bob is at a constant distance from side of carriage but string is inclined to the vertical, in a plane parallel to the track, with the bob behind the vertical.

Exercise 15e – p. 322

In each question the object or person is treated as a particle and it is assumed that there is no air resistance.
1. 0.63 rad s^{-1}
2. 4.4 rad s^{-1}; assume constant speed
3. (a) 4800 N
   (b) 32 m s^{-1}; possible overturning ignored and constant speed assumed (not very likely)
4. (a) 91 N (b) 1.9 (c) No
5. Horiz: 570 N; vert: 390 N
6. (a) $mr\omega^2$
   (b) mg; assume man exactly rm from axis
   (c) $\sqrt{g/\mu r}$ (d) 3.13 rad s^{-1}; 7.83 m s^{-1}
7. (a) 1.5ω m s^{-1}, 3ω m s^{-1}, 4.5ω m s^{-1}
   (b) (i) $225\omega^2$ N (ii) $375\omega^2$ N
   (iii) $450\omega^2$ N
   People modelled as particles is a *very* rough assumption.
8. (a) 490 N (b) $37°$ (c) 3.0 m; the rope does not stretch and lies in the vertical plane containing the girl's weight and the pole.

Exercise 15f – p. 328

People and objects modelled as particles.
Answers based on $g = 10$

1. (a) 11 kN   (b) outer
2. (a) outer   (b) 620 m
3. 11 m s^{-1} (40 km h^{-1})
4. (a) $\frac{1}{2}$ (c) 22°
5. 44 m s^{-1} (160 km h^{-1});
   77 m s^{-1} (280 km h^{-1})
6. 8200 N; 0.015 m
7. Between 8.9 m s^{-1} and 38 m s^{-1}
8. (a) $\dfrac{V^2}{3g}$ (b) $\dfrac{V\sqrt{3}}{3}$
9. 6.6 m s^{-1}; treat motor cyclist and machine
   as a particle; assume wall to be uniformly
   rough.
10. (a) When the aircraft is banked, the lift
    force has a component towards the
    centre of the circle.
    (b) 49°. No wind, lift force constant.
11. (b) $\dfrac{V_2{}^2}{rg} = \dfrac{1 - \mu}{1 + \mu}$ (c) $r = \dfrac{V_1 V_2}{g}$
    (d) $\sqrt{V_1 V_2}$

CHAPTER 16

Exercise 16a – p. 336

1. (a) 2.65 m s^{-1} (b) 2.25 m s^{-1}
   (c) 1.08 m s^{-1}
2. (a) 40.1 N towards O
   (b) 25.4 N towards O
   (c) 4 N away from O
3. (a) 1.4   (b) 1.98   (c) 2.8   (d) 0
4. 113°
5. (a) 4.87   (b) 5.24
6. (a) 25.5 N, tension   (b) 29.4 N, tension
7. (a) 41.9 N   (b) 14.7 N
8. 27.9 m s^{-2} rad., 9.8 m s^{-2} tang.
9. (a) 204°; 1.22 cm   (b) 218°; 1.84 cm
10. (a) 2.63 m s^{-2} rad.; 4.9 m s^{-2} tang.
    (b) 19.6 m s^{-2} rad.; 9.8 m s^{-2} tang.
    (c) 36.6 m s^{-2} rad.; 4.9 m s^{-2} tang.
11. $u > 2\sqrt{ga}$
12. $u > 4\sqrt{2gl}$
13. a; $\frac{1}{4}a$

Exercise 16b – p. 343

1. 4.48 m s^{-1}
2. (a) 4.18 m s^{-1} (b) 8.69 N
3. (a) $u > \sqrt{3g}$
   (b) oscillation through 180°

4. 4.81 m s^{-1}; 13.2 N
5. $3\sqrt{ga}$
6. (a) $\sqrt{5ga}$ (b) $4mg$
7. (a) $\sqrt{ga/2}$ (b) $\sqrt{7ga/2}$
8. 0.544 m; 2.31 m s^{-1}
9. 62.6 m s^{-1}
10. 3.13
11. (a) $3\frac{1}{3}$ m   (b) 3.61 m s^{-1} (c) 8.85 m s^{-1}
12. (a) $\frac{1}{3}m(5v^2 - 3g)$ (b) 2.42
13. (a) 10.2 m   (b) 9.82 m s^{-1}

Exercise 16c – p. 350

1. (a) 12 m s^{-1}
   (c) (i) 1.16 N, tension;
       (ii) 0.830 N, thrust
   (d) 165°
2. (a) 2.8 m s^{-1}, 29.4 N
   (b) 1.2 m s^{-1} (c) 20.1 N   (d) 0.073 m
3. (a) 1600 N   (b) 1300 N
   (c) 7600 N   (d) 15.7 m s^{-1}
4. (a) $J \gg 7.42$ N s
   (b) $\frac{3}{4}\sqrt{7g}$ N s (6.2 N s)
5. $\frac{1}{2}l$; 11 : 5
6. Assume catcher and flier each to be a
   particle distant 7 m from point of
   attachment of ropes.
   (a) 9.5 m s^{-1} (b) 4.6 m s^{-1}
   (c) 1700 N   (d) 32°
7. (a) 1.01 m s^{-1} (b) 57.6 N   (c) 505 m s^{-1}
8. (a) 350 N, 6.9 m s^{-1} (b) 17 m s^{-1}
   (c) 1700 N, 29 m s^{-2} (d) 18 m s^{-1}
   (e) 1500 N, decrease
9. (a) $\frac{4}{3}\pi$ m s^{-1}
   (b) (i) $\frac{2}{9}\pi^2$ m s^{-2} (ii) 0
   (c) $R = 60g \cos\theta + \frac{40}{3}\pi^2$; $S = 60g \sin\theta$
   (d) (i) 720 N vert. upwards
       (ii) 460 N vert. upwards
10. (a) 12 m s^{-2} (b) 15 m s^{-2}
11. (a) 4.0 m s^{-1} (b) 2000 N s
    (c) (i) 6200 N   (ii) 4900 N
12. (a) 2.1 m s^{-1} (b) 5.6 m s^{-1}
    (c) $\vartriangleleft$— 1.05 m s^{-1}; ↑ 1.05$\sqrt{3}$
    (d) 0.17 m   (e)

13. (a) 28°   (b) 1.9 m   (c) 2.9 m s^{-1}
    (d) —▷ 2.58 m s^{-1}, ↓ 1.4 m s^{-1}
    (e) (i) 6.2 m s^{-1} (ii) 0.49 s   (iii) 1.3 m

CONSOLIDATION D

Miscellaneous Exercise D – p. 356

1. $37.6\,\mathrm{N\,s}$

2. (a) $4\,\mathrm{m\,s^{-1}}$ (b) $3\,\mathrm{J}$

3. $5.88\,\mathrm{N\,s}$

5. (a) $11\,\mathrm{cm}$ (b) $12\,\mathrm{cm}$

6. (a) $3\,mg$ (b) $\frac{7}{3}\,mg$

7. 6.49

8. 0.392; $21.4°$; $35.5\,\mathrm{m\,s^{-1}}$

9. (a)

$30g$

(b) $375\,\mathrm{N}$

(d) It is light and inextensible.

(e) $4\,\mathrm{s}$

10. (a)

$0.25g$

(b) $0.392\,\mathrm{N}$ (c) $12.2\,\mathrm{m\,s^{-2}}$

(d) $19.5\,\mathrm{m\,s^{-1}}$ (e) $\omega = \sqrt{g/h}$

(f) If T is horizontal it cannot balance the weight of the particle.

11. B **12.** C

13. C **14.** D

15. B **16.** (a) $\frac{3}{16}a$

17. $\frac{2}{3}m\sqrt{6gh}$

18. (a) $4.5\,\mathrm{m\,s^{-1}}$ (b) $6750\,\mathrm{J}$ (c) $2250\,\mathrm{N\,s}$

19. $0.036\,\mathrm{N\,s}$

20. $58°$ (nearest degree);
tan θ is independent of m and M

21. (i) F (ii) F $(e \neq 1)$
(iii) F; impulses are equal (iv) T

22. B **23.** C

24. D **25.** C

26. (i) $2.8u$ (ii) $6mu$ (iii) 0.909

27. (b) $\frac{1}{4}u(3e - 1)$ (d) $\frac{27}{16}mu$

28. (i) $\sqrt{2h/g}$ (ii) $2e\sqrt{2h/g}$

30. (a) $\frac{3}{8}u$

31. (a) $3.20\,\mathrm{m\,s^{-1}}$ (b) $6.81\,\mathrm{m\,s^{-1}}$

32. (b) $\frac{1}{2}Mg$

33. (a) $v = \sqrt{[ag(2\cos\theta - 1)]}$
(b) $mg(3\cos\theta - 1)$

34. (i) $14\,\mathrm{m\,s^{-1}}$ (iii) $2.45\,\mathrm{N\,s}$
(iv) $4.29\,\mathrm{J}$
(v) Each upward speed is
$e \times$ the previous one.
(vi) 18th

35. (i) $7.0\,\mathrm{m\,s^{-1}}$ (ii) $6.5\,\mathrm{m\,s^{-1}}$; $2.2\,\mathrm{m}$
All resistance reduces speed so a greater safe height is possible.

CHAPTER 17

Exercise 17a – p. 369

1.

$\triangle ABC$
$F = 6\,\mathrm{N}$, $T = 8\,\mathrm{N}$

2.

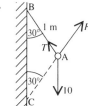

$\triangle ABC$
$F = 10\frac{\sqrt{3}}{3}\,\mathrm{N} = T$ $(5.77\,\mathrm{N})$

3.

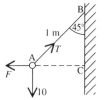

$\triangle ABC$
$F = 10\,\mathrm{N}$, $T = 10\sqrt{2}\,\mathrm{N}$ $(14.1\,\mathrm{N})$

4.

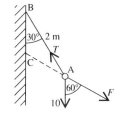

$\triangle ABC$
$F = 10\,\mathrm{N}$, $T = 10\sqrt{3}\,\mathrm{N}$ $(17.3\,\mathrm{N})$

5.

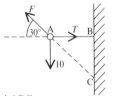

△ABC

$F = 20$ N, $T = 10\sqrt{3}$ N (17.3 N)

6.

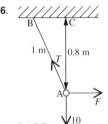

△ABC

$F = 7\frac{1}{2}$ N, $T = 12\frac{1}{2}$ N

7.

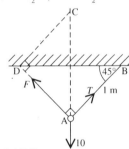

△ADC

$F = 5\sqrt{2}$ N $= T$ (7.07 N)

8.

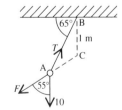

△ABC

$T = 18.1$ N, $F = 10$ N

Exercise 17b – p. 371

	P	Q
1.	34.6	40
2.	11.3	7.99
3.	3.54	4.34
4.	8.51	2.91
5.	67.2	82.4
6.	17.1	11.1

Exercise 17c – p. 372

1. $P = 12.5$, $Q = 7.5$
2. $P = 8\sqrt{3}$ (13.9), $\theta = 30°$
3. $\theta = 154°$, $P = 18.4$

Exercise 17d – p. 377

1. (a) 12 N m↺   (b) 19.2 N m↻
   (c) 10.5 N m↻   (d) 22 N m↻
2. (a) 11 N m↻   (b) 1.5 N m↻
   (c) 3 N m↻   (d) 0
   (e) 2.5 N m↻   (f) 0
   (g) 32 N m↻
3. (a) 1 N m↺   (b) 2.53 N m↺
4. (a) $\sqrt{3}$ N m (1.73 N m)
   (b) $(3\sqrt{3} + 2)$ N m (7.20 N m)
5. (a) 5.5 N m   (b) 2.75
6. 10 N m
7. (a) 23 N m↻   (b) 6 N m↻

Exercise 17e – p. 381

1. F along $\overrightarrow{DA}$; $F\sqrt{2}$ along $\overrightarrow{CA}$;
   $4Fa$ in direction of ABC
2. $F\sqrt{5}$ at $27°$ to DA (nearest degree)
3. 10 units anticlockwise
4. 26 units clockwise
5. $p = -\frac{5}{4}$, $q = -\frac{9}{4}$
   Whatever the values of p and q there is
   always a turning effect about the centre,
   therefore equilibrium is impossible.
6. $P = -1$, $Q = 6$

Exercise 17f – p. 384

1. A: $11\frac{2}{3}$ N, B: $13\frac{1}{3}$ N
2. A: $13\frac{3}{4}$ N, B: $21\frac{1}{4}$ N
3. A: 26 N, B: 6
4. A: 36 N, B: 0
5. 0.43 m; 500 N
6. $T_1 = \frac{2}{3}W$, $T_2 = \frac{1}{3}W$
7. $T_1 = \frac{3}{2}W$, $T_2 = 0$
8. $T_1 = 2\frac{1}{3}W$, $T_2 = 4\frac{2}{3}W$
9. $T_1 = 0$, $T_2 = 3W$
10. About 1 m
11. (a) 150 N, 570 N
    (b) 312 N, 408 N (3 sf)
    (c) each 360 N
12. (a) 54.3 kg   (b) 10.9 kg; too risky
13. 1.375 m
14. W
15. (a) 0.875 m   (b) 0.8 m
16. (a) $\frac{3}{4}a$ (b) $\frac{3}{4}$
17. (a) 0.96 m
18. (a) 6.14 kg   (b) 0.011 m
19. (a) $53°$ (b) 233

CHAPTER 18

Exercise 18a – p.394
1. $34°$
2. (a) 200 N   (b) 100 N   (c) $\frac{1}{2}$
3. $28°$
4. $300\sqrt{3}$ N $(520$ N$)$; $\mu = \frac{3}{10}\sqrt{3}$ (0.520)
5. To the top   6. $21°$
7. 12 N;   12 N↑
8. (a) 12 N   (b) 20.8 N
9. $12\sqrt{3}$ N (20.8)
10. (a) $53\frac{1}{3}$ N   (b) 44.3 N at $16°$ to PQ
11. (a) $W\frac{\sqrt{3}}{3}$ (b) $W\frac{\sqrt{3}}{6}$
12. (a) 9.8 N
    (b) 35.3 N at $76°$ to
    horizontal

13. $\mu = \frac{9}{8}$; $F = 26\frac{2}{3}$ N
14. (a) 81.3 N
    (b) 49.4 N at $81°$ to the downward
    vertical.
15. (a) $\frac{1}{2}W$ (b) $\frac{1}{3}\sqrt{3}$ (0.577)
16. 67.5; each man modelled as a particle on
    the rung. Not very sensible as each man's
    weight acts somewhere in the middle of
    his body. Also his weight may not be
    vertically over the rung.
17. (a) 20.4 N
    (b) 40.8 N at $37°$ to the horizontal.

Exercise 18b – p.400
1. (a)

    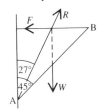
    (b) $R = \frac{1}{2}W\sqrt{5}$ $(1.12W)$
2. Reactions at A and C both 1200 N
    (a)
    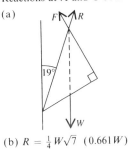
    (b) $R = \frac{1}{4}W\sqrt{7}$ $(0.661W)$

3. (a)
    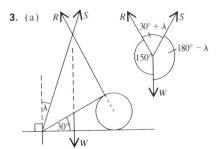
    (b)
    $$\frac{R}{\sin(180° - \lambda)} = \frac{S}{\sin 150°} = \frac{W}{\sin(30° + \lambda)}$$
5. $56°$; $\frac{3}{4}W$
6. $\sqrt{2} - 1$
7. $53°$
8. There must be friction at the ground to
   counteract the reaction perpendicular to
   the wall.
9. (a)
    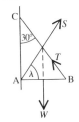
    (b) $\lambda = 60°$; $\mu = \sqrt{3}$ (c) $\frac{1}{3}W\sqrt{3}$
10. (a) The reaction R and the weight meet
    at the centre O; the tension must be
    concurrent with R and W.
    (b) AO $= 3$ m, CO $= 5$ cm, CA $= 4$ cm
    (c) $\triangle$CAO; $T = 25$ N, $R = 15$ N

Exercise 18c – p.404
1. AB: $40\sqrt{3}$ N, tie
   CB: $20\sqrt{3}$ N, strut
2. AB or BC: 2400 N, strut
   AC: $1200\sqrt{3}$ N, tie
3. AB or CD: $\frac{2}{3}W\sqrt{3}$, strut
   AD: $\frac{1}{3}W\sqrt{3}$, tie
   BC: $\frac{1}{3}W\sqrt{3}$, strut
   BE and CE: no force – these members are
   not needed
4. AB or DE: $75\sqrt{2}$ kN, strut
   BC or CD: $50\sqrt{2}$ kN, strut
   BF or FD: $25\sqrt{2}$ kN, strut
   AF or FE: 75 kN, tie
   CF: 50 kN, tie

CHAPTER 19

Exercise 19a – p. 408

1. (a) 16 cm   (b) 6 cm
   (c) 17.5 cm   (d) 9.6 cm
   (e) 18 cm   (f) 10.3 cm
2. $\frac{40}{9}a$ $(4.44a)$
3. $\frac{5}{2}a$ $(2.5a)$
4. $\frac{11}{5}a$ $(2.2a)$
5. $\frac{13}{5}a$ $(2.6a)$
6. $\frac{7}{4}a$ $(1.75a)$
7. $\frac{34}{11}a$ $(3.09a)$
8. $3.9a$
9. $3.5a$
10. $5a$
11. (a) 3   (b) $\left(\frac{10}{3}, 0\right)$
12. (a) 34.7 cm   (b) 0.8 kg
   (c) 36.7 cm

Exercise 19b – p. 411

1. $\left(\frac{13}{5}, \frac{16}{5}\right)$ 2. $\left(\frac{6}{7}, \frac{16}{7}\right)$
3. $\left(\frac{1}{5}, \frac{11}{5}\right)$ 4. $\left(-\frac{4}{5}, \frac{6}{5}\right)$
5. $\left(5, \frac{23}{5}\right)$ 6. $\left(\frac{1}{3}, 1\right)$

Exercise 19c – p. 417

1. (a) $(2, 2)$ (b) $(4, 3)$
   (c) $(2, 1)$ (d) $(2, 1)$
2. (a) $(3.5, 5.5)$, $(7, 4)$
   (b) $(1.5, 3.5)$, $(5, 1.5)$
   (c) $(3, 3)$, $(6, 4)$
   (d) $(2, 4)$, $(4, 2)$, $(6, 4)$

3.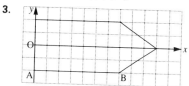
$\bar{x} = \frac{73}{17}, \quad \bar{y} = 0$

4.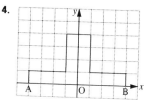
$\bar{x} = 0, \quad \bar{y} = \frac{19}{14}$

5.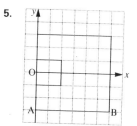
$\bar{x} = 3.25, \quad \bar{y} = 0$

6.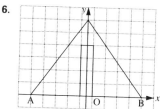
$\bar{x} = 0, \quad \bar{y} = 2$

7. $\left(\frac{13}{3}, \frac{7}{3}\right)$ 8. $\left(\frac{113}{24}, \frac{79}{24}\right)$
9. $(5.6, 4.3)$ 10. $\left(\frac{16}{3}, \frac{14}{3}\right)$
11. $(5.5, 3)$ 12. $\left(\frac{119}{26}, \frac{58}{13}\right)$
13. $(12a, 7.2a)$ 14. $\left(\frac{10}{7}a, \frac{6}{7}a\right)$
15. 4 cm, 5 cm
16. 13.75 cm from AB; 6.25 cm from BC
17. 4330 km

Exercise 19d – p. 422

1. 22.1 cm
2. 97.4 g
3. $1\,\text{g/cm}^2$

4. (a) $x = 16$ cm;
   cannot be done

   (b) $x = 13.6$ cm;
   can be done

CHAPTER 20

Exercise 20a – p. 431

1. (a) $\frac{8}{3}$ (b) $\frac{3}{2}$ 2. (a) $\frac{16}{3}$ (b) $\frac{12}{5}$
3. $\frac{45}{14}$ 4. $\frac{45}{28}h$
5. $\frac{5}{3}$ 6. $\frac{95}{166}$ (0.572)
7. $\frac{14}{5}$ 8. $\frac{18}{11}$ (1.64)
9. $\frac{1}{2}\left(\frac{e^2 + 1}{e^2 - 1}\right)$ 10. $\frac{5}{\ln 6} - 1$ (1.79)

11. $\dfrac{4}{\ln 5}$ **12.** 3

13. $\dfrac{4}{3}$ **14.** $\dfrac{525}{152} a \,(3.45a)$

15. $\dfrac{27}{40} a$

16. (a) $\dfrac{4a^2 x^2}{h^2}$ (b) $\dfrac{4a^2 x^2}{h^2} \rho \,\delta x$

  (c) $a^2 h^2 \rho$ (d) $\tfrac{3}{4} h$ from O on Ox

Exercise 20b – p. 438

1. 4.14 cm

2. (a) $\tfrac{5}{3}$ cm $(1.67\,\text{cm})$

  (b) $\tfrac{15}{4}$ cm $(3.75\,\text{cm})$

4. $\dfrac{h^2 - 24}{2(h+4)}$ cm

5. 4.7 cm

6. $\tfrac{9}{13} a$

7. (a) $\tfrac{65}{24}$ cm $(2.71\,\text{cm})$

  (b) $\tfrac{115}{32}$ cm $(3.59\,\text{cm})$

8. $30°$

9. (a) $\dfrac{3\,|2h^2 - r^2|}{4\,(3h + 2r)}$ (b) $r = h\sqrt{2}$

10. $\tfrac{93}{88} a \,(1.06a)$

11. 20

12. (a) 13 cm, 57 cm   (b) 22 500 cm^3

  (c) (i) 11 cm   (ii) 14 400 cm^3

  (iii) 9425 cm^3 (iv) 55.2 cm

  (d) (i) B   (ii) B

CHAPTER 21

Exercise 21a – p. 445

1. (a) $25°$ (b) $19°$ (c) $43°$

2. (a) $\bar{x} = \dfrac{4a^2 + 3ad}{4a + 2d}$, $\bar{y} = \dfrac{2a^2 + 2ad + d^2}{2a + d}$

  (b) $d = \tfrac{1}{3} a$

3. $38°$ **4.** $74°$ **5.** $88°$

6. (a) $2\sqrt{6}$ cm (4.90) (b) 8.79 cm

Exercise 21b – p. 451

1. (a) 1.75 cm   (b) Yes

2. (a) 9.11 cm   (b) rest in equilibrium

3. It topples $(\bar{x} = 8.94\,\text{cm which} > \text{AB})$

4. (a) $\dfrac{3 + 8d}{2(3 + 2d)}$ (b) Yes, $\tfrac{19}{14} > 1$

5. $d \leqslant \tfrac{3}{4}$

6. (a) $\dfrac{k^2 + 20k + 250}{2k + 50}$

  (b) (i) Yes, $\tfrac{775}{80} < 10$ (ii) No, $\tfrac{1050}{90} > 10$

  (c) $k \leqslant \sqrt{250} \ (\leqslant 15.8)$

7. will topple

8. (a) $\tfrac{120}{13}$ cm

  (b) will stay in equilibrium

Exercise 21c – p. 457

1. (a) rest in equilibrium   (b) topple

2. (a) $59°$; $\tfrac{5}{3}$ (b) $31°$; $\tfrac{3}{5}$

3. 0.8

4. will topple

5. will not topple

6. is on the point of toppling

7. $20\sqrt{3}$ cm $(34.6\,\text{cm})$

8. (a) $27°$ (b) 0.1

9. 4.08 cm, 8.17 cm

10. (a) $29°$

11. $10°$

Exercise 21d – p. 464

1. (b) (i) 5.39 N   (ii) 219 N

2. (iii)

3. (ii)

4. (a) $h = 5\sqrt{3}$ cm $(8.66\,\text{cm})$

  (b) it will remain in equilibrium (this is true for any angle)

5. $\tfrac{1}{3} \tan \theta$

6. (a) $\tfrac{3}{4} W$ (b) $\tfrac{1}{2} W$ (c) by toppling

7. By sliding when $\tan \theta = \tfrac{1}{4}$ (for toppling $\tan \theta = 2$)

8. (a) $\tfrac{50}{3} M$ (b) $\tfrac{127}{33} M$

CONSOLIDATION E

Miscellaneous Exercise E – p. 468

1. $R = 80 - 50 \sin^2\theta \cos\theta$,
  $F = 50 \sin\theta \cos^2\theta$,
  $N = 50 \sin\theta \cos\theta$; $\tfrac{100}{9} \sqrt{3}$ N; 30.4 N

2. (a) 94 N   (b) 0.47   (c) 133 N

3. (a) 49 N

  (b) vertical: 123 N; horizontal: 42.4 N

  (c) 0.35

4. D   **5.** A   **6.** D

7. (i) F   (ii) T   (iii) T   (iv) F

8. (b) $38°$

9. (i) There must be a horizontal component to balance that of T.

  (ii) $T \sin 35° = R \sin \alpha°$,
    $T \cos 35° + R \cos \alpha° = 80°$

  (iii) about 457 N

  (iv) about 563 N along the string away from A

  (v) about 563 N

  (vi) $2T \cos 17.5° \approx 1070$ N

10. B: 3100 N; A: 3500 N; 1800 N, 1300 N

11. 7.3 cm; $\mu \geqslant \tan 20°$, i.e. $\mu \geqslant 0.36$

12. (b) 9°

13. (i) must be concurrent (ii) $\frac{1}{10} W\sqrt{5}$

(iii) $\frac{1}{10} W\sqrt{85}$ (iv) $\frac{1}{6} W\sqrt{2}$; 45°

14. Uniform bridge, light rope, smooth pulley.
(i) 87 kg (ii) 1400 N

16. (b) $2\frac{2}{3}$ cm

17. $k = 0.71$; the normal reaction is vertical and passes through C – therefore is collinear with the weight.

18. (b) 38°

19. $3\mathbf{i} + 2.5\mathbf{j}$

20. (a) 0.34 m (b) 0.24 m (c) 16°

21. 0.109 m (3 sf)

22. (b) AB: $20\sqrt{3}$ N, strut
BC: $20\sqrt{3}$ N, tie
CA: $10\sqrt{3}$ N, strut
(c) $10\sqrt{21}$ N (46 to 2 sf)

CHAPTER 22

Exercise 22a – p. 480

1. 85 m s^{-1}

2. 50 m s^{-2}

3. 26 m s^{-2}

4. $82\frac{2}{3}$ m

5. 17 m s^{-1}; 12 m

6. $t = 2$ and 4

7. $t = \frac{2}{3}$ and $\frac{1}{3}$

8. $v = t^2$; $r = \frac{1}{3}t^3$

9. $29\frac{1}{3}$ m s^{-1}; $47\frac{1}{3}$ m

10. (a) $t = 3.1$, $t = -0.43$
(b) 21 m, -0.90 m

11. $v = 3(t + \frac{1}{3})^2 + \frac{35}{3}$ which is always positive

12. (a) $31\frac{1}{2}$ m s^{-1} (b) 27 m

13. (a) $\frac{1}{2}(3t^2 + 1)$ (b) 10 m s^{-1}

14. (a) 2.53 m s^{-1} (b) 2.5 m s^{-1}

15. (a) $10\frac{2}{3}$ m s^{-1} (b) $t = 2$
(c) $13\frac{1}{3}$ m

16. $s = \dfrac{32t^2}{15(t + 1)}$

17. $v = u + at$, $s = ut + \frac{1}{2}at^2$

18. (a) 8 m s^{-1} (b) $5\frac{1}{3}$ m
(c) $2\sqrt{\frac{7}{3}}$ (3.06) (d) 9.85 m

19. (a) $2\left(1 + \dfrac{1}{t^3}\right)$ (b) 6.048 N

20. $\frac{1}{2}t^3 + \frac{5}{2}t^2 + 6t - 26$

21. $66\frac{2}{3}$ m

Exercise 22b – p. 486

1. (a) (i) $\mathbf{v} = 6t^2\mathbf{i} + 6t\mathbf{j}$, $\mathbf{a} = 12t\mathbf{i} + 6\mathbf{j}$
(ii) When $t = 2$, $\mathbf{v} = 24\mathbf{i} + 12\mathbf{j}$,
$\mathbf{a} = 24\mathbf{i} + 6\mathbf{j}$
When $t = 3$, $\mathbf{v} = 54\mathbf{i} + 18\mathbf{j}$,
$\mathbf{a} = 36\mathbf{i} + 6\mathbf{j}$
(b) (i) $\mathbf{v} = (2t + 1)\mathbf{i} - 2t\mathbf{j}$, $\mathbf{a} = 2\mathbf{i} - 2\mathbf{j}$
(ii) When $t = 1$, $\mathbf{v} = 3\mathbf{i} - 2\mathbf{j}$,
$\mathbf{a} = 2\mathbf{i} - 2\mathbf{j}$
When $t = 4$, $\mathbf{v} = 9\mathbf{i} - 8\mathbf{j}$,
$\mathbf{a} = 2\mathbf{i} - 2\mathbf{j}$
(c) (i) $\mathbf{v} = -\dfrac{2}{t^2}\mathbf{i} - \dfrac{12}{t^3}\mathbf{j}$, $\mathbf{a} = \dfrac{4}{t^3}\mathbf{i} + \dfrac{36}{t^4}\mathbf{j}$
(ii) When $t = 1$, $\mathbf{v} = -2\mathbf{i} - 12\mathbf{j}$,
$\mathbf{a} = 4\mathbf{i} + 36\mathbf{j}$
When $t = 2$, $\mathbf{v} = -\frac{1}{2}\mathbf{i} - \frac{3}{2}\mathbf{j}$,
$\mathbf{a} = \frac{1}{2}\mathbf{i} + \frac{9}{4}\mathbf{j}$

2. (a) $4y = 3x^2$ (b) $(y + 4)^2 = 16x$
(c) $xy = 6$

3. (a) (i) $2 - t$ (ii) 2, 1, 0
(b) (i) $\frac{3}{2}t$ (ii) $\frac{1}{2}$, 1, $\frac{3}{2}$

4. $v = t\mathbf{i} - 2t\mathbf{j}$; $\mathbf{r} = (\frac{1}{2}t^2 + 3)\mathbf{i} + (1 - t^2)\mathbf{j}$

5. (a) $-\mathbf{j}$ (b) $18\mathbf{i} + \mathbf{j}$ (c) $8\mathbf{i}$

6. $\mathbf{r} = 16\mathbf{i} + 8\mathbf{j}$; $8\sqrt{5}$ m s^{-1}

7. $\mathbf{v} = t(3\mathbf{i} - 2\mathbf{j})$; $3\sqrt{13}$ m s^{-1}

8. $2\sqrt{10}$ m s^{-2}

9. $\mathbf{v} = 9\mathbf{i} + 14\mathbf{j}$; $\mathbf{r} = 9\mathbf{i} + \frac{57}{2}\mathbf{j}$

10. (a) $\mathbf{v} = 9\mathbf{i} - \dfrac{16}{t^2}\mathbf{j}$ (b) $\mathbf{r} = 9t\mathbf{i} + \dfrac{16}{t}\mathbf{j}$
(c) $\frac{4}{3}$ (d) 9 m s^{-1} parallel to $\mathbf{i}$

11. (a) (i) parallel to $\mathbf{i}$
(ii) parallel to $(\mathbf{i} - t\mathbf{j})$
(b) The direction of motion always has a horizontal component so the direction can never be vertical.
(c) 2 m s^{-2} $(-2\mathbf{j})$

12. (a) (i) $\mathbf{v} = -gt\mathbf{j} + V\cos\alpha\,\mathbf{i} + V\sin\alpha\,\mathbf{j}$
(ii) $\mathbf{r} = -\frac{1}{2}gt^2\mathbf{j} + Vt\cos\alpha\,\mathbf{i}$
$+ Vt\sin\alpha\,\mathbf{j}$
(b) $y = x\tan\alpha - \dfrac{gx^2}{2V^2\cos^2\alpha}$

13. (a) (i) $6\mathbf{i}$ (ii) $4\mathbf{j}$
(iii) $\mathbf{v} = 6\mathbf{i} + 4t\mathbf{j}$, $\mathbf{r} = 6t\mathbf{i} + 2t^2\mathbf{j}$
(b) $18y = x^2$

14. (a) $8\mathbf{i} + 4\mathbf{j}$; $4\mathbf{i} + 2\mathbf{j}$
(b) $\mathbf{r} = 2t^2\mathbf{i} + (t^3 + 3)\mathbf{j}$
15. (a) $\mathbf{F}_1 = 8\mathbf{i} + 6\mathbf{j}$, $\mathbf{F}_2 = 6\mathbf{i} + 3\mathbf{j}$,
$\mathbf{F}_1 + \mathbf{F}_2 = 14\mathbf{i} + 9\mathbf{j}$
(b) $\mathbf{a} = 7\mathbf{i} + \frac{9}{2}\mathbf{j}$, $\mathbf{v} = 7t\mathbf{i} + \frac{9}{2}t\mathbf{j}$
(c) $14\mathbf{i} + 9\mathbf{j}$
16. (a) $2\mathbf{i} + \frac{3}{2}\mathbf{j}$ (b) 6.25 J
17. (a) $\mathbf{v} = 6\mathbf{i} - 8t\mathbf{j}$, $\mathbf{a} = -8\mathbf{j}$
(b) $\mathbf{F} = -32\mathbf{j}$
(c) 2120 J
(d) 2048 J
18. (a) $2\mathbf{i} - 3\mathbf{j}$ (b) $t(2\mathbf{i} - 3\mathbf{j})$ (c) 52 J
(d) 52 J (e) 26 W

Exercise 22c – p. 492

1. (a) $v^2 = 2(s^2 + 5s + 2)$
(b) $4\,\mathrm{m\,s^{-1}}$ (c) 1 (or −6)
2. (a) $v^2 = 6s^2 + 8s + 9$
(b) $\pm 7\,\mathrm{m\,s^{-1}}$ (c) 1.10, −2.43
3. (a) $v^2 = s^2 - 8s + 160$
(b) ± 12.6, $\pm 13\,\mathrm{m\,s^{-1}}$
(c) $s = 4$ (d)

（graph with v-axis showing 13, 12 dashed lines and −12, −13 dashed lines; s-axis marked at 4 and 9, origin O）

4. (a) $v = 2s^2$ (b) 5
5. (a) $v = 9s^2$ (b) $a = 162s^3$
(c) 10
6. (a) $v^2 = 25s^2 - 576$ (b) 5.2 (c) 4.8
7. (a) $v^2 = 72s - 8s^3 = 8s(9 - s^2)$
(b) $8\,\mathrm{m\,s^{-1}}$
(c) −3, 0, 3
(e) (i) $36\,\mathrm{m\,s^{-2}}$ (ii) $-72\,\mathrm{m\,s^{-1}}$
(f) $\sqrt{3}$; $\sqrt{48\sqrt{3}}\,\mathrm{m\,s^{-1}}$ ($9.12\,\mathrm{m\,s^{-1}}$)
(g) Oscillates between $s = 0$ and $s = 3$
8. (a) $v^2 = 2(e^s + 1)$
(b) 10.5 (c) 5.29
9. (a) $v^2 = 144 - 80e^{-s}$
(b) 1.25 (c) 12
10. (a) $v^2 = 20\ln(s + 1) + 16$
(b) 6.94 (c) 601
11. (a) $8\,\mathrm{m\,s^{-1}}$
12. (a) $v^2 = 21 + \dfrac{8 \times 10^5}{x}$
(b) $10.0\,\mathrm{km\,s^{-1}}$
(c) $\sqrt{21}\,\mathrm{km\,s^{-1}}$ ($4.58\,\mathrm{km\,s^{-1}}$)

Exercise 22d – p. 497

1. (a) $a = -\dfrac{2}{s^5}$ (b) −64
2. 18
3. $a = \dfrac{-50}{(1 + 2s)^3}$; −0.4
4. (a) $s = \frac{1}{4}t^2$
(b) (i) 10 seconds
(ii) $10(\sqrt{2} - 1)$ seconds (4.14)
(c) 2.5
5. (a) $s = 5e^{-\frac{1}{4}t}$ (b) $5e^{-\frac{1}{2}}$ (3.03) (c) 0
6. (a) $c = -2$, $d = 12$
(c) $s = 6(1 - e^{-2t})$
(d) 6
7. (b) $t = \dfrac{1}{kp}(e^{-3k} - e^{-ks})$
(c) $p = 20$, $k = \frac{1}{3}\ln\frac{1}{2}$ (or $-\frac{1}{3}\ln 2$)
(d) 6 m
8. (a) $s = 2\sqrt{(t + 4)}$ (b) $v = \frac{2}{s}$
(c) $\frac{1}{6}\sqrt{6}$ (0.408)
9. (a) 0.2 (b) $9.8e^{-0.4s}$
10. (c) $\lambda = 0.08$, $\mu = 0.0283$
(d) 1st, $7.35s$; 2nd, $6.03s$
(e) 1st, $27.5s$; 2nd, $20.8s$; strengthens case for 1st model.

Exercise 22e – p. 500

1. (a) $s = (2t + 6)^{1/3}$ (b) 29
2. (a) $s = 8t + 3e^{-2t} - 3$ (b) 7.19
3. (a) $a = -36\cos 3t$ (b) −2.55
(c) $s = 4\cos 3t$ (d) 3
4. (a) (i) $v = 5t^2 - 40t$
(ii) $t = 0$ and $t = 8$
(iii) $+100\,\mathrm{m\,s^{-1}}$, $t = 10$;
$-80\,\mathrm{m\,s^{-1}}$, $t = 4$
(iv)

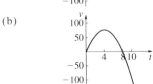

(b)

5. (a) $10\frac{2}{3}\,\mathrm{m\,s^{-1}}$ (b) $13\frac{1}{3}\,\mathrm{m}$
6. (a) $\mathbf{a} = 6t\mathbf{i} - 4\mathbf{j}$ (b) $\mathbf{R} = 8\mathbf{j}$

7. (a) $\mathbf{v} = 2e^{2t}\mathbf{i} + 2t\mathbf{j}$
    (b) $\mathbf{a} = 4e^{2t}\mathbf{i} + 2\mathbf{j}$
    (c) $2\mathbf{i}$; $4\mathbf{i} + 2\mathbf{j}$
    (d) $t = \frac{1}{2}\ln 5$, $\mathbf{v} = 10\mathbf{i} + \ln 5\mathbf{j}$
8. (a) $\mathbf{r} = 5\sin 2t\,\mathbf{i} + 5\cos 2t\,\mathbf{j}$
    (b) $\mathbf{a} = -20\sin 2t\,\mathbf{i} - 20\cos 2t\,\mathbf{j}$
    (c) $5\mathbf{i}$ (d) 5   (e) 10   (f) -4
    (g) $x^2 + y^2 = 25$
9. (a) $12.5\,\mathrm{m\,s^{-1}}$
    (b) $25.8\,\mathrm{m\,s^{-1}}$, 200 m
    (c) 400 m
10. (a) $12\sqrt{5}\,\mathrm{m\,s^{-1}}$ (b) $6\mathbf{i} - \frac{3}{2}\mathbf{j}$
11. $8\ln 8 - 7$ (9.64)
12. (b) $s = k\{\frac{1}{4}t^4 - 6t^3 + 36t^2\}$; $k = \frac{1}{4}$
    (c) Yes
    (d) $t = 2.53$, $20.8\,\mathrm{m\,s^{-1}}$;
        $t = 9.46$, $20.8\,\mathrm{m\,s^{-1}}$

CHAPTER 23
Exercise 23a – p. 513
1. (a) $\frac{2}{3}\pi\,\mathrm{s}$ (b) $15\,\mathrm{m\,s^{-1}}$
    (c) $3\sqrt{21}\,\mathrm{m\,s^{-1}}$ $(13.7\,\mathrm{m\,s^{-1}})$
2. (a) $\frac{8}{5}\pi\,\mathrm{s}$ (b) $6\frac{2}{5}\,\mathrm{m}$
    (c) $\sqrt{39}\,\mathrm{m\,s^{-1}}$ $(6.24\ \mathrm{m\,s^{-1}})$
3. (a) 2.5 m   (b) $\pi\,\mathrm{s}$ (c) $5\,\mathrm{m\,s^{-1}}$
    (d) $10\,\mathrm{m\,s^{-2}}$
4. (a) 5 m
    (b) $\frac{4}{5}\sqrt{6}\,\mathrm{m\,s^{-1}}$ (1.96), $\frac{4}{25}\,\mathrm{m\,s^{-2}}$ (0.16)
5. (a) 4 s   (b) 0.376 m
    (c) $0.541\,\mathrm{m\,s^{-1}}$, $0.370\,\mathrm{m\,s^{-2}}$
6. (a) (i) $-12\sin 3t$ (ii) $-36\cos 3t$
    (b) $\ddot{x} = -9x$ (c) $\frac{2}{3}\pi$
7. (a) $x = \frac{1}{2}\cos\left(\frac{1}{4}\pi t\right)$
    (b) $v = -\frac{1}{8}\pi\sin\left(\frac{1}{4}\pi t\right)$
    (c)

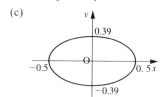

8. 1.33 s
9. $a = 0.023\,\mathrm{m}$, $\ddot{x}_{max} = 3.65\,\mathrm{m\,s^{-2}}$,
    $v = 0.262\,\mathrm{m\,s^{-1}}$
10. (a) $x = 2\cos\left(t + \frac{1}{6}\pi\right)$; $\ddot{x} = -x \Rightarrow$ SHM
    (b) 2 m; 2π s

11. (a) $\frac{1}{12}\pi\sqrt{3}\,\mathrm{s}$ (b) $\frac{1}{8}\pi\sqrt{3}\,\mathrm{s}$
    (c) $\frac{1}{12}\pi\sqrt{3}\,\mathrm{s}$
12. (a) 1 m   (b) $0.274\,\mathrm{m\,s^{-2}}$
    (c) $0.453\,\mathrm{m\,s^{-1}}$
13. (a) 38.7 s   (b) 9.16 s
14. (a) $0.161\,\mathrm{m\,s^{-1}}$,
        $259\,\mathrm{m\,s^{-2}}$
    (b) $0.139\,\mathrm{m\,s^{-1}}$,
        $129\,\mathrm{m\,s^{-2}}$

Exercise 23b – p. 516
1. (a) $v^2 = \frac{g}{l}(2lx - 3x^2)$
    (b) 0, $\frac{2}{3}l$ (c) $\frac{1}{3}l$
2. (b) 7.35×10^6 kJ lost
3. (a) $v = \lambda t - \frac{1}{2}\mu t^2 + k$,
        $s = \frac{1}{2}\lambda t^2 - \frac{1}{6}\mu t^3 + kt$
    (b) Measure the time taken to reach three
        measured distances from the start and
        use $s = f(t)$
    (c) Check further corresponding values of
        s and t
4. (b) $k = 2 \times 60^3$ (432 000)
    (c) (i) 1800 m   (ii) 1920 m
    (d) (i) $15\,\mathrm{m\,s^{-1}}$ (ii) $16.9\,\mathrm{m\,s^{-1}}$
    (e) (i) 225 m   (ii) 199 m
    (f) Second model correlates very well.
5. (a) 3 m, 750 minutes
    (b) $7\frac{1}{2}$ minutes past 3, 0.025 m/minute
    (c) 3.48 pm   (d) 8.42 pm
6. (a) 2550 minutes
    (b) 2.47×10^{-3} rad/minute
    (c) 423 000 km

CHAPTER 24
Exercise 24a – p. 522
1. (a) $6(5 - e^{3t})$ (b) 24 N
2. (a) $-\dfrac{1}{(t+1)^2}$ (b) $\dfrac{8}{(t+1)^3}$
3. (a) $8 - 2e^{-2t}$ (b) $8t + e^{-2t} + 4$
4. (a) $F = 3s\left(1 - \frac{1}{s^4}\right)$ (b) $b = 1$
5. (a) $F = \dfrac{120}{v}$
    (b) (i) $v^2 = 80t + 36$
        (ii) $v^3 = 120s + 344$
6. (a) $v = -6\sin 3t$ (b) $x = 2\cos 3t$
    (c) $\ddot{x} = -9x$, SHM
7. (a) 50 J   (b) $\frac{125}{3}$ J

8. $66\frac{2}{3}$ m

9. $\frac{1}{20}d^2$; $\frac{1}{20}(20mgd - d^2)$

10. $T = \sqrt{2mu/k}$

11. (a) $8mk$ (b) $(V^2 - 16k)^{1/2}$

12. (a) $38.3\,\text{m s}^{-1}$

(b) $s = 75\ln\left(\dfrac{150g}{150g - v^2}\right)$

(c) $71.1\,\text{m}$

Exercise 24b – p. 527

1. (a) $p = 20q$

(b) $v = \frac{1}{800}(pt - \frac{1}{2}qt^2)$

(c) $p - 10q = 600$

(d) $p = 1200$, $q = 60$

(e) $11.3\,\text{m s}^{-1}$; agrees very well with the value from the graph

(f) No, because the character of the motion changes after 20 seconds.

2. (a) $10v\,\text{N}$

(b) $v = 8(1 - e^{-\frac{1}{8}t})$

(c) $3.72\,\text{m s}^{-1}$

(d) (i) ≈ 17 s   (ii) ≈ 35 s   (iii) infinite time

(e) $s = 64(\frac{1}{8}t + e^{-\frac{1}{8}t} - 1)$; $10.3\,\text{m}$

(f) Results in (d) are not realistic and the result in (e) is not as big as would be expected in practice.

3. (a) $J = 1.1mv$

(b) 20   40   80   100   150   200

     66   132   264   330   495   660

(c) The correlation is now much closer but all the predicted values are still lower than those measured.

(d) From (c) we deduce that the coefficient of restitution has been underestimated and should be increased. Trying $e = 0.14$ gives very close correlation.

4. The two values of k given by the second model are quite close, at approximately 4.75×10^6. The two values given by the first model differ by too great a margin.

5. (a)   (i) $50\,\text{m s}^{-1}$

(iii) $t = 50\ln\dfrac{50}{50 - v} - v$

(iv) 5.5 s, 16 s

(b) (i) 16

(ii) $t = 25\ln\left(\dfrac{2500}{2500 - v^2}\right)$

(iii) 4.4 s, 11 s

(c) Model 1 does not agree very well, giving values that are too large. Model 2 has reasonable agreement.

(d) (i) 1st: 25.2 s   2nd: 16.8 s

(ii) Model 2 agrees better than Model 1, but at this speed even Model 2 is not very good.

Exercise 24c – p. 533

1. $6.63 \times 10^{-11}\,\text{m}^3\,\text{kg}^{-1}\,\text{s}^{-2}$

2. 6.0×10^{24} kg

3. (a) $2.7 \times 10^{-6}\,\text{rad s}^{-1}$ (b) 3.8×10^8 m

4. 3.42×10^8 m

5. (a) 1.75×10^6 m

(b) $9.6 \times 10^{-4}\,\text{rad s}^{-1}$

(c) 7.4×10^{22} kg

6. (a) $140\,\text{N}$ (b) $1.7\,\text{m s}^{-2}$

(c) $2.2\,\text{m s}^{-1}$

7. (b) $u^2 - \dfrac{2k}{R}$ (c) $\sqrt{\dfrac{2k}{R}}$

(d) $1.1 \times 10^4\,\text{m s}^{-1}$

8. (a) $3.7\,\text{m s}^{-2}$

(b) $4.3 \times 10^{13}\,\text{m}^3\,\text{s}^{-2}$

(c) $k = g_1 R^2$ (d) $\dfrac{2g_1 R^2}{(2g_1 R - u^2)}$

(e) (i) $\sqrt{2g_1 R}$ (ii) $5000\,\text{m s}^{-1}$

Exercise 24d – p. 537

1. (a) $9.79\,\text{m s}^{-2}$ (b) $1.70\,\text{m s}^{-2}$

2. (a) $0.635\,\text{m}$, $1.43\,\text{m}$ (b) 4.8 s

3. $1.5\,\text{m}$ **4.** $0.0018\,\text{m}$

5. 9.803

6. (a) $429\,\text{s/day}$ (b) $435\,\text{s/day}$

Exercise 24e – p. 545

1. (a) $72\,\text{N}$ (b) $90\,\text{N}$

2. (a) $\frac{2}{15}\pi$ seconds   (b) $0.4\,\text{m}$

3. (a) $\frac{1}{5}\pi$ seconds   (b) $0.3\,\text{m}$ (c) $6\,\text{N}$

4. (a) $\frac{1}{3}\pi$ seconds   (b) $5\,\text{m}$

5. (a) $3550\,\text{N}$ (b) $18.8\,\text{m s}^{-1}$

6. (a) No $\left(ma\left(\dfrac{2\pi}{T}\right)^2 < \mu mg\right)$

(b) Yes $\left(ma\left(\dfrac{2\pi}{T}\right)^2 > \mu mg\right)$

7. 2.59 s

8. (a) $R - mg = m\ddot{x}$

(b) $R = m(g - \omega^2 x)$

(c) (i) $m(g - a\omega^2)$ (ii) $m(g + a\omega^2)$

(d) $\dfrac{g}{\omega^2}$

9. (a) 2π seconds   (b) 880 N

10. (a) $\frac{1}{5}\pi$ seconds   (b) 1 m   (c) Yes

11. (b) 1.6 m   (c) $\frac{1}{3}\pi$ seconds
     (d) (i) 1.34 m s^{-1} (ii) 1.8 m s^{-1}

12. (a) (i) 0.7 m; complete oscillations if the
         spring obeys Hooke's Law in the
         range $0.3 \leqslant \text{AP} \leqslant 0.7$
        (ii) 1.1 m; incomplete oscillations
         as the position when $\text{AP} = -0.1$ m
         is impossible to reach
    (b) $\frac{2}{5}\pi$ seconds, 0.2 m

13. (a) $\text{AO} = 3l,\ \text{OB} = 2l$
    (b) (i) $2l - x,\ l + x$
        (ii) $\ddot{x} = -\frac{3g}{l}x$
        (iii) $2\pi\sqrt{\frac{l}{3g}},\ \frac{1}{2}l$

14. (a) 3.4 m
    (b) (i) $\ddot{x} = -7x$ (ii) $\frac{2}{7}\pi\sqrt{7}$ seconds
        (iii) Incomplete SHM oscillations for
        $1.4 \leqslant x \leqslant 1.8$, vertical motion
        under gravity for $\text{AP} < 2$ m

15. (a) $2mg$ (b) $3l$
    (c) $2\pi\sqrt{\frac{l}{g}},\ l$
    (d) (i) SHM with smaller amplitude but
        same period
       (ii) Incomplete SHM

16. (a) 23.7 N   (b) 0.105 kg

CONSOLIDATION F

Miscellaneous Exercise F – p. 551

1. C   **2.** B   **3.** D

4. C   **5.** D

6. (a) 2 m s^{-1} (b) $(3 - e)$ m

7. (a)

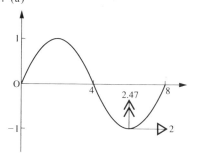

    (b) $\mathbf{v} = 2\mathbf{i},\ \mathbf{a} = 2.47\mathbf{j}$

8. (a) $2\mathbf{i} - e^{-t}\mathbf{j}$ (b) 4.32 m s^{-1}

9. (a) 56 m s^{-2} (b) 156 m
    (c) $6\mathbf{i} - 8\mathbf{j}$ (d) $\mathbf{i} + 8\mathbf{j}$
    (e) $\sqrt{65}$ m s^{-1} (8.06)   (f) 82.9°

10. (a) $\dfrac{d\mathbf{v}}{dt} = (2t - 1)\mathbf{i} + (2 - 2t)\mathbf{j}$

    (b) $\dfrac{d^2\mathbf{v}}{dt^2} = 2\mathbf{i} - 2\mathbf{j}$

    (c) $8t^2 - 12t + 5$

11. A: $(40t^2 + 0.5)\mathbf{i}$;
    $_A\mathbf{r}_B = (40t^2 - 0.25)\mathbf{i} - 30t^2\mathbf{j}$;
    $T = \frac{1}{50}\sqrt{10}$ s
    $F_1 : 4\,\text{J}, \quad F_2 : 2.4\,\text{J}$

12. (a) $6\mathbf{j}$ (b) $\frac{3}{2}(9 + 4t^2)$
    (c) $7\mathbf{i} + 5\mathbf{j}$

13. (a) $-(2\sin 2t)\mathbf{i} - (2\cos 2t)\mathbf{j}$
    (b) 2 m s^{-1}

14. (a) $v\frac{dv}{dx} = x^2 + 4$ (c) 55.8 J

15. (a) (ii)

       v becomes constant

    (b) The initial acceleration is zero so
       the ball descends at constant
       velocity.

17. (a) $0.3 - 0.6v = 0.03\dfrac{dv}{dt}$
    (d) 0.035 s (2 sf)

18. (i) $\left(\dfrac{P}{Mk}\right)^{1/3}$ (ii) $\dfrac{P - Mkv^3}{Mv}$

    (iii) $\left[\dfrac{P}{Mk}(1 - e^{-3kx})\right]^{\frac{1}{3}}$

19. $u^2 = \frac{2}{3}k \ln 3$

20. (a) 6.4×10^{12} (b) 1.46×10^3 m s^{-1}

21. (a) 1.62 m s^{-2}

22. $v = \sqrt{\left[u^2 + 2k\left(\dfrac{1}{x} - \dfrac{1}{a}\right)\right]}$

23. C   **24.** D   **25.** B

26. D   **27.** C   **28.** C

29. F Only if acceleration and displacement
      have opposite signs.

30. F String may go slack near the top.

31. T

32. F Not true if F is variable.

33. T

34. T

35. (a) $2\sqrt{3}$ m s^{-1} (b) 1.9 N

36. (i) 8 m   (ii) $\frac{1}{18}\pi$ seconds
    (iii) $108\sqrt{3}$ W

37. (ii) 4π m s^{-1} (iii) $80\pi^2$ m s^{-2} (iv) $\frac{2}{3}$

38. (a) $g_1 : g_2 = (3601)^2 : (3599)^2$
    (b) different heights above sea level.

39. (b) 3.8×10^{-3} N   (c) 0.34 s

40. (a) $a = 3$, $b = -\frac{3}{2}$

(b) $\frac{3}{4}$

(c) Acceleration $= \frac{3}{2}t - 3$

Deceleration $= 3 - \frac{3}{2}t$ which decreases as $t \to 2$

(d) 2 m

41. $7\,\mathrm{m\,s^{-1}}$

42. (a) 2190   (c) 1460

43. (a) $\frac{3}{2}L$ (c) $0 \leqslant h \leqslant \frac{1}{2}L$

(d) $\pi\sqrt{L/2g}$

(e) $\frac{1}{3}\sqrt{2gL}$

44. (b) (i) 13.22×10^3 seconds

(ii) $1.41 \times 10^3\,\mathrm{m\,s^{-1}}$

(c) (ii) $10^4 v = 1.41 \times 10^7\,\mathrm{N\,s}$

(iii) $\theta = 35.2°$

45. $\dfrac{\mathrm{d}v}{\mathrm{d}t} = kv - g$;   0.28 V

46. (a) (i) Resultant force is zero.

(ii) Resultant force is perpendicular to the tangent on the bends.

(b) $R = kv$ with $k = 6$

(d) $10(1 - e^{-t/15})$

(e) Resistance when just starting unlikely to be zero. At high speeds resistance quite likely to be almost constant.

47. (a) Assume smooth joints and light members.

(b) 50 N

(c) AB and CD: $25\sqrt{2}\,\mathrm{N}$; DA: $75\sqrt{2}\,\mathrm{N}$; AC: 50 N

(d) AB and AC

INDEX